MEDICAL RADIOLOGY
Diagnostic Imaging

Editors:
A. L. Baert, Leuven
K. Sartor, Heidelberg

A. Jackson · D. L. Buckley · G. J. M. Parker (Eds.)

Dynamic Contrast-Enhanced Magnetic Resonance Imaging in Oncology

With Contributions by

M. W. Ah-See · R. C. Brasch · D. L. Buckley · F. Calamante · J. L. Evelhoch · L. S. Fournier
I. S. Gribbestad · K. I. Gjesdal · M. H. B. Hjelstuen · A. Jackson · G. C. Jayson
M. W. Knopp · M. O. Leach · K. L. Li · S. Lundgreen · N. A. Mayr · D. S. McMeekin
J. F. Montebello · C. T. W. Moonen · D. A. Nicholson · G. Nilsen · M. D. Noseworthy
A. R. Padhani · G. J. M. Parker · M. Pedersen · W. E. Reddick · T. P. L. Roberts
J. S. Taylor · P. van Gelderen · J. C. Waterton · D. H. Wu · W. T. C. Yuh · X. P. Zhu

Foreword by

A. L. Baert

With 110 Figures in 295 Separate Illustrations, 85 in Color and 20 Tables

Springer

Alan Jackson, MBChB (Hons), PhD, FRCP FRCR
Professor, Imaging Science and
Biomedical Engineering
The Medical School
University of Manchester
Stopford Building
Oxford Road
Manchester M13 9PT
UK

Geoffrey J. M. Parker, PhD
Research Fellow in Imaging Science
Imaging Science and Biomedical Engineering
University of Manchester
Stopford Building
Oxford Road
Manchester M13 9PT
UK

David L. Buckley, PhD
Lecturer in Image Science
Imaging Science and Biomedical Engineering
University of Manchester
Stopford Building
Oxford Road
Manchester M13 9PT
UK

Medical Radiology · Diagnostic Imaging and Radiation Oncology
Series Editors: A. L. Baert · L. W. Brady · H.-P. Heilmann · M. Molls · K. Sartor

Continuation of Handbuch der medizinischen Radiologie
Encyclopedia of Medical Radiology

ISBN 978-3-642-07607-7 e-ISBN 978-3-540-26420-0

Library of Congress Cataloging-in-Publication Data

Dynamic contrast-enhanced magnetic resonance imaging in oncology / A. Jackson, D. Buckley, G.J.M. Parker, eds. ; with contributions by R.C. Brasch ... [et al.] ; foreword by A. L. Baert.
p. ; cm. -- (Medical radiology)
Includes bibliographical references and index.

1. Cancer--Magnetic resonance imaging. I. Jackson, Alan, Ph.D. II. Buckley, D. (David), 1969- III. Parker, G. J. M. (Geoffrey J. M.) 1971- IV. Series.
[DNLM: 1. Magnetic Resonance Imaging--methods. 2. Neoplasms--diagnosis. 3. Contrast Media. QZ 241 D997 2003]
RC270.3.M33D96 2003
616.99'407548--dc22 2003067408

Springer-Verlag is part of Springer Science+Business Media

http//www.springeronline.com

Softcover reprint of the hardcover 1st edition 2005

Medical Editor: Dr. Ute Heilmann, Heidelberg
Desk Editor: Ursula N. Davis, Heidelberg
Production Editor: Kurt Teichmann, Mauer
Cover-Design and Typesetting: Verlagsservice Teichmann, 69256 Mauer

21/3150xq – 5 4 3 2 1 0 – Printed on acid-free paper

To our wifes

Anna and *Susan*

and our children

Joseph and *Jessica*

Theodore and *Alexandra*

Foreword

Dynamic contrast enhancement techniques using iodinated contrast media have been employed routinely in computed tomography for many years and can help the radiologist considerably in narrowing down the differential diagnosis of tumors by adding functional information to the anatomically detailed morphological images.

One of the many advantages of magnetic resonance imaging is the possibility to use not one but several types of contrast media, each with its specific composition and particular properties.

The functional imaging capabilities of dynamic contrast-enhanced MRI are for that reason substantially greater than those of dynamic contrast-enhanced CT.

Contrast-enhanced MR is now well established as a high-performance imaging modality for the diagnosis and management of patients with solid tumors.

This book is the first volume to cover comprehensively these new and fascinating techniques and represents an excellent up-to-date review of our knowledge in this field. It deals with the theoretical and technical principles of the methods, their practical implementation and their application in tumors of the brain, the breast, the musculoskeletal system, the liver, the prostate and the cervix. It also discusses the actual and possible future applications of the method in assessing tumor treatment including anti-angiogenesis.

The editors are world-renowned experts in the field with a long-standing interest in functional MRI. The authors of the individual chapters were invited to contribute because of their outstanding experience and their major contributions to the literature.

I would like to thank both the editors and the authors and congratulate them sincerely for their innovative approach and their superb efforts in compiling this excellent volume.

This book will be of great interest to radiologists, oncologists and oncological surgeons but also to scientists involved in more basic research. I am confident that it will meet the same success with readers as the previous volumes published in this series.

Leuven

Albert L. Baert

Preface

Radiologists have long been interested in imaging and measuring blood flow and in characterising the structure of blood vessels within pathological tissues. In 1927 Egas Moniz described the technique of cerebral angiography using injection of radio-opaque contrast agent combined with conventional radiology. His earlier published work had focused on the physiology of sex, so the fact that dynamic contrast imaging was able to divert him is a testament to the excitement intrinsic to the topic. Perhaps it is worrying that he later went on to invent the technique of prefrontal lobotomy – one can only hope that it was not his frustration with dynamic contrast imaging that drove him to it.

More seriously, it soon became obvious that dynamic contrast imaging using conventional radiography (i.e. angiography) was an enormously powerful clinical tool for the identification, localisation and characterisation of a wide range of pathological tissues. This was especially true in the head, where alternative radiological tools were not available, and in vascular and malignant lesions. The angiographer became a mainstay in many areas of diagnosis, and numerous articles were subsequently published on the characterisation of tumours based on their angiographic appearances. Even in these early days before the Second World War it was clear that classifying tumours based on the number of blood vessels, the density of capillaries, the rate of blood flow and the extent of contrast leakage provided valuable, and at the time unique, information, although these classification methods remained purely subjective.

The development of axial imaging techniques, particularly CT and SPECT, in the second half of the twentieth century offered a range of alternative imaging approaches for lesion diagnosis and classification, and the usage of diagnostic angiography for tissue classification declined. With the widespread adoption of CT in neuroradiology and in oncological diagnosis and grading, the use of vascular patterns for tissue classification became of principally historical interest. In retrospect, the obvious biological importance of vascular structure in pathology meant that it was inevitable this situation would change. This book reflects the change that has occurred over the past 10 years due to the convergence of a number of important factors. First, and probably most important, was the realisation less than 20 years ago that the development of tumours is entirely dependent on the induction of growth of new blood vessels from existing host vasculature, a process known as angiogenesis. One important concomitant of this was the realisation that histological estimates of vascular structure, such as microvascular density, could provide prognostic information over and above formal histological grading strategies in a number of tumours. Even more important was the gradual elucidation of mechanisms controlling the angiogenic process and the demonstration, in animal models, that anti-angiogenic therapies that target the mechanism or the endothelial cell itself can be highly effective in a wide range of tumours. This led to a demand for surrogate markers of the angiogenic process that could be used in drug development trials, and the use of imaging to characterise the angiogenic vasculature was a clear candidate. At the same time as this revolution in the understanding of tumour biology there were dramatic leaps in the technology of MRI that made it ideal for this type of study. The ability of MRI to produce images without radiation, using a selectable range of contrast mechanisms and in any arbitrary plane were already providing it with a high profile in clinical practice. The development of strategies for faster image acquisition offered the opportunity to study time-course changes in vascular beds. This convergence of demand and capability has provided fertile soil for the growth of a virtual subspecialty of angiogenic imaging, of which dynamic contrast-enhanced MRI has become the mainstay.

The technique of dynamic contrast-enhanced MRI is now firmly established in drug development, clinical research and clinical practice. Despite this there is a wide choice of image-acquisition protocols, contrast agents and analysis methods which can be overwhelming to the newcomer. In the main this reflects the fact that, as with so many things in life, all dynamic contrast-enhanced MRI studies represent a series of compromises. The contrasting demands of the clinical or biological study and the limitations of the image-acquisition equipment and analysis techniques can lead to the use of study protocols that seem incompatible or contradictory to the uninitiated. This book is designed to provide the reader with the insight to understand this range of techniques and to allow them to tackle dynamic contrast-enhanced MRI studies with confidence.

The contributors to this book are an outstanding sample of clinicians and scientists working with dynamic contrast-enhanced MRI and all have international reputations in their field. We feel fortunate that they agreed to contribute to this work and would like to offer our thanks for their efforts and dedication. Finally, we hope that our readers will find this field as exciting as we all do. We look forward to meeting many of you and reading your work over the coming years.

Manchester

Alan Jackson
David L. Buckley
Geoffrey J. M. Parker

Contents

Principles of
Dynamic Contrast-Enhanced MRI

1 An Introduction to Dynamic Contrast-Enhanced MRI in Oncology

Ingrid S. Gribbestad, Kjell I. Gjesdal, Gunnar Nilsen, Steinar Lundgren, Mari H. B. Hjelstuen, and Alan Jackson

CONTENTS

1.1 Background

The diagnosis, grading and classification of tumours has benefited considerably from the development of magnetic resonance imaging (MRI) which is now essential to the adequate clinical management of many tumour types. The ability of MRI to demonstrate tumour morphology and the relationships of malignant lesions to neighbouring structures provides essential clinical information for both clinical management and surgical planning. Magnetic resonance has innate advantages in these applications enabling clear delineation of normal anatomical structures and organs and, in most cases clearly delineating and identifying pathological change. The ability to acquire multiplanar images or even volume acquisitions is extremely valuable and provides the clinician with a true three dimensional appreciation of tumour and tissue morphology. The development of small molecular weight paramagnetic contrast agents has had a major impact on the application of magnetic resonance in oncology. Many tumours exhibit distinctive enhancement patterns which may increase their conspicuity and provide useful diagnostic or staging information.

This book addresses the introduction of dynamic contrast enhanced magnetic resonance imaging in oncology. Improvements in magnetic resonance technology and sequence design have produced a series of fast and ultrafast image acquisition techniques which allow the collection of sequential sets of morphological data. These "MRI movies" represent a potentially major development in the management of a wide range of diseases but have particular applicability in oncology (Verstraete et al. 1994; Barentsz et al. 1999; Knopp et al. 1999; Kuhl et al. 1999; Mayr et al. 2000; Taylor and Reddick 2000). Variations in microvascular structure and pathophysiology give rise to temporospatial variations in enhancement patterns which can provide valuable information on tumour characteristics (Fig. 11). Dynamic contrast enhanced MRI (DCE-MRI) is now widely used in the diagnosis of cancer and is becoming a promising tool for monitoring tumour response to treatment (Hayes 2002; Ross 2002). Dynamic contrast enhancement patterns can be affected by a wide range of physiological factors which include vessel density, blood flow, endothelial permeability and the size of the extravascular extracellular space in which contrast is distributed (Kvistad et al. 2000; Taylor and Reddick 2000; Hayes 2002).

I. S. Gribbestad, MD
SINTEF Unimed, MR Center, 7465 Trondheim *and* Cancer Clinic, St. Olavs University Hospital, 7006 Trondheim, Norway
K. I. Gjesdal, MD
Ullevaal University Hospital, 0407 Oslo, Norway
G. Nilsen, MD
Molde Hospital, 6400 Molde, Norway
S. Lundgren, MD
Cancer Clinic, St. Olavs University Hospital, 7006 Trondheim, Norway
M. H. B. Hjelstuen, MD
Central Hospital of Stavanger, 4068 Stavanger, Norway
A. Jackson, MBChB (Hons), PhD, FRCP, FRCR
Professor, Imaging Science and Biomedical Engineering, The Medical School, University of Manchester, Stopford Building, Oxford Road, Manchester, M13 9PT, UK

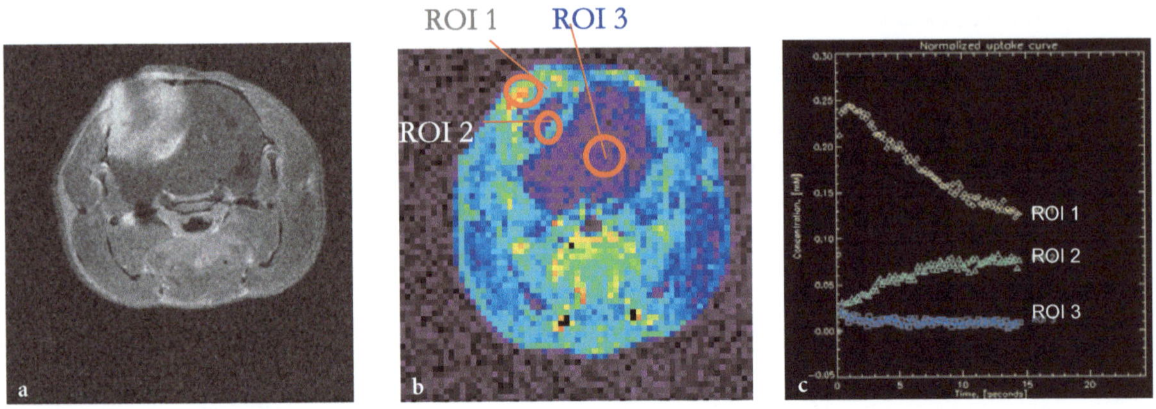

Fig. 1.1. a,b A T1-weighted MR slice through a rat brain with a glioblastoma tumour (**a**) and corresponding postcontrast image from the dynamic MR sequence (**b**). **c** Signal intensity curves for three selected regions of interest (ROIs) demonstrate different contrast enhancement in the chosen brain and tumour regions, where *ROI 1* = viable part of tumour, *ROI 2* = tumour necrosis and *ROI 3* = normal brain with intact blood--brain barrier. (**a** FOV 3.5×3.5cm, matrix 256×256, slice thickness 1 mm; **b** FOV 3.5×3.5cm, matrix 64×64, slice thickness 3 mm)

Many different methods for image acquisition and data analysis have been described for use in DCE-MRI. The analysis models are designed to derive the optimal biologically relevant components from the dynamic MR signal changes and to relate these to the underlying pathophysiological processes taking place in the tissue. In particular dynamic contrast enhanced MRI combined with physiological model-based analysis has been widely used in the study of tumour angiogenesis and in the development and trial of anti angiogenic drugs. The derivation of physiological data from dynamic contrast MRI relies on the application of appropriate pharmacokinetic models to describe the distribution of contrast media following its systemic administration. A range of modelling techniques are available which will be discussed in details in Chap. 6. However, these techniques are complex and are not widely available outside specialist centres (Tofts 1997; Tofts et al. 1999). In response to this many quantitative or semi-quantitative approaches for the classification of enhancement curve shapes have been described and are now in relatively common use in clinical settings (Daniel et al. 1998). As we shall see even these simplistic approaches for data analysis can provide extremely valuable information for clinical management.

1.2 Pathophysiological Basis of Contrast Enhancement

Numerous studies using dynamic contrast enhanced MRI have demonstrated that malignant tumours generally show faster and higher levels of enhancement than is seen in normal tissue (Padhani 1999, 2002). This enhancement characteristic reflects the features of the tumour microvasculature which in general will tend to demonstrate increased proportional vascularity and higher endothelial permeability to the contrast molecules than do normal or less aggressive malignant tissues.

Cancer can develop in any tissue of the body that contains cells capable of division (Ruddon 1987) The earliest detectable malignant lesions, referred to as cancer in situ, are often a few millimetres or less in diameter and at an early stage are commonly avascular. In avascular tumours cellular nutrition depends on diffusion of nutrients and waste materials and places a severe limitation on the size that such a tumour can achieve (Delorme and Knopp 1998). The maximum diameter of an avascular solid tumour is approximately 150–200 µm, and is governed effectively by the maximum diffusion distance of oxygen. Avascular tumours of this nature are not detectable by MRI (Knopp et al. 2001).

Conversion of a dormant tumour in situ to a more rapidly growing invasive neoplasm, may take several years and is associated with vascularization of the tumour. The development of neovascularization within a tumour results from a process known as angiogenesis (Carmeliet and Jain 2000). Why an avascular tumour at a certain stage should recruit its own vascular supply is not clearly known, but Rak et al. (1996) assume a stepwise accumulation of genetic alteration during tumour progression which is parallel to stepwise increases in the angiogenic competence of tumour cells. These angiogenically competent cells have the ability to induce neovascularization through the release of angiogenic factors. There are positive and negative regulators of angiogenesis. Release of a promoter substance stimulates the endothelial cells of the existing vasculature close to the neoplasia to initiate the formation of solid endothelial sprouts that grow toward the solid tumour [see CARMELIET and JAIN (2000) and references therein; Fig. 12]. FOLKMAN (1976) was the first to isolate such a tumour angiogenic factor in 1971.

The morphology of the neovascular network in tumours can differ significantly from that seen in normal tissue. Tumour vasculature is often highly heterogeneous, and the capillaries are extremely coarse, irregularly constricted or dilated, and distorted with twisting and sharp bends (JAIN 1987, 1988; KNOPP et al. 2001). The vessels might be acutely or transiently collapsed, they are often poorly differentiated and fragile. The capillary walls have numerous 'openings', widened inter-endothelial junctions, and a discontinuous or absent basement membrane. These defects make tumour capillaries extremely leaky (CARMELIET and JAIN 2000). Mural defects may even be large enough to allow extravasation of red blood cells (DELORME and KNOPP 1998). Such structural abnormalities can increase the transcapillary permeability many fold. The degree of abnormality seen within the tumour or the vascular bed appears to depend on whether structural maturation can occur at a rate sufficient to keep in step with the

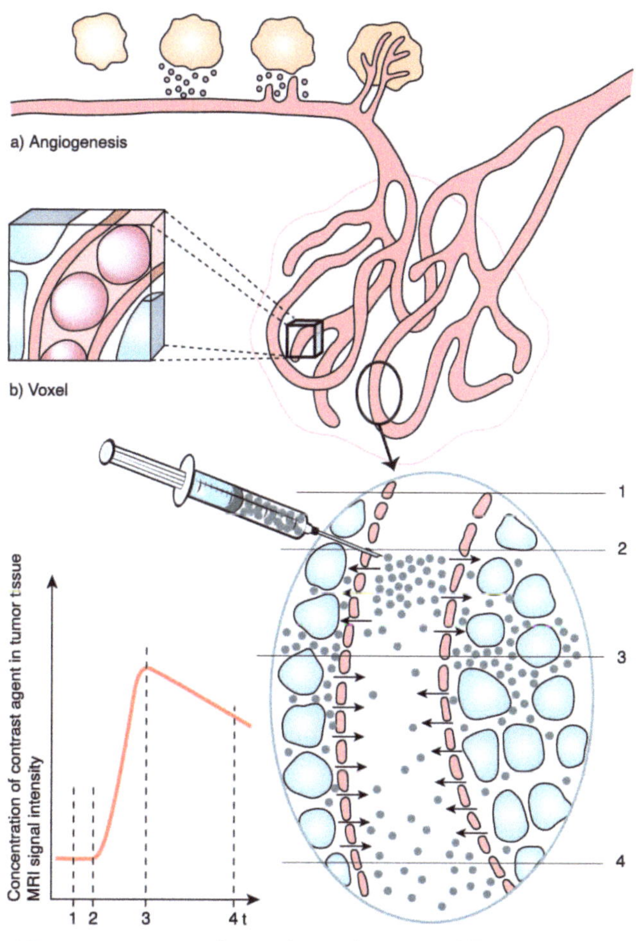

Fig. 1.2. a Growth of a malignant tumour depends upon its ability to stimulate neighbouring vasculature to initiate formation of new blood vessels that can grow into the tumour and supply it with oxygen and nutrients. This process is called angiogenesis. Angiogenesis starts with cancerous tumour cells releasing molecules, angiogenic promoter substances that send signals to surrounding normal host tissue. These signals activate certain genes in the host tissue that, in turn, make proteins to encourage growth of new vessels. A new blood capillary can form by sprouting of endothelial cells from the wall of an existing small vessel. The cells at first form a solid sprout, which then hollows out to form a tube. This process continues until the sprout encounters another vessel, with which it connects, allowing blood to circulate (CARMELIET and JAIN 2000). **b** The resolution of an MR image is determined by the field of view (FOV) and the matrix size. The pixel size and the thickness of the image slice give the volume of the voxel shown in the figure. One voxel contains many different cells even when using the smallest FOV and the largest matrix size possible. This means that the MR signal obtained from one voxel is the average of the proportion of tissue covered by the voxel. **c** The zoomed region shows a cross section through a blood vessel and the surrounding extravascular tissue consisting of tumour cells, extracellular matrix components and normal cells. The vessel wall is mainly made up of endothelial cells. The small *grey circles* indicate contrast agent molecules. The contrast agent is administered as a single intravenous bolus injection at point *2*. The contrast agent leaks into the extravascular-extracellular space (EES), also called the leakage space, through VVOs and widened inter-endothelial junctions (*line 2* to *line 3*). How fast the contrast agent extravasates is determined by the permeability of the microvessels, their surface area, and the blood flow [see PADHANI (2002) and references therein]. At first the contrast agent accumulates in the extravascular tissue before it diffuses back into the vasculature from which it is excreted (usually by the kidneys, although some contrast media have significant hepatic excretion; *line 3* to *line 4*). In an MR image the accumulation and wash-out of contrast agent is observed as changes in the MR signal intensity which is proportional to the concentration of contrast media. The time--intensity curve to the *left* in the figure shows the intensity of the MR signal from the zoomed region before (*line 1* to *2*) and after injection of contrast agent (*line 2* to line *4*)

angiogenic process (Delorme and Knopp 1998). Thus slow growing and relatively benign tumours such as intracranial meningiomas tend to show a relatively regular vascular morphology whereas aggressive rapidly growing tumours such as glioblastoma multiforme show some of the most disordered and bizarre vascular morphology seen in any tissue.

Current research into the structural basis of tumour microvascular hyperpermeability suggests that the vesiculo-vascular organelles (VVOs) provide the major pathway for the extravasation of circulating macromolecules across endothelia (Dvorak et al. 1996). VVOs are grape-like clusters of interconnecting uncoated vesicles and vacuoles that span the entire thickness of vascular endothelium, therefore providing a potential transendothelial connection between the vascular lumen and the extravascular space. The number and leakage rate of VVOs are regulated by local mediators especially the multifunctional cytokine vascular endothelial growth factor (VEGF) (Knopp et al. 1999). VEGF also known as vascular permeability factor (VPF), induces angiogenesis and strongly increases microvascular permeability to plasma proteins (Senger et al. 1983). Under hypoxic conditions, which are typical for most solid tumours, VEGF expression is regulated by pH and glucose concentration. VEGF expression can also be modulated by oncogenes and by the tumour suppressor gene p53 (Mukhopadhyay 1995). The characteristic increase of permeability in tumour capillaries is probably attributable to upregulation of VVO function (Knopp et al. 1999). The VVOs might also be the most important pathway for the Gd-chelate MRI contrast agent leakage (Dvorak et al. 1996; Delorme and Knopp 1998). After intravenous administration of a contrast agent it travels through the vascular system (Fig. 1.1) reaching the neoplastic tissues and will immediately start to leak from the tumour vasculature, accumulating in the extracellular extravascular space by a passive diffusion process driven by the contrast concentration differences. As the concentration in plasma drops due to tissue leakage and renal excretion it will eventually be lower than the concentration in the EES and backflow of contrast will occur. As the plasma concentration falls due to renal excretion backflow from the EES to plasma will continue until all contrast has been eliminated (Knopp et al. 1999).

The mechanisms underlying the signal enhancement patterns seen on dynamic MRI include variations in regional blood flow, proportional blood vessel density, regional variations in haematocrit, proportional vascularization of existing blood vessels and variations in the surface area permeability of the endothelial membranes as well as the concentration difference which exists between plasma and the EES (Bhujwalla 1999; Knopp et al. 1999). In many tumour types including breast, lung, prostate, and head and neck cancer, measurements of microvascular density made on histopathological samples correlate closely with clinical stage and act as an independent prognostic factor of considerable sensitivity (Weidner 1996). The rationale for this relationship appears to be that rapid tumour growth can be supported only in the presence of highly active angiogenesis and more aggressive tumours are therefore associated with increased evidence of angiogenesis-related microvasculature abnormalities. On the basis of this histopathological evidence it has been suggested that dynamic contrast enhanced MRI may also be able to provide independent indices of angiogenic activity and therefore act as a prognostic indicator in a broad range of tumour types. Clearly, if this is substantiated then the non-invasive technique of magnetic resonance imaging would have significant advantages over other methods which rely on tissue sampling and secondary histopathology.

1.3 Dynamic MR Imaging

1.3.1 Patient Examination

Dynamic contrast enhanced MR can be satisfactorily performed on the majority of currently available clinical scanners. Because of the requirements for high temporal resolution there will be restrictions on spatial resolution and spatial coverage dependent on gradient performance. However, it should be stressed that useful clinical information can be obtained in many applications using even a single slice of dynamic data. The majority of currently available clinical scanners will comfortably allow multislice acquisitions to be performed with adequate temporal resolution for most analysis techniques.

Contrast administration is performed through a peripheral vein. A large antecubital vein is commonly employed and the injections can be given through a small cannula which should be inserted and secured in place prior to the investigation. The injection technique is of considerable importance. Most dynamic imaging methodologies use a bolus injection of contrast and it is important that this be administered in a consistent manner. Most centres now use an automated pressure injection system to ensure reproducibility. Protocols for contrast administration vary depending on the technique in use. Typically however, a single dose of contrast (0.1 mmol/Kg) of a standard gadolinium chelate will be administered at a rate in the region of 4 ml/s. Some centres prefer to vary the injection rate so that the overall period of contrast administration is kept constant rather than having a constant injection rate of different volumes in different patients. It is important that the contrast bolus remain coherent in its passage through the body and in order to achieve this a chaser injection of normal saline is given immediately after the contrast. The chaser injection must be given at the same flow rate as the contrast and must be of adequate volume to empty the draining veins, typically 20–30 ml, so that the contrast passes into the systemic circulation as a coherent bolus. The venous injection should be placed into the right arm if possible since variations in venous anatomy can lead to significant jugular reflux on the left side which can impair the coherence of the contrast bolus (Fig. 1.3).

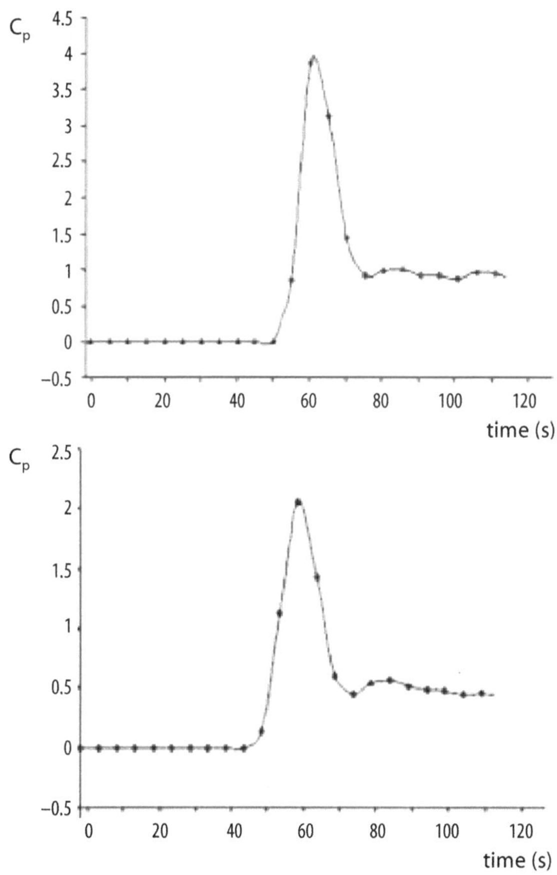

Fig. 1.3. Showing the change in contrast concentration over time within the middle cerebral artery following a bolus injection of contrast media into a peripheral vein. The two graphs represent different patients and show clear differences in the spread of the bolus despite the use of the same injection technique and contrast dose

The imaging of the dynamic sequence is usually performed following initial anatomical and localisation scans. Localisation is of prime importance particularly if hardware restrictions limit the number of slices or the spatial resolution of the images that can be obtained. There are a number of important considerations concerning slice localisation. Clearly it is essential that the pathology is included in the slice but also, for most analysis techniques it is important to include an appropriate large blood vessel. This allows the measurement of the contrast concentration changes in the plasma over time, which is commonly referred to as the arterial (AIF; Fig. 1.3) or vascular input function (VIF). This is used in many analysis techniques to represent the contrast changes occurring in the blood vessels within the tumour and to allow calculation of the contrast concentration gradient between blood and the tumour extravascular extracellular space. A number of other technical complications will be encountered depending on the choice of acquisition sequence and anatomical location. Problems associated with respiratory and other physiological motion and with the presence of inflow artefacts distorting the dynamic contrast signal in blood vessels must be considered. These will be discussed in more detail in Chap. 5. If it is intended to use a simple subjective or semi-quantitative analysis of enhancement curve then the dynamic image series provides adequate data for this approach. If it is intended to use a pharmacokinetic analysis then it is necessary to calculate contrast concentration in each image in the dynamic series. Unfortunately the relationship between contrast concentration and signal intensity is non linear (see Chaps. 5 and 6) and will be affected by the underlying native T1 of the tissues. For pharmacokinetic analyses it is therefore necessary to add additional imaging sequences to the investigation before the dynamic run is performed. These sequences are designed to allow calculation of quantitative T1 maps to enable subsequent calculation of contrast concentration. The approaches taken for T1 mapping and the imaging methods employed are also described in detail in Chap. 6 (Figs. 1.4, 1.5).

The dynamic sequence is performed following preliminary anatomical imaging and T1 mapping. The choice of image acquisition technique, injection rate and temporal resolution will be entirely dependent on the analysis method to be employed. This in turn will be chosen to optimise the amount of biologically relevant information that can be extracted from the data which will depend on the organ system

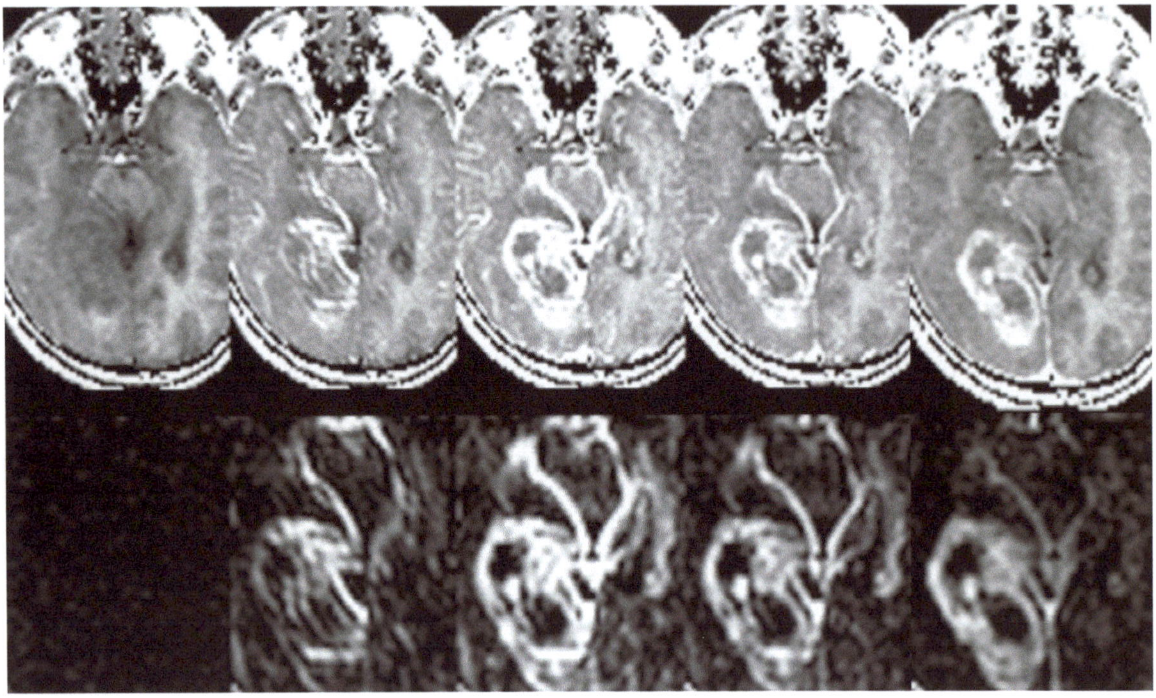

Fig. 1.4. A series of dynamic MR images (*top*) showing contrast enhancement and passage of the contrast agent into the interstitial tissues using a 3D T1-weighted gradient echo acquisition. Illustrated images are spaced approximately 10 s apart. The *lower row* of images shows the calculated concentration of contrast agent which is derived from the images in the *top row* and which can be used as the basis for pharmacokinetic analysis of enhancement patterns

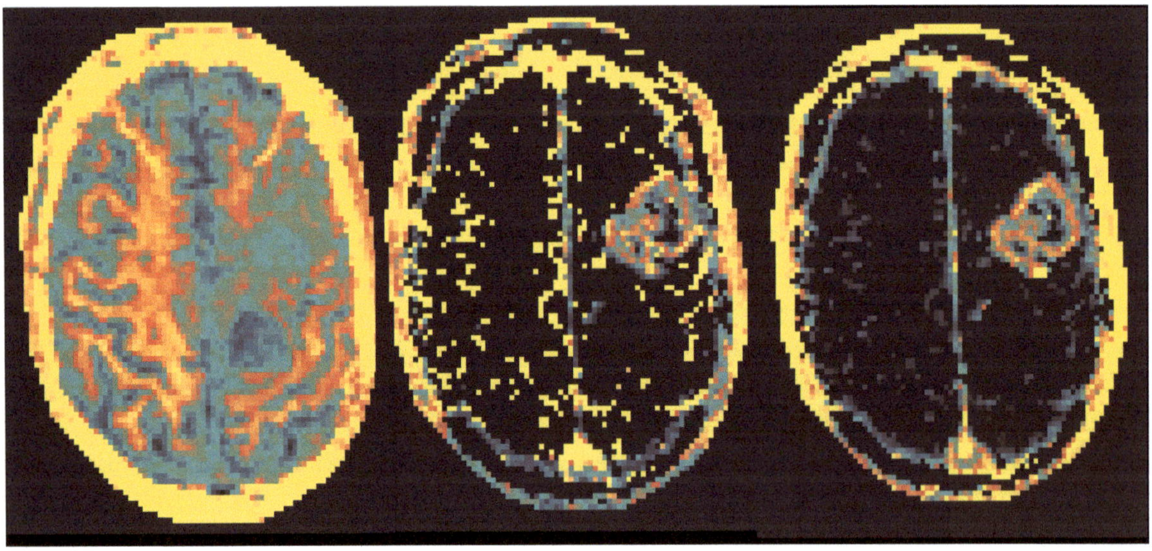

Fig. 1.5. Parametric images from a dynamic susceptibility contrast enhanced study in a patient with a grade four glioma. The *left-hand image* shows a map of the baseline relaxivity (R10) with areas of significant prolongation of T1 (*blue*) in the location of peri-tumoral oedema. The *centre image* shows a map of K^{trans} calculated using a simple two compartment model. Note the presence of elevated values in the region of normal cerebral arteries. The *right-hand image* shows the distribution of ve (the size of the extravascular extracellular space)

and the pathology being studied. At one extreme, measurements of blood flow in the brain require a temporal resolution in the region of 2 s or less in order to adequately demonstrate the first passage of the contrast bolus through the cerebral vasculature (Barbier 2001) and lower temporal resolutions will introduce progressively greater errors into the calculated parameters. Some T1-weighted dynamic techniques demand accurate measurement of the arterial input function and a temporal resolution of at least 5 s is necessary for these techniques. Where these high temporal resolutions cannot be achieved satisfactorily then the analysis method can be compromised to some extent and the use of surrogate or averaged arterial input functions may still allow meaningful pharmacokinetic analysis to be performed (Tofts 1997; Tofts et al. 1999). As an extreme example, the temporal resolution may be reduced to a period of minutes as in the case of multi-phasic imaging of the liver where the dynamic sequence consists only of one pre-contrast image and 3–4 post-contrast images performed over a period of several minutes (Bartolozzi et al. 1999). Clearly pharmacokinetic analysis of this data is impossible but nonetheless it provides extremely valuable clinical diagnostic data for showing variations in temporal enhancement patterns.

Because of the complexity of these issues the design of dynamic contrast enhanced MRI protocols can seem daunting and is indeed complex. The design process must begin with clear identification of the biological parameters available from dynamic contrast enhanced MRI which are likely to be of clinical or research value. This will in turn govern the choice of analysis techniques, which will define the characteristics necessary within the dataset and the design of the image acquisition protocol. It is important that this design process is followed appropriately and it must be stressed that there is no recipe for dynamic contrast enhanced MRI that works in all clinical cases or all applications.

In order to try and provide an understanding of these mechanisms and the problems associated with them this book will provide detailed overviews of the potential acquisition techniques and analysis methods that are currently in use or being explored by groups working with dynamic contrast enhanced MRI in oncological applications. The early chapters will deal with the technical aspects of acquisition and analysis and the later chapters will focus on the clinical benefits of dynamic contrast enhanced MRI and will review in detail the methods that have been applied in specific clinical areas.

1.3.2 Dynamic MR Acquisition Techniques

There are two generic approaches for the acquisition of dynamic contrast enhanced MRI data. Relaxivity-based methods use T1-weighted acquisitions whilst susceptibility-based techniques use T2 or more commonly $T2^*$-based sequences. Both methods have specific advantages and disadvantages and historically they have tended to be used in different applications. Dynamic susceptibility contrast enhanced MRI (DSC-MRI) was first used to study the microcirculation of the brain (Rosen et al. 1991). Susceptibility contrast agents act by changing the local magnetic environment around the contrast molecule. This effect extends some significant distance which may be in the order of several millimetres depending on the sensitivity of the image acquisition sequence used. This contrast mechanism will therefore affect water molecules lying outside the blood vessels for some considerable distance. The effect of this is that susceptibility-based imaging sequences will be more sensitive to small amounts of contrast that are distributed through a large tissue volume such as is seen when contrast is present in diffuse capillary beds (Rosen et al. 1991). When contrast molecules are collected together in a single large vessel there will still be some amplification effect but only around the vessel periphery. DSC-MRI was therefore used to study cerebral blood flow because blood passes through the brain within diffuse grey and white matter capillary beds which are areas of relatively low blood volume (in the order of 2%–5% by volume) (Figs. 1.6, 1.7). Susceptibility-based imaging techniques have a number of significant disadvantages (Calamante 2000; Thacker 2003). The paramagnetic recruitment effect described above leads to distortion of anatomy particularly where contrast lies within large vessels which appear larger than they are in reality. Susceptibility-based sequences are also liable to image distortion from a variety of other paramagnetic effects particularly those that occur at air tissue interfaces and chemical shift boundaries. This can lead to significant distortion in the hind brain around the petrous temporal bones and the base of the forebrain adjacent to the paranasal sinuses. Perhaps more importantly most T2- and $T2^*$-weighted images will exhibit some degree of T1 sensitivity and will show signal changes resulting from relaxivity effects particularly from contrast that has leaked into the EES. As a consequence imaging of normal brain parenchyma with an intact blood brain barrier, where no effective contrast leakage occurs, offers no technical problems. Suscep-

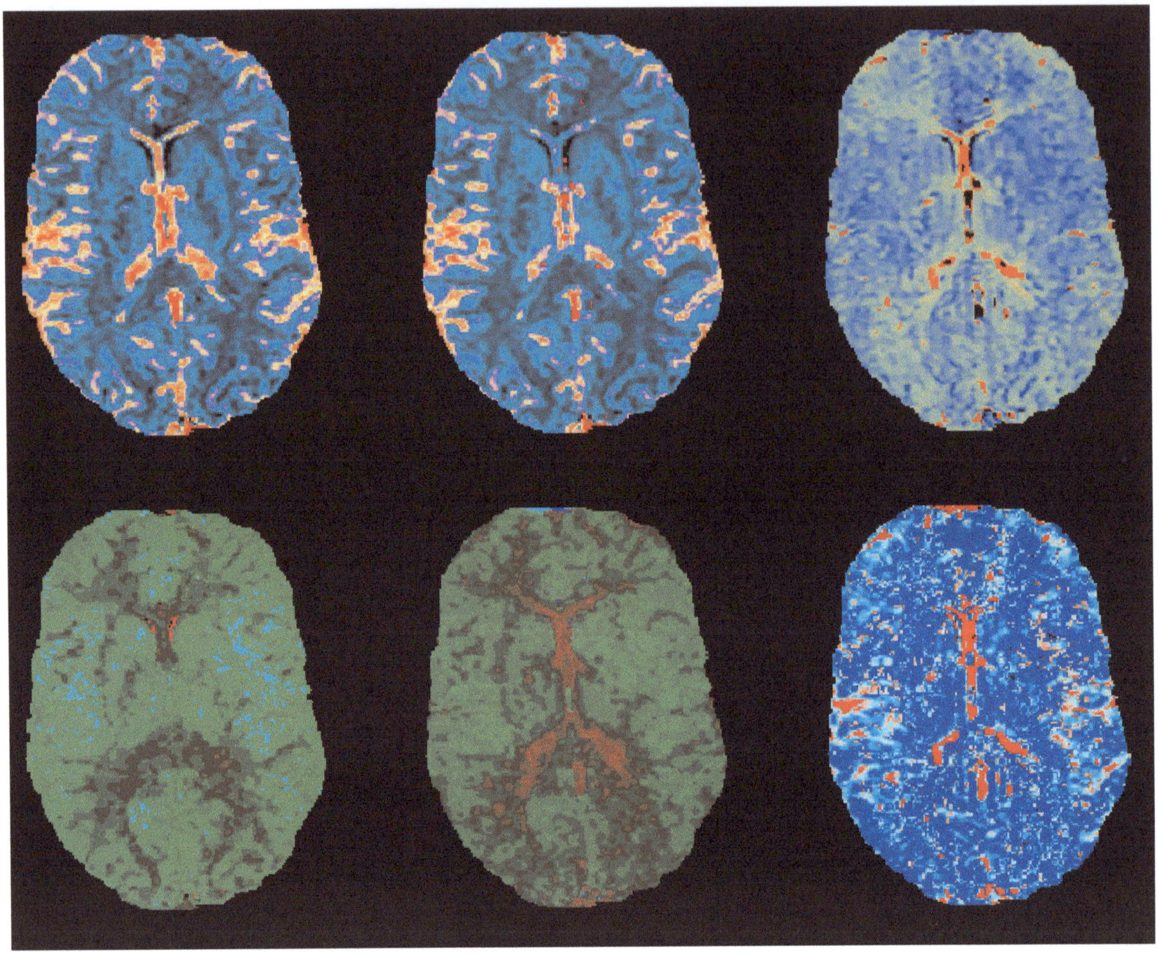

Fig. 1.6. Parametric maps of cerebral blood volume (*top left*), cerebral blood flow (*top centre*), mean transit time (*top right*), time to contrast arrival (*bottom left*), time to maximum contrast concentration (*bottom centre*) and fitting error (*bottom right*). All images were derived from the single dynamic susceptibility contrast enhanced experiment in a normal volunteer

tibility-based dynamic imaging of enhancing tissues such as tumours is far more difficult and requires the use of contrast pre-enhancement or low T1 sensitivity-based sequences (see Chap. 5) (Kassner et al. 2000). Perhaps more importantly when considering application to tumours the effects of extravascular leakage of contrast material on the T2 or T2* signal are difficult to quantify and, although estimates of contrast transfer coefficient can be made these are less reliable than those obtained with comparable relaxivity-based techniques (Weisskoff et al. 1994).

Dynamic relaxivity contrast enhanced MRI (DCE-MRI) uses T1-weighted images to detect the relaxivity effects of contrast agents during dynamic data collection (Tofts et al. 1999). It must be realised that there is no difference in the administration or nature of the contrast agent only in the image acquisition sequences used to record the data. The choice of a relaxivity-based acquisition significantly changes the observed behaviour of the contrast as it is distributed through the vasculature and extracellular spaces. Relaxivity contrast mechanisms are strongly localised to the contrast molecule itself and therefore effects are only seen in the immediate vicinity of the contrast and no spread of the contrast mechanism effects is seen as is the case in T2* acquisitions. As a result the signal intensity changes seen in areas of diffuse intravascular contrast such as capillary beds, is proportionally less on relaxivity weighted sequences. In addition there will be comparable effects seen from contrast within the extravascular extracellular space where leakage has occurred through the endothelial membrane. These effects are generally greater in magnitude compared to the intravascular signal on T1-weighted images than is seen on T2 or T2* acquisitions. Unfortunately, both intra and extravascular contrast will cause a rise

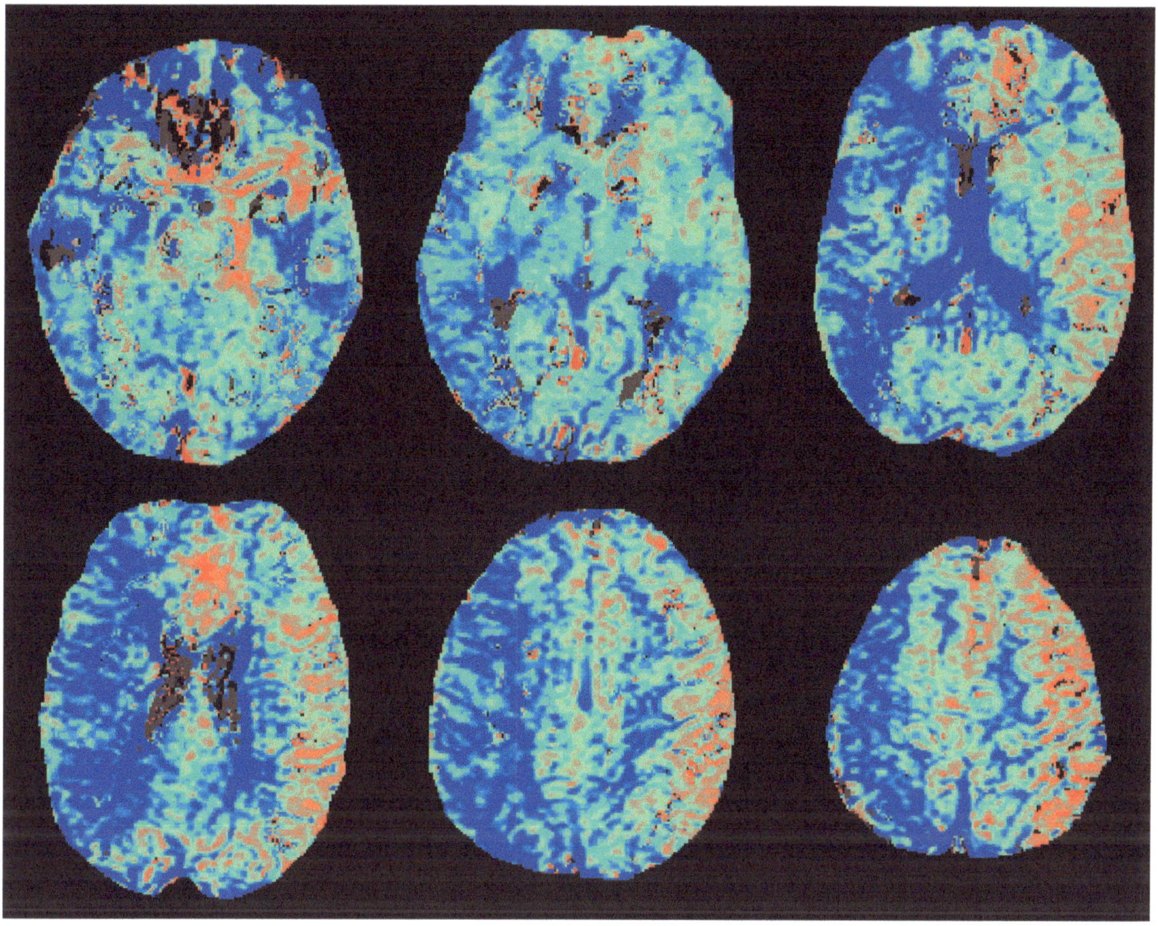

Fig. 1.7. Images of time to peak contrast concentration from a patient with a severe right sided carotid artery stenosis. Note the severe prolongation (*blue*) in the time of contrast arrival in the affected hemisphere

in signal intensity occurring with a similar time course making the separation of intra and extravascular contrast contributions to the signal time course data difficult to distinguish (Li et al. 2003).

1.3.3 Image Acquisition Sequences

Typical modern clinical MR scanners can offer a variety of choices of pulse sequence which are suitable for dynamic contrast enhanced magnetic resonance applications. Despite this the sequence chosen will almost always be a compromise between a number of imaging quality factors including time, spatial resolution, anatomical coverage, sensitivity to artefacts, image signal to noise ratio (SNR) and degree of contrast weighting.

T1-weighted DCE-MRI is most commonly acquired using gradient echo-based sequences which can be broadly divided into three groups: the steady state, the transient state and the echo planar-based sequences. The "standard" gradient echo sequences include GRASS, FISP or FFE (Haase 1990), a range of T2 weighted sequences called T2-FFE or PSIF (Gyngell 1988), specific T1-weighted sequences (Spoiled GRASS, T1 FFE or FLASH) (Wang and Riederer 1990) and methods which applied balanced gradients (FIESTA, balanced FFE or True FISP) (Oppelt et al. 1986). Standard, balanced, gradient echo sequences have high sensitivity to T2 effects which is undesirable as the signal decreases once contrast agent arrives at the tissue. Many dynamic studies have therefore used spoiled gradient echo sequences which are more specifically sensitive to T1 effects and therefore show signal increases. The main problem with these techniques is their low signal to noise ratio which can be compensated to some extent by the use of 3D acquisitions. The transient gradient echo sequences combine a contrast preparation pulse with a sequence for image readout which allows control over the contrast sensitivity. The T1 sensitive versions of transient gradient echo techniques have been used extensively in DCE-MRI. Echo planar imaging (EPI) allows single shot image acquisition and a temporal resolution of 4 images per second with a 256^2 matrix are easily achievable. However EPI methods are highly sensitive to susceptibility changes and the technique is difficult to use in the abdomen or thorax because of these susceptibility effects. Even in anatomically favourable areas such as the brain, distortion will occur and will be most marked in areas of paramagnetic effect including areas with contrast enhancement. This can be reduced by the use of segmented EPI sequences with consequent loss in temporal resolution.

$T2^*$-weighted DSC-MRI acquisition techniques are more problematic. The need for heavy $T2^*$ weighting demands a long echo time (TE) which in turn severely limits the available temporal resolution. The commonest application for $T2^*$-weighted DSC-MRI is to study perfusion in the brain and for these methods a temporal resolution of 2 s or below is required. At the present time the most common approach to this problem has been the use of echo planar image acquisitions with sequential single slice EPI. This however retains the problems of spatial distortion, which have been described above. Another approach to this problem has been the development of echo shifting techniques such as PRESTO (principles of echo shifting with a train of observations) where the echo is shifted into the subsequent data collection period allowing the use of TEs which are longer than the sequence TR (Liu et al. 1993). This approach can be combined with segmented EPI to produce a fast heavily $T2^*$-weighted volume acquisition with good anatomical covering and spatial resolution.

At the present time the development in analysis methodologies particularly the T1-weighted DCE-MRI continues to place major demands on the speed and spatial resolution required from image acquisitions. These are being addressed by many workers who are developing optimised keyhole or parallel imaging techniques. Keyhole imaging takes advantage of the fact that central lines of case space are sensitive to image contrast whereas peripheral lines are more sensitive to spatial resolution. Acquisition of a high resolution image can therefore be followed by multiple central K-space samples which are far faster and which can be combined in some way with a spatial information from the first image (Oesterle et al. 2000). The application of these K-space sub-sampling techniques can produce an increase in temporal resolution of three- to four-fold but does have specific problems particularly in areas where there is physiological movement. A large number of modified partial K-space sampling techniques are currently available each addressing particular pitfalls of the approach (Jones et al. 1993; van Vaals et al. 1993). For a more detailed discussion of the benefits and pitfalls of keyhole imaging see (Oesterle et al. 2000). Parallel imaging uses multiple receiver coils to simultaneously receive separate spatial components of the signal and produce proportional decreases in acquisition time (Sodickson and Manning 1997; Pruessmann et al. 1999; Griswold et al. 2000). Commercial systems can increase temporal resolution by a factor of 6 in the 3D scan making this technique very attractive for DCE-MRI (Pruessmann et al. 1999; Tsao 2003). There are however problems associated with loss of signal-to-noise ratio in these scans.

1.3.4 Contrast Agents

An increasing range of contrast agents is available for MRI studies. By far the most commonly used group are the paramagnetic gadolinium chelates of which 6 are currently available (see Table 1.1). These agents are similar showing similar pharmacokinetics and predominantly renal excretion. Gadolinium BOPTA to Gd-BOPTA (MultiHance) differs from the other available agents in that it shows weak protein binding and therefore some hepatic excretion. This is discussed further in Chaps. 2 and 14. Macromolecular agents such as Gadomer-17 or albumin-GdDTPA remain in the intravascular space with a half life of several hours due to their large molecular weights (35 and 65 kDa, respectively) (Gillies et al. 2000). The use of these blood pool agents allows accurate measurement of local blood volume using low temporal resolution scans. However, it is clear that these blood pool agents will leak in areas of extreme endothelial permeability such as are seen in a number of aggressive tumours. They can therefore be used as agents for the measurement of endothelial permeability and indeed have some significant advantages over the gadolinium chelates, which are more commonly used (see Chap. 2). At the present time however none

Table 1.1. Some of the contrast agents available and approved for the market in 2002

Paramagnetic		Superparamagnetic	
Non-protein binding	Weak protein binding	SPIO	USPIO
Gd-DTPA (Magnevist) (Magnevist enteral)	Gd-BOTA (MultiHance)	AMI-25 (Feridex I.V. Enorem)	AMI227 (Combidex, Sinerem)
Gd-DTPA-BMA (Omniscan)		GastroMARK	NC100150 (Clariscan)
Gd-DOTA (Dotarem)		Ferucarbotran (Resovist)	
Gd-HP-DO3A (ProHance)			
Gd-DO3A-butriol (Gadovist 1.0)			

of these are marketed for clinical use although it is clear that this development will occur in the near future and that these compounds are likely to have broad application.

1.4 Analysis of Dynamic Data

A large range of techniques have been applied to the analysis of the signal enhancement curves observed in DCE-MRI which range from simple visual inspection to complex quantification using applications of pharmacokinetic models. Most of these methods will be discussed in detail in subsequent chapters. Many analysis techniques are based on measurements taken from user-defined regions of interest (ROI). This has the advantage of ease of use but also has the disadvantage in that it produces a wide degree of variability and potential intraobserver errors into the technique. More importantly it is incapable of identifying or quantifying significant heterogeneity within the tumour microvascular which may occur within the region of interest. Inappropriate selection of the ROI so that it includes both enhancing and necrotic or non-enhancing components of tumour would give rise to misleading interpretation (Gribbestad et al. 1994). These shortcomings can be addressed by the production of calculated parametric images which allow pixel by pixel analysis of the calculated microvascular components. This pixel by pixel analysis deals specifically with tumour heterogeneity and potentially provides a far wider range of information concerning tumour behaviour than is available from region of interest analysis. Unfortunately, the use of parametric images imposes significant further demands on the acquisition and analysis techniques. In particular, the use of pixel by pixel analysis assumes that there is no significant motion at the spatial resolution of the individual voxel. Since a typical voxel may be below 1 mm in size even small physiological motions can have significant impact on the calculated parameters.

Whatever approach to analysis is selected careful inspection of the original dynamic contrast series images is important. Visual review of these images will allow identification of significant patient motion or unexpected artefacts which may have occurred during the dynamic acquisition and will also allow appreciation of the distribution of enhancement that has occurred in various tissues. This initial visual inspection is greatly aided by the use of simple subtraction techniques which can usually be performed on the scanner console or a standard clinical workstation.

1.4.1 Visual Inspection of Enhancement Curves

A commonly used analysis of dynamic enhancement patterns is based upon a subjective evaluation of the time-signal intensity curve, in which each curve is classified in accordance with the evaluation system shown in Fig. 1.8. Classification of signal intensity curves according to this scheme achieved very good diagnostic performance in differentiating malignant from benign breast lesions as described by Daniel et. al. (Daniel et al. 1998). The high numbered curves are interpreted as representing more aggressive tumour types.

1.4.2 Enhancement Curve Analysis

A broad range of approaches have been taken to assess the properties of enhancement curves in various tumours (Kuhl and Schild 2000). Most of these techniques are designed to deal with baseline variations in signal intensity and with the inherent differences in signal intensity that would be observed due to changes in tuning and scaling factors between scanners or even between sessions on the same scanner. Heywang and co-workers (1989) proposed a classification scheme for breast lesions using nor-

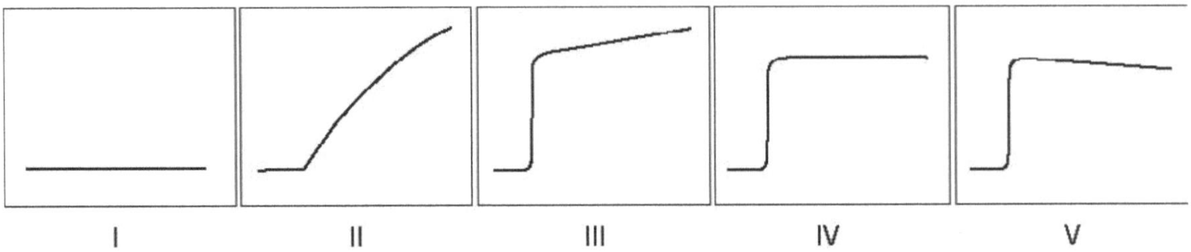

Fig. 1.8. The diagrams show the classification system for visual evaluation of the enhancement curves. *I*, no enhancement; *II*, slow sustained enhancement; *III*, rapid initial and sustained late enhancement; *IV*, rapid initial and stable late enhancement; *V*, rapid initial and decreasing late enhancement. [Adapted from Daniel et al. (1998)]

malised signal intensities (NU) where lesions beyond the threshold of 300 were classified as malignant, 250–300 as borderline and below 250 NU as non-malignant. Kaiser et al. (1989) normalised the signal enhancement using the intensity of the lesion before contrast injection and defined malignant lesions as those showing a signal enhancement of at least 100% during the first minute. Other workers (Boetes et al. 1994; Gilles et al. 1994) have identified the presence of very early enhancement in the tumour after the arrival of the arterial bolus as an indicator of malignancy.

These studies illustrate the potential problems in the use of signal enhancement curves to study malignant vasculature. Firstly, signal intensity changes are non-specific and will vary according to a wide range of scanner parameters. Because of the non-linear relationship between contrast concentration and signal intensity changes and the wide variation in baseline T1 values and signal intensity seen in the tissues being studied, attempts to calibrate or produce standard images which could be used in multiple sites have not been successful. Subjective enhancement curve analysis must therefore include some form of curve normalisation to account for these variations which are principally seen as differences in the amplitude of the enhancement curve. Secondly, the enhancement curves will contain important information not only in terms of amplitude but also in terms of time of arrival and curve slope. The choice of normalisation technique and of the most representative curve shaped parameter is difficult and has led to the development of many semi-quantitative or descriptive systems. Different methods have been applied in different applications and in different tissues and many of these will be described in more detail in the clinical applications chapters (see specifically Chaps. 9–14). Despite these shortcomings subjective or semi-quantitative curve shape analysis techniques can be extremely valuable, particularly in clinical applications for the grading or classification of tumours. This reflects the fact that differences in tumour vasculature between benign and malignant tumours are large and maybe demonstrated by relatively crude analysis techniques.

Despite their clear clinical utility the shortcomings of these subjective and semi-quantitative techniques have led many workers to develop more robust quantitative approaches to analysis. There are several reasons why more quantitative approaches may be beneficial. Firstly, the ability to produce measurements which reflected the physiological anatomical structure of the tumour microvasculature and which are truly independent of scanner acquisition and tumour type is highly attractive when compared to the use of a range of tissue or scanner specific semi-quantitative methods. Secondly, the development of precise and reproducible quantitative measures is highly desirable for use in longitudinal or multi-centre studies. This is of particular importance in clinical trials of new therapeutic agents where the ability to test the hypothesis that an agent affects tumour microvasculature will depend entirely on the accuracy and reproducibility of the measure used. Thirdly, it must be appreciated that signal enhancement curves are a crude indicator of the contrast distribution mechanisms that are occurring within the voxel. Even a small imaging voxel is large by biological standards and may contain varied proportions of blood vessels, cells and extracellular extravascular space. A signal enhancement curve may therefore represent contrast which is principally in blood vessels, contrast which has principally leaked into the extracellular extravascular space or any combination of these. Contrast enhancement curves cannot therefore differentiate between voxels which contain few vessels but have rapid leakage of contrast medium into the EES and voxels in which there is little leakage but a large vascular fraction. Fourthly, simple analyses of contrast enhancement curves are unable to compensate for variations in the contrast delivery to the tumour, which might occur due to poor injection technique, anatomical variations or abnormal physiological features in the individual patients such as poor cardiovascular function. Since these techniques are widely used in cancer patients where venous access can be difficult and cardiovascular and renal functions are commonly compromised this can introduce significant variation into the results of quantitative analysis techniques.

1.4.3 Quantitative Approaches Using Pharmacokinetic Models

The optimal analysis of DCE-MRI data would be designed to identify specific quantitative physiological parameters which describe the tissue microvasculature being observed. We have discussed previously that the distribution of contrast material will be governed by regional blood flow, blood volume, vessel shape and size, endothelial permeability, endothelial surface area, and the size of the EES. In theory an optimal analysis should allow independent assessment of each of these descriptive variables. In practice this is extremely difficult to achieve but many groups have described approaches which try to derive some of these parameters in a principled manner from the changes in signal intensity that occur during contrast passage (Larsson et al. 1990; Brix et al. 1991; Tofts 1996, 1997; Tofts et al. 1995; St Lawrence and Lee 1998). This type of analysis requires the application of a quantitative mathematical model which describes the pharmacokinetics of the contrast agent. The model must include a description of each of the parameters listed above and describe their potential relationships in terms of the effect that they will have on the flux of contrast between compartments and the relative contrast concentrations seen within them.

In order to apply pharmacokinetic models of contrast distribution to imaging-based data the first essential step is to use the signal changes observed in the dynamic acquisition to calculate quantitative parametric images of contrast concentration at each time point. Since the relationship between signal intensity and contrast concentration may be non-linear this adds an additional complication and often requires the measurement of the pre-contrast T1 values for each of the voxels to be studied. There are several ways to achieve this and these are discussed in detail in Chap. 6. All these methods produce a parametric image of T1 which corresponds to the dynamic time course which is to be obtained (Fig. 1.5). Using these quantitative T1 images together with the observed signal change it is possible to transform each of the images within the dynamic time course series to a parametric image of contrast concentration. It is these contrast concentration time course curves for individual voxels that are then used as the substrate for pharmacokinetic models.

A wide range of pharmacokinetic models have been described and applied to the analysis of DCE-MRI data. It should be appreciated that this is not because the pharmacokinetics of the contrast agent are in doubt but rather because the true pharmacokinetics are complex and applications of an idealised pharmacokinetic model lead to logistical problems which make the analysis unstable or unreliable. Almost all of the pharmacokinetic analysis techniques use curve fitting methods to estimate the parameters of the pharmacokinetic models being studied. For the non mathematician this can be visualised as an algorithm that changes the parameters you are trying to measure until it finds the best combination of parameters to describe the relationship that has been observed between the vascular input function curve and the tumour voxel curve. The inherent problem of these curve fitting methods is that the more complex the description of the curve and the more unknown parameters that have to be estimated the more likely it is that a range of different solutions can be found. The less specific the solution the less accurate and reliable the estimate of the underlying parameters will be. In basic terms it can therefore be seen that the use of a complex multiparametric model describing all of the physiological features that we have listed above will lead to instabilities in the analysis and increasing number of errors in the estimated parameters. This has led to the development of a range of simplified models which combine the effects of several parameters into one in order to reduce the number of variables used in the curve fit analysis.

Most workers have concentrated on the calculation of the contrast transfer coefficient K^{trans} (Larsson et al. 1990; Brix et al. 1991; Tofts 1996, 1997; Tofts et al. 1995; St Lawrence and Lee 1998; Li et al. 2003). Papers describing variations or changes in K^{trans} are now commonplace but it is often not appreciated that the meaning of K^{trans} will differ depending on the analytical model that has been applied. A simple model such as that described by Tofts and Kermode (1991) will estimate only two parameters, the first of these is the size of the EES (v_e) and the second is K^{trans}. Where this model is used therefore K^{trans} will be affected by flow, endothelial permeability, endothelial surface area product and by the proportional blood volume of the voxel (Fig. 1.9). High values of K^{trans} will therefore be seen where there is high flow, high permeability, high capillary surface area or a large proportion of intravascular contrast within the voxel. Although this measurement is highly non-specific it is relatively reproducible, provides a quantitative measurement of microvascular structure and function and has been widely used in many studies.

A further level of complexity in the analysis is introduced in other models which attempt to separate out the effect of the vascular fraction which is one of the main problems with the simple Tofts and Kermode type approach. These models will calculate vascular fraction (v_p), v_e and K^{trans} which will now be affected by flow, capillary surface area and endothelial permeability (Fig. 1.9). In practice it is also extremely desirable to separate out the effects of flow from those of capillary endothelial permeability and surface area (St Lawrence and Lee 1998; Li et al. 2003). There are two main reasons for this. Firstly, many pharmaceutical agents currently in development directly affect the angiogenic process. These imaging techniques are widely used in the study of such compounds and there is a specific interest in the changes in vascular endothelial permeability which occur in response to antiangiogenic drugs. The use of measurements of K^{trans}, which are unable to differentiate these parameters from flow-based effects, is therefore potentially confusing. Secondly, it must be appreciated that measurements of K^{trans} with this type of model are essentially flow limited. This is an important concept which can be illustrated as follows. If a tumour has very high capillary endothelial surface area and capillary endothelial permeability then the leakage of contrast from the plasma into the EES will be rapid and large. However, if the blood flow to this tumour is extremely low then this fast leakage out of the vascular system will deplete the contrast in the vascular tree which will not be replenished sufficiently quickly to keep up with the rate of leakage. Under these circumstances the measured K^{trans} will not reflect the capillary endothelial permeability or surface area but will reflect only the flow. A measurement of K^{trans} which is low may therefore represent an area of relatively high flow and low permeability or could represent low flow and high permeability. The value K^{trans} is commonly considered to be synonymous with endothelial permeability and it can be seen that this is clearly not true unless and appropriate model is used. Such a model has been described and is available for analysis but as described above the presence of multiple parameters makes it relatively unstable and highly sensitive to poor signal-to-noise ratio in the underlying data (St Lawrence and Lee 1998).

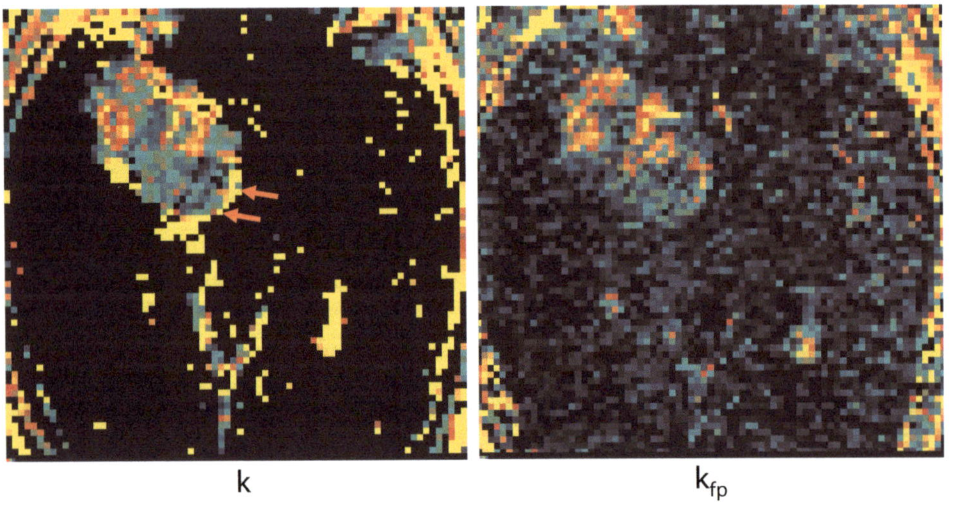

Fig. 1.9. Maps of K^{trans} calculated using a standard two compartment model (*left*) and a modified first past model designed to exclude erroneous high values at the sight of blood vessels. Note that the high value seen in the *left image* (*arrows*) corresponding to a large draining vein adjacent to the tumour is not seen on the *right*

1.5 Applications of Dynamic Contrast Enhanced MRI

1.5.1 Tumour Detection

Early detection and treatment of malignant disease offers the opportunity to improve outcome and survival. Early detection and treatment has a major impact on the survival of patients with many different cancers including breast, cervix, endometrium, ovary; rectum, prostate and lung. Early detection of metastatic spread is also of considerable importance of therapeutic planning (Sobin and Fleming 1997) and will strongly affect the treatment options which will be offered to individual patients. DCE-MRI has been shown to be more sensitive than mammography in demonstrating multifocal breast disease (Drew et al. 1999) and it has been suggested that there is probably a place in screening high-risk women for breast cancer (Boetes and Stoutjesdijk 2001; Kneeshaw 2003). A role for the detection of early recurrence after radiotherapy has also been identified (Drew et al. 1998; Belli et al. 2002). The identification of direct clinical roles of this type is steadily increasing in a wider range of tumours including breast, head and neck neoplasms and prostate cancer (Gribbestad et al. 1992; Baba et al. 1997; Padhani et al. 2000; Ogura et al. 2001; Aronen and Perkio 2002; Kremer et al. 2003).

1.5.2 Tumour Characterisation

The ability of DCE-MRI to quantify a range of characteristics of the tumour microvasculature has encouraged many investigators to use this technique as a basis for "in-vivo staging" of tumours. Correlation of DCE-MRI-based features of the tumour microvasculature with pathology, therapeutic response and prognosis remains the main goal for future work with these techniques. Even early studies were able to demonstrate a clear relationship between rapid and large increases in signal enhancement and high grade malignant behaviour in a range of tumours. Several studies however have also shown significant overlapping of contrast enhancement patterns between benign and malignant tumours (Fig. 1.10). This has been described in breast cancer and fibroadenoma (Kvistad et al. 2000) and prostate cancer and benign prostatic hyperplasia (Barentsz et al. 1999) (Figs. 1.11, 1.12). It is hoped that increased specificity and accuracy in the identification of microvascular characterisation parameters will progressively improve lesion characterisation possible with non-invasive in-vivo techniques. Improved differentiation for breast tumours has certainly been obtained using fast T2*-weighted first-pass perfusion imaging after the standard T1-weighted DCE-MRI examination (Kuhl et al. 1999; Kvistad et al. 2000) and increased speed and model specificity for T1-weighted DCE-MRI improves the characterisation of prostate and bladder cancer (Barentsz et al. 1999).

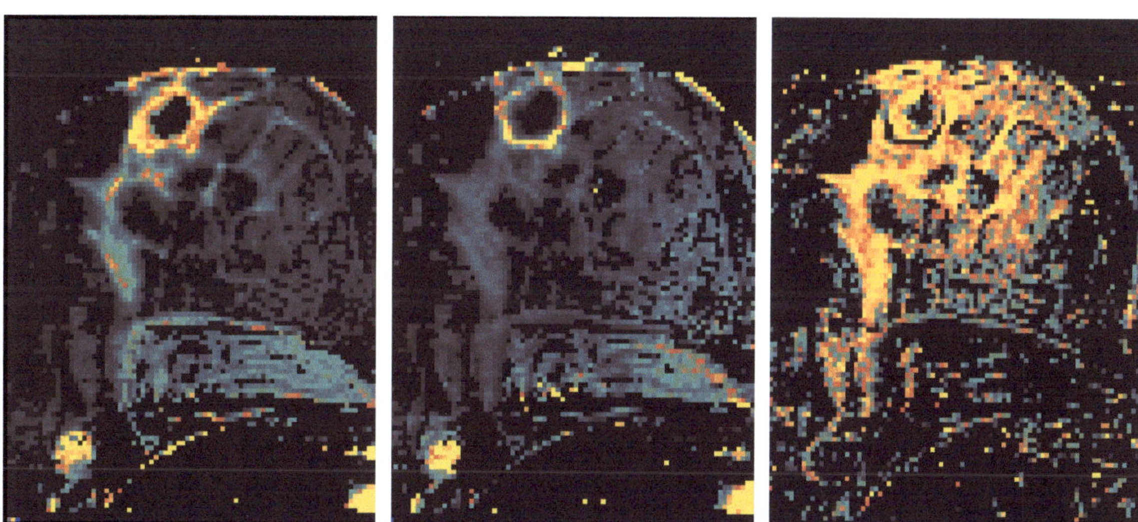

Fig. 1.10. Images of tumour blood volume (*right*), K^{trans} (*centre*) and ve in a patient with a malignant breast carcinoma

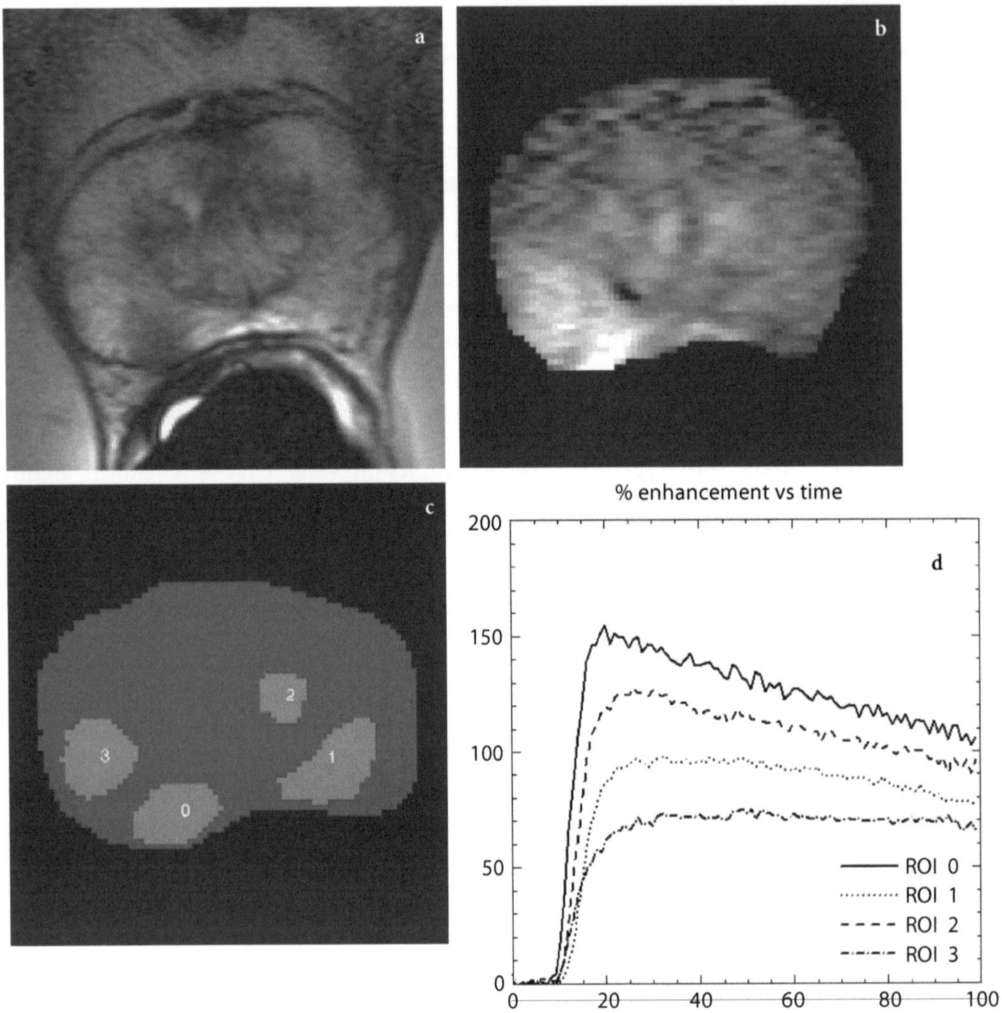

Fig. 1.11. a,b T2-weighted MR image (**a**) and corresponding 3D T1-weighted EPI (**b**) of a patient with prostate cancer (Gleason score 8). The time resolution is 2.8 s/ten slices, slice thickness ~6 mm for DCE-MRI examinations. **c** Four regions of interest (ROIs) were evaluated, the area marked as *ROI 0* has been proven by histopathology to be cancer. **d** The enhancement curves demonstrated highest signal intensity in this region of the prostate with a clear wash-out phase. Also, *ROI 2* demonstrates enhancement but consists of benign hyperplasia

DCE-MRI also offers the opportunity for monitoring tumour treatment. For the operable cancer patient verification and complete resection of the tumour is of primary importance and governs the decision as to whether to use adjuvant therapeutic techniques such as chemotherapy, endocrine therapy or radiotherapy. Treatment monitoring in these patients consists of regular clinical examinations using endpoints such as survival or time-to-progression (TTP). The clinical evaluation of tumour response to treatment is based on changes in tumour size using the WHO criteria (WHO 1979) or the revised version (Therasse et al. 2000) under the acronym RECIST (response evaluation criteria in solid tumours). A complete response (CR) is the complete disappearance of tumour, partial response (PR) is defined as 50% (WHO) or 30% (RECIST) decrease of tumour diameters, progressive disease (PD) is increase of tumour size by 25% (WHO) or 30% (RECIST). A stable disease (SD) is defined as neither CR, PR nor PD.

New evaluation strategies are needed to follow individual treatment response in detail. Individual treatment plans depend on tumour characteristics and results of histopathological staging. The need for early assessment of response is increasing in order to optimise individual treatment regimes and to avoid over- as well as under treatment. Methods that not only demonstrate the anatomical outline but also

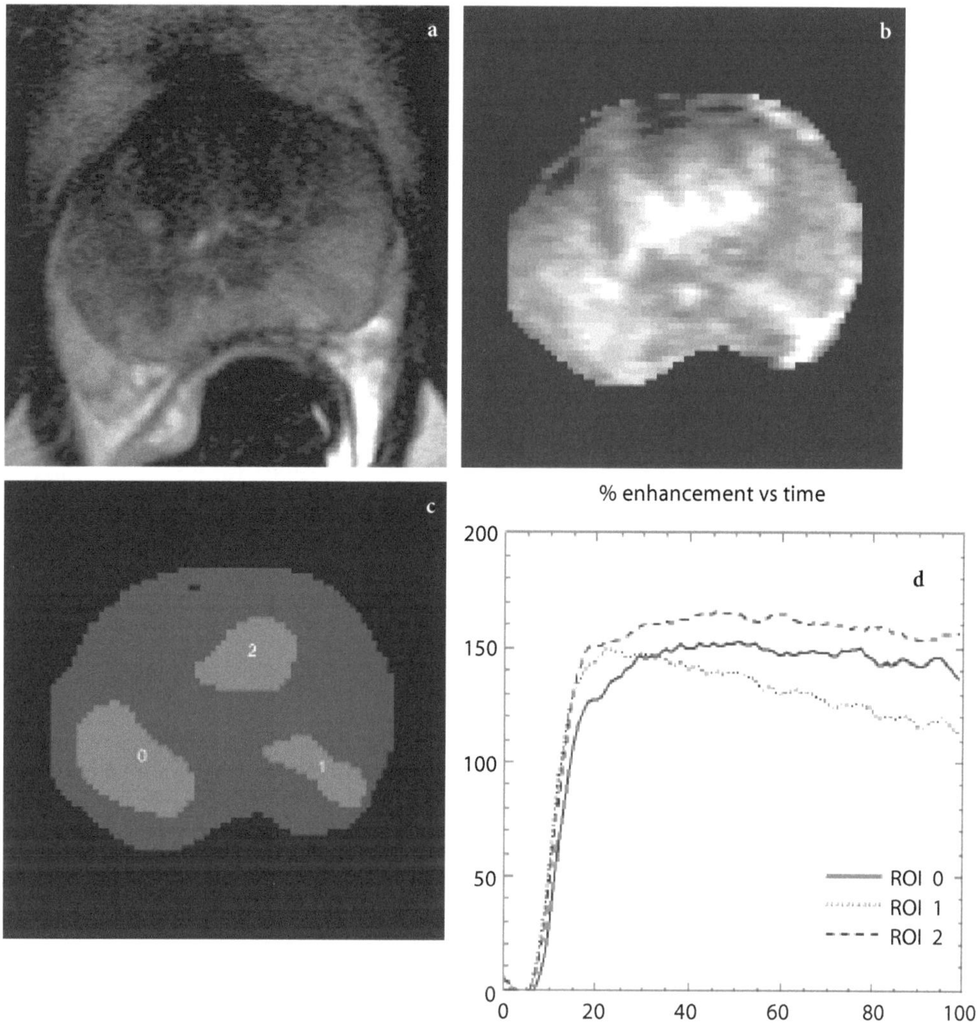

Fig. 1.12a–d. MR images from another prostate cancer patient (Gleason score 8). **a** T2-weighted MR image, (**b**) corresponding dynamic 3D T1-weighted EPI image, (**c**) selected ROIs and corresponding enhancement curves (**d**). Again, the area *ROI 0* was shown by histopathology to be cancer. The *ROI 1* was interpreted as cancer, but proved to be adenomatous hyperplasia with prostatitis

functional or metabolic changes would provide valuable information for clinical decisions. Several studies have demonstrated the possible role of DCE-MRI as a tool for early detection of treatment response (Drew et al. 1998, 2001; Barentsz et al. 1999; Padhani 2002). It has been proposed that DCE MRI during the first 2 weeks of radiotherapy may provide early prediction of tumour regression rate, and therefore be of value in designing treatment schedules for cervical carcinoma (Gong et al. 1999). Also, DCE-MRI has been used to quantify the extent of poor vascularity regions within the tumour to predict long-term tumour control and treatment outcome in cervical cancer (Mayr et al. 2000).

1.6 Conclusions

Dynamic contrast enhanced MRI is a relatively new technique with a wide range of applications. The importance of the technique lies in its ability to produce surrogate markers which describe features of the anatomy or physiology of the tumour microvasculature. Modern scanners are more than capable of producing appropriate imaging sequences for a wide range of DCE-MRI applications. Although the design of image acquisition and analysis protocols is complex it is clear that even simple approaches can produce valuable clinical information in some settings and that the development of more complex pharmacokinetic-based modelling approaches have the facility to provide exquisite characterisation of tumour with a wide range of potential applications in diagnosis management treatment planning and follow up. Subsequent chapters in this book will describe in detail the technical applications of sequence design and analysis methods which have been introduced here and will illustrate specific clinical applications in a wide range of tumour types.

References

Aronen HJ, Perkio J (2002) Dynamic susceptibility contrast MRI of gliomas. Neuroimaging Clin N Am 12(4):501–523
Baba Y, Furusawa M, Murakami R, Yokoyama T, Sakamoto Y, Nishimura R, Yamashita Y, Takahashi M, Ishikawa T (1997) Role of dynamic MRI in the evaluation of head and neck cancers treated with radiation therapy. Int J Radiat Oncol Biol Phys 37:783–787
Barbier EL, Lamalle L, Decorps M (2001) Methodology of brain perfusion imaging. J Magn Reson Imaging 13(4):496–520
Barentsz JO, Engelbrecht M, Jager GJ, Witjes JA, de LaRosette J, van Der Sanden BP, Huisman HJ, Heerschap A (1999) Fast dynamic gadolinium-enhanced MR imaging of urinary bladder and prostate cancer. J Magn Reson Imaging 10:295–304
Bartolozzi C, Lencioni R, Donati F, Cioni D (1999) Abdominal MR: liver and pancreas. Eur Radiol 9:1496–1512
Belli P, Pastore G, Romani M, Terribile D, Canade A, Costantini M (2002) Role of magnetic resonance imaging in the diagnosis of recurrence after breast conserving therapy. Rays 27(4):241–257
Bhujwalla ZM, Artemov D, Glockner J (1999) Tumor angiogenesis, vascularization, and contrast-enhanced magnetic resonance imaging. Top Magn Reson Imaging 10:92–103
Boetes C, Stoutjesdijk M (2001) MR imaging in screening women at increased risk for breast cancer. Magn Reson Imaging Clin N Am 9:357-72, vii
Boetes C, Barentsz JO, Mus RD, van der Sluis RF, van Erning LJ, Hendriks JH, Holland R, Ruys SH (1994) MR characterization of suspicious breast lesions with a gadolinium-enhanced TurboFLASH subtraction technique. Radiology 193:777–781
Brix G, Semmler W, Port R, Schad LR, Layer G, Lorenz WJ (1991) Pharmacokinetic parameters in CNS Gd-DTPA enhanced MR imaging. J Comput Assist Tomogr 15:621–628
Calamante F, Gadian DG, Connelly A (2000) Delay and dispersion effects in dynamic susceptibility contrast MRI: simulations using singular value decomposition. Magn Reson Med 44:466–473
Carmeliet P, Jain RK (2000) Angiogenesis in cancer and other diseases. Nature 407:249–257
Daniel BL, Yen YF, Glover GH, Ikeda DM, Birdwell RL, Sawyer-Glover AM, Black JW, Plevritis SK, Jeffrey SS, Herfkens RJ (1998) Breast disease: dynamic spiral MR imaging. Radiology 209:499–509
Delorme S, Knopp MV (1998) Non-invasive vascular imaging: assessing tumour vascularity. Eur Radiol 8:517–527
Drew PJ, Kerin MJ, Turnbull LW, Imrie M, Carleton PJ, Fox JN, Monson JR (1998) Routine screening for local recurrence following breast-conserving therapy for cancer with dynamic contrast-enhanced magnetic resonance imaging of the breast. Ann Surg Oncol 5(3):265–270
Drew PJ, Chatterjee S, Turnbull LW, Read J, Carleton PJ, Fox JN, Monson JR, Kerin MJ (1999) Dynamic contrast enhanced magnetic resonance imaging of the breast is superior to triple assessment for the pre-operative detection of multifocal breast cancer. Ann Surg Oncol 6:599–603
Drew PJ, Kerin MJ, Mahapatra T, Malone C, Monson JR, Turnbull LW, Fox JN (2001) Evaluation of response to neoadjuvant chemoradiotherapy for locally advanced breast cancer with dynamic contrast-enhanced MRI of the breast. Eur J Surg Oncol 27:617–620
Dvorak AM, Kohn S, Morgan ES, Fox P, Nagy JA, Dvorak HF (1996) The vesiculo-vacuolar organelle (VVO): a distinct endothelial cell structure that provides a transcellular pathway for macromolecular extravasation. J Leukoc Biol 59:100–115
Gilles R, Guinebretière JM, Toussaint C, Spielman M, Rietjens M, Petit JY, Contesso G, Masselot J, Vanel D (1994) Locally advanced breast cancer: contrast-enhanced subtraction MR imaging of response to preoperative chemotherapy. Radiology 191:633–638
Gillies RJ, Bhujwalla ZM, Evelhoch J, Garwood M, Neeman M, Robinson SP, Sotak CH, Van Der Sanden B (2000) Applications of magnetic resonance in model systems: tumor biology and physiology. Neoplasia 2:139–151
Gong QY, Brunt JN, Romaniuk CS, Oakley JP, Tan LT, Roberts N, Whitehouse GH, Jones B (1999) Contrast enhanced dynamic MRI of cervical carcinoma during radiotherapy: early prediction of tumour regression rate. Br J Radiol 72(864):1177–1184
Gribbestad IS, Nilsen G, Fjosne H, Fougner R, Haugen OA, Petersen SB, Rinck PA, Kvinnsland S (1992) Contrast-enhanced magnetic resonance imaging of the breast. Acta Oncol 31(8):833–842
Gribbestad IS, Nilsen G, Fjøsne HE, Kvinnsland S, Haugen OA, Rinck PA (1994) Comparative signal intensity measurements in dynamic gadolinium-enhanced MR mammography. J Magn Reson Imaging 4:477–480
Griswold MA, Jakob PM, Nittka M, Goldfarb JW, Haase A (2000) Partially parallel imaging with localized sensitivities (PILS). Magn Reson Med 44:602–609
Gyngell ML (1988) The application of steady-state free precession in rapid 2DFT NMR imaging: FAST and CE-FAST sequences. Magn Reson Imaging 6:415–419
Haase A (1990) Snapshot FLASH MRI. Applications to T1, T2, and chemical-shift imaging. Magn Reson Med 13:77–89
Hayes C, Padhani AR, Leach MO (2002) Assessing changes in tumour vascular function using dynamic contrast-enhanced magnetic resonance imaging. NMR Biomed 15:154–163
Jain RK (1987) Transport of molecules across tumor vasculature. Cancer Metastasis Rev 6:559–593
Jain RK (1988) Determinants of tumor blood flow: a review. Cancer Res 48:2641–2658
Jones RA, Haraldseth O, Müller TB, Rinck PA, Oksendal AN (1993) K-space substitution: a novel dynamic imaging technique. Magn Reson Med 29:830–834

Kassner A, Annesley DJ, Zhu XP, Li KL, Kamaly-Asl ID, Watson Y, Jackson A (2000) Abnormalities of the contrast re-circulation phase in cerebral tumors demonstrated using dynamic susceptibility contrast-enhanced imaging: a possible marker of vascular tortuosity. J Magn Reson Imaging 11(2):103–113

Kneeshaw PJ, Turnbull LW, Drew PJ (2003) Current applications and future direction of MR mammography. Br J Cancer 88:4–10

Knopp MV, Weiss E, Sinn HP, Mattern J, Junkermann H, Radeleff J, Magener A, Brix G, Delorme S, Zuna I, van Kaick G (1999) Pathophysiologic basis of contrast enhancement in breast tumors. J Magn Reson Imaging 10:260–266

Knopp MV, Giesel FL, Marcos H, von Tengg-Kobligk H, Choyke P (2001) Dynamic contrast-enhanced magnetic resonance imaging in oncology. Top Magn Reson Imaging 12:301–308

Kremer S, Grand S, Berger F, Hoffmann D, Pasquier B, Remy C, Benabid AL, Bas JF (2003) Dynamic contrast-enhanced MRI: differentiating melanoma and renal carcinoma metastases from high-grade astrocytomas and other metastases. Neuroradiology 45(1):44–49

Kuhl CK, Schild HH (2000) Dynamic image interpretation of MRI of the breast. J Magn Reson Imaging 12:965–974

Kuhl CK, Mielcareck P, Klaschik S, Leutner C, Wardelmann E, Gieseke J, Schild HH (1999) Dynamic breast MR imaging: are signal intensity time course data useful for differential diagnosis of enhancing lesions?. Radiology 211:101–110

Kvistad KA, Rydland J, Vainio J, Smethurst HB, Lundgren S, Fjosne HE, Haraldseth O (2000) Breast lesions: evaluation with dynamic contrast-enhanced T1-weighted MR imaging and with T2*-weighted first-pass perfusion MR imaging. Radiology 216(2):545–553

Larsson HB, Stubgaard M, Frederiksen JL, Jensen M, Henriksen O, Paulson OB (1990) Quantitation of blood-brain barrier defect by magnetic resonance imaging and gadolinium-DTPA in patients with multiple sclerosis and brain tumors. Magn Reson Med 16:117–131

Li KL, Zhu XP, Checkley DR, Tessier JJ, Hillier VF, Waterton JC, Jackson A (2003) Simultaneous mapping of blood volume and endothelial permeability surface area product in gliomas using iterative analysis of first-pass dynamic contrast enhanced MRI data. Br J Radiol 76:39–50

Liu G, Sobering G, Duyn J, Moonen CT (1993) A functional MRI technique combining principles of echo-shifting with a train of observations (PRESTO). Magn Reson Med 30:764–768

Mayr NA, Yuh WT, Arnholt JC, Ehrhardt JC, Sorosky JI, Magnotta VA, Berbaum KS, Zhen W, Paulino AC, Oberley LW, Sood AK, Buatti JM (2000) Pixel analysis of MR perfusion imaging in predicting radiation therapy outcome in cervical cancer. J Magn Reson Imaging 12:1027–1033

Mukhopadhyay D, Tsiokas L, Sukhatme VP (1995) Wild-type p53 and v-Src exert opposing influences on human vascular endothelial growth factor gene expression. Cancer Res 55(24):6161–6165

Oesterle C, Strohschein R, Köhler M, Schnell M, Hennig J (2000) Benefits and pitfalls of keyhole imaging, especially in first-pass perfusion studies. J Magn Reson Imaging 11:312–323

Ogura K, Maekawa S, Okubo K, Aoki Y, Okada T, Oda K, Watanabe Y, Tsukayama C, Arai Y (2001) Dynamic endorectal magnetic resonance imaging for local staging and detection of neurovascular bundle involvement of prostate cancer: correlation with histopathologic results. Urology 57:721–726

Oppelt A, Graumann R, Fisher H, Hertl W, Schajor W (1986) FISP: a new fast MRI sequence. Electromedica 3:15–18

Padhani AR (1999) Dynamic contrast-enhanced MRI studies in human tumours. Br J Radiol 72(857):427–431

Padhani AR (2002) Dynamic contrast-enhanced MRI in clinical oncology: current status and future directions. J Magn Reson Imaging 16(4):407–422

Padhani AR, Gapinski CJ, Macvicar DA, Parker GJ, Suckling J, Revell PB, Leach MO, Dearnaley DP, Husband JE (2000) Dynamic contrast enhanced MRI of prostate cancer: correlation with morphology and tumour stage, histological grade and PSA. Clin Radiol 55:99–109

Pruessmann KP, Weiger M, Scheidegger MB, Boesiger P (1999) SENSE: sensitivity encoding for fast MRI. Magn Reson Med 42:952–962

Rosen BR, Belliveau JW, Buchbinder BR, McKinstry RC, Porkka LM, Kennedy DN, Neuder MS, Fisel CR, Aronen HJ, Kwong KK et al (1991) Contrast agents and cerebral hemodynamics. Magn Reson Med 19:285–292

Ross BD, Chenevert TL, Rehemtulla A (2002) Magnetic resonance imaging in cancer research. Eur J Cancer 38(16):2147–2156

Ruddon RW (1987) Cancer Biology.

Senger DR, Galli SJ, Dvorak AM, Perruzzi CA, Harvey VS, Dvorak HF (1983) Tumor cells secrete a vascular permeability factor that promotes accumulation of ascites fluid. Science 219:983–985

Sobin LH, Fleming ID (1997) TNM Classification of Malignant Tumors, fifth edition (1997). Union Internationale Contre le Cancer and the American Joint Committee on Cancer. Cancer 80:1803–1804

Sodickson DK, Manning WJ (1997) Simultaneous acquisition of spatial harmonics (SMASH): fast imaging with radiofrequency coil arrays. Magn Reson Med 38:591–603

St Lawrence KS, Lee TY (1998) An Adiabatic Approximation to the Tissue Homogeneity Model for Water Exchange in the Brain: I. Theoretical Derivation. J Cereb Blood Flow Metab 18:1365–1377

Taylor JS, Reddick WE (2000) Evolution from empirical dynamic contrast-enhanced magnetic resonance imaging to pharmacokinetic MRI. Adv Drug Deliv Rev 41:91–110

Thacker NA, Scott ML, Jackson A (2003) Can dynamic susceptibility contrast magnetic resonance imaging perfusion data be analyzed using a model based on directional flow?. J Magn Reson Imaging 17:241–255

Therasse P, Arbuck SG, Eisenhauer EA, Wanders J, Kaplan RS, Rubinstein L, Verweij J, Van Glabbeke M, van Oosterom AT, Christian MC, Gwyther SG (2000) New guidelines to evaluate the response to treatment in solid tumors. European Organization for Research and Treatment of Cancer, National Cancer Institute of the United States, National Cancer Institute of Canada. J Natl Cancer Inst 92(3):205–216

Tofts PS (1996) Optimal detection of blood-brain barrier defects with Gd-DTPA MRI-the influences of delayed imaging and optimised repetition time. Magn Reson Imaging 14(4):373–380

Tofts PS (1997) Modeling tracer kinetics in dynamic Gd-DTPA MR imaging. J Magn Reson Imaging 7:91–101

Tofts PS (1997) Modeling tracer kinetics in dynamic Gd-DTPA MR imaging. J Magn Reson Imaging 7:91–101

Tofts PS, Berkowitz B, Schnall MD (1995) Quantitative analysis of dynamic Gd-DTPA enhancement in breast tumors using a permeability model. Magn Reson Med 33:564–568

Tofts PS, Brix G, Buckley DL, Evelhoch JL, Henderson E, Knopp MV, Larsson HB, Lee TY, Mayr NA, Parker GJ, Port RE, Taylor J, Weisskoff RM (1999) Estimating kinetic parameters from dynamic contrast-enhanced T(1)-weighted MRI of a diffusable tracer: standardized quantities and symbols. J Magn Reson Imaging 10(3):223–232

Tsao J, Boesiger P, Pruessmann KP (2003) k-t BLAST and k-t SENSE: dynamic MRI with high frame rate exploiting spatiotemporal correlations. Magn Reson Med 50(5):1031–1042

van Vaals JJ, Brummer ME, Dixon WT, Tuithof HH, Engels H, Nelson RC, Gerety BM, Chezmar JL, den Boer JA (1993) "Keyhole" method for accelerating imaging of contrast agent uptake. J Magn Reson Imaging 3:671–675

Verstraete KL, De Deene Y, Roels H, Dierick A, Uyttendaele D, Kunnen M (1994) Benign and malignant musculoskeletal lesions: dynamic contrast-enhanced MR imaging-parametric "first-pass" images depict tissue vascularization and perfusion. Radiology 192(3):835–843

Wang HZ, Riederer SJ (1990) A spoiling sequence for suppression of residual transverse magnetization. Magn Reson Med 15:175–191

Weidner N (1996) Intratumoral vascularity as a prognostic factor in cancers of the urogenital tract. Eur J Cancer 32A:2506–2512

Weisskoff RM, Zuo CS, Boxerman JL, Rosen BR (1994) Microscopic susceptibility variation and transverse relaxation: theory and experiment. Magn Reson Med 31:601–610

WHO (1979) Handbook for reporting results of cancer treatment.

Zhu XP, Li KL, Kamaly-Asl ID, Checkley DR, Tessier JJ, Waterton JC, Jackson A (2000) Quantification of endothelial permeability, leakage space, and blood volume in brain tumors using combined T1 and T2* contrast-enhanced dynamic MR imaging. J Magn Reson Imaging 11:575–585

2 Contrast Agents for Magnetic Resonance Imaging

Timothy P. L. Roberts and Michael D. Noseworthy

CONTENTS

2.1 Overview

One of the distinguishing characteristics of magnetic resonance imaging (MRI) compared with other cross sectional imaging modalities, such as X-ray computed tomography (CT), is the potential sensitivity to a wide range of microenvironmental physico-chemical properties. Sometimes viewed as signal intensity confounds, these sensitivities can be harnessed by focusing the MR imaging experiment (or "pulse sequence") on an isolated or limited subset of such parameters. These physico-chemical microenvironmental characteristics include among others: (1) tissue relaxation times (T_1 and T_2), which reflect local macromolecule/membrane density and water mobility due to magnetic interactions between water molecules and their environment, (2) water molecule diffusion, (3) tissue perfusion and fluid flow, (4) magnetization transfer, which is indicative of chemical exchange between bound and free proton environments, and thus provides information concerning macromolecule density and exchange kinetics. Indeed with such an impressive array of physiologically-specific sensitivities to choose from, the exquisite soft tissue contrast of MRI can be exploited and interpreted in a range of physiological and pathophysiological tissue states.

However, despite this range, there are situations in which such endogenous contrast mechanisms are insufficient. The apparent physiological specificity alluded to above, is simply inadequate to optimally delineate and, importantly, characterize pathological tissue. Often this may be the result of the dynamic range and measurement error in the parameters being calculated: for example, the entire dynamic range of physiologically-relevant T_1 measurements may go from only 250 ms to 2500 ms, or only a factor of 10 difference from minimum to maximum. With the complication of measurement error, the ability to discriminate normal from pathological tissues within such a narrow range becomes problematic. As in nuclear medicine and indeed X-ray CT, this has motivated the development of exogenous contrast media (tracers or "magnetic dyes") with properties suited to yet more specific physiological interpretation. It is desirable, for example, to develop and explore newer agents that can act to change tissue signal from specific proton pools, or tissue "compartments" (e.g., extravascular vs. intravascular), or to exhibit antigenic site directed contrast enhancement or contrast modulation due to the action of some metabolic "switch".

The purpose of this chapter is to provide an overview of the mechanisms of contrast agent-mediated signal changes, to describe contrast agent categories (clinically available and under research development), to introduce dynamic contrast-enhanced MRI (dMRI or DCE-MRI) and some forms of kinetic time series analysis (which will be expounded upon in subsequent chapters), to discuss attempts to validate the physiological inferences drawn from DCE-MRI data, and finally to direct the reader's attention towards current and future applications of these

T. P. L. Roberts, PhD
Professor, Department of Medical Imaging, University of Toronto, Canada Research Chair in Imaging Research, Fitzgerald Building, 150 College St., Room 88, Toronto, Ontario, M5S 3E2, Canada
M. D. Noseworthy, PhD
Departments of Medical Imaging and Medical Biophysics, University of Toronto, Hospital for Sick Children Research Scientist, The Hospital for Sick Children, 555 University Ave., Toronto, Ontario, M56 1X8, Canada

methodologies in research and clinical practice: the primary goal of contrast agents is to increase the signal difference between normal and abnormal tissues. In theory, this difference offers the promise of improved discrimination of pathological areas and allows one to monitor the progression of disease and therapeutic efficacy.

2.2 Relaxation Enhancement

Biological tissues are associated with characteristic relaxation time constants T_1 and T_2, which reflect processes of longitudinal and transverse relaxation (often referred to as spin–lattice and spin–spin relaxation, respectively). These relaxation time-constants may indeed be directly interpreted as indicative of the nature of the microenvironment: longitudinal relaxation is facilitated by (although not solely by) the presence of macromolecular or microstructural entities – hence water proton T_1 may be short in white matter, and rather longer in cerebrospinal fluid in which such moieties are less prevalent. Similarly, spin–spin relaxation rates are facilitated by interactions between protons of water molecules. In the cerebrospinal fluid (CSF) environment, such interactions are brief in duration and so remain relatively ineffective (leading to a rather long T_2 value). By comparison in more physically constrained biological tissues, in which interactions are more pronounced, the T_2 times are consequently generally shorter.

T_1 and T_2 values can be selectively probed using different MR imaging pulse sequences designed to achieve either "T_1-weighting" or "T_2-weighting". By varying the "T_1-" or "T_2-" sensitivity, it is further possible to produce parametric estimates (on a regional basis) of the T_1 and T_2 values themselves. T_1-weighting is generally achieved either using a short TR, short TE spin-echo or gradient-echo pulse sequence (in which relaxation occurs during the period TR, and therefore signal is elevated in rapidly relaxing – short T_1 – species), or using an inversion-recovery (IR) prepared imaging sequence, in which magnetization is inverted a time TI prior to onset of the "imaging portion" of the pulse sequence. Two widespread applications of this approach occur at short and long TI extrema: "short tau inversion recovery" (STIR) and "fluid attenuated inversion recovery" (FLAIR), for the suppression of short T_1 species (predominantly lipids), and long T_1 species (predominantly CSF), respectively.

Sensitivity or "weighting" towards T_2 is commonly achieved with a long TR, long TE spin-echo approach in which spin-spin or transverse relaxation occurs over the time period TE and leads to signal loss. In the case of $T_2 << TE$, signal will be lost rapidly and signal intensity will be low. For species with T_2 values ~ TE or longer, signal loss will be less appreciable and relative hyperintensity will be observed in the resulting image. In spin-echo experiments the effect of TR and TE is most easily understood by evaluating the general signal equation:

$$S_{SE} \approx N[H] \cdot \left(e^{-TE/T_2}\right) \cdot \left(1 - e^{-TR/T_1}\right) \quad (1)$$

where S_{SE}, the relative spin echo signal is proportional to proton density, N[H]. When TE and TR are long the middle term grows while the last term approaches 1, resulting in signal dominated by T_2, and is therefore "T_2-weighted". T_1-weighting, with short TE and short TR, results in dominance of the final term with the middle term approaching 1. Finally if short TE is used with a long TR the N[H] term dominates resulting in proton density weighting. In general it should be appreciated that T_1, T_2 and proton density contribute to the signal intensity and contrast in any image. Changing the contrast weighting of a pulse sequence will change the relative contributions from these factors. However, all sequences are, to a degree, of "mixed weighting".

Beyond T_1 and T_2, a further apparent relaxation mechanism warrants introduction. In general less tissue specific, apparent transverse relaxation can be described by a constant T_2^*. Although this might initially be considered similar to T_2, T_2^* in fact incorporates other mechanisms of transverse signal loss, that are eliminated in the spin-echo based construction of T_2-weighting. These signal losses arise from "dephasing" or loss of precessional coherence of water proton spins as their transverse magnetization evolves following excitation. The origins of such loss of coherence are multiple, but can be considered simply as reflecting different local magnetic fields at different spatial positions included in the imaging volume. This magnetic field inhomogeneity may be attributable to magnet construction limits, endogenous magnetic susceptibility differences between tissues (leading to varying magnetic fields, particularly across bone–tissue and air–tissue interfaces), or to the presence of contaminates (e.g., particulate iron) which disturb the local magnetic field. Since these forms of dephasing are "reversed" in the formation of the spin-echo (which indeed motivated its inception), to achieve T_2*-sensitivity, a gradient-recalled echo imaging sequence is commonly employed. Similar to T_2, contrast agents may act to shorten T_2*, primarily through enhancement of spin incoherence mechanisms, or introduction of spatial inhomogeneity in magnetic field.

In general, exogenous contrast agents operate by enhancing or facilitating longitudinal and transverse relaxation (shortening T_1 or T_2), and/or by disturbing local magnetic field homogeneity (consequently shortening T_2*). The presence of the contrast agent can then be revealed on appropriately selected (and weighted) MR images. The added value of such relaxation enhancement is that it is in some way "tissue-specific", "tissue-compartment" specific, or even "biologically-targeted" (typically to probe expression of a specific receptor to which the agent may bind).

While T_1 and T_2 relaxation enhancement occurs via interaction with the outer shell electrons of the paramagnetic (e.g., Gadolinium, Gd) or ferromagnetic (e.g., Iron, Fe) species within the exogenous agent, T_2*-shortening occurs via field disruption associated with the paramagnetic, superparamagnetic or ferromagnetic nature of the agent itself. An important aspect of this is that T_1- and T_2-shortening contrast mechanisms are "short-range" phenomena, demanding close approach of water molecules to the contrast agent (~nm scale), whereas T_2*-shortening is a "field effect" and often significantly extends several mm beyond the physical location of the contrast agent itself. In both cases and in important contradistinction to nuclear medicine and X-ray CT tracer methodologies, MRI does not image the tracer concentration directly but rather observes the "effect of contrast agent presence on local water proton signal intensity." As such, there is a somewhat indirect link between degree of signal change and concentration of tracer. This impacts subsequent simple physiological modeling based on standard tracer-kinetic approaches.

2.3 Categories of Contrast Agents

MRI contrast agents may be divided into four broad categories: non-selective, partially selective, targeted, and activated. The most common agents are those of the non-selective group as they are those that are currently used clinically. Often referred to as extracellular fluid (ECF), or extravascular agents, these are typically small molecules (~500 Da) administered intravenously and leading to both T_1 and T_2* shortening. This subcategory of contrast media includes those agents routinely used in clinical applications such as the lanthanide chelates: Gadolinium diethylene-triaminepentaacetic acid (gadopentetate, Gd-DTPA, Magnevist, Schering AG, Berlin, Germany), Gd-diethylene-triaminepentaacetic acid bis-methylamide (gadodiamide, Gd-DTPA-BMA, Omniscan, Amersham Health, Oslo, Norway) and gadolinium 10-(2-hydroxypropyl)-1,4,7,10-tetraazacyclododecane-1,4,7-triacetic acid (gadoteridol, Gd-HP-DO3A, Prohance, Bracco s.p.A, Milan, Italy). Additionally, in Europe Gd-DOTA, gadolinium 1,4,7,10-tetraazacyclododecane-1,4,7,10-tetraacetic acid (gadoterate, Dotarem, Guerbet, Paris, France) is approved for clinical use. In general, these agents are administered intravenously and, while they do not cross an intact blood–brain barrier, they have a high (~50%) first pass extraction fraction, a quantitative measure of their relatively rapid extravasation from the intravascular to extravascular extracellular spaces in the body. In general, however, they do not subsequently enter tissue cells.

There are four main categories of normal microvasculature, based on endothelial structure: continuous (e.g., skeletal muscle), discontinuous (e.g., liver sinusoids), fenestrated (e.g., kidney glomerular microvasculature), and tight junctional (blood–brain barrier, BBB). The molecular size, shape, and charge of an agent determine whether it will be diffusible in a specific microvascular environment. As mentioned above, the most restrictive environment is the brain due to the presence of the BBB. Certain vascular midline brain regions, collectively known as circumventricular organs, are exceptions, however, and are void of a BBB. These include the posterior pituitary (neurohypophysis), median eminence, area postrema, pineal gland, subcommissural organ, and subfornical organ. Due to the vascular nature of these areas they often enhance with routine contrast applications (especially the larger brain regions such as posterior pituitary and pineal gland). Other microvascular barriers exist (e.g., blood–testicular, blood–retinal, blood–CSF); however, these are not as restrictive to passive molecular passage as the BBB.

Many diseases and pathologies can affect the permeability and hence enhancement characteristics of the microvasculature. The most widely studied is cancer which is known to produce an uncontrolled increase in microvascular density (hypervascularity). Often the tissue is also hyperpermeable to freely diffusing agents. However, this may not *always* be the case (Noseworthy et al. 1998) and permeability may be differently accentuated to tracers of different size (Su et al. 1999; Roberts et al. 2002a). Some theories of cancer biology have suggested that tumor microvascular permeability to blood borne proteins, fibrin and fibrinogen, may be a necessary part of new vessel formation (angiogenesis) critical for sustained tumor growth (Harris et al. 1996; Goede et al. 1998). Furthermore, microvascular permeability to tumor cells themselves may be a mechanism of distant metastatic spread. Many other diseases have a microvascular component and have been shown to cause increase in microvascular permeability to diffusible agents. For example, oxidative stress and/or low antioxidant status leads to elevated brain microvascular permeability (Noseworthy and Bray 1998, 2000). Inflammation, hypoxia, seizures, certain toxins, diabetes, and stroke are examples of other insults that alter the microvasculature and hence the nature of dynamic contrast enhancement (Brasch 1991; Demsar et al. 1997; van Dijke et al. 1997). On the other hand, wound healing and pro-angiogenesis in, for example, myocardial ischemia represent two conversely positive aspects of angiogenesis, similarly visualizable as demonstrating elevated microvascular permeability (Helbich et al. 2002).

The clinically approved Gd chelates differ in their biochemistry in that some are referred to as linear agents (Gd-DTPA, Gd-DTPA-BMA) while others are macrocyclic (Gd-HP-DO3A, Gd-DOTA). These agents differ in their osmolarity (630-1950 mOsm/kg water), ionic nature (for example, Gd-DTPA is ionic, whereas Gd-DTPA-BMA and Gd-HP-DO3A are non-ionic), and stability. Due to the toxicity of free Gd it is most desirable to have it tightly bound to the chelate (i.e., high stability constant). Gd-DTPA is more thermodynamically stable than the non-ionic Gd-DTPA-BMA, but also has a higher osmolarity and carries a +2 charge. The macrocyclic agents are typically more stable. These are newer and not yet as clinically widespread as Gd-DTPA which is the oldest approved MRI contrast agent.

Gadolinium has also been successfully bound to porphyrin structures. Porphyrins are well know to be effective chelates of metals (e.g. iron in hemoglobin, cobalt in vitamin B_{12}, and magnesium in chlorophyll). Gadolinium- texaphyrin, also called Xcytrin (Pharmacyclics, CA) is one such agent. Gd-texaphyrin can enhance tumor contrast and also has the property of increasing the sensitivity of hypoxic regions to radiation therapy (Rosenthal et al. 1999).

Other lanthanide chelates have been used with success under experimental conditions. For example Dysprosium (as Dy-DTPA-BMA) has been used to assess brain perfusion. Due to the higher T_2^* relaxivity (greater susceptibility effect) of the Dy ion (which has a greater magnetic moment than the Gd ion) the contrast with this agent is greater per unit dose (Moseley et al. 1991). However, toxicity issues have limited the usefulness to experimental studies, although a phase I clinical trial of a Dy-based contrast agent for MR perfusion of the brain showed promising imaging efficacy (Roberts et al. 1994). Other uses of dysprosium include Dy-bis-tripolyphosphate ($Dy[PPP]_2$) which has been used in vivo as a shift reagent to allow differentiation between intra and extracellular sodium in ^{23}Na-MRI. Although this agent produces a significant separation of the intra and extracellular ^{23}Na resonances it is cardiotoxic as a result of interference with Ca^{2+} and thus decreasing cardiac contractility (Miller et al. 1991).

Non-lanthanide contrast agents including manganese chelates, oxygen (O_2), and nitroxide spin labels have also been explored as putative diagnostic contrast agents for MRI. Manganese is the most successful, and most often clinically used, non-lanthanide based MRI contrast agent, most commonly seen as a chelate of dipyridoxyl diphosphate (Mn-DPDP), a derivative of vitamin B_6. Mn-DPDP is also known as Mangafodipir trisodium and is marketed as Teslascan by Nycomed-Amersham. In oncology Mn contrast reagents have been successfully used in the analysis of pancreatic lesions (Diehl et al. 1999; Sahani et al. 2002). In addition, Mn-DPDP has been hailed as a liver-specific contrast agent as it is sequestered there by Kupffer cells. There has even been some suggestion that Mn-DPDP can be useful in detecting lymph node involvement from hepatocellular carcinoma (Burkhill et al. 2001).

Some MRI applications have made use of the fact that ground state oxygen (O_2), the redox state that we are constantly breathing, is paramagnetic and contains two unpaired electrons (Karczmar et al. 1994; Oikawa et al. 1997; Noseworthy et al. 1999; Rijpkema et al. 2002). However, high resolution image contrast rarely changes as O_2 predominantly changes the T_2* through change in the oxy/deoxyHb ratio in the microvasculature. However, as a method, applied together with blood-oxygen level-dependent (BOLD) imaging, MRI with hyperoxia provides functional information on tumor biology such as whether a tumor may have greater probability of responding to radiation therapy (Rijpkema et al. 2002). The enhancement of T_1 relaxation via molecular oxygen has also been used as a contrast mechanism for studying lung ventilation (wherein local partial pressure of oxygen gas influences parenchymal proton relaxation). Alternately breathing room air (20% O_2) and 100% O_2, during inversion recovery (T_1-weighted) imaging allows delineation of local oxygen abundance (EDELMAN et al. 1996).

In recent years there has been a tremendous interest in non-diffusible contrast agents, particularly for application in MR angiography. One difficulty in MR angiography, when using diffusible agents, is the 'washing-out' of blood vessels of interest as the agent leaks from the circulation to tissue extracellular spaces. This facilitated a great push in research into non-diffusible agents which carried over into successful applications of cancer assessment. In some experimental and clinical circumstances differentiation between tumor grade has been made possible by using macromolecular and particulate agents as the extravascular leakage rate is reduced in general compared with low molecular weight agents, and may in fact not be detectable in healthy tissues (even outside the brain) (DALDRUP et al. 1998; TURETSCHEK et al. 2001b). Thus only pathologically hyperpermeable tissues (e.g., angiogenically active tumors) show extravasation of such tracers and thus progressive enhancement.

To date, most of the non-diffusible agents are not clinically useful due to potentially severe adverse effects. Toxicity may be due to the prolonged half-life of the agent and therefore probability of subsequent metabolism and release of the free metal. Alternatively, some agents induce potentially lethal immunologic reactions. Even though these agents catalyze well documented adverse events, thereby preventing widespread clinical utility, they have been very useful experimentally as prototypes. These agents include polymers of lysine chelated with Gd (Gd-polylysine) and derivatives (e.g., PEG-Gd-polylysine), the final size being in the hundreds of kDa (compared with clinically approved Gd chelates which are roughly 500 Da in size) as well as Gd-containing dendrimer structures of various sizes (or "generations") represented, for example, by Gadomer-17 (Schering AG, Berlin, Germany), which has shown considerable experimental promise in experimental models of anti-angiogenic therapy (ROBERTS et al. 2002a). In particular, for molecules of the order of 20–30 kDa, there may be a wide "dynamic range" of microvascular permeability rendering this measure potentially more sensitive to changes associated with disease progression or response than smaller molecules, such as Gd-DTPA (ROBERTS et al. 2002a). Gd has also been successfully bound to proteins such as albumin (65 kDa) and fibrinogen (150–200 kDa) and used as an intravascular or "true blood pool" agent (BRASCH 1991). It is worthy of note that at least at current clinical field strengths (~1.5 T), the enhancement potency or r_1 relaxivity (s^{-1} $mM[Gd]^{-1}$) of gadolinium increases significantly when the Gd moiety is bound to macromolecular structures compared with the low molecular weight agents.

Arguably the most clinically promising class of non-diffusible agent is the iron-oxide particle (Ultrasmall Particulate Iron Oxides, USPIOs) or monocrystalline iron oxide nanoparticle (MION). Examples include Clariscan (Amersham), Resovist (Schering), and Sinerem (Guerbet). In oncology these have been used in the evaluation of lymph node metastasis (SIGAL et al. 2002; STETS et al. 2002) and in one study for estimation of microvascular permeability in breast cancer (DALDRUP-LINK et al. 2002). Lymph nodes and liver seem to enhance well with particulate iron oxides, which is likely to be due to the sequestering of these agents within cells of the reticuloendothelial system (RES), in macrophages and Kupffer cells, which are plentiful in these tissues. Although not rich in macrophages, brain tumor evaluation has been reported to be superior with the use of these large particles (ENOCHS et al. 1999).

To avoid immunologic reactions due to the injection of foreign proteins like covalently-bound Gd-HSA (human serum albumin), EPIX medical (Cambridge, MA, USA) have developed a gadolinium based contrast agent (MS-325), a small molecule, which when injected binds to native albumin in the subject's own blood. The result is an intravascular agent that is safe for patient use. This has proved to be excellent in MR angiography studies (Grist et al. 1998). Preliminary studies in tumors, however, have demonstrated less success than using covalently-bound albumin-Gd-DTPA for permeability-based assessment of tumor malignancy (TURETSCHEK et al. 2001a). Additionally, as the agent in vivo is 85% bound and 15% free, the models describing signal change and its relationship with microvascular parameters such as capillary permeability would need to be more complex than ones currently used. A similar transient protein-binding agent, gadocoletic acid (B22956/1, Bracco) with different binding affinity has, however, shown considerable utility in experimental oncology therapy models (ROBERTS et al. 2002b)

Recently, there has been a reconsideration of ultrasound (US) microbubble contrast agents and their application in an MR environment. Microbubbles are typically filled with a gas, but may include chemotherapeutic agents. They have been successfully applied in US evaluation of tumor microvascular perfusion. And, in addition, by using focused US one can disrupt the microbubbles at sites of interest and thereby allow site directed delivery of therapeutic agents. Due to the size of the microbubbles (~ µm diameter) they remain intravascular, and thereby act as a non-diffusible contrast agent. When filled with gas there is a tremendous T_2* effect due to the microbubble gas–tissue interface. Interestingly, as the T_2* effect is related to the size of the microbubble and this size is affected by local microvascular pressure there have been some attempts in the use of these agents for the evaluation of intravascular pressure (Alexander et al. 1996). Such an application in tumors would prove invaluable as higher pressure is thought to be related to tumor severity.

One intriguing approach to non-diffusible agents has been demonstrated by Johnson et al. (1998). In this method, red blood cells (RBCs) were extracted from rabbits. They were then subjected to osmotic shock and subsequently loaded with high concentrations of Gd-DTPA. These were then re-introduced into the same rabbits, producing a large intravascular contrast effect without the risk of immunologic reactions. Although novel this method would be difficult to reproduce in a clinical setting and as the Gd biological half-life would be increased there would be increased risk of toxicity.

The final classes of contrast agents, targeted and activated, are experimental but have shown the most diverse applications of contrast agents in MRI. Targeted agents are those that contain a paramagnetic or ferromagnetic center bound to an antibody (Ab) which is directed at a specific in vivo antigen. As the concentration is typically lower (limited to the number of available epitopes) usually these experiments are done with large macromolecular complexes (to maximize the T_2* effect). Site directed MR imaging of the $\alpha_v\beta_3$ integrin, associated with angiogenesis has been demonstrated using a large liposome covered with Gd and linked to a biotinylated antibody (Sipkins et al. 1998). This experiment showed that MRI could be used to 'image' angiogenesis (indirectly).

Metabolically activated agents have recently been developed for model systems in which a particular biochemical event or condition results in a conformational change of a molecular "basket" containing a paramagnetic center. Some of these agents have been compared to "waste bins with lids": a metabolic switch opens the lid which is usually closed. Opening the lid allows water to interact with the contained paramagnetic nucleus and hence accelerates relaxation. Some examples include pH sensitive switches (Zhang et al. 1999) and Ca^{2+} sensitive agents (Li et al. 2002). For oncology, tumor pH is related to hypoxia. Therefore in vivo pH measurement could prove useful in tumor grading.

Imaging of gene therapy through metabolic switching has been shown to be possible (Weissleder et al. 1997). By adding the gene for tyrosine kinase into a vector containing a gene to be inserted for gene therapy purposes, contrast modulation can result and therefore potentially show whether the gene therapy was successful. The mechanism is through the polymerization of tyrosine into melanin by tyrosine kinase. This macromolecule subsequently binds native iron. When the insert has been successful there is a concurrent increase in local iron which therefore produces a T_2* effect, visualizable on MRI.

These are just a sample of the many potential contrast agents and clever biochemical ways to modulate MRI signal contrast in a functionally-dependent manner, for the assessment of tumors. The rest of this book describes these and many others in finer detail.

2.4 Dynamic Contrast-Enhanced MRI

A critically important extra-dimension is brought to MRI by the use of exogenous contrast agents: namely the time domain. In general, the process of selective accumulation of the tracer is non-instantaneous and thus its dynamics can be interrogated by rapidly repeated imaging. Parametric analysis of this dynamic process often allows physiologically specific inference. Specifically, application of this approach permits the non-invasive evaluation of tissue microvasculature in vivo (Tofts 1997; Griebel et al. 1997; Brix et al. 1991; Tofts and Kermode 1991; Larsson et al. 1990). Such microvascular parameters have been used to assess, for example, cancer (Griebel et al. 1997; Reddick et al. 1999) and cerebrovascular disease (Edelman et al. 1990; Tsuchida et al. 1997), amongst other pathologies.

There are two approaches used in DCE-MRI, based on whether the acquired signal is T_1-weighted or T_2* (or T_2)-weighted (the choice of method is often dictated by the vascular parameters that are being investigated). The primary differences between the two approaches are:

(1) The T_1-weighted methods (sometimes called "pseudo-steady-state") typically take longer to

acquire the data (5–7 min or more are usually needed to model the data sufficiently accurately to give meaningful results, while T_2* methods typically take less than 1–2 min, focusing on "first-pass" kinetics).

(2) The T_2* methods usually have complete coverage of the tissue of interest (at least over multiple 2D sections) while T_1 methods typically give limited spatial coverage if high temporal resolution is demanded[1].

(3) T_1-weighted methods result in an increase in local signal while T_2* methods give a decrease.

(4) T_2* methods are usually derived from echo-planar imaging methodologies and consequently suffer from intrinsic susceptibility sensitivity – manifest not only as (desirable) contrast agent-mediated signal changes, but also as (undesirable) spatial geometric distortions and signal abnormalities.

(5) T_2* methods make assessment of a vascular reference function (reflecting local intravascular tracer concentration) difficult compared with T_1 methods, because of lower resolution and inflow sensitivity (also the T_2* signal dependence on blood contrast agent concentration is different between large vessels and the microvasculature).

For both T_1 and T_2* approaches, it is possible that improvement in quantitation can be achieved by converting signal intensity changes observed directly on MRI into dynamic changes of contrast agent concentration to provide the source data for descriptive or kinetic modeling[2] (Schwickert et al. 1995; Roberts 1997). In contrast to nuclear medicine and CT, signal changes in MRI are not directly related to tracer concentration, in vivo. Common practice is to convert observed signal intensities into local changes in the pertinent relaxation rate ($1/T_1$, or $1/T_2$*). The implied assumption subsequently is that contrast agent potency (or relaxivity, r_1 or r_2* – the amount of relaxation enhancement per unit concentration of tracer) is the same across all tissues and compartments. This assumption is likely to be flawed in vivo, and may be especially inappropriate for transverse relaxivity, r_2*, where local magnetic susceptibility field gradients are a function of local compartmentalized distribution of the tracer, which remains beyond the resolution of current MR imaging. Converting signal intensity changes to relaxation rates, despite the above limitation, has been shown both theoretically (Roberts 1997) and experimentally (Schwickert et al. 1995) to offer considerable improvement in the accuracy of target parameter estimation.

Within the class of T_2 methodologies, there are two variants: T_2-weighted and T_2*-weighted. The T_2*-weighted method is the most routinely used clinical approach to DCE-MRI, where it is most often used to assess cerebral perfusion and blood volume. It has also proven efficacious in clinical breast cancer scanning (Kuhl et al. 1997). Both techniques monitor the passage of a bolus through the microvascular environment; however, the T_2*-weighted method, being based on a gradient echo technique, is more sensitive to blood vessels with larger diameters. T_2-weighting, based on spin echoes, gives sensitivity predominantly to smaller vessels (which arguably therefore provides a closer measure of perfusion at the functional level) and consequently produces a smaller signal change during bolus passage. The advantage of the T_2-weighted approach is in assessment of tissues that are typically plagued by susceptibility artifacts due to gas (or bone)–tissue interfaces (e.g., frontal lobes of the brain, abdomen).

The observed signal change in DCE-MRI reflects a complex association of tissue perfusion, microvascular permeability, blood volume, and extravascular volume (Tofts 1997; Henderson et al. 2000). Many assumptions are made when modeling the data in order to calculate these parameters. Of particular significance is tissue contrast agent location. In healthy brain it is fairly well understood that the contrast agent perfuses the tissue, causing a change in MR signal, and this signal returns to, or close to, baseline as the bolus passes into the draining veins. In this case, the intact BBB does not allow the agent to leak into the extravascular space. However, it is assumed the agent perfuses the tissue of interest equally through all capillaries. In brain tumors, and indeed other tissues where ultrastructural barriers are absent, it is assumed that microvascular leakage plays a highly significant role in maintaining the signal change, thus preventing the signal from returning to baseline following the passage of the bolus (Fig. 2.1). In some

---

[1] While T_1-weighted methods have historically been too slow to allow rapid volume coverage, developments in scanner hardware, in particular parallel imaging with multiple radiofrequency coils, have reduced the temporal sampling limits to the point where a T_1-weighted volume series may be acquired at a rate of 2–3 s. It is therefore likely that the differential in T_2*- and T_1-weighted methods will become less significant.

[2] It should be noted that doing so affects the distribution of errors in the observation, making subsequent estimates of parametric uncertainties more difficult.

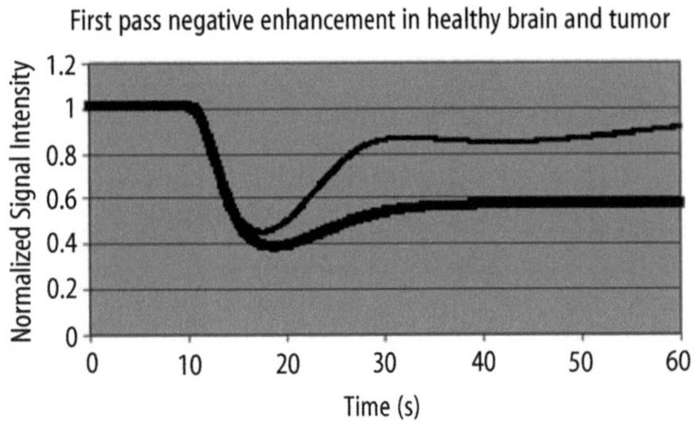

Fig. 2.1. First pass signal intensity during passage of a bolus of gadolinium-based contrast agent imaged using T_2*-weighted magnetic resonance imaging. In healthy brain (*thin line*) a transient drop in signal intensity associated with the contrast agent recovers towards baseline, although residual recirculation effects limit full recovery of signal. In tumor (*solid line*) rapid extravasation of the contrast agent as a result of blood–brain barrier damage (hyperpermeability) leads to persistence of signal loss beyond the first pass

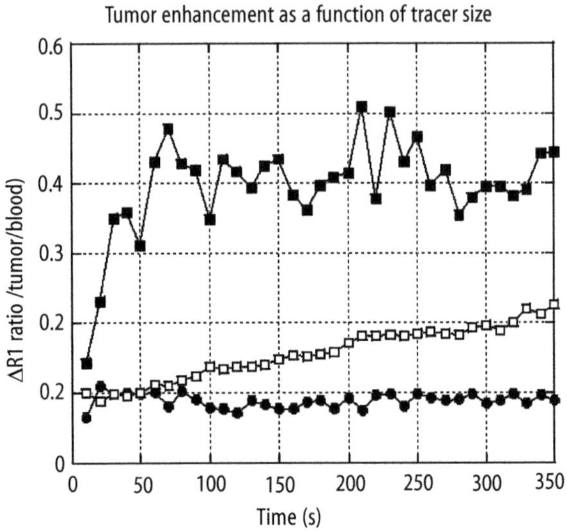

Fig. 2.2. Contrast media behavior as a function of molecular weight. In a rodent model of human breast cancer (Roberts et al. 2002a), different contrast media were used in an attempt to characterize the tumor microvasculature. Shown are curves of the ratio of tumor enhancement to reference vascular enhancement (from the inferior vena cava) for Gd-DTPA (~500 Da) (*solid squares*), Gadomer-17 (~17 kDa) (*open squares*) and covalently-bound albumin-Gd-DTPA (~92 kDa) (*solid circles*). It can be readily appreciated that progressive enhancement is extremely rapid with the Gd-DTPA, with signal enhancement rapidly indicating the distribution volume (the sum of intravascular and extravascular extracellular space). This may not be ideally sensitive to changes in permeability associated with disease progression or therapy. On the other hand, progressive enhancement with the much larger albumin-Gd-DTPA is difficult to resolve due to its small magnitude in comparison to noise (note, the estimation of fractional blood volume from the intercept of such plots is robustly made with albumin-Gd-DTPA due to effective oversampling). Finally, it can be appreciated from the intermediate sized agent, Gadomer-17, that progressive enhancement can be observed clearly within a 5-min scanning period, with potential for clear resolution of increases or decreases associated with disease progression or treatment

DCE-MRI models the content of tracer in the blood is assumed negligible in the mathematical model, and change in MRI signal is suggested to be due entirely to permeability between the plasma and the extravascular space (Tofts et al. 1995; Larsson et al. 1990; Brix et al. 1991). To eliminate the need for this type of assumption some investigators have used blood-pool (intravascular) agents, such as ultrasmall iron oxide particles (USPIOs), which are macromolecular complexes assumed too large to get out of the microcirculation into the interstitial space (Loubeyre et al. 1999). Whether with small or macromolecular contrast agents there are many investigators who have attempted to model MRI signal changes following contrast administration. The influence of molecular weight (and indeed other physical properties, such as electrical charge) of the tracer on vascular permeability additionally has a practical impact if extrapolation to a clinical setting is desired; while low molecular weight agents may be "too leaky" for physiological modeling, rapidly attaining a "distribution volume" in extravascular extracellular space, much larger agents may require long scan times to reveal discernible enhancement (associated with their much lower leak rates). Intermediate-sized agents may offer an appealing compromise (Fig. 2.2). However, an accurate understanding of the spatiotemporal distribution kinetics of MRI contrast agents is still not completely available. This is due both to the nature of the modeling process (i.e., one that describes the available data using a concise set of explanatory parameters) and to the limitations of the dynamic time series in terms of temporal resolution and signal-to-noise ratio (SNR).

2.5 Diagnostic and Prognostic Inferences

A number of parameters can potentially be extracted using DCE-MRI and used to improve the specificity of the MR exam in fully characterizing the vascularity of the tumor. It may be possible to improve both diagnostic and grading accuracy, as well as offer some prognostic indicator and a measure of tumor response to putative therapy. Parameterization of the dynamic time series typically takes one of two forms. In one approach, the dynamic time series data is interpreted in terms of the signal-intensity vs. time plot, and described in terms of "measured parameters," such as the "peak signal change," or the "time to peak," or some form of "integral, or area under the curve," or "slope, or rate of enhancement." These parameters are most certainly affected by the vascular properties of the tissue but often not directly. They retain some composite sensitivity to perfusion, fractional blood volume, vascular permeability and extravascular extracellular fraction, but are also sensitive to the imaging method chosen.

A second school of thought seeks to extract from the dynamic data parameters more specifically representing each of the above tissue microvascular characteristics, rather than a convenient but potentially confounded signal intensity-based measurement. Such approaches rely on tracer kinetic modeling subsequent to dynamic data acquisition, producing parameters that are "modeled" or "estimated" in contradistinction to the above "measured" quantities. In determining the most suitable approach for practical use one has to weigh up the advantages and disadvantages of physiological specificity for a particular microvascular characteristic versus random error and computational complexity required by modeling. Furthermore, each model is itself a simplified representation of the tissue it seeks to describe and thus its limitations impose an additional source of error on the target parameters derived. Further still, the development of more sophisticated and complex physiological models incurs the involvement of more estimated or target parameters, each additional parameter increasing the complexity and random noise in the fitting process.

Such microvascular characterizations have been used in experimental animal models for their predictive value in non-invasively assessing tumor histological grade, or pathological scoring (Daldrup et al. 1998). Furthermore, in the developing field of anti-angiogenic pharmaceuticals, such microvascular characterization has been used to monitor the immediate effects of novel anti-angiogenic therapies (Pham et al. 1998; Gossmann et al. 2002), which may have a mode of action that involves reducing tumor microvascular hyperpermeability, via blocking of the action of VEGF (vascular endothelial growth factor), alternatively known as VPF (vascular permeability factor), a molecule that seems integral in the development of many tumor types.

Extending these approaches to the clinical environment, it has been shown that similar analysis of dynamic enhancement studies in terms of microvascular characteristics can be used in human gliomas as a non-invasive correlate of tumor grade (Roberts et al. 2000) (Fig. 2.3). In fact, microvascular permeability also showed a strong correlation with tumor labeling index, MIB-1 (Roberts et al. 2001). Outside the brain, clinically available (ECF) contrast media have high permeabilities even in healthy tissue, rendering quantitative physiological modeling of hyperpermeability in tumors difficult to resolve. Consequently most of the successful implementations of dynamic contrast-enhanced MRI in peripheral tumors have been restricted to morphological descriptions of time-series data (Kuhl et al. 1997; Muller-Schimpfle et al. 1997).

2.6 Validation Studies of Dynamic Contrast-Enhanced MRI

There are numerous imaging and modeling strategies available for the application of DCE-MRI. The results of so many studies are difficult to compare due to the wide variety of existing mathematical models, each having their own assumptions and constraints (Tofts 1997).

The DCE-MRI approach is widely used to draw inferences into microvascular parameters such as microvascular permeability, blood volume, and tissue perfusion. Confirmed insight into the whereabouts, and amount of the tracer, at the tissue level would unquestionably help to determine the most accurate and precise way to model the signal changes in DCE-MRI. This can be accomplished by correlating DCE-MRI with quantitative transmission electron microscopy (TEM) methods. TEM can be used, in combination with energy dispersive X-ray spectrometry (EDXS) microanalysis to assess the subcellular content and location of heavy metals like gadolinium and iron (Elster 1989; LoPachin and Saubermann 1990; Taherzadeh et al. 1998). This method analyses the characteristic X-ray patterns, produced from the heavy metal based MRI contrast agents when an electron beam passes through the tissue. Where the concentration is high enough elemental distribution maps can be produced (LoPachin and Saubermann 1990). A previous study has shown the feasibility and necessity of combined MRI/EDSX studies (Hawkins et al. 1990). This was done to examine whether Gd-DTPA leaked out of the cerebral vasculature of Guinea pigs with experimental allergic encephalomyelitis, a multiple sclerosis-like disorder that causes breakdown of the BBB. Neither ultrastructural MRI contrast agent quantification nor elemental map-

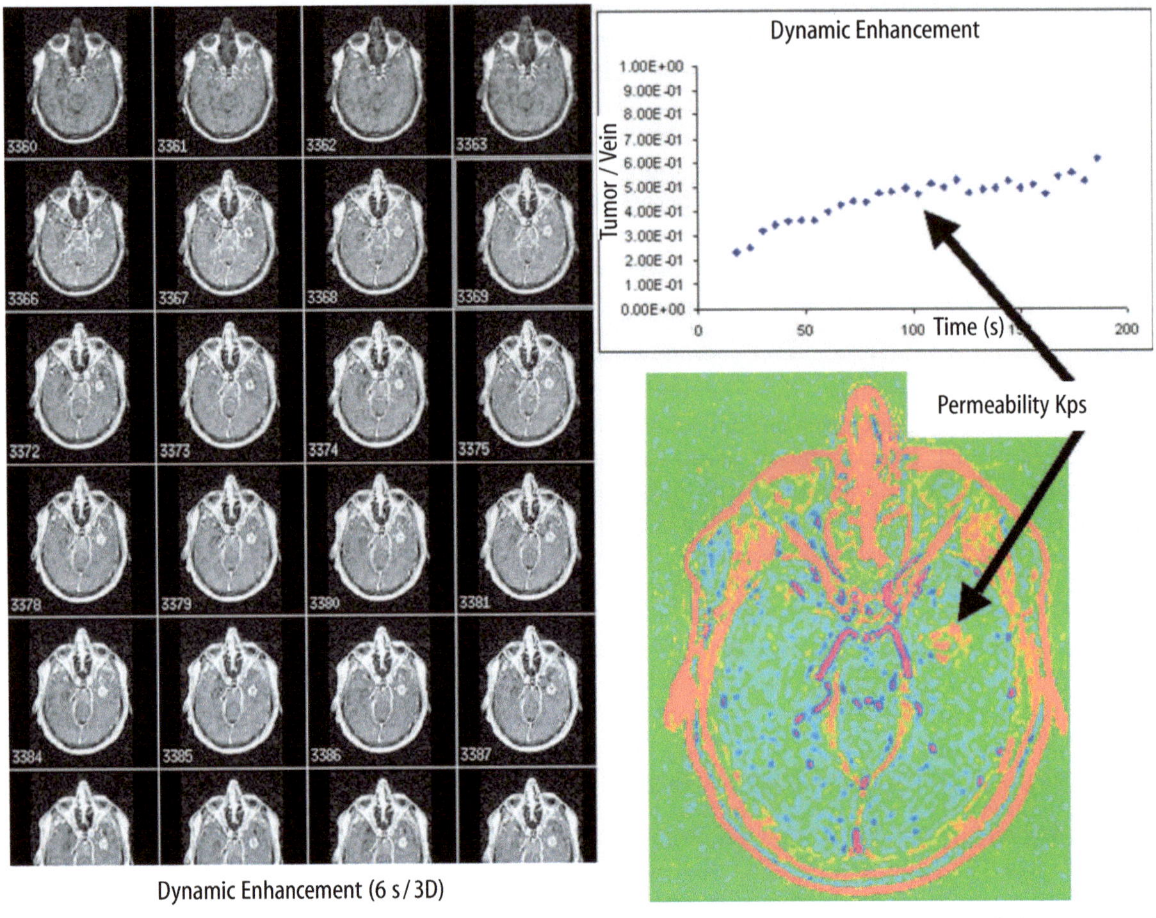

Fig. 2.3. Clinical imaging of brain tumor permeability. T_1-weighted 3D imaging is used dynamically during a bolus injection of paramagnetic contrast agent (gadolinium-based). A representative slice from a 3D data set is shown at 6-s intervals (*left*) revealing signal enhancement after the fourth frame, which progresses in a metastasis (*right*, anterior portion of brain). Constructing the ratio of tumor enhancement to reference blood vessel enhancement (from the sagittal sinus) and plotting as a function of time indicates the rate of such progressive enhancement (*right, upper*). Kinetic modeling allows construction of a pixel-by-pixel map of the permeability, Kps (*right, lower*), clearly delineating the tumor (*arrow*).

ping were performed, due to the equipment limitations. However, this study showed the feasibility of combining both methods in the study of microvasculature. With the rapid advances made in MR and EDSX imaging technologies over the last 5 years, the merging of the macro and microscopic technologies is feasible.

There are cheaper and simpler microscopic imaging methods (e.g., light microscopy), other than the EDSX electron microscopy methods. Macromolecular contrast agents containing Gd have been histochemically assessed using light microscopy (Saeed et al. 1998). Van Dijke et al. (2002) demonstrated via light microscopy post-in vitro streptavidin staining that even macromolecular contrast agents permeate out of the intravascular space over a period of 1 h in a breast cancer model. However, no current microscopic imaging method can image heavy metals such as gadolinium in small chelates. There are more sensitive biochemical methods, other than EDXS, for measuring heavy metal content. For example, inductively coupled plasma (ICP) atomic emission spectrometry, polarized X-ray fluorescence excitation analysis (FEA), and HPLC. However, these procedures lack the ability to localize the contrast agent precisely within the tissue architecture (Elster 1989). The tissue specimen must be first homogenized, digested, or vaporized, before they can be analyzed by the ICP-AES, FEA, or HPLC instruments.

In an experiment to localize the subcellular location of Gd-DTPA, New Zealand white rabbits implanted with vx2 tumor cells (Park et al. 1998) were evaluated using a combination of DCE-MRI and analytical electron microscopic methods (Noseworthy et al. 2002). Following DCE-MRI and parametric mapping of microvascular permeability, portions of muscle and tumor were examined under a field emission scanning electron microscope (FESEM). Using quantitative energy dispersive X-ray spectroscopy gadolinium was detected in the lumen of blood vessels in both tumor and muscle tissues, as expected, at concentrations similar to that observed in plasma collected prior to euthanasia. Interestingly at 5 min post-euthanasia, a time when the extra and intravascular spaces are theoretically in equilibrium, with respect to Gadolinium concentration, there was a four-fold larger Gd accumulation in the tumor extravascular space (Fig. 2.4). And, contrary to all reports on the in vivo behavior of small molecular weight Gd-based contrast agents, Gd was detected within the endothelial cells lining blood vessels of both tumor and muscle. This study showed that the current mathematical models for fitting DCE-MRI data may be oversimplified. The nature of the intracellular signal was speculated to arise from Gd being transported within vesiculo-vacuolar organelles (VVOs), which are known to transport molecules across vascular endothelial cells between intra- and extravascular spaces, or within transendothelial cell channels (Dvorak and Feng 2001). With the existence of an intracellular pool of Gd it is possible, therefore, that current DCE-MRI models could be overestimating blood volume or extravascular volume. In other experiments examining high molecular weight iron (Fe) particles (Ultrasmall superparamagnetic iron oxide particles, USPIOs), Fe was localized in only the intravascular space (unpublished observation).

A form of validation is also offered by comparing imaging derived parameters of microvascular characterization with histological assessment of microvascular density (MVD), assessed by staining with Factor VIII or CD-31. Critics of this approach point to sampling errors in comparing "hot spots" of immunohistochemical staining under light micros-

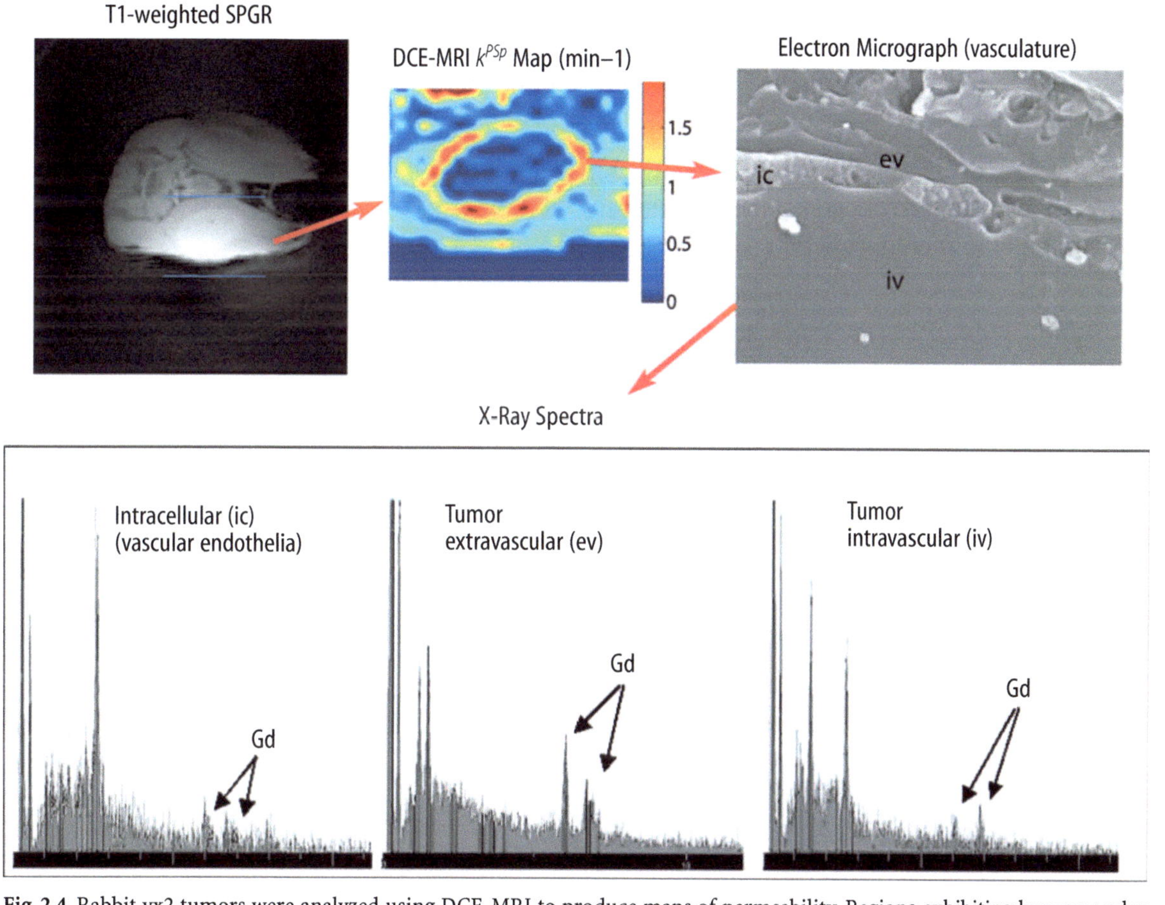

Fig. 2.4. Rabbit vx2 tumors were analyzed using DCE-MRI to produce maps of permeability. Regions exhibiting hypervascular permeability were biopsied, cryopreserved, and analyzed using energy dispersive X-ray microanalysis. Gd concentration in intravascular (*iv*), extravascular (*ev*), and endothelial intracellular (*ic*) spaces were assessed at 5 min post injection. Gd was found to be four times higher in the extravascular space, relative to intravascular. In addition, Gd was localized within vascular endothelial cells (*ic*)

copy with regional assessments derived from macroscopic imaging. Furthermore, the MVD assessment will provide a count of all visible vessels, not simply "functional" or perfused vessels. Nonetheless, coarse agreement between microvascular characterization derived from imaging with MVD is usually demonstrated (e.g., van Dijke et al. 1996).

Other attempts at validating dynamic DCE-MRI-derived parameters have taken advantage of permeability reducing pharmaceuticals such as Avastin (a monoclonal antibody to the VEGF molecule) (Ferrara 2002). Demonstrating "sensitivity to change," or reduction of permeability post-treatment with such an agent has been interpreted as confirming both the mode of action or biological effect of the pharmaceutical as well as providing some form of experimental utility validation for the microvascular characterization afforded by DCE-MRI (Pham et al. 1998; Brasch et al. 1997; Gossmann et al. 2002; Roberts et al. 2002a,b).

2.7 Summary

In summary, dynamic contrast-enhanced MRI is becoming more and more widely used in both preclinical and clinical settings. Improving hardware is allowing higher and higher temporal and spatial resolutions, while preserving signal to noise. Kinetic analysis is becoming more robust and target parameters derived from such analyses are being suggested as surrogate markers for assessing the efficacy of novel anti-tumor pharmaceuticals as well as for providing non-invasive correlates of histopathological assessment. Practically, the parametric maps can also be used to guide, or direct, surgical biopsy to target the most aggressive portion of a heterogeneous tumor, as well as to more appropriately define the margins of therapy zone for radiation treatment.

Dynamic contrast-enhanced MRI and its subsequent analysis, either based on signal measurements directly or physiological-modeling offers attractive opportunities to both preclinical and clinical trials of novel anti-cancer pharmaceuticals, especially anti-angiogenics. The role of such physiological imaging of the microvasculature can be considered in clinical trials at several levels:

(1) Patient selection: Imaging-based criteria can be used to identify and select patients who, for example, demonstrate abnormalities of the vasculature as a criterion for entry into a trial of a putative anti-angiogenic agent
(2) Disease state characterization: Imaging should allow quantitative and physiologically-specific description of the nature and degree of the disease-related pathophysiology at baseline
(3) Therapeutic efficacy assessment: Use similar quantitation to assess the effect of an interventional pharmaceutical therapy on a physiologically-specific property of tissue (e.g., blood volume, permeability)

In particular, with regard to the last two facets, it is important that descriptive estimators (directly measured or modeled) exhibit not only precision and reproducibility but also "sensitivity to change" (wide dynamic range) in response to either disease progression or response to therapy.

The ongoing development of contrast media of increasing molecular size for clinical use may be of considerable benefit to the interpretation of dynamic contrast-enhanced MRI, as has been demonstrated in preclinical studies, primarily as a result of lower intrinsic permeability in healthy tissues. This potentially offers the dynamic range for improved detection of hyperpermeability in tumors and the possible reduction thereof with novel and emerging antiangiogenic therapy (Roberts et al. 2002a,b).

References

Alexander AL, McCreery TT, Barrette TR, Gmitro AF, Unger EC (1996) Microbubbles as novel pressure-sensitive MR contrast agents. Magn Reson Med 35:801–806
Brasch RC (1991) Rationale and applications for macromolecular Gd-based contrast agents. Magn Reson Med 22:282–287
Brasch R, Pham C, Shames D, Roberts T, van Dijke K, van Bruggen N, Mann J, Ostrowitzki S, Melnyk O (1997) Assessing tumor angiogenesis using macromolecular MR imaging contrast media. J Magn Reson Imaging 7:68–74
Brix G, Semmler W, Port R, Schad LR, Layer G, Lorenz WJ (1991) Pharmacokinetic parameters in CNS Gd-DTPA enhanced MR imaging. J Comp Assist Tomogr 15:621–628
Burkill GJ, Mannion EM, Healy JC (2001) Technical report: lymph node enhancement at MRI with MnDPDP in primary hepatic carcinoma. Clin Radiol 56:67–71
Daldrup H, Shames DM, Wendland M, Okuhata Y, Link TM, Rosenau W, Lu Y, Brasch RC (1998) Correlation of dynamic contrast-enhanced MR imaging with histologic tumor grade: comparison of macromolecular and small-molecular contrast media. AJR Am J Roentgenol 171:941–949
Daldrup-Link HE, Kaiser A, Link TM, Settles M, Helbich T, Werner M, Roberts TP, Rummeny EJ (2002) Comparison between gadopentetate and feruglose (Clariscan)-enhanced MR-mammography: preliminary clinical experience. Acad Radiol 9:S343–S347

Demsar F, Roberts TP, Schwickert HC, Shames DM, van Dijke CF, Mann JS, Saeed M, Brasch RC (1997) A MRI spatial mapping technique for microvascular permeability and tissue blood volume based on macromolecular contrast agent distribution. Magn Reson Med 37:236–242

Diehl SJ, Lehmann KJ, Gaa J, McGill S, Hoffmann V, Georgi M (1999) MR imaging of pancreatic lesions. Comparison of manganese-DPDP and gadolinium chelate. Invest Radiol 34:589–595

Dvorak AM, Feng D (2001) The vesiculovacuolar organelle (VVO): a new endothelial cell permeability organelle volume. J Histochem Cytochem 49:419–431

Edelman RR, Mattle HP, Atkinson DJ, Hill T, Finn JP, Mayman C, Ronthal M, Hoogewoud HM, Kleefield J (1990) Cerebral blood flow: assessment with dynamic contrast-enhanced T2*- weighted MR imaging at 1.5 T. Radiology 176:211–220

Edelman RR, Hatabu H, Tadamura E, Li W, Prasad PV (1996) Noninvasive assessment of regional ventilation in the human lung using oxygen-enhanced magnetic resonance imaging. Nature Med 2:1236–1239

Elster AD (1989) Energy-dispersive x-ray microscopy to trace gadolinium in tissues. Radiology 173:868–870

Enochs WS, Harsh G, Hochberg F, Weissleder R (1999) Improved delineation of human brain tumors on MR images using a long-circulating, superparamagnetic iron oxide agent. J Magn Reson Imaging. 9:228–232

Ferrara N (2002) Role of vascular endothelial growth factor in physiologic and pathologic angiogenesis: therapeutic implications. Semin Oncol 29 [Suppl 16]:10–14

Goede V, Fleckenstein G, Dietrich M, Osmers RG, Kuhn W, Augustin HG (1998) Prognostic value of angiogenesis in mammary tumors. Anticancer Res 18:2199–2202

Gossmann A, Helbich TH, Kuriyama N, Ostrowitzki S, Roberts TP, Shames DM, van Bruggen N, Wendland MF, Israel MA, Brasch RC (2002) Dynamic contrast-enhanced magnetic resonance imaging as a surrogate marker of tumor response to anti-angiogenic therapy in a xenograft model of glioblastoma multiforme. J Magn Reson Imaging 15:233–240

Griebel J, Mayr NA, de Vries A, Knopp MV, Gneiting T, Kremser C, Essig M, Hawighorst H, Lukas PH, Yuh WT (1997) Assessment of tumor microcirculation: a new role of dynamic contrast MR imaging. J Magn Reson Imaging 7:111–119

Grist TM, Korosec FR, Peters DC, Witte S, Walovitch RC, Dolan RP, Bridson WE, Yucel EK, Mistretta CA (1998) Steady-state and dynamic MR angiography with MS-325: initial experience in humans. Radiology 207:539–544

Harris AL, Zhang H, Moghaddam A (1996) Breast cancer angiogenesis–new approaches to therapy via antiangiogenesis, hypoxic activated drugs, and vascular targeting. Breast Cancer Res Treat 38:97–108

Hawkins CP, Munro PM, MacKenzie F, Kesselring J, Tofts PS, du Boulay EP, Landon DN, McDonald WI (1990) Duration and selectivity of blood-brain barrier breakdown in chronic relapsing experimental allergic encephalomyelitis studied by gadolinium-DTPA and protein markers. Brain 113:365–378

Helbich TH, Roberts TP, Rollins MD, Shames DM, Turetschek K, Hopf HW, Muhler M, Hunt TK, Brasch RC (2002) Noninvasive assessment of wound-healing angiogenesis with contrast-enhanced MRI. Acad Radiol 9:S145–S147

Henderson E, Sykes J, Drost D, Weinmann HJ, Rutt BK, Lee TY (2000) Simultaneous MRI measurement of blood flow, blood volume, and capillary permeability in mammary tumors using two different contrast agents. J Magn Reson Imaging 2:991–1003

Johnson KM, Tao JZ, Kennan RP, Gore JC(1998) Gadolinium-bearing red cells as blood pool MRI contrast agents. Magn Reson Med 40:133–142

Karczmar GS, River JN, Li J, Vijayakumar S, Goldman Z, Lewis MZ (1994) Effects of hyperoxia on T2* and resonance frequency weighted magnetic resonance images of rodent tumours. NMR Biomed 7:3–11

Kuhl CK, Bieling H, Gieseke J, Ebel T, Mielcarek P, Far F, Folkers P, Elevelt A, Schild HH(1997) Breast neoplasms: T2* susceptibility-contrast, first-pass perfusion MR imaging. Radiology 202:87–95

Larsson HB, Stubgaard M, Frederiksen JL, Jensen M, Henriksen O, Paulson OB (1990) Quantitation of blood-brain barrier defect by magnetic resonance imaging and gadolinium-DTPA in patients with multiple sclerosis and brain tumors. Magn Reson Med 6:117–131

Li WH, Parigi G, Fragai M, Luchinat C, Meade TJ (2002) Mechanistic studies of a calcium-dependent MRI contrast agent. Inorg Chem 41:4018–4024

LoPachin RM Jr, Saubermann AJ (1990) Disruption of cellular elements and water in eurotoxicity: studies using electron probe X-ray microanalysis. Toxicol Appl Pharmacol 106:355–374

Loubeyre P, de Jaegere T, Miao Y, Landuyt W, Marchal G (1999) Assessment of iron oxide particles (AMI 227) and a gadolinium complex (Gd-DOTA) in dynamic susceptibility contrast MR imagings (FLASH and EPI) in a tumor model implanted in rats. Magn Reson Imaging 17:627–631

Miller SK, Chu WJ, Pohost GM, Elgavish GA (1991) Improvement of spectral resolution in shift-reagent-aided NMR spectroscopy in the isolated perfused rat heart system. Magn Reson Med 20:184–195

Moseley ME, Vexler Z, Asgari HS, Mintorovitch J, Derugin N, Rocklage S, Kucharczyk J (1991) Comparison of Gd- and Dy-chelates for T2 contrast-enhanced imaging. Magn Reson Med :259–264

Muller-Schimpfle M, Ohmenhauser K, Sand J, Stoll P, Claussen CD (1997)Dynamic 3D-MR mammography: is there a benefit of sophisticated evaluation of enhancement curves for clinical routine? J Magn Reson Imag 7:236–240

Noseworthy MD, Bray TM (1998) Effect of oxidative stress on brain damage detected by MRI and in vivo 31P-NMR. Free Radic Biol Med 24:942–951

Noseworthy MD, Bray TM (2000) Zinc deficiency exacerbates loss in blood-brain barrier integrity induced by hyperoxia measured by dynamic MRI. Proc Soc Exp Biol Med 223:175–182

Noseworthy MD, Morton G, Wright GA (1998) Dynamic MR imaging for assessment of prostate carcinoma. ISMRM workshop on MR in experimental and clinical cancer research, p 110

Noseworthy MD, Stanisz GJ, Kim JK, Stainsby JA, Wright GA (1999) Tracking oxygen effects on MR signal in blood and skeletal muscle during hyperoxia exposure. J Magn Reson Imag 9:814–820

Noseworthy MD, Ackerley C, Qi X, Wright GA (2002) Correlating subcellular contrast agent location from dynamic contrast enhanced magnetic resonance imaging (dMRI) and analytical electron microscopy. Acad Radiol 9 [Suppl 2]:S514–S518

Oikawa H, al-Hallaq HA, Lewis MZ, River JN, Kovar DA, Karczmar GS (1997) Spectroscopic imaging of the water resonance with short repetition time to study tumor response to hyperoxia. Magn Reson Med 38:27–32

Park KS, Choi BI, Won HJ, SEO JB, Kim SH, Kim TK, Han JK, Yeon KM (1998) Intratumoral vascularity of experimentally induced VX2 carcinoma. Invest Radiol 33:39–44

Pham CD, Roberts TP, van Bruggen N, Melnyk O, Mann J, Ferrara N, Cohen RL, Brasch RC (1998) Magnetic resonance imaging detects suppression of tumor vascular permeability after administration of antibody to vascular endothelial growth factor. Cancer Invest 16:225–230

Reddick WE, Taylor JS, Fletcher BD (1999) Dynamic MR imaging (DEMRI) of microcirculation in bone sarcoma. J Magn Reson Imaging 10:277–285

Rijpkema M, Kaanders JH, Joosten FB, van der Kogel AJ, Heerschap A (2002) Effects of breathing a hyperoxic hypercapnic gas mixture on blood oxygenation and vascularity of head-and-neck tumors as measured by magnetic resonance imaging. Int J Radiat Oncol Biol Phys 53:1185–1191

Roberts HC, Roberts TP, Brasch RC, Dillon WP (2000) Quantitative measurement of microvascular permeability in human brain tumors achieved using dynamic contrast-enhanced MR imaging: correlation with histologic grade. AJNR Am J Neuroradiol 21:891–899

Roberts HC, Roberts TP, Bollen AW, Ley S, Brasch RC, Dillon WP (2001) Correlation of microvascular permeability derived from dynamic contrast-enhanced MR imaging with histologic grade and tumor labeling index: a study in human brain tumors. Acad Radiol 8:384–391

Roberts TPL (1997) Physiologic measurements by contrast-enhanced MR imaging: expectations and limitations. J Magn Reson Imaging 7:82–90

Roberts TPL, Kucharczyk J, Cox I, Moseley ME, Prayer L, Dillon WP, Harnish P (1994) Sprodiamide injection-enhanced MR imaging of cerebral perfusion: phase I clinical trial results. Invest Radiol 29:S24–S26

Roberts TPL, Turetschek K, Preda A, Novikov V, Moeglich M, Shames DM, Brasch RC, Weinmann HJ (2002a) Tumor microvascular changes to anti-angiogenic treatment assessed by MR contrast media of different molecular weights. Acad Radiol 2:S511–S53

Roberts TPL, Preda A, Turetschek K, Novikov V, Moeglich M, Shames DM, Brasch RC, Cavagna FM (2002b) Permeability of B22956/1, a novel protein-binding contrast agent, resolves anti-angiogenic therapy in human breast cancer model. Proc Int Soc Magn Reson Med 319

Rosenthal DI, Nurenberg P, Becerra CR, Frenkel EP, Carbone DP, Lum BL, Miller R, Engel J, Young S, Miles D, Renschler MF (1999) A phase I single-dose trial of gadolinium texaphyrin (Gd-Tex), a tumor selective radiation sensitizer detectable by magnetic resonance imaging. Clin Cancer Res 5:739–745

Saeed M, van Dijke CF, Mann JS, Wendland MF, Rosenau W, Higgins CB, Brasch RC (1998) Histologic confirmation of microvascular hyperpermeability to macromolecular MR contrast medium in reperfused myocardial infarction. J Magn Reson Imaging 8:561–567

Sahani D, Prasad SR, Maher M, Warshaw AL, Hahn PF, Saini S (2002) Functioning acinar cell pancreatic carcinoma: diagnosis on mangafodipir trisodium (Mn-DPDP)-enhanced MRI. J Comput Assist Tomogr 26:126–128

Schwickert HC, Roberts TP, Shames DM, van Dijke CF, Disston A, Muhler A, Mann JS, Brasch RC. (1995) Quantification of liver blood volume: comparison of ultra short TI inversion recovery echo planar imaging (ULSTIR-EPI), with dynamic 3D-gradient recalled echo imaging. Magn Reson Med 6:845–852

Sigal R, Vogl T, Casselman J, Moulin G, Veillon F, Hermans R, Dubrulle F, Viala J, Bosq J, Mack M, Depondt M, Mattelaer C, Petit P, Champsaur P, Riehm S, Dadashitazehozi Y, de Jaegere T, Marchal G, Chevalier D, Lemaitre L, Kubiak C, Helmberger R, Halimi P (2002) Lymph node metastases from head and neck squamous cell carcinoma: MR imaging with ultrasmall superparamagnetic iron oxide particles (Sinerem MR) – results of a phase-III multicenter clinical trial. Eur Radiol 12:1104–1113

Sipkins DA, Cheresh DA, Kazemi MR, Nevin LM, Bednarski MD, Li KC (1998) Detection of tumor angiogenesis in vivo by alphaVbeta3-targeted magnetic resonance imaging. Nat Med 4:623–626

–Stets C, Brandt S, Wallis F, Buchmann J, Gilbert FJ, Heywang-Kobrunner SH (2002) Axillary lymph node metastases: a statistical analysis of various parameters in MRI with USPIO. J Magn Reson Imaging 16:60–68

Su MY, Wang Z, Carpenter PM, Lao X, Muehler A, Nalcioglu O (1999) Characterization of N-ethyl-N-Nitrosourea-induced malignant and benign breast tumors in rats by using 3 MR contrast agents. J Magn Reson Imaging 9:177–186

Taherzadeh M, Das AK, Warren JB (1998) Nifedipine increases microvascular permeability via a direct local effect on postcapillary venules. Am J Physiol 275:H1388–H1394

Tofts PS (1997) Modeling tracer kinetics in dynamic Gd-DTPA MR imaging. J Magn Reson Imag 7:91–101

Tofts PS, Kermode AG (1991) Measurement of the blood-brain barrier permeability and leakage space using dynamic MR imaging. 1. Fundamental concepts. Magn Reson Med 17:357–367

Tofts PS, Berkowitz B, Schnall MD (1995) Quantitative analysis of dynamic Gd-DTPA enhancement in breast tumors using a permeability model. Magn Reson Med 33:564–568

Tsuchida C, Yamada H, Maeda M, Sadato N, Matsuda T, Kawamura Y, Hayashi N, Yamamoto K, Yonekura Y, Ishii Y (1997) Evaluation of peri-infarcted hypoperfusion with T2*-weighted dynamic MRI. J Magn Reson Imaging 7:518–522

Turetschek K, Floyd E, Helbich T, Roberts TP, Shames DM, Wendland MF, Carter WO, Brasch RC (2001a) MRI assessment of microvascular characteristics in experimental breast tumors using a new blood pool contrast agent (MS-325) with correlations to histopathology. J Magn Reson Imaging 14:237–242

Turetschek K, Huber S, Floyd E, Helbich T, Roberts TP, Shames DM, Tarlo KS, Wendland MF, Brasch RC (2001b) MR imaging characterization of microvessels in experimental breast tumors by using a particulate contrast agent with histopathologic correlation. Radiology 218:562–569

Van Dijke CF, Brasch RC, Roberts TP, Weidner N, Mathur A, Shames DM, Mann JS, Demsar F, Lang P, Schwickert HC, (1996) Mammary carcinoma model: correlation of macromolecular contrast-enhanced MR imaging characterizations of tumor microvasculature and histologic capillary density. Radiology 198:813–818

Van Dijke CF, Kirk BA, Peterfy CG, Genant HK, Brasch RC,

Kapila S (1997) Arthritic temporo-mandibular joint: correlation of macromolecular contrast-enhanced MRimaging parameters and histopathologic findings. Radiology 204:825–832

Van Dijke CF, Mann JS, Rosenau W, Wendland MF, Roberts TP, Roberts HC, Demsar F, Brasch RC (2002) Comparison of MR contrast-enhancing properties of albumin-(biotin)10-(gadopentetate)25, a macromolecular MR blood pool contrast agent, and its microscopic distribution. Acad Radiol Suppl 1:S257–S260

Weissleder R, Simonova M, Bogdanova A, Bredow S, Enochs WS, Bogdanov A Jr (1997) MR imaging and scintigraphy of gene expression through melanin induction. Radiology 204:425–429

Zhang S, Wu K, Sherry AD (1999) A novel pH sensitive MRI contrast agent. Angew Chem Int Ed Engl 38:3192–3194

3 The Role of Blood Pool Contrast Media in the Study of Tumor Pathophysiology

Laure S. Fournier and Robert C. Brasch

CONTENTS

3.1 Introduction

Characterizing the biological properties of individual tumors has become a major goal for non-invasive imaging. With growing enthusiasm, the use of serial MRI to monitor the pharmacokinetics of paramagnetic contrast media is being explored to functionally characterize tumors, particularly tumor vascularity. Analysis and kinetic modeling of the dynamic tumor enhancement response after contrast medium administration is being applied to generate quantitative estimates of microvascular characteristics, particularly the fractional blood volume and the microvascular permeability of tumor vessels. Understandably, it is commonly hoped by many MR imagers that the currently available and widely used gadolinium-based contrast media, all belonging in the class of small molecular contrast media (SMCM <1000 Daltons) will be satisfactory for characterization of microvessels with dynamic contrast-enhanced imaging. But it should be emphasized that endothelial permeability to solutes of substantially different size cannot be equated. Unfortunately, SMCM are too small to optimally exploit the well-recognized hyperpermeability of neoplastic microvessels; such hyperpermeability, with rare exception, has been consistently demonstrated in relation to larger macromolecular solutes. Smaller solutes in the size range of commercially available gadolinium chelates are known to diffuse across vascular endothelium in both normal and neoplastic tissues; a notable exception being the blood–brain barrier. Thus, the differentiation of normal from neoplastic tissue by the MRI assay of tissue SMCM leakage is problematic. A macromolecular contrast media (MMCM) is favored to probe and quantitatively estimate by MRI the elevated macromolecular permeability of tumor microvessels. Another obstacle to quantitative MRI tumor characterization pursued with SMCM is the highly variable and unpredictable vascular extraction fraction, both in normal and in neoplastic tumor tissues.

MMCM are also referred to as blood pool contrast media (BPCM). While BPCM/MMCM are now being clinically developed and positioned for governmental approval, prototype MMCM are being applied with considerable success in a host of experimental tumor models. The pre-clinical results show strongly positive and significant correlations between MMCM-assayed tumor vascular permeability and tumor blood volume with pathologic tumor grade, tumor angiogenesis as assayed by the histologic microvascular density, and with tumor response to multiple forms of anti-angiogenesis therapy. In groups of tumors characterized both by SMCM and MMCM-enhanced MRI, the correlations between MRI assays and histopathologic endpoints are consistently superior using the macromolecular-enhanced imaging.

3.2 What Are Blood Pool Contrast Media and What Is Their Appeal?

Blood pool contrast media (BPCM) are not strictly defined; which compounds should be included within

L. S. Fournier, MD
R. C. Brasch, MD, Professor of Radiology and Pediatrics
Center for Pharmaceutical and Molecular Imaging, Department of Radiology, University of California San Francisco, 513 Parnassus Avenue, San Francisco, CA 94143-0628, USA

this category may depend upon whom one asks. Yet, a qualitative definition is universally accepted: a blood pool contrast medium for MRI is an enhancing, typically paramagnetic, formulation, that after intravascular administration remains largely in the vascular space for an extended time. For instance, one could choose to define a BPCM as an agent with a plasma half-life of more than 60 min. Typically, such plasma retention would be associated with molecules having molecular weights greater than 50,000 Daltons (Vexler et al. 1994). However, plasma half-life not only varies with molecular weight but also depends on molecular shape and charge, and on the animal species used for the evaluation. Other MRI scientists might wish to include relatively smaller molecules in the category of BPCM and might define "prolonged vascular retention" as a net plasma half-life greater than that observed for typical small molecular gadolinium contrast media (SMCM) represented by gadopentetate (MW=547 Daltons). For reference, the plasma half-life of gadopentetate in rats is 13 min and approximately 20 min in humans (Weinmann et al. 1984). Using such a broad definition of BPCM would permit inclusion of contrast-enhancing formulations ranging from 5000-50,000 Daltons and larger. Obviously, the pharmacokinetic properties and thus the clinical utility of BPCM could vary substantially across this broad range of molecular sizes. Importantly for the reader, all BPCM should not be considered equivalent or interchangeable. The specific agent and its properties must be considered in any discussion of kinetics, or diagnostic application (see "Appendix").

Appendix

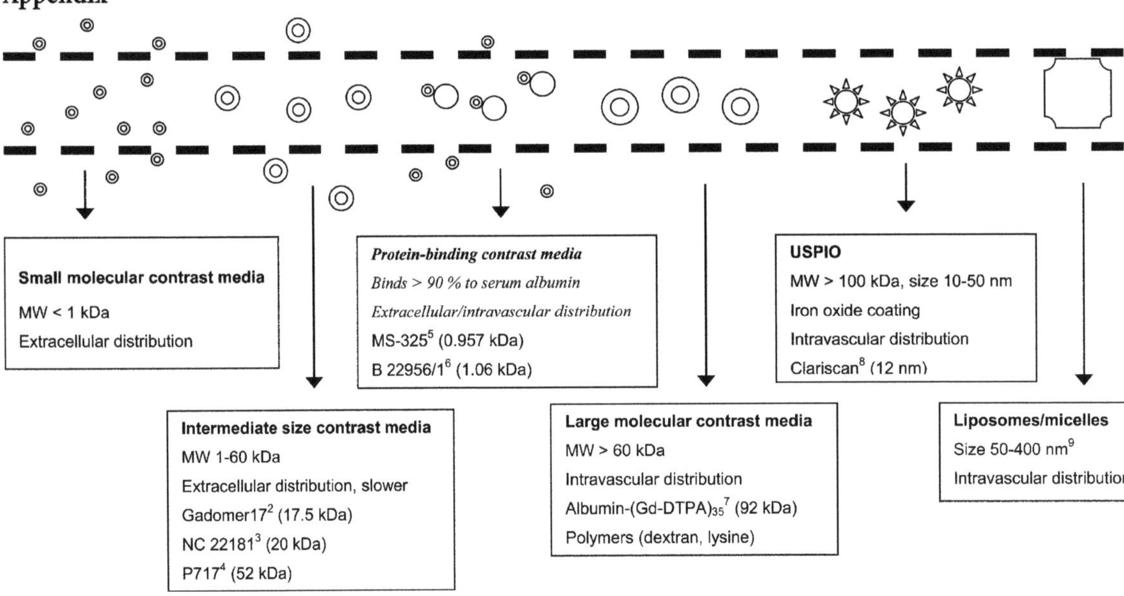

Different classes of contrast media (manufacturers and references). [1]Weinmann et al. (1984); [2]Schering AG, Berlin, Germany (Henderson et al. 2000); [3]Amersham Health, Oslo, Norway (Bonk et al. 2000); [4]Guerbet, Aulnay-sous-Bois, France (Daldrup-Link et al. 2001); [5]Epix, Cambridge, MA, USA (Kroft and de Roos 1999); [6]Bracco, Milan, Italy (Cavagna et al. 2001); [7]Ogan (1988); [8]Amersham Health, Oslo, Norway (Hoffmann et al. 2002); [9]Okuhata et al. (1999); Weissig et al. (2000)

3.3 Biological Aspects of Tumor Vessel Hyperpermeability

Although it is commonly taught in biological sciences to "never say never" and conversely, nothing is "always" true, it can be stated with some certainty that tumor microvessels when compared to normal non-tumor vessels are more permeable to the transendothelial diffusion of large molecular solutes. Macromolecular hyperpermeability of tumor vessels has been demonstrated consistently over more than 50 years by numerous, but generally invasive assays; published assays have utilized macromolecular dyes like Evan's blue, radio-labeled proteins such as fibrinogen and albumin, and fluorescent-labeled proteins detected with video microscopy (Gerlowski and Jain 1986; Nagy et al. 1989; Dvorak 1990; Sevick and Jain 1991; Yuan et al. 1993; Jain 1994). Our group at UCSF first detected and quantitatively monitored by MRI this macromolecular hyperpermeability of tumors in 1989 when we performed dynamic contrast-enhanced (DCE) MRI in an experimental mouse fibrosarcoma model using a prototype macromolecular contrast agent, albumin-$(Gd\text{-}DTPA)_{35}$ (Aicher et al. 1990). Albumin-$(Gd\text{-}DTPA)_{35}$ is a highly paramagnetic bio-probe with a molecular weight of 92,000 Daltons and a hydrodynamic radius of approximately 6 nm (Ogan 1988). It is interesting to note that even this early attempt to quantitatively assess tumor microvascular permeability previewed the later confirmed potential of the MMCM-enhanced DCE MRI for monitoring tumor response to therapy. This relatively early investigation showed a significant reduction in tumor vessel leakiness, measured by both MRI and Evan's blue assays after only 1-2 h of treatment with tumor necrosis factor-alpha (Aicher et al. 1990). Jain (1994), in an excellent review article appearing in *Scientific American*, summarized more than a decade of work from his laboratory and from others explaining, from an engineering perspective, the nature of macromolecular diffusion from blood within tumor microvessels into the tumor interstitium. Physiologically, diffusion dominates the transendothelial exchange of macromolecules in the tumor periphery where interstitial pressure is not as high as that often to be found in the tumor core. Jain (1994) notes that within the tumor core, there is an inhibitory effect from high interstitial pressure on the extravascular leakage/diffusion of solutes. This pathophysiological property of tumor hyperpermeability with respect to macromolecular solutes can be exploited by MMCM-enhanced MRI to characterize this consistent biological feature of malignant tumors.

3.4 Shortcomings of Small Molecular Contrast Media for Characterizing Tumor Vessels

To understand the biology underlying the kinetics of MRI contrast media, it is essential to recognize that tumor microvascular hyperpermeability is relatively specific to macromolecular solutes. Small molecular solutes such as insulin, glucose, or paramagnetic gadolinium chelates are known to be readily diffusible across the endothelial barrier of both normal and neoplastic vessels (Crone and Levitt 1984; Jain 1987). There is no basis from previous invasive studies to anticipate that tumor vessels will leak SMCM while normal vessels in the same organ will not be permeable to the same bio-probes. Two notable exceptions are found in the microvessels of the brain and the testes; normal vessels in these organs have unusually tight junctions between endothelial cells, limiting diffusion of even SMCM, while tumor vessels in these same organs allow extravascular accumulation of contrast agents.

The degree of transendothelial diffusion for any substance is reflected in its extraction fraction (E) or by the more complex functional parameter termed the 'permeability surface area product'' (PS) (Renkin 1959; Crone 1963). As implied by the name, the PS parameter depends on both the localized permeability of the vessel and the surface area of the vessel available for transendothelial diffusion. Intuitively, one can appreciate that diffusion of solutes across the endothelium will also depend on the rate of blood flow in the leaky microvessel. One could reasonably predict that with slower flow there would be more time for permeable solute molecules to actually escape the blood compartment and diffuse into the extravascular space.

3.4.1 What Are the Determinants of PS?

Physiologists have been fascinated by vascular permeability for generations and more than one kinetic model has been proposed; Renkin (1959) and Crone (1963) described a useful and widely accepted model of this process. The Renkin/Crone equation is as follows:

$$FLR = \frac{F}{PV}\left(1 - e^{\frac{-PS}{F}}\right) \quad (1)$$

where FLR represents fractional leak rate, F flow, PV plasma volume, and PS the permeability surface area product.

In the limiting situation in which the value for PS is much smaller than the value for flow (F), the Renkin/Crone equation can be simplified by substituting

$$\frac{PS}{F} \text{ for } 1 - e\frac{-PS}{F} \quad (2)$$

$$\text{then } FLR = \frac{F}{PV} \times \frac{PS}{F} \quad (3)$$

$$\text{and } FLR = \frac{PS}{PV} \quad (4)$$

In this specific situation, realized with macromolecular blood pool contrast agents for which PS and the extraction fraction are quite small (less than 0.001 per passage through the vascular bed), the complex exponential Renkin/Crone equation reduces to

$$PS = PV \times FLR \quad (5)$$

Thus, for macromolecular contrast agents, the PS value is relatively flow independent and can be derived without knowledge of the flow term (F). This is most fortunate, because currently there are no easily employed techniques using MRI for measurement of flow in tissue microvessels (St. LAWRENCE and LEE 1998a,b).

Unfortunately for those wishing to estimate PS using SMCM, the extraction fraction and the PS are not small compared to F. The flow term cannot be ignored, cannot be easily estimated on a gadolinium-enhanced MRI examination, and cannot reasonably be assumed to be constant in different tissues, tumors, or even in all regions of a given tumor.

Not withstanding the fact that estimating flow in tumor microvessels by MRI methods is problematic, tumor microvascular flow can be estimated invasively using labeled microspheres. DALDRUP and colleagues (1998) used a combination of invasive microsphere assays and non-invasive dynamic contrast-enhanced MRI acquisitions in a group of human breast tumors (MDA-MB-435) implanted in the mammary fat pads of athymic rats (DALDRUP et al. 1998b). By estimating the fractional leak rate (FLR) and the fractional plasma volume (fPV) from the MRI tumor enhancement response to gadopentetate and by measuring the microvascular blood flow (F) from the microsphere accumulation, they were able to quantitatively estimate the extraction fractions in this series of tumors with respect to gadopentetate. The tumor gadopentetate extraction fractions were neither consistent nor small; E values were all in the range of 20%–48%. In fact, those experienced with DCE MRI using this contrast agent and the trend for strong rapid tumor enhancement, might predict such a relatively high rate of transendothelial diffusion for gadopentetate. The E of gadopentetate can be even higher than in these tumors, for example 55% in normal myocardium (SVENDSEN et al. 1992; HAUNSO et al. 1980). Returning to the Renkin/Crone equation, for gadopentetate the extraction fraction is clearly not small compared to the PS. Thus, for gadopentetate (and similar SMCM), the equation cannot be simplified algebraically to eliminate the flow (F) term, and the PS value cannot be reliably estimated for this class of small molecular MRI contrast media. These considerations argue for the development and use of MMCM/BPCM for assessing the permeability of tumor microvessels. The same agents can also be used to advantage for blood volume measurements and angiographic imaging.

3.5 Applications of Blood Pool Contrast Media

The potential applications of blood pool contrast media for tumor evaluations are still being defined, but no less than three clinically feasible and attractive goals are emerging. To be discussed separately will be angiography, characterizing individual tumor biology including differentiation of benign from malignant tumors and tumor grading, and monitoring of tumor response to treatment.

3.5.1 Angiography

First, it is useful to consider angiography with conventional small molecular contrast media (SMCM). For these agents having molecular weights less than 1000 Daltons, there is a fast transendothelial diffusion resulting in a rapidly declining vessel-to-tissue contrast (TAUPITZ et al. 2000). Bolus tracking is also recommended when using SMCM to ensure that the image acquisitions are obtained with optimal timing, ideally just after the bolus arrives in the vessels (BONK et al. 2000). However, the requirement for relatively fast acquisitions, within the short duration of high vessel-to-background ratio, excludes the possibility of using higher spatial resolution pulse sequences, which entail longer acquisition periods. MMCM/BPCM, on the other hand, provide a prolonged enhancement of blood within vessels related to their low transendothelial extraction fraction and slower blood clearance rates (KROFT and DE ROOS 1999; KAUCZOR and KREITNER 2000).

Several macromolecular BPCM formulations have been tested for their potential to angiographically define both large vessels and to characterize tumors generally. A 20-kDa hexamethylene diamine co-polymer (BONK et al. 2000), and a carboxymethyl-dextran (P717, Guerbet) with a molecular weight of 52 kDa (DALDRUP-LINK et al. 2001), when evaluated in rabbits showed stable blood enhancement for 40–60 min in the abdominal aorta and pelvic vessels. In comparison, liposomes filled with gadolinium chelates, much larger than the polymeric molecules, produced a substantial shortening of blood T1 for over 4 h (WEISSIG et al. 2000). Although the vessel definition was considered good with all three of these BPCM, the MRI seemed to slightly underestimate the vascular diameter compared to measurements by conventional radiographic angiography, for example in pelvic vessels (BONK et al. 2000).

Of course, there are additional important considerations, beyond blood half-life, in the choice among BPCM. For example, when using the 20-kDa carboxymethyl-dextran formulation (p717), the enhancement in the liver did not clear until 10 days, considered to be an undesirable long retention. Similarly, the whole body retention of albumin-$(Gd\text{-}DTPA)_{35}$ is considered undesirably high and long, 18% at 3 weeks (WHITE et al. 1989) for this much-used prototype BPCM; such retention discouraged the development of this compound as a clinical pharmaceutical. However, this 92-kDa prototype compound has several near ideal pharmacokinetic properties including a volume of distribution equal to the plasma volume and a blood half-life of 90 min (SCHMIEDL et al. 1987). Albumin-$(Gd\text{-}DTPA)_{35}$ has been used frequently to demonstrate the range of potential benefits for MMCM-enhanced MRI (Fig. 3.1); SCHWICKERT and coworkers (1995) showed detailed rodent angiograms using albumin-$(Gd\text{-}DTPA)_{35}$ which persisted essentially unchanged visually for 88 min after administration.

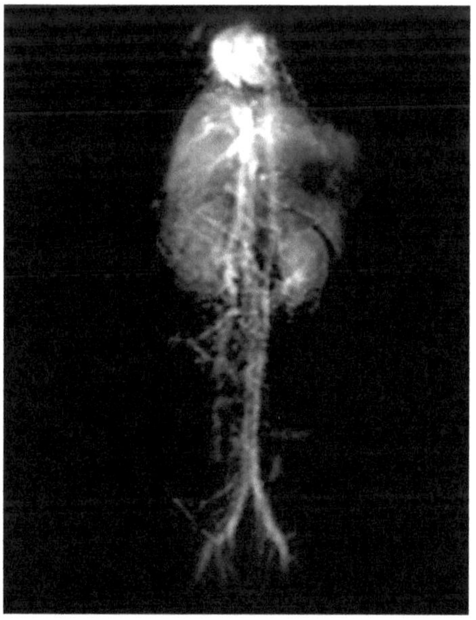

Fig. 3.1. Anterior view of a rat MR angiography (maximum intensity projection), performed 10 min after injection of albumin-$(Gd\text{-}DTPA)_{35}$ at a dose of 0.03 mmol Gd/kg, on a 2T MRI system. Large and relatively small veins and arteries can be defined

Additional intermediate size gadolinium-based blood pool contrast agents have been tested for magnetic resonance angiography. Gadomer-17 (Schering AG, Berlin), for example, has a molecular weight of 17.5 kDa, and has been used successfully in pigs to angiographically study of coronary and pulmonary arteries, with a noticeable increase of signal-to-noise (SNR) and contrast-to-noise (CNR) ratios between pre- and post-contrast images (LI et al. 2001; ABOLMAALI et al. 2002). However, such intermediately sized molecules serve as reliable markers of the blood pool only during the first passes. Being smaller than serum proteins, they diffuse progressively into the interstitial space, and their kinetics tend to resemble those of extracellular agents with a bi-exponential plasma disappearance. Though this plasma clearance characteristic might not diminish MR angiography, it tends to limit the usefulness of these intermediate-sized agents for tissue characterization, since they would tend to overestimate tissue blood volume (KROFT and DE ROOS 1999).

MS-325 (Epix, Cambridge, MA) is a small gadolinium chelate (957 Da), which binds reversibly to circulating albumin, forming a macromolecular complex (KROFT and DE ROOS 1999). The percentage of binding varies with species but is approximately 95% at equilibrium in humans and rabbits. Despite the presence of some small unbound molecules diffusing into the interstitial space and cleared by glomerular filtration, MS-325 produces a strong vascular enhancement in patients for a length of time sufficient to accurately depict stenoses and ulcerations in carotid arteries (BLUEMKE et al. 2001; GRIST et al. 1998; LI et al. 1998; STUBER et al. 1999), with an SNR and CNR in the carotid vessels which decreased by only 10% between 5 and 50 min post-contrast.

Another serum protein-binding gadolinium chelate (94% binding at equilibrium), B-22956/1 (Bracco, Milan, Italy) with a molecular weight of 1060 Da also has shown promise in MR angiography (CAVAGNA et al. 2001, 2002), due to its strong binding to albumin and the slow elimination from the plasma in humans. Although it is not an issue for MR angiography, the presence of unbound molecules may limit the usefulness of this type of agent for tumor microvascular characterization. The unbound molecule, though in small proportion, presents a much higher extraction fraction and transendothelial transport rate than the albumin-bound complex, and its very fast kinetics can mask the slower exchanges of the macromolecular form.

Other even larger macromolecules have shown potential utility as angiographic contrast agents.

Clariscan (Feruglose, Amersham, UK), representative of ultrasmall superparamagnetic iron oxide (USPIO) particles, has been used extensively in pilot studies as a vascular imaging agent in animals and patients. Other preparations of iron oxide particles have also been evaluated for their potential in angiographic studies; for example, TAUPITZ and colleagues (2000) tested a VSOP (very small superparamagnetic iron oxide particles), with hydrodynamic diameter of 8 nm, which yielded a highly contrasted depiction of even small thoracic and abdominal vessels in rats and rabbits for up to 50 min. Improved MR angiography is a goal attainable with virtually all of the examined blood pool contrast media, and may contribute to the definition of tumor vascular characteristics, with a better delineation of the feeding vessels of the tumor and their relationship with host vasculature. For example, BPCM-enhanced angiography during biopsies would highlight the large vessels so that they could be avoided during biopsy, and would ease the time pressure inherent to rapidly deteriorating vascular contrast typically observed with SMCM (KAUCZOR and KREITNER 2000).

3.5.2
Characterizing Individual Tumor Biology

Malignant tumors differ from benign lesions in several regards, notably in having a more active recruitment of neovascularity. This acceleration of angiogenesis is essential for the exponential growth and metastasis of the tumor cells (FOLKMAN 1992). Tumor vessels differ from normal tissue vessels by their structural irregularity (abnormal endothelial cell contours and peculiar branching patterns), heterogeneity (flow, diameter, and spacing), and leakiness to macromolecular solutes (JAIN 1988; LESS et al. 1991; BAISH and JAIN 2000; EBERHARD et al. 2000). Recent data from scanning electron microscopy (EM) show that tumor endothelial cells overlap one another and are loosely interconnected, leaving gaps ranging from 0.3 to 4.7 μm (Fig. 3.2) (MCDONALD and FOSS 2000; HASHIZUME et al. 2000). These gaps likely account for the macromolecular hyperpermeability that leads to extravasation of plasma proteins, considered necessary for angiogenesis, as well as transendothelial passage of tumor cells, required for hematogenous metastases (HEUSER and MILLER 1986; DVORAK et al. 1988; HASHIZUME et al. 2000).

This endothelial characteristic of macromolecular hyperpermeability in malignant tumors can be exploited by the use of dynamic contrast-enhanced imaging to identify and grade the abnormal tumor microvessels. Stated in the simplest of terms, malignant tumors should leak macromolecular contrast media and benign tumors should not leak.

In clinical oncology, a neoplastic lesion is usually evaluated for its aggressiveness by performing a biopsy followed by histopathologic microscopic examination. However, this method is invasive and can only sample a small percentage of the entire lesion, possibly missing the most aggressive part of the tumor and leading to an erroneous evaluation.

Non-invasive imaging can complement the histopathologic information and may in some respects surpass it as a means to grade tumor properties. Imaging can be performed on living tissues, on multiple occasions, allows for evaluation of the entire (and often heterogeneous) tumor, and provides both morphologic and physiologic data.

Biopsy tumor specimens are graded by their specific histopathologic characteristics. In breast tumors, for example, the Scarff-Bloom-Richardson (SBR) grading method is used to define the presence and degree of malignant characteristics for a tumor (BLOOM and RICHARDSON 1957; SCARFF and TORLONI 1968). The SBR score sums the microscopic evaluations of three separate morphologic elements: frequency of mitotic figures, nuclear polymorphism, and glandular/tubular formation, each scored from '1' to '3'. Benign tumors, were they scored like the malignant lesions, would have an SBR score of '3' (summing the minimum score of '1' in each category), whereas malignant tumors can have scores ranging from '3' to '9'.

In a rodent model of mammary tumors induced by the single intraperitoneal administration of a chemical carcinogen, *N*-ethyl-*N*-nitrosourea (ENU), a spectrum of tumors develop over months, paralleling the spectrum of breast tumors encountered clinically in women. Several research groups have used this ENU model to evaluate and compare different contrast media for DCE MRI in the characterization of mammary neoplasms (DALDRUP et al. 1998a; SU et al. 1998; HELBICH et al. 2000; TURETSCHEK et al. 2001a).

DALDRUP and coworkers (1998a) used this model to determine if DCE MRI enhanced with either an MMCM [albumin-(Gd-DTPA)$_{35}$] or an SMCM (gadopentetate dimeglumine) could differentiate benign from malignant tumors, and furthermore, if MRI results could predict the histopathologic SBR score. MRI-assayed microvascular parameters including the coefficient of endothelial permeability, defined as K^{trans} (reflecting leakiness), and the fractional plasma

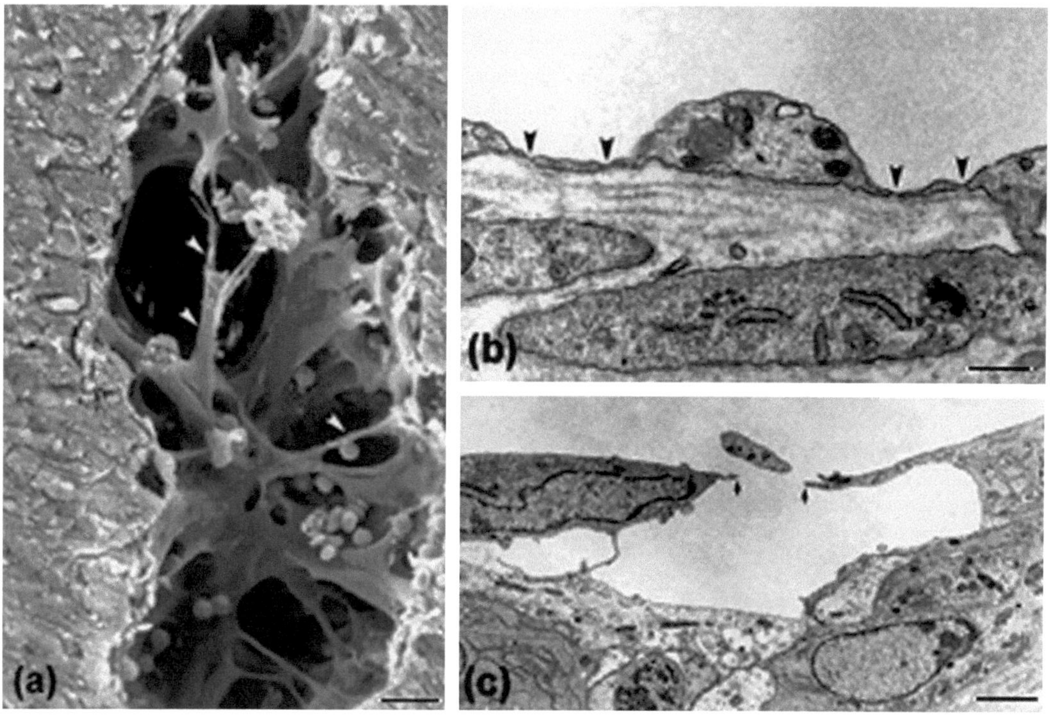

Fig. 3.2a-c. **a** Scanning electron micrograph of vascular endothelial cells from a mouse mammary tumor, showing irregular cell structure such as cellular overlap, bridges (*arrowheads*), tunnels, and wide openings in the vessel wall (scale bar represents 15 µm). **b** Transmission electron micrograph of a mouse mammary tumor blood vessels showing transcellular fenestrae (*arrowheads*) of 50-80 nm in diameter visible in the endothelial cell (scale bar represents 0.5 µm). **c** Transmission electron microscopy of mammary tumor endothelium lining showing intercellular gaps, which may account for the characteristic leakiness of tumor blood vessels (scale bar represents 3 µm). (All images courtesy of Donald M. McDonald).

volume, fPV, (reflecting richness of vascularity), were estimated in each tumor for each contrast agent using a simple two compartment tissue model comprising the blood and interstitial water of the tumor tissue. Correlations were sought between MRI-assayed characteristics and pathologic status including assignment to benign or malignant status and SBR scores. MRI-assayed permeability (K^{trans}) estimated using the macromolecular albumin-(Gd-DTPA)$_{35}$, showed a significant difference between benign fibroadenomas and malignant carcinomas ($p<0.05$). All ten benign tumors had K^{trans} values of zero, whereas all tumors showing measurable permeability to this macromolecular contrast agent, ($K^{trans}>0$), were diagnosed pathologically as carcinomas. There was a slight overlap in pathology for tumors having no measurable MRI leakiness to macromolecules; other than the ten benign tumors, five of 23 carcinomas had no MRI measurable macromolecular permeability, but these five tumors also had the lowest possible SBR scores ('3'–'4'). In this series of 33 tumors, a positive MRI-assayed endothelial permeability value, as estimated with macromolecular albumin-(Gd-DTPA)$_{35}$, was a consistent sign of malignancy with an observed specificity of 100%. Regarding MRI tumor grading, microvascular permeability to albumin-(Gd-DTPA)$_{35}$ (K^{trans}) showed a strong positive correlation with histological tumor grade ($r^2=0.76$; $p<0.001$) (Fig. 3.3).

By comparison, in the same tumors, when using gadopentetate dimeglumine as the contrast agent, there was a broad overlap and no significant difference in K^{trans} values observed between benign fibroadenomas and carcinomas (K^{trans} of 13.2 versus 13.3, respectively; $p>0.99$). No correlation between gadopentetate-assayed K^{trans} or fractional plasma volume and histologic SBR grade was found ($r^2=0.01$ and $p>0.95$ for K^{trans}; $r^2=0.03$ and $p>0.15$ for fPV) (Fig. 3.4). This initial report of positive results using macromolecular contrast media and DCE MRI to characterize and grade ENU-induced mammary tumors was reconfirmed in multiple studies despite sometimes differing methods of kinetic analysis, but all using albumin-(Gd-DTPA)$_x$ (Su et al. 1998; Turetschek et al. 2001a,b).

The same chemically induced mammary tumor model has been used to evaluate other BPCM for-

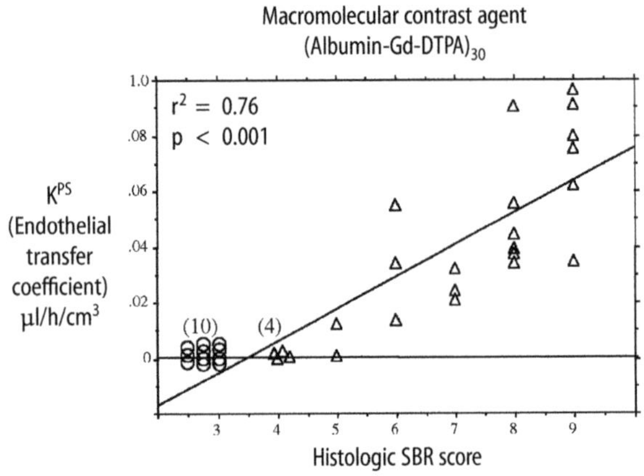

Fig. 3.3. Plot showing strong positive and significant correlation between the endothelial transfer coefficient (K^{PS}=K^{trans}) after intravenous injection of albumin-$(Gd\text{-}DTPA)_{35}$ and histologic tumor grade in benign (*circle*) and malignant tumors (*triangle*). (Adapted from Daldrup et al. 1998a)

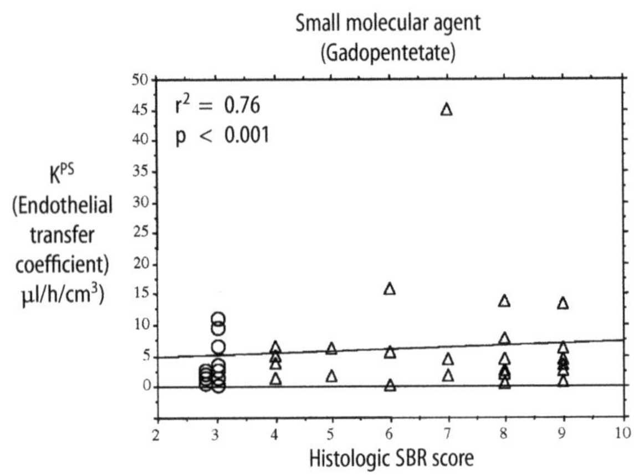

Fig. 3.4. Plot showing the lack of correlation between the endothelial transfer coefficient (K^{PS}=K^{trans}) after intravenous injection of gadopentetate, and histologic tumor grade quantified according to the Scarff-Bloom-Richardson method in benign (*circle*) and malignant tumors (*triangle*). (Adapted from Daldrup et al. 1998a)

mulations and provides a useful means to compare performance of different agents. Using Gadomer-17 (Schering AG, Berlin) with an apparent molecular weight of 17.5 kDa, Daldrup-Link and coworkers (2000) showed a significant difference in MRI-estimated permeability between benign and malignant tumors. However, apparently due to a high variability within both fibroadenoma (benign) and carcinoma (malignant) groups, there was no significant correlation between K^{trans} or fPV, and histopathologic tumor grade. Similarly, Su and colleagues (1998) reported limited specificity when using the intermediately-sized Gadomer-17 being able to differentiate between high grade and low-grade carcinomas, but not between low-grade carcinomas and benign tumors.

Turetschek and colleagues (2001a) evaluated MS-325, a small molecule that spontaneously associates with albumin, for characterization of tumor microvessels assaying plasma volume and permeability in the ENU rodent tumor model. No significant correlations were found between MRI-estimated characteristics and pathologic tumor grade or microvascular count, a marker of angiogenesis; the lack of correlation was attributed in part to the inability to resolve the kinetics of the small-unbound MS-325 (25% in rats) and the larger protein-bound complex.

Yet another class of potential BPCM, the ultrasmall superparamagnetic iron oxide (USPIO) particles, has been evaluated in the ENU-induced mammary tumor model (Turetschek et al. 2001d). NC100150 injection (Clariscan, Amersham, UK) yielded a strongly positive correlation between MRI-derived K^{trans} estimates and histological SBR tumor grade (r=0.82; p<0.001). K^{trans} also correlated significantly with histologically assessed microvascular density (MVD). In this study of 19 total tumors, five were benign fibroadenomas, all with non-measurable (zero) leakiness to the USPIO. Nine of 14 carcinomas did show measurable permeability to the MRI probe; the other five carcinomas without leakiness showed low aggressive potential as reflected in low SBR scores. Overall, the tumors that were leaky to USPIO were all carcinomas. The significance of these results in animal tumor models is accentuated because USPIO particles have already been tested extensively in human clinical trials as angiographic and lymph node enhancers with favorable results; governmental approval for USPIO is anticipated soon (Kernstine et al. 1999; Hudgins et al. 2002; Varallyay et al. 2002).

A preliminary study was performed in women to evaluate the capacity of USPIO particles for breast tumor characterization. Although the qualitative tumor enhancement evident to the eye was relatively low with the USPIO at a dose of 2 mg Fe/kg, making tumor detection more difficult than with gadopentetate, Daldrup-Link and colleagues (2002) found a significant difference in enhancement patterns and kinetic analyses between carcinomas (n=9), and benign lesions including fibroadenomas and mastopathic lesions (n=10) (Daldrup-Link et al. 2002). However, there was no significant difference in enhancement profiles between these same two groups using the small-molecular gadopentetate. The lack of significance was attributed to the broad overlap in gadopentetate enhancement patterns between the two groups, results similar to those observed in the animal mammary models. The results of this clinical study in women with breast tumors supports a unique role of blood pool contrast media for tumor characterization, allowing the differentiation of benign from malignant lesions. Initial tumor detection may be best accomplished by other means such as radiographic mammography or SMCM-enhanced MRI.

MMCM-enhanced MRI can also non-invasively assay tumor angiogenesis, the process by which cancers recruit new vessels growing in from the non-tumor host tissue. Although there is no single "gold standard" assay for angiogenesis, the counting of immunohistochemically stained endothelial cell clusters within a given area of the tumor to yield the microvascular density (MVD) has been used widely as a surrogate marker of the angiogenesis process (WEIDNER 1995). Clinical series have shown that MVD correlates with the presence of metastases at time of diagnosis and with decreased patient survival in numerous types of malignancies including breast, lung, prostate, bladder, ovary, and head and neck carcinomas. Of note, the status of tumor cell differentiation, for example, the SBR score, and that of concomitant tumor angiogenesis do not necessarily correlate. In fact, tumor grade and angiogenesis are considered independent biological characteristics.

Using albumin-$(Gd\text{-}DTPA)_{35}$ in two groups of xenograft human breast carcinomas grown in athymic rats, VAN DIJKE and coworkers (1996) evaluated the potential of MMCM-enhanced MRI to assay angiogenesis. The first group of tumors was angiogenically less active and showed a slower growth rate with distinctly and significantly lower MVD values (MVD<50; $p<0.001$). In contrast, the more aggressive and rapidly growing group of tumors had MVD values ranging from 80 to 305. MRI-derived tumor plasma volumes and permeability estimates increased exponentially with increasing microvascular density, and there was a strong positive correlation between MRI-assayed microvascular characteristics and MVD ($r^2=0.8$; $p<0.001$) (Fig. 3.5).

The authors discussed that MRI estimated angiogenesis might be superior to the pathologic MVD assay, which reflects only the number of vessels, because the MRI technique reflects both vessel number and size through the plasma volume assay, and the functional characteristics of the vessels through the permeability essay. Indeed, MRI samples the entire tumor, is non-invasive, is not operator-dependent, and can be used repeatedly in the same subject.

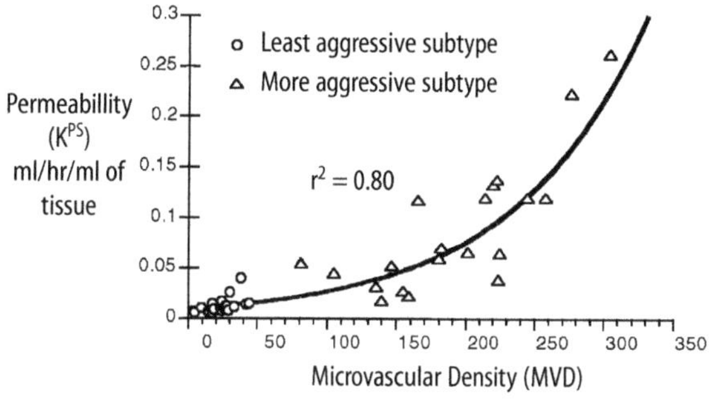

Fig. 3.5. Correlation between MRI-assayed permeability ($K^{PS}=K^{trans}$) and pathologically determined microvascular density in two populations of xenograft breast tumors, a slow-growing group (*circles*) and a more aggressive rapidly-growing group (*triangles*). (Adapted from VAN DIJKE et al. 1996)

3.5.3 Monitoring Tumor Response to Treatment

Beyond assessing individual tumor biology at time of diagnosis, MMCM-enhanced MRI with quantitative estimates of microvascular blood volume and permeability have been shown effective to define the responses of individual tumors to various forms of treatment. In this sense, MMCM-derived MRI measurements are treatment response biomarkers.

In current clinical practice, monitoring of tumor treatment response is typically assessed on imaging examinations by the measurement of tumor size. Generally such size evaluations are performed at 6- or 8-week intervals. Adding to the problem of a long delay, tumor size is a rather indirect and imprecise morphologic sign of treatment effectiveness. Defining additional and more direct biological signs of response to treatment would be highly desirable, particularly if that response were detectable non-invasively and before other measurable changes such as tumor shrinkage or necrosis. However, not only do clinicians need a means to define biological effectiveness soon after the initiation of a given treatment, there is also an urgent requirement in the field of oncologic pharmaceutical development for a sensitive treatment response marker for a broad spectrum of therapeutic agents undergoing development.

The need for MRI treatment biomarkers is in no place more evident than for angiogenesis inhibitors. This class of anti-cancer drugs is known to generally retard tumor growth, but not to cure or totally eradicate lesions; therefore, the drug may be effective but the tumor persists, albeit inhibited from attracting additional blood vessels (FENTON et al. 2001). An imaging biomarker may be superior to tumor sizing for documentation of angiogenesis inhibitory effect. These biomarker changes may be detectable hours after treatment initiation, rather than in weeks or months needed typically to define substantial changes in tumor size. An early evaluation of therapeutic efficacy would allow physicians to adapt treatment regimens and doses based on individual response, optimally associate anti-angiogenic with other anti-tumor therapies, and interrupt an eventually inefficient treatment, reducing morbidity.

Several reports indicate that MMCM-enhanced MRI-derived microvascular characteristics can be used effectively to monitor the biological effectiveness of angiogenesis inhibitory treatment (Cohen et al. 1995; Schwickert et al. 1996; Pham et al. 1998; Su et al. 1999; Gossmann et al. 2000; Clement et al. 2001; Roberts et al. 2001, 2002; Turetschek et al. 2001c; Allegrini et al. 2002; Gossmann et al. 2002; Petrovsky et al. 2002). Vascular endothelial growth factor (VEGF), also known as vascular permeability factor (VPF) is considered a central signaling molecule in the complex process of tumor angiogenesis (Dvorak 2000). VEGF/VPF has multiple stimulatory effects, all tied to angiogenesis, including endothelial cell mitogenesis, endothelial cell migration, cell survival, and increased endothelial permeability. The VEGF-induced hyperpermeability of cancer microvessels, 50,000 times stronger than that induced by histamine, leads to an extravasation of macromolecular proteins that form a favorable substrate in the tumor interstitium into which the new vessels grow. It was therefore relevant to design an experiment which would probe the potential of MRI-assayed microvascular responses for the detection of anti-angiogenic effect in tumors.

The first anti-angiogenic drug tested in our center was a human antibody directed against VEGF (Avastin, Genentech, South San Francisco, CA) (Pham et al. 1998). A significant ($p<0.01$) suppression (>75%) of MRI-assayed microvascular hyperpermeability, expressed as the coefficient of permeability surface area product (K^{trans}), was observed in a human breast cancer (MB-MDA-435) grown in athymic rats following a 1-week course of three 1-mg doses of human anti-VEGF antibody. With appropriate controls, the anti-VEGF antibody was shown to reduce tumor weight, growth rate (Fig. 3.6), and MRI-assayed permeability to the blood pool contrast agent (Fig. 3.7).

In a follow-up experiment (Brasch et al. 2000), MMCM-enhanced MRI demonstrated a reduction in permeability, induced by anti-VEGF antibody as early as 24 h after only a single dose. A significant ($p<0.01$) reduction in tumor microvascular macromolecular permeability to levels less than 40% of baseline was recorded.

In a parallel manner, subsequent studies showed anti-VEGF antibody induced reductions in MRI-assayed permeabilities in models of human ovarian carcinoma and cerebral glioblastoma multiforme (Gossmann et al. 2000, 2002). In the case of intraperitoneal human ovarian cancers (SKOV-3) grown in athymic rats, after five 1-mg doses of anti-VEGF antibody administered every 3 days, permeability assayed by MMCM-enhanced MRI was seen to decrease significantly in treated tumors, while saline-treated control ovarian tumors exhibited an increase in permeability to MMCM (Fig. 3.8).

Consistent with the hypothesis that elaboration and deposition of VEGF/VPF by ovarian cancer cells into the peritoneal cavity leads to microvascular hyperpermeability, macromolecular extravasation,

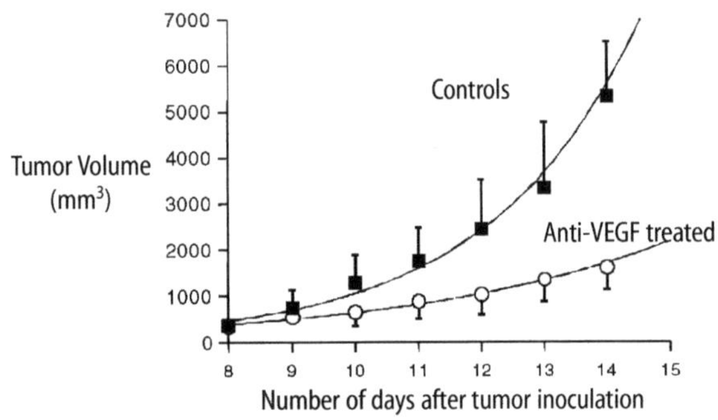

Fig. 3.6. Tumor growth for controls and animals treated with 1-mg doses of anti-VEGF antibody every third day over a period of 1 week. Notice the substantial slowing of tumor growth in the anti-VEGF antibody-treated group. (Adapted from Pham et al. 1998)

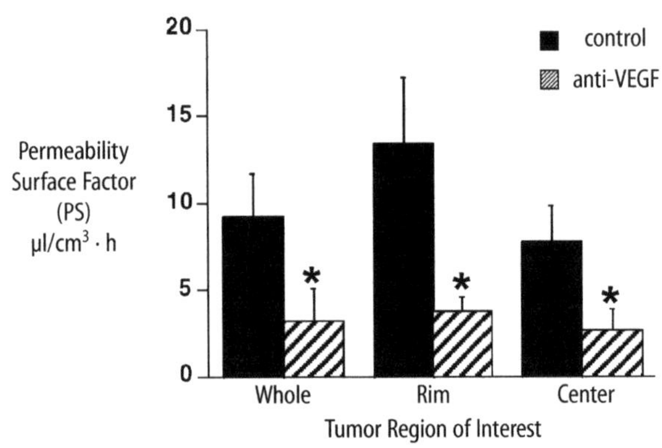

Fig. 3.7. MRI-assayed permeability for controls (inactive immunoglobin) and animals treated with anti-VEGF antibody following a 1-week course of treatment. Notice a significant reduction in MRI-estimated PS whether assayed for the whole tumor or exclusively in the tumor rim or center. (Adapted from Pham et al. 1998)

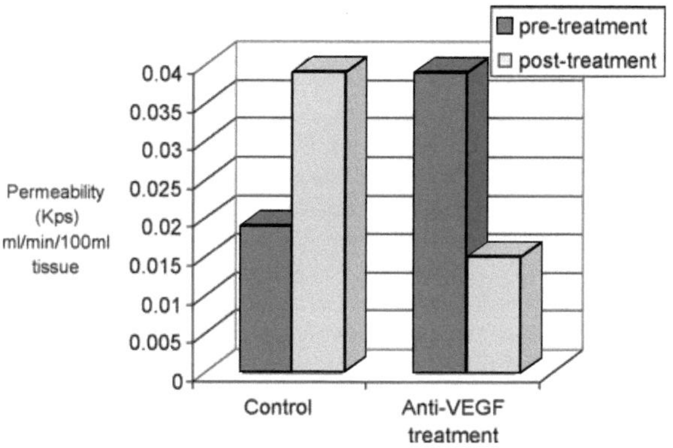

Fig. 3.8. Mean coefficients of endothelial transport in human ovarian carcinomas implanted in nude rats, estimated by dynamic albumin-(Gd-DTPA)$_{35}$ enhanced MRI, before and after treatment in control animals receiving saline solution and treated animals receiving five 1-mg doses of anti-VEGF antibody every third day. (Adapted from GOSSMANN et al. 2000)

and ascites (NAGY et al. 1989), there was a significant reduction in the measured ascites accompanying the reduced permeability in the angiogenically-inhibited tumors (Fig. 3.9).

Other contrast agents have been shown to be potentially useful for the definition of changes in tumor microvessels induced by anti-VEGF antibody treatment. For example, in a human breast cancer rodent model the intermediately sized molecule, gadomer-17, showed significant changes both in permeability (K^{trans}) and in fractional blood volume (fPV) after treatment (ROBERTS et al. 2001). In a similar study, the serum protein-binding molecule B22956/1 also demonstrated a decrease in vascular permeability (K^{trans}) reflecting the biological effectiveness of the antibody on tumor vessels (ROBERTS et al. 2002).

Dynamic MMCM-enhanced MRI has examined other anti-angiogenic drugs known to diminish tumor growth and metastatic spread for measurable effects on microvessels. TURETSCHEK and coworkers (2001c) studied the effect of an inhibitor of the tyrosine kinase VEGF receptor (PTK787/ZK222584, Novartis, Basel, Switzerland) in athymic rats bearing human MB-MDA-435 breast adenocarcinomas. MRI-assayed microvascular characteristics were evaluated to determine whether they could reflect treatment efficacy, and were compared to tumor size and microvessel density, respectively the clinical and pathological methods used to evaluate biological response. Two macromolecular contrast agents were tested, albumin-(Gd-DTPA)$_{35}$ and USPIO (SHU555C, Schering AG, Berlin, Germany), to generate quantitative estimates of tumor blood volume and microvascular permeability.

With both albumin-(Gd-DTPA)$_{35}$ and USPIO, a decrease in estimated permeabilities was observed in the treated group after VEGF receptor inhibition, whereas there was an increase in MRI-estimated permeabilities for the control group (Fig. 3.10). These microvascular responses correlated with the observed slowing in the treatment group of tumor growth and with the significantly reduced microvascular density.

PETROVSKY and colleagues (2002) tested macromolecular contrast-enhanced MRI for detection of vascular changes after treatment with another but similar tyrosine kinase inhibitor. Treated animals received this agent, VEGF-RTKI AG013925 (Pfizer, San Diego, CA) at the dose of 25 mg/kg b.i.d. for 12 days. Using a large polylysine contrast agent, shielded by methoxy polyethylene glycol (MPEG) chains, and labeled with gadolinium chelates, they showed a significant decrease (>50%) in the vascular volume fraction of MV-522 human colon carcinoma

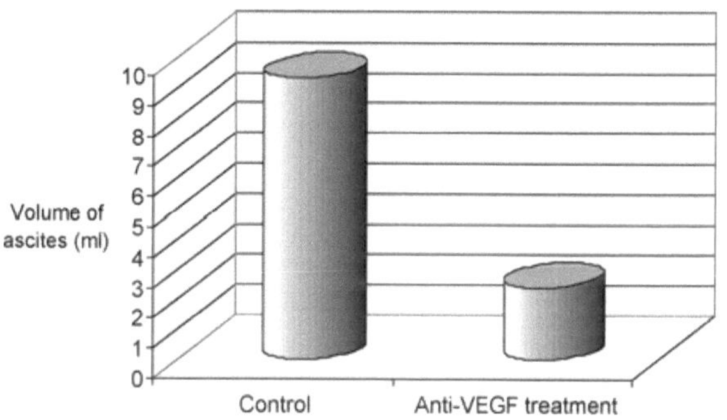

Fig. 3.9. Volume of ascites at necropsy in rodents bearing intraperitoneal human ovarian SKOV-3 carcinomas measured for saline-treated control animals and those receiving five 1-mg doses of anti VEGF antibody every third day. These differences in volume of ascites corresponded to anti-VEGF antibody-induced reductions in MRI-estimated microvascular permeability. (Adapted from GOSSMANN et al. 2000)

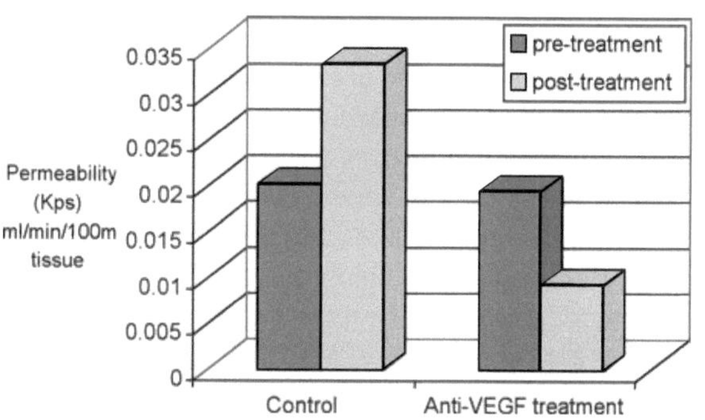

Fig. 3.10. Coefficient of endothelial transport ($K^{PS}=K^{trans}$) before and after treatment in control animals receiving saline solution and treated animals receiving two daily doses of 50 mg/kg of tyrosine kinase inhibitor, both by oral gavage for 7 days. (Adapted from TURETSCHEK et al. 2001c)

xenografts in mice after only three inhibitor doses (1.5 days) of the treatment, predictive of a 2.5-times decrease in tumor volume visible after 2 weeks.

In a similar experiment, yet another tyrosine kinase inhibitor, ZD4190 (AstraZeneca Pharmaceuticals, Macclesfield, UK), was tested on the human PC-3 prostate carcinoma implanted in nude mice. A significant decline (>40%) in MRI-estimated permeability was detected after only two inhibitor doses administered in 24 h, when comparing the treated group and the control group (Clement et al. 2001).

MMCM-assayed microvascular status has been applied with success for monitoring responses to other forms of therapy, in addition to anti-angiogenic drugs, and for other diseases in addition to malignancies. For example, acute and highly significant increases in microvascular blood volume and permeability to macromolecular albumin-(Gd-DTPA)$_{35}$ were reported following a single administration of gamma radiation (30 Gy) to a mammary tumor model (Cohen et al. 1995). The diseased tissue response to treatment for conditions as diverse as arthritis (Jiang et al. 2002), spinal cord injury (Philippens et al. 2002), and oxygen-induced pulmonary fibrosis can be monitored by MMCM-enhanced MRI (Brasch et al. 1993).

3.6 Conclusion

In summary, there is a growing body of evidence supporting the development and use of quantitative MRI tumor microvascular bioassays, achievable through the application of macromolecular blood pool contrast media, to characterize individual tumors and non-neoplastic disease processes that affect microvessels. It is reasonable to anticipate that the clinical introduction of a dynamic MMCM-enhanced MRI bioassay could lead to better definition of tumor biological properties including the more precise identification of malignancy, the grading of malignancy, the angiographic demarcation of tumor feeding vessels, the presence and degree of accelerated angiogenesis, and perhaps clinically most important, a means to rapidly and non-invasively evaluate tumor response to treatment. Critical to the realization of the full potential of MMCM-enhanced MRI will be the identification, development, and ultimately the governmental approval of macromolecular blood pool contrast agents that combine favorable characteristics for strong image enhancement, prolonged intravascular retention, complete and timely bodily elimination, and safety.

References

Abolmaali ND, Hietschold V, Appold S et al (2002) Gadomer-17-enhanced 3D navigator-echo MR angiography of the pulmonary arteries in pigs. Eur Radiol 12:692-697

Aicher KP, Dupon JW, White DL et al (1990) Contrast-enhanced magnetic resonance imaging of tumor-bearing mice treated with human recombinant tumor necrosis factor alpha. Cancer Res 50:7376-7381

Allegrini P, Rudin M, Wood J et al (2002) Nvp-laf389 reduces tumor blood volume and vascular permeability in ca20948 pancreatic tumor model as measured in vivo by dynamic contrast enhanced MRI - putative surrogate markers for efficacy. In: International Society for Magnetic Resonance in Medicine Tenth Scientific Meeting and Exhibition. Honolulu, Hawaii, USA

Baish JW, Jain RK (2000) Fractals and cancer. Cancer Res 60:3683-3688

Bloom H, Richardson W (1957) Histologic grading and prognosis in breast cancer. Br J Cancer 11:359-377

Bluemke DA, Stillman AE, Bis KG et al (2001) Carotid MR angiography: phase ii study of safety and efficacy for ms-325. Radiology 219:114-122

Bonk RT, Schmiedl UP, Yuan C et al (2000) Time-of-flight MR angiography with Gd-DTPA hexamethylene diamine copolymer blood pool contrast agent: comparison of enhanced MRA and conventional angiography for arterial stenosis induced in rabbits. J Magn Reson Imaging 11:638-646

Brasch RC, Berthezene Y, Vexler V et al (1993) Pulmonary oxygen toxicity: Demonstration of abnormal capillary permeability using contrast-enhanced mri. Pediatr Radiol 23:495-500

Brasch RC, Li KC, Husband JE et al (2000) In vivo monitoring of tumor angiogenesis with MR imaging. Acad Radiol 7:812-823

Cavagna F, Lorusso V, Anelli P et al (2001) Preclinical profile and clinical potential of gadocoletic acid trisodium salt (b-22956/1), a new intravascular contrast medium. Contrast Media Research, Capri, Italy

Cavagna F, La Noce A, Maggioni F et al (2002) Mr coronary angiography with the new intravascular contrast agent b-22956/1: first human experience. International Society for Magnetic Resonance in Medicine Tenth Scientific Meeting and Exhibition. Honolulu, Hawaii, USA

Clement O, Pradel C, Siauve N et al (2001) Assessing perfusion and capillary permeability changes induced by a VEGF inhibitor in human tumor xenografts using macromolecular MR imaging contrast media. Contrast Media Research, Capri, Italy

Cohen F, Kuwatsuru R, Shames D et al (1995) Contrast enhanced MRI estimation of altered capillary permeability in experimental mammary carcinomas following x-irradiation. Invest Radiol 29:970-977

Crone C (1963) The permeability of capillaries in various organs determined by the use of the "indicator diffusion" method. Acta Physiol Scand 58:292-305

Crone C, Levitt DG (1984) Capillary permeability to small solutes. American Physiological Society, Bethesda

Daldrup H, Shames DM, Wendland M et al (1998a) Correlation of dynamic contrast-enhanced MR imaging with histologic tumor grade: comparison of macromolecular and small-molecular contrast media. AJR Am J Roentgenol 171:941-949

Daldrup HE, Shames DM, Husseini W et al (1998b) Quantification of the extraction fraction for gadopentetate across breast cancer capillaries. Magn Reson Med 40:537-543

Daldrup-Link HE, Shames DM, Wendland M et al (2000) Comparison of gadomer-17 and gadopentetate dimeglumine for differentiation of benign from malignant breast tumors with MR imaging. Acad Radiol 7:934-944

Daldrup-Link HE, Link TM, Moller HE et al (2001) Carboxymethyldextran-a2-Gd-DOTA enhancement patterns in the abdomen and pelvis in an animal model. Eur Radiol 11:1276-1284

Daldrup-Link HE, Kaiser A, Link TM et al (2002) Quantification of breast tumor microvascular permeabilities with Feruglose (Clariscan) enhanced MR-mammography: initial clinical trial. Ecr 2002. Eur Radiol [Suppl] 1:158

Dvorak HF (1990) Leaky tumor vessels: consequences for tumor stroma generation and for solid tumor therapy. Prog Clin Biol Res

Dvorak HF (2000) VPF/VEGF and the angiogenic response. Semin Perinatol 24:75-78

Dvorak HF, Nagy JA, Dvorak JT et al (1988) Identification and characterization of the blood vessels of solid tumors that are leaky to circulating macromolecules. Am J Pathol 133:95-109

Eberhard A, Kahlert S, Goede V et al (2000) Heterogeneity of angiogenesis and blood vessel maturation in human tumors: implications for antiangiogenic tumor therapies. Cancer Res 60:1388-1393

Fenton BM, Beauchamp BK, Paoni SF et al (2001) Characterization of the effects of antiangiogenic agents on tumor pathophysiology. Am J Clin Oncol 24:453-457

Folkman J (1992) The role of angiogenesis in tumor growth. Semin Cancer Biol 3:65-71

Gerlowski LE, Jain RK (1986) Microvascular permeability of normal and neoplastic tissues. Microvasc Res 31:288-305

Gossmann A, Helbich T, Mesiano S et al (2000) Magnetic resonance imaging in an experimental model of human ovarian cancer demonstrating altered microvascular permeability after inhibition of vascular endothelial growth factor. Am J Obstet Gynecol 183:956-963

Gossmann A, Helbich TH, Kuriyama N et al (2002) Dynamic contrast-enhanced magnetic resonance imaging as a surrogate marker of tumor response to anti-angiogenic therapy in a xenograft model of glioblastoma multiforme. J Magn Reson Imaging 15:233-240

Grist TM, Korosec FR, Peters DC et al (1998) Steady-state and dynamic mr angiography with ms-325: initial experience in humans. Radiology 207:539-544

Hashizume H, Baluk P, Morikawa S et al (2000) Openings between defective endothelial cells explain tumor vessel leakiness. Am J Pathol 156:1363-1380

Haunso S, Paaske WP, Sejrsen P et al (1980) Capillary permeability in canine myocardium as determined by bolus injection, residue detection. Acta Physiol Scand 108:389-397

Helbich TH, Gossmann A, Mareski PA et al (2000) A new polysaccharide macromolecular contrast agent for MR imaging: biodistribution and imaging characteristics. J Magn Reson Imaging 11:694-701

Henderson E, Sykes J, Drost D et al (2000) Simultaneous MRI measurement of blood flow, blood volume, and capillary permeability in mammary tumors using two different contrast agents. J Magn Reson Imaging 12:991-1003

Heuser LS, Miller FN (1986) Differential macromolecular leakage from the vasculature of tumors. Cancer 57:461-464

Hoffmann U, Loewe C, Bernhard C et al (2002) MRA of the lower extremities in patients with pulmonary embolism using a blood pool contrast agent: Initial experience. J Magn Reson Imaging 15:429-437

Hudgins PA, Anzai Y, Morris MR et al (2002) Ferumoxtran-10, a superparamagnetic iron oxide as a magnetic resonance enhancement agent for imaging lymph nodes: a phase 2 dose study. AJNR Am J Neuroradiol 23:649-656

Jain R (1987) Transport of molecules across tumor vasculature. Cancer Metast Rev 6:559-593

Jain R (1988) Determinants of tumor blood flow: a review. Cancer Res 48:2641-2658

Jain R (1994) Barriers to drug delivery in solid tumors. Sci Am 271:58-65

Jiang Y, Zhao JJ, Tang H et al (2002) Blood pool MR contrast media ms-325 improves contrast and disease characterization of rheumatoid arthritis for longitudinal quantification of inflamed synovium and joint fluid. International Society for Magnetic Resonance in Medicine Tenth Scientific Meeting and Exhibition. Honolulu, HI, USA

Kauczor HU, Kreitner KF (2000) Contrast-enhanced MRI of the lung. Eur J Radiol 34:196-207

Kernstine KH, Stanford W, Mullan BF et al (1999) PET, CT, and MRI with combidex for mediastinal staging in non-small cell lung carcinoma. Ann Thorac Surg 68:1022-1028

Kroft LJ, de Roos A (1999) Blood pool contrast agents for cardiovascular MR imaging. J Magn Reson Imaging 10:395-403

Less JR, Skalak TC, Sevick EM et al (1991) Microvascular architecture in a mammary carcinoma: branching patterns and vessel dimensions. Cancer Res 51:265-273

Li D, Dolan RP, Walovitch RC et al (1998) Three-dimensional MRI of coronary arteries using an intravascular contrast agent. Magn Reson Med 39:1014-1018

Li D, Zheng J, Weinmann HJ (2001) Contrast-enhanced MR imaging of coronary arteries: comparison of intra- and extravascular contrast agents in swine. Radiology 218:670-678

McDonald DM, Foss AJ (2000) Endothelial cells of tumor vessels: abnormal but not absent. Cancer Metastasis Rev 19:109-120

Nagy J, Brown L, Senger D et al (1989) Pathogenesis of tumor stroma generation: a critical role for leaky blood vessels and fibrin deposition. Biochim Biophys Acta 948:305-326

Ogan MD (1988) Albumin labeled with Gd-DTPA: an intravascular contrast-enhancing agent for magnetic resonance blood pool imaging: preparation and characterization. Invest Radiol 23:961

Okuhata Y, Brasch RC, Pham CD et al (1999) Tumor blood volume assays using contrast-enhanced magnetic resonance imaging: regional heterogeneity and postmortem artifacts. J Magn Reson Imaging 9:685-690

Petrovsky A, Weissleder R, Hu-Lowe D et al (2002) Non-invasive mr imaging of anti-angiogenic effects induced by a VEGF-RTKI in a human xenograft model. International Society for Magnetic Resonance in Medicine Tenth Scientific Meeting and Exhibition. Honolulu, HI, USA

Pham C, Roberts T, van Bruggen N et al (1998) Magnetic resonance imaging detects suppression of tumor vascular permeability after administration of antibody to vascular endothelial growth factor. Cancer Invest 6:224-230

Philippens M, Pikkemaat J, Schellekens S et al (2002) USPIO contrast enhanced MRI of irradiated rat spinal cord, monitoring macrophages and blood volume changes. International Society for Magnetic Resonance in Medicine Tenth Scientific Meeting and Exhibition. Honolulu, HI, USA

Renkin EM (1959) Transport of potassium-42 from blood to tissue in isolated mammalian skeletal muscles. Am J Physiol 197:1205-1210

Roberts T, Kuretschek K, Preda A et al (2001) Tumor microvascular changes to anti-angiogenic treatment assessed by MR contrast media of different molecular weights. Contrast Media Research, Capri, Italy

Roberts T, Preda A, Turetschek K et al (2002) Permeability of b22956/1, a novel protein-binding contrast agent, resolves anti-angiogenic therapy in human breast cancer model. International Society for Magnetic Resonance in Medicine Tenth Scientific Meeting and Exhibition. Honolulu, HI, USA

Scarff R, Torloni, H. (1968) Histological typing of breast tumors. World Health Organization, Geneva, pp 13-20

Schmiedl U, Moseley ME, Ogan MD et al (1987) Comparison of initial biodistribution patterns of Gd-DTPA and albumin-(Gd-DTPA) using rapid spin echo MR imaging. J Comput Assist Tomogr 11:306-313

Schwickert H, Stiskal M, van Dijke CF et al (1995) Tumor angiography using high-resolution, three-dimensional magnetic resonance imaging: comparison of gadopentetate dimeglumine and a macromolecular blood-pool contrast agent. Acad Radiol 2:851-858

Schwickert H, Stiskal M, Roberts T et al (1996) Contrast-enhanced MRI assessment of tumor capillary permeability: the effect of pre-irradiation on the tumor delivery of chemotherapy. Radiology 198:893-898

Sevick EM, Jain RK (1991) Measurement of capillary filtration coefficient in a solid tumor. Cancer Res 51:1352-1355

St Lawrence KS, Lee TY (1998a) An adiabatic approximation to the tissue homogeneity model for water exchange in the brain. II. Experimental validation. J Cereb Blood Flow Metab 18:1378-1385

St Lawrence KS, Lee TY (1998b) An adiabatic approximation to the tissue homogeneity model for water exchange in the brain. I. Theoretical derivation. J Cereb Blood Flow Metab 18:1365-1377

Stuber M, Botnar RM, Danias PG et al (1999) Contrast agent-enhanced, free-breathing, three-dimensional coronary magnetic resonance angiography. J Magn Reson Imaging 10:790-799

Su MY, Muhler A, Lao X et al (1998) Tumor characterization with dynamic contrast-enhanced MRI using MR contrast agents of various molecular weights. Magn Reson Med 39:259-269

Su MY, Wang Z, Nalcioglu O (1999) Investigation of longitudinal vascular changes in control and chemotherapy-treated tumors to serve as therapeutic efficacy predictors. J Magn Reson Imaging 9:128-137

Svendsen JH, Efsen F, Haunso S (1992) Capillary permeability of 99mtc-dtpa and blood flow rate in the human myocardium determined by intracoronary bolus injection and residue detection. Cardiology 80:18-27

Taupitz M, Schnorr J, Abramjuk C et al (2000) New generation of monomer-stabilized very small superparamagnetic iron oxide particles (VSOP) as contrast medium for MR angiography: preclinical results in rats and rabbits. J Magn Reson Imaging 12:905-911

Turetschek K, Floyd E, Helbich T et al (2001a) MRI assessment of microvascular characteristics in experimental breast tumors using a new blood pool contrast agent (ms-325) with correlations to histopathology. J Magn Reson Imaging 14:237-242

Turetschek K, Floyd E, Shames DM et al (2001b) Assessment of a rapid clearance blood pool MR contrast medium (p792) for assays of microvascular characteristics in experimental breast tumors with correlations to histopathology. Magn Reson Med 45:880-886

Turetschek K, Preda A, Floyd E et al (2001c) MRI monitoring of tumor response to a novel VEGF tyrosine kinase inhibitor in an experimental breast cancer model. Contrast Media Research, Capri, Italy

Turetschek K, Roberts TP, Floyd E et al (2001d) Tumor microvascular characterization using ultrasmall superparamagnetic iron oxide particles (USPIO) in an experimental breast cancer model. J Magn Reson Imaging 13:882-888

Van Dijke CF, Brasch RC, Roberts TP et al (1996) Mammary carcinoma model: correlation of macromolecular contrast-enhanced MR imaging characterizations of tumor microvasculature and histologic capillary density. Radiology 198:813-818

Varallyay P, Nesbit G, Muldoon LL et al (2002) Comparison of two superparamagnetic viral-sized iron oxide particles ferumoxides and ferumoxtran-10 with a gadolinium chelate in imaging intracranial tumors. AJNR Am J Neuroradiol 23:510-519

Vexler V, Clèment O, Schmitt-Willich H et al (1994) Effect of varying molecular weight of the MR contrast agent Gd-DTPA-polylysine on blood pharmacokinetics and enhancement patterns. J Magn Reson Imag 4:381-388

Weidner N (1995) Current pathologic methods for measuring intratumoral microvessel density within breast carcinoma and other solid tumors. Breast Cancer Res Treat 36:169-180

Weinmann H-J, Brasch RC, Press WR et al (1984) Characteristics of gadolinium-DTPA complex: a potential MRI contrast agent. Am J Roentgenol 142:619-624

Weissig VV, Babich J, Torchilin VV (2000) Long-circulating gadolinium-loaded liposomes: Potential use for magnetic resonance imaging of the blood pool. Colloids Surf B Biointerfaces 18:293-299

White D, Wang S-C, Aicher K et al (1989) Albumin-(DTPA-Gd)15-20: Whole body clearance, and organ distribution of gadolinium. Society of Magnetic Resonance in Medicine, 8th Annual Meeting. Amsterdam, p 807

Yuan F, Leunig M, Berk DA et al (1993) Microvascular permeability of albumin, vascular surface area, and vascular volume measured in human adenocarcinoma ls174t using dorsal chamber in SCID mice. Microvasc Res 45:269-289

4 Quantification of Dynamic Susceptibility Contrast T_2^* MRI in Oncology

Fernando Calamante

CONTENTS

4.1 Introduction

Within the last decade, magnetic resonance imaging (MRI) has become a very powerful technique for the assessment of perfusion, and perfusion-related parameters in many different tissues, such as brain, breast, heart, etc. MRI techniques for measuring perfusion can be divided in two main groups depending on the type of contrast agent used (Calamante et al. 1999; Barbier et al. 2001): arterial spin labelling techniques (using magnetically labelled blood as an endogenous contrast agent), and dynamic susceptibility contrast MRI techniques (using an exogenous MR contrast agent). These latter techniques, despite the need of an exogenous agent, are the most common methodologies in clinical studies so far [see Calamante et al. (1999) for a recent review]. When a bolus of an exogenous contrast agent (such as gadolinium-DTPA) is injected into a vein, its paramagnetic properties introduce a decrease in both the T_1 and T_2^* (or T_2)[a] relaxation properties of water. This relaxation enhancement can be exploited to quantify haemodynamic parameters using either T_1-weighted (detected as signal enhancement) or T_2^*-weighted (detected as signal loss) sequences. However, the effect of the paramagnetic contrast agent on these relaxation times is based on different mechanisms. The shortening of T_1 relaxation is due to dipole-dipole interactions, which are only effective at very short distances (Farrar and Becker 1971). Therefore, the signal enhancement on T_1-weighted images will be observed only in areas that have direct access to the contrast agent. On the other hand, the reduction in T_2^* is due to the signal dephasing associated with the susceptibility-induced gradients surrounding the paramagnetic contrast agent. This effect is more significant in areas where the contrast agent is compartmentalised since this increases the induced gradients (Villringer et al. 1988). In this sense, the T_1 and T_2^* effects can be considered complementary: an area with uniformly distributed contrast agent would have a very large T_1 effect (direct access to the contrast agent in the whole area) but a very small T_2^* effect (no induced gradients due to the homogeneous distribution of the contrast); in contrast, an area with contrast agent within a small compartment would have negligible T_1 effect but produce very significant T_2^* decrease. Any intermediate situation, as discussed below, will complicate the quantification of perfusion using MRI because both effects are present, and compete against each other.

This chapter describes some of the more common models involved in the quantification of MRI data in the presence of an exogenous contrast agent using T_2^*-weighted sequences. For a description of the models involved using T_1-weighted sequences the reader is referred to Chaps. 5 and 6.

F. Calamante, MD, PhD
Radiology and Physics Unit, Institute of Child Health, University College London, 30 Guilford Street, London, WC1N 1EH, UK

[a] For the remainder of this chapter, all the statements referring to T_2^* are also applicable to T_2, except where otherwise stated

4.2 Measurement of Perfusion Using Bolus Tracking T_2^* MRI

As mentioned in Sect. 4.1, the effect of the contrast agent on the signal intensity on T_2^*-weighted images can be used to measure perfusion (Villringer et al. 1988; Rosen et al. 1990). This technique, usually referred to as dynamic susceptibility-contrast (DSC) MRI, is also known as *bolus tracking*. It involves the rapid injection of a bolus of paramagnetic contrast agent and the measurement of the transient signal loss during its passage through the tissue of interest (Fig. 4.1). Since the transit time of the bolus through the tissue is only a few seconds, a rapid MRI technique is required to properly characterise the signal intensity time course. The more commonly used sequences are echo-planar imaging (EPI) (Stehling et al. 1991) and fast low-angle single-shot (FLASH) (Haase 1990). Although EPI is faster (therefore allowing the acquisition of more slices for a given time resolution), the increased image distortions in regions of high magnetic susceptibility gradients (such as close to tissue–air interfaces) and hardware demands (Fischer and Ladebeck 1998; Bowtell and Schmitt 1998) favour FLASH sequences for many applications.

As discussed in Sect. 4.1, T_2^*-based techniques for measuring perfusion are primarily used in cases where there is a significant compartmentalisation of the contrast agent (to observe a significant decrease in T_2^* relaxation). Therefore, they

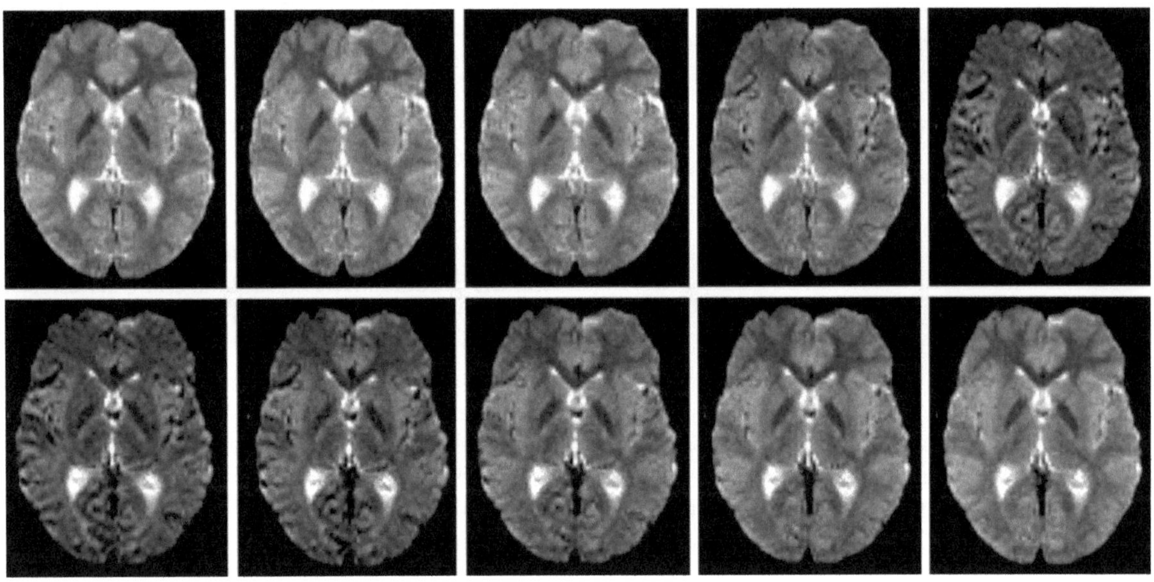

a

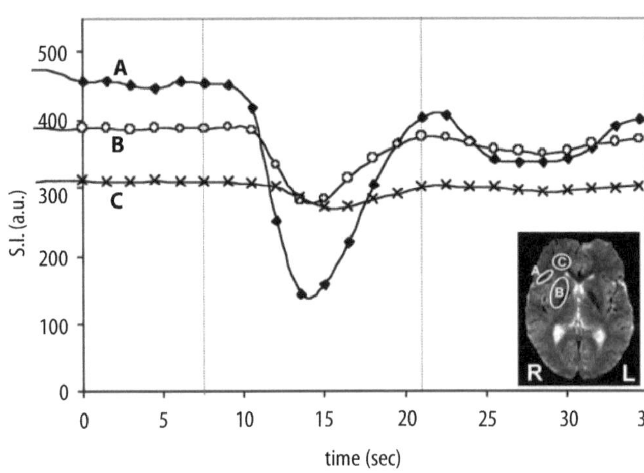

b

Fig. 4.1a,b. MR perfusion data in a 13-year-old boy with sickle-cell disease. The patient had no infarctions and was asymptomatic at the time of the MR examination. **a** Sequential spin-echo planar images during the passage of a bolus of contrast agent (TR=1.5 s). Timed sequence of images runs from *top left* to *top right* and then from *bottom left* to *bottom right*. These images show the signal intensity decrease associated with the passage of the bolus. **b** Signal intensity time course plots for three regions of interest (ROI) (see *inset*): ROI *A*, peripheral branch of the right MCA; ROI *B*, right basal ganglia; ROI *C*, right frontal white matter. Three different periods can be identified in the time course data: the baseline (before the arrival of the bolus approximately from t=0 s to t=9 s), the first passage of the bolus (approximately from t=9 s to t=21 s) and the recirculation period (in this case a second smaller peak more clearly seen in the arterial region (ROI *A*), approximately for times t>21 s. Note that the grey matter region ROI *B* shows a signal drop larger than that from the white matter region ROI *C* consistent with the higher CBV in grey matter. *SI*, signal intensity; *a.u.*, arbitrary units are ideal for using in combination with blood pool contrast agents, since the tracer remains intravascular. In these cases, although the vascular space is a small fraction of the total tissue volume (e.g. 3%–5% in the human brain), the susceptibility effect extends beyond the vascular space (Gillis and Koenig 1987; Villringer et al. 1988), leading to a significant transient signal drop. Due to the presence of the blood–brain barrier (BBB), the investigation of brain perfusion is one of the main applications of bolus tracking, since even the contrast agents that are not blood pool agents (such as gadolinium-DTPA) behave as such. Therefore, the kinetic models used in the quantification of DSC-MRI data in this chapter will be described in the context of cerebral perfusion. However, these models are not restricted to quantification in the brain, but they can also be applied to other organs if the appropriate model is chosen. Section 4.2.1 describes the model for the case of an MR contrast agent that remains intravascular (such as gadolinium-DTPA in brain areas where the BBB is intact). Section 4.2.2 describes some of the modifications proposed to extend the model to areas where the contrast agent does not remain intravascular (such as in brain areas where the BBB is disrupted). Section 4.2.3 describes a more commonly used approach for the analysis of bolus tracking data which has the benefits of simplicity, but which provides only indirect information about perfusion. Finally, Sect. 4.3 discusses the different sequence types for measuring perfusion (e.g. T_2^*- or T_2-weighted sequences).

4.2.1 Perfusion Model for Intravascular MR Contrast Agents

Quantification of bolus tracking data using an intravascular agent is based on the principles of tracer kinetics for non-diffusible tracers (Zierler 1965; Axel 1980). These are used to model the time-dependent concentration of the contrast agent in the tissue ($C_t(t)$) as a function of the injected bolus, the cerebral blood flow (CBF), and the fraction of the contrast remaining in the tissue at a given time for an ideal instantaneous injection [see Eq. (2) below]. One of the differences of MRI compared with other imaging modalities is that MRI does not measure the concentration of the contrast agent directly. The concentration has to be inferred from its effect on the relaxation time (Villringer et al. 1988; Rosen et al. 1990). Despite some recent concerns regarding its validity (Kiselev 2001), it has been shown empirically (Villringer et al. 1988; Rosen et al. 1990) and with numerical simulations (Weisskoff et al. 1994b; Boxerman et al. 1995) that, for the doses commonly used in clinical studies, the change (compared with the pre-injection baseline value) in the relaxation rate R_2^* ($R_2^*=1/T_2^*$) is linearly proportional to the concentration of the contrast agent:

$$C_t(t) = \kappa_t \Delta R_2^*(t) = -\frac{\kappa_t}{TE}\ln\left(\frac{S(t)}{S_0}\right) \quad (1)$$

where $S(t)$ is the signal intensity in the tissue at time t (see Fig. 4.2), S_0 is the signal intensity during the baseline period before the arrival of the tracer, TE is the echo-time of the MRI sequence, and κ_t is a proportionality constant that depends on the tissue, the contrast agent, the field strength and the pulse sequence parameters (Weisskoff et al. 1994b; Boxerman et al. 1995). Equation (1) assumes negligible T_1 effects, which is a reasonable assumption for a T_2^*-weighted sequence using an intravascular contrast agent. The second equality is obtained from the single exponential relationship between R_2^* and the signal intensity.

According to the indicator dilution theory for intravascular contrast agents, the concentration of tracer in the tissue can be described (Østergaard et al. 1996a; Calamante et al. 1999) by:

$$C_t(t) = \alpha \cdot CBF \cdot (C_a(t) \otimes R(t)) = \alpha \cdot CBF \cdot \int_0^t C_a(\tau)R(t-\tau)d\tau \quad (2)$$

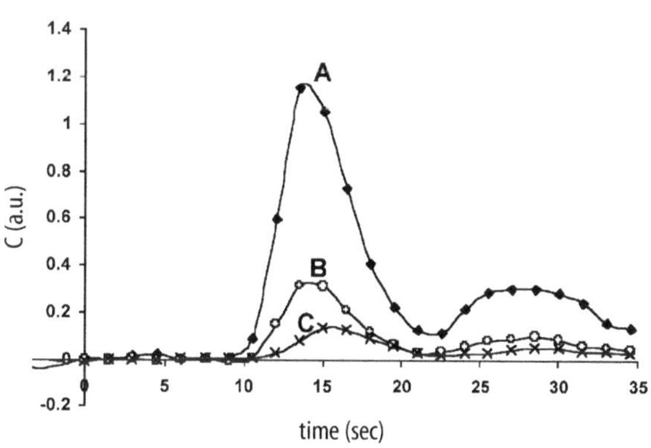

Fig. 4.2. Concentration time-curves [see Eq. (1)] corresponding to the signal intensity time-courses of the three ROIs shown in Fig. 4.1b. *C*, concentration of contrast agent; *a.u.*, arbitrary units. As expected, the concentration of the contrast agent is zero before the arrival of the bolus (during the baseline), and the concentration in the tissue (ROIs *B* and *C*) is much smaller than that in the vessel (ROI *A*)

where $C_a(t)$ is the arterial input function (AIF), i.e. the concentration of tracer entering the tissue at time t, $R(t\text{-}\tau)$ is the residue function, which describes the fraction of contrast agent remaining in the tissue at time t, following the injection of an ideal instantaneous bolus (i.e. a delta function) at time τ[a], and ⊗ indicates the mathematical convolution operation [whose explicit integral expression is given by the last term in Eq. (2)]. The proportionality constant α depends on the density of brain tissue, and the difference in haematocrit levels between capillaries and large vessels (to compensate for the fact that only the plasma volume is accessible to the contrast agent) (Calamante et al. 1999). The convolution operation in Eq. (2) accounts for the fact that for a non-ideal bolus, part of the spread in the concentration time curve is due to the finite length of the actual bolus (i.e. the wider the injected bolus, the wider the $C_t(t)$, regardless of the CBF). It is possible to interpret the integral expression for $C_t(t)$ in Eq. (2) by considering the AIF as a superposition of consecutive ideal boluses "$C_a(\tau)d\tau$" injected at time τ. For each ideal bolus, based on the definition of the residue function, the concentration still present in the tissue at time t will be proportional to "$C_a(\tau)R(t\text{-}\tau)d\tau$", and the total concentration $C_t(t)$ will be given by the sum (or integral) of all these contributions. The AIF is commonly calculated from the change in R_2^* [similar to Eq. (1), with a corresponding proportionality constant κ_a] on a major artery present in one of the images [typically the internal carotid artery (van Osch et al. 2001), or the middle cerebral artery (Østergaard et al. 1996b)]. Therefore, in order to calculate CBF, Eq. (2) must be deconvolved to isolate "CBF·$R(t)$", and the flow obtained from its value at time t = 0 (since $R(0)$ = 1).

The mathematical deconvolution operation of Eq. (2) is very sensitive to noise, and different deconvolution methods have been proposed (see for example (Rempp et al. 1994; Østergaard et al. 1996a; Schreiber et al. 1998; Vonken et al. 1999). At present, one of the most commonly used methods is to rewrite Eq. (2) as a matrix equation, and invert the resulting system by means of singular value decomposition (Østergaard et al. 1996a; Press et al. 1992). This method has been compared with some other common methods, and was found to be the most accurate for the quantification of perfusion, independent of the underlying vascular structure [$R(t)$] and volume [cerebral blood volume (CBV)], even for the low signal-to-noise ratio of typical DSC-MR images (Østergaard et al. 1996a,b). A detailed description of the different deconvolution methods is beyond the scope of this chapter, and the reader is referred to the relevant literature for a full description [see for example Østergaard et al. (1996a)].

Bolus tracking MRI can provide information not only about perfusion but also about CBV and the mean transit time (MTT: the average time for a molecule of contrast agent to pass through the tissue vasculature following an ideal instantaneous bolus injection). These three physiological parameters are not independent, but they are related through the central volume theorem [Stewart (1894): MTT=CBV/CBF].

Due to the compartmentalisation of the contrast agent within the intravascular space, the CBV is proportional to the normalised total amount of tracer:

$$CBV = \alpha^{-1} \frac{\int C_t(t)dt}{\int C_a(t)dt} \tag{3}$$

The proportionality factor α^{-1} is the inverse of the factor in Eq. (2). The normalisation to the integral of AIF accounts for the fact that, the more tracer is injected the greater concentration will reach the tissue, regardless of the CBV. In practice, the measured $C_t(t)$ has a contribution from recirculation of the tracer (usually seen as a second, smaller and more spread peak; see for example Fig. 4.2 between 25 s and 35 s). However, the model previously described is only valid for the first pass of the bolus. Therefore, although the integrations performed on Eq. (3) should continue until the contrast concentration has returned to zero, in practice this is not possible and the calculation is performed up to the time point where contrast recirculation begins. This can lead to an underestimation of CBV (the tail of the peak is not included) and, therefore, an alternative approach involves removing the recirculation contribution by fitting the first pass of $C_t(t)$ to an assumed bolus shape function, typically a gamma-variate function (see Fig. 4.3):

$$C_t(t) = C_0 \cdot (t - BAT)^r \cdot e^{-b\cdot(t-BAT)} \tag{4}$$

where the coefficient BAT is the bolus arrival time, r and b the coefficients that determine the shape of the function, and C_0 a scaling factor (Thompson et al. 1964).

Once CBF and CBV are calculated, MTT can be obtained through the central volume theorem

[a] By definition, $R(0)$ = 1, i.e. no tracer has left the tissue at time t = τ after an ideal instantaneous bolus injected at time τ, and $R(\infty)$ = 0, i.e. at very long times, no tracer remains in the tissue since the BBB is intact and the contrast is washed out by perfusion

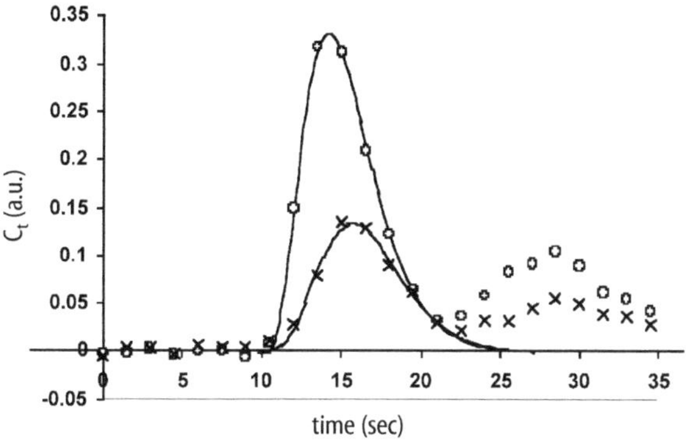

Fig. 4.3. Gamma-variate fitting to the concentration time curves. The graph shows the concentration time curves for an ROI in grey matter (*open circles*) and an ROI in white matter (*crosses*) (see ROI *B* and ROI *C* in Figs. 4.1 and 4.2). The solid lines correspond to the associated gamma-variate fitting, according to Eq. (4). As can be seen in the figure, the fitted curves can be used to eliminate the contribution from recirculation to the concentration time curves. C_t, concentration of contrast agent in the tissue, *a.u.*, arbitrary units

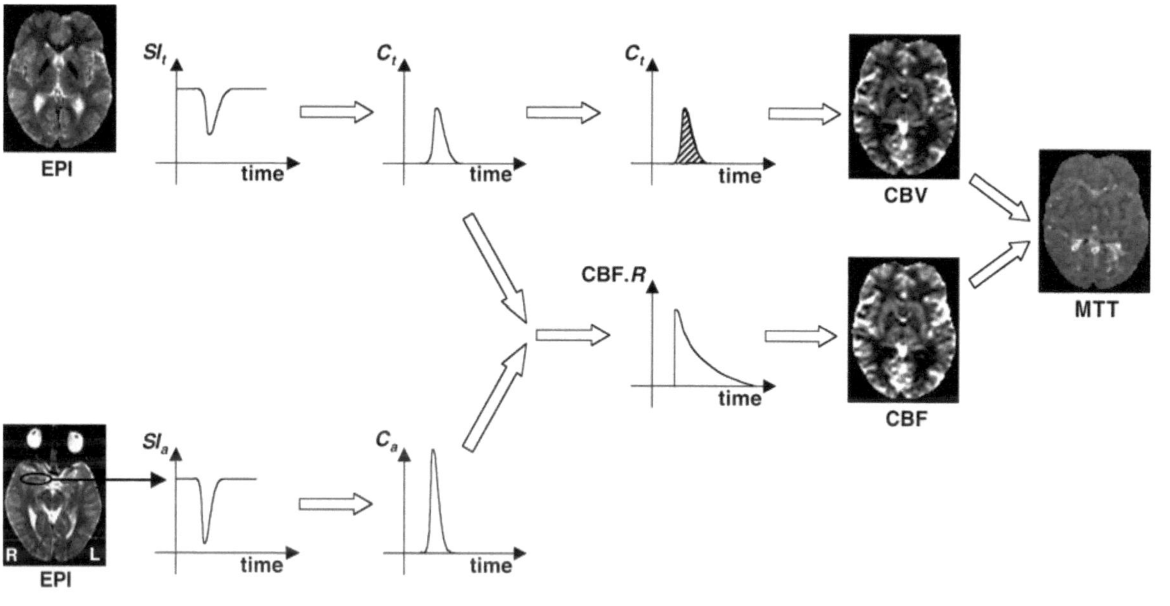

Fig. 4.4. Schematic representations of the steps involved in the quantification of CBF, CBV and MTT using intravascular contrast agents. From each pixel in the echo-planar image (spin-echo, in this example), the signal intensity time course is obtained (SI_t). Using Eq. (1), the SI_t is converted to the concentration time course (C_t). From the area under the peak [see Eq. (3)], a map of CBV can be calculated. From pixels on a major artery (right middle cerebral artery, in this example), the arterial signal intensity time course is measured (SI_a). The AIF (C_a) is obtained by converting SI_a to concentration of contrast agent [using Eq. (1)]. The AIF information is then used to deconvolve C_t [see Eq. (2)] to obtain the product of CBF times the residue function (R). From the initial value of this resulting curve, a map of CBF can be generated. By using the central volume theorem, the MTT can be calculated as the ratio of CBV to CBF

(Stewart 1894). Figure 4.4 shows a schematic description of the quantification of bolus tracking data using intravascular contrast agents.

Bolus tracking MRI can provide, in principle, CBF, CBV and MTT in absolute units (typically ml/100g/min, ml/100g, and s, respectively). However, this requires an estimation of the proportionality constants mentioned above (κ_t, κ_a, and α) (Rempp et al. 1994; Gückel et al. 1996; Schreiber et al. 1998; Vonken et al. 1999; Smith et al. 2000). Alternatively, a cross correlation to a 'gold standard' technique (e.g. PET) has been proposed to calibrate the DSC-MRI measurement by calculating a scaling factor (Østergaard et al. 1998). Although the values obtained in normal subjects are consistent with expected CBF values, there are still some concerns regarding the accuracy under various physiological conditions (Sorensen 2001; Lin et al. 2001; Calamante et al. 2002). The main issues are potential errors due to: the presence of delay and dispersion on the estimated AIF (Østergaard et al. 1999; Calamante et al. 2000), partial volume effects in the quantification of the AIF (Wirestam et al. 2000; van Osch et al. 2001; Lin et al. 2001), change in the proportionality constants or scaling factors (JOHNSON et al. 2000; LIN et al. 2001; KISELEV 2001), and changes in haematocrit levels (CALAMANTE et al. 2002). Work is currently under way to address many of these issues, and therefore accurate absolute measurements of CBF may be possible in the near future.

As mentioned before, the perfusion model described in this section is primarily used to quantify CBF in areas of the brain with an intact BBB. The most common application by far is in the investigation of acute stroke (SORENSEN et al. 1999; GRANDIN et al. 2001; WU et al. 2001; RØHL et al. 2001; FIEHLER et al. 2002).

Although the description of the kinetic models in this section have been based on cerebral perfusion, the future development of true blood pool agents (e.g. SIMONSEN et al. 1999) will allow the widespread use of these models to other organs.

4.2.2 Perfusion Models Including Extravascular Exchange

When the BBB is disrupted (or when a contrast agent that is not a blood pool agent is used in other organs), the kinetic model described in the previous section is no longer valid. First, the fundamental assumption (i.e. the tracer remains intravascular) is incorrect. Furthermore, with the distribution of the contrast throughout the extravascular space, the assumption of a negligible T_1 effect is inappropriate, and the signal enhancement (due to T_1 shortening) will mask the signal loss due to T_2^* (KASSNER et al. 2000). In these cases, a common approach consists of eliminating (or in practice, minimising) this effect and using the model described in Sect. 4.2.1. It has been suggested that, for tissues with moderate contrast extravasation, a small pre-loading dose of contrast agent can produce enough T_1-related signal enhancement, such that this effect is negligible when the second (main) bolus is used to quantify perfusion. In this way, the extra signal enhancement can be neglected, and the only remaining effect is the T_2^* shortening. KASSNER et al. (2000) has recently compared this approach using 2D and 3D sequences with imaging sequences that are insensitive to T_1-enhancement. They concluded that the use of pre-enhancement techniques were effective in eliminating the T_1-contamination in the tumours they investigated. Furthermore, they pointed out that that approach does not suffer from some of the drawbacks of the T_1-insensitive MR sequences, such as low signal-to-noise ratio (for low flip angle gradient-echo based sequences) or reduced maximum number of slices (for dual-echo techniques). This approach has become a common option when analysing DSC-MRI data in the presence of a disrupted BBB (SORENSEN and REIMER 2000; DONAHUE et al. 2000; SUGAHARA et al. 2001; JACKSON et al. 2002).

The problem of T_1 contamination when the BBB is disrupted has been long recognised and, in the early 1990s, WEISSKOFF et al. (1994a) modelled the MR signal in terms of the combined T_1 and T_2 effects of the contrast agent, replacing Eq. (1) by:

$$\Delta R^*_{2,eff}(t) = K_1(x,y)\cdot\overline{\Delta R^*_2}(t) + K_2(x,y)\cdot\int_0^t \overline{\Delta R^*_2}(\tau)d\tau \quad (5)$$

where $\Delta R^*_{2,eff}(t)$ represents the contaminated (by T_1-enhancement) estimate of $\Delta R^*_2(t)$, $K_1(x,y)$ and $K_2(x,y)$ are the spatially varying ΔR^*_2 and permeability weighting factors, and $\overline{\Delta R^*_2}$ is the average ΔR_2. The weighting factors can be calculated using a linear least-square fit, and an estimation of the corrected CBV [e.g. Eq. (3), which is only valid for an intravascular tracer] can be obtained from the uncorrected CBV (using) plus the time integral of the second term in Eq. (5) [for a detailed description of the method, see WEISSKOFF et al. (1994a) or the appendix in DONAHUE et al. (2000)]. WEISSKOFF et al. (1994a) suggested that the parameter $K_2(x,y)$ could be used to create a map of a measure of permeability. However, it should be noted that $K_2(x,y)$ is not only dependent on the vascular permeability, but also on the tissue T_1 (DONAHUE et al. 2000). The method by WEISSKOFF et al. (1994a) relies on the assumption that the T_1-based enhancement is small, and therefore they suggested the use of a small pre-loading dose and a relatively long TR. Furthermore, DONAHUE et al. (2000) showed that the permeability weighting factor tends to zero with increasing pre-loading dose, suggesting that with a sufficient large pre-dose the correction to CBV may not be necessary. However, with a very high dose some of the assumptions in Weisskoff's model may be no longer valid [e.g. negligible reflux of contrast agent from the extravascular space into the vascular system; DONAHUE et al. (2000)]. Therefore, it is recommended to use a moderate pre-dose and to perform the correction suggested by WEISSKOFF et al. (1994a). This approach has been used in many studies to quantify CBV in tumours (e.g. see ØSTERGAARD et al. 1999; SORENSEN and REIMER 2000; DONAHUE et al. 2000). Figure 4.5 shows an example of a corrected CBV map from a 58-year-old woman with a primary CNS lymphoma.

It should be noted that the use of a pre-loading dose does not overcome the decreased T_2^* effect due

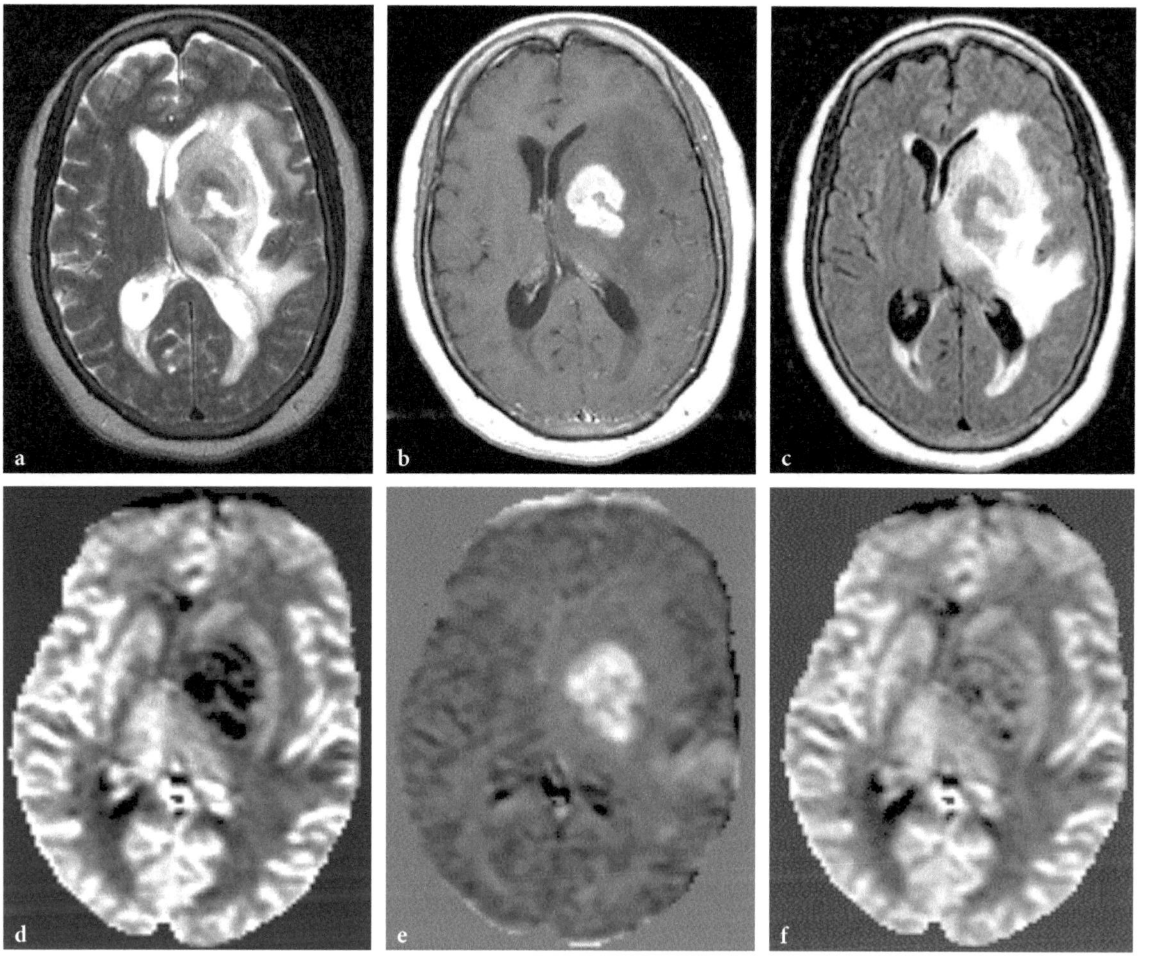

Fig. 4.5a–f. Primary CNS lymphoma. T_2 images (**a**), T_1-post contrast images (**b**), and FLAIR images (**c**) demonstrate an enhancing mass in the right putamen with surrounding oedema. There is central necrosis and a small amount of central haemorrhage. Uncorrected CBV map (**d**), K_2 map (**e**), and CBV corrected for BBB breakdown (**f**) demonstrate heterogeneous volume throughout the lesion. Note the extensive BBB breakdown and how the corrected CBV map shows CBV within the lesion. [Image kindly provided by Dr. A.G. Sorensen, and previously published in Sorensen and Reimer (2000)]

to the loss of tracer compartmentalisation, and the use of a modified kinetic model is still preferable (Sorensen and Reimer 2000).

An alternative approach to avoid the contamination from T_1-enhancement is to use a dual-echo T_2^*-weighted sequence (and therefore directly quantify ΔR_2^* without assumptions regarding T_1):

$$\Delta R_2^*(t) = \frac{\ln\left(S_1(t)/S_2(t)\right) - \ln\left(S_{1,pre}/S_{2,pre}\right)}{TE_2 - TE_1} \qquad (6)$$

where $S_i(t)$ is the signal intensity for the i-th echo (echo-time TE_i) at time t, and $S_{i,pre}$ is the corresponding signal intensity before contrast administration (Miyati et al. 1997; Barbier et al. 1999). Although this approach effectively removes the T_1 contribution, a significant leakage to the extravascular space can be expected even when only the first pass of the bolus is considered (Heiland et al. 1999). Therefore, a modification of the kinetic model to include leakage of the contrast agent is also necessary to quantify CBF (Vonken et al. 2000). Vonken et al. (2000) have recently extended the tracer kinetic model to include a two-compartment model: the intravascular space, and the extravascular space. Assuming negligible back-diffusion of the tracer during the first pass, the concentration in the two compartments are related according to:

$$\frac{dC_{ex}(t)}{dt} = k \cdot C_{in}(t) \qquad (7)$$

where $C_{ex}(t)$ is the concentration in the extravascular space, $C_{in}(t)$ the concentration in the intravascular space, and k the transfer constant into the extravascular space. This equation assumes a simple concentration-difference transfer. Extending Eq. (1) to include the contributions from both compartments, the change in R_2^* (assuming fast exchange) becomes:

$$\Delta R_2^*(t) = \beta \cdot CBV \cdot C_{in}(t) + \beta \cdot (1 - CBV) \cdot C_{ex}(t) = \beta \cdot C(t) \quad (8)$$

where β is a proportionality constant, and $C(t)$ is the total contrast concentration in the voxel[a].

In the presence of BBB disruption, the deconvolution of $C(t)$ in Eq. (2) provides information about the *total* residue function $R_{tot}(t)$ (including possible extravasation of the tracer). This function is related to the residue function $R(t)$ for an ideal intravascular tracer by VONKEN et al. (2000):

$$R_{tot}(t) = R(t) + \lambda \int_0^t R(\tau) d\tau \quad (9)$$

The integrative term represents the fact that, unlike $R(t)$, the total residue function will reach a non-zero value for long t (i.e. due to the extravasation of the contrast agent $R_{tot}(\infty) = \lambda \int_0^\infty R(\tau) d\tau \neq 0$). The coefficient λ determines the magnitude of this term, and it is therefore related to the extraction fraction[b]. VONKEN et al. (2000) proposed an iterative method to calculate $R(t)$, therefore allowing the quantification of CBF, CBV and MTT as previously described in Sect. 4.2.1 (since $R(t)$ corresponds to the residue function for an intravascular tracer). By this iterative method, the coefficient λ is also determined, which allows the calculation of the extraction fraction (see VONKEN et al. 2000 for a detailed description). Figure 46 shows an example of this method for a 55-year-old patient with a glioblastoma multiforme. This method is very promising, although a validation of the modified kinetic model is still needed.

4.2.3 Quantification of Bolus Tracking Using Summary Parameters

Although the methods described in the previous two sections in principle allow the quantification of CBV and MTT, direct calculation of summary parameters from the concentration time curve (or even from the signal intensity time course)[a] is the most commonly used technique in clinical practice. Many such parameters have been proposed (see for example Fig. 4.7), such as: bolus arrival time (BAT), time to peak (TTP), i.e. time until the maximum value of $C(t)$, maximum peak concentration (MPC)[b], i.e. maximum value of $C(t)$, area under the peak (AUP), full width at half maximum (FWHM), first moment of the peak (FMP; FMP= $\int tC(t)dt$)[c], etc.

The use of summary parameters is common since the analysis of the data is fast and straightforward, and does not require measurement of the AIF. However, none of the parameters provides direct measures of perfusion (WEISSKOFF et al. 1993) since the $C(t)$ depends not only on CBF but also on the AIF, and the residue function [see Eq. (2)]. Therefore, the values of the summary parameters can be affected not only by perfusion, but also by the injection conditions (volume injected, injection rate, cannula size, etc.), the vascular structure, and the cardiac output of the patient (PERTHEN et al. 2002). For this reason, the interpretation of summary parameters in terms of physiological information is not straightforward. The only parameter that can be unequivocally interpreted (and only for an intravascular tracer) is the area under the peak: this parameter is proportional to CBV [see Eq. (3)]. Despite this concern, many of the e.g. TTP, FWHM and FMP[c] are used as a measure of MTT; and MPC and the ratio AUP/FMP as a measure of CBF. However, recent numerical simulations have shown that this can lead to very large errors (PERTHEN et al. 2002). Therefore, although summary parameter maps can provide very relevant clinical information, they should be interpreted with caution. Studies using summary parameters include a large range of applications, such as the investigation of acute stroke (SCHLAUG et al. 1999; NEUMANN-HAEFELIN et al. 1999), the assessment of patients with

[a] Note that for a blood pool agent, $C_{ex}(t) = 0$ and Eq. (8) reduces to Eq. (1).

[b] The extraction fraction (E_f) is given by $E_f = 1\text{-}e^{-E}$, where $E = k/CBF$ is the extraction ratio. The transfer constant k is obtained as VONKEN et al. (2000):

$$k = \frac{\lambda \cdot CBV}{(1 - CBV)}$$

[a] It should be noted that due to the logarithm relationship between the signal intensity and the concentration of contrast agent [see Eq. (1)], the relationship between the summary parameters in two regions will be in general different if they are calculated on $S(t)$ or in $C_t(t)$

[b] The parameter MPC is proportional to CBF for the case of an ideal instanteneous bolus injection, i.e. when the AIF is a delta function (ZIERLER 1965). The ratio AUP/FMP is commonly used as a CBF index based on the central volume theorem and the assumption of FMP as a measure of MTT.

[c] The parameter FMP is proportional to MTT when the $C(t)$ is measured in the output of the tissue (AXEL 1995). However, FMP depends also on the vasculature when $C(t)$ is measured in the tissue, as in bolus tracking MRI (WEISSKOFF et al. 1993).

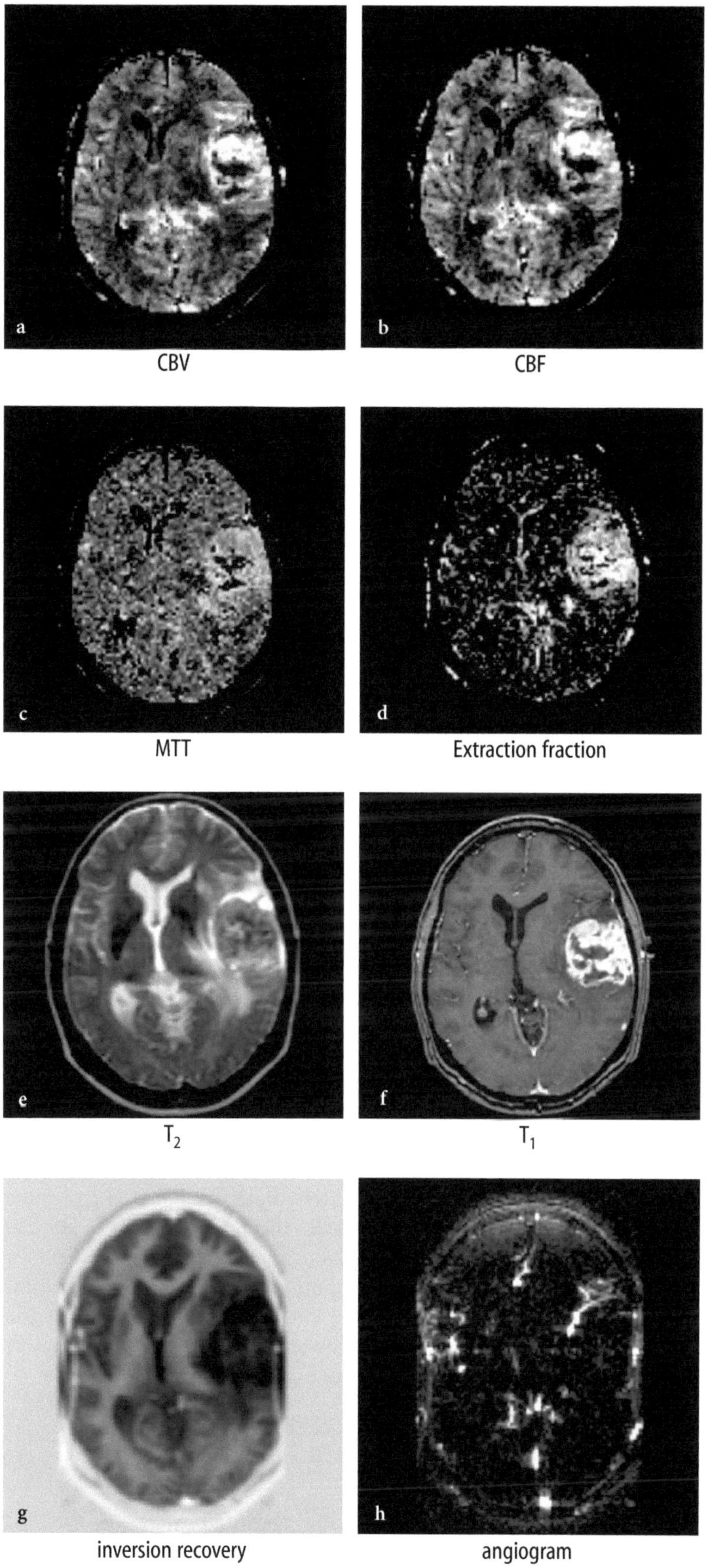

Fig. 4.6a–h. Clinical case of a glioblastoma multiforme. **a** High-resolution post contrast T_1-weighted image. **b** T_2-weighted image showing the CSF and oedema. **c** Inversion recovery image of the perfusion slice, delineating grey and white matter (acquired with a smaller FOV). **d** Angiogram, to distinguish high perfusion from macrovasculature. **e** CBV map after extravasation correction. **f** CBF map. **g** MTT map, which shows increased MTT (while CBV and CBF are increased). **h** Extraction fraction map. [Image kindly provided by Dr. E.P.A. Vonken, and previously published in Vonken et al. (2000)]

a

b

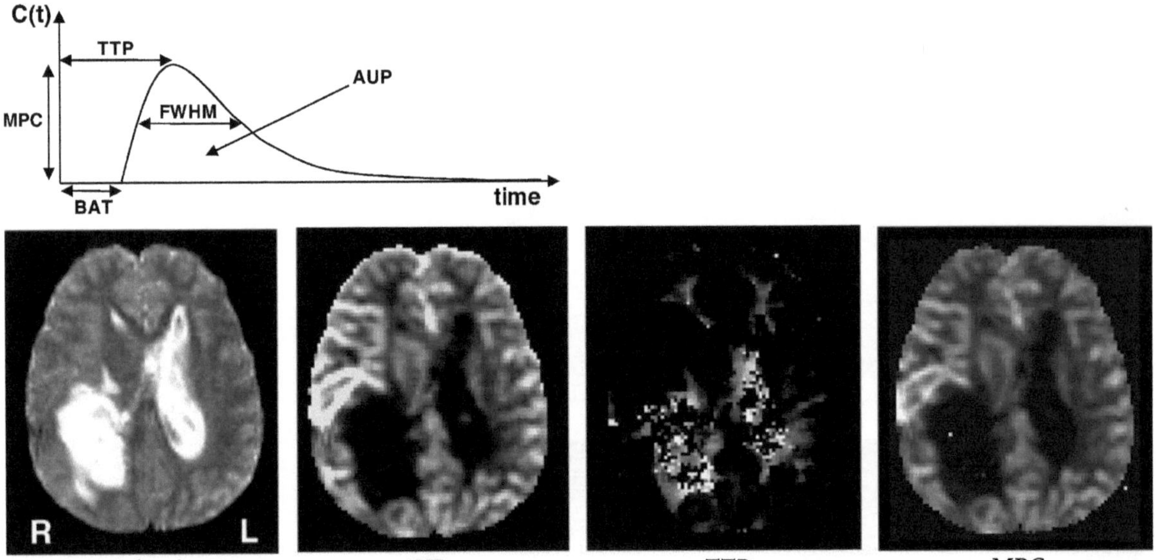

Fig. 4.7a,b. a Schematic graph of a concentration time course (i.e. peak), and some of the more common summary parameters. *TTP*, time to peak; *BAT*, bolus arrival time; *FWHM*, full width at half maximum; *MPC*, maximum peak concentration; *AUP*, area under the peak. **b** Some examples from a 6-year-old child, with an enhancing solid and cystic tumour in the right ventricle (*left side* of the image). From *left* to *right*: T_2-weighted echo-planar image (baseline image before arrival of the bolus), *AUP* map, *TTP* map, and *MPC* map

chronic cerebrovascular abnormalities (Kluytmans et al. 1998; Calamante et al. 2001), the investigation of the relationship between functional impairment and hypoperfusion in acute stroke (Hillis et al. 2001), and the assessment of patients with various brain lesions showing evidence of BBB disruption (Heiland et al. 1999).

The interpretation of summary parameters is further complicated in the presence of contrast extravasation, since the parameters can also be influenced by the extraction fraction. In this case, therefore, it is recommended to use one of the methods described in Sect. 4.2.2 to eliminate or minimise the T_1-contamination in the images (Kassner et al. 2000).

For a more accurate quantification of the summary parameters (due to the relatively low time resolution used, and the presence of recirculation), it is common practice to fit the concentration time course to a gamma-variate function [see Eq. (4)], and to calculate the parameters from this function. However, it has been shown that this model does not accurately describe the first pass of the tracer when a pre-dose is used (Levin et al. 1995; Kassner et al. 2000), and a different model may be necessary in this situation.

All the summary parameters described above depend only on the first pass of the tracer. The recirculation part is generally considered as a 'contamination' to the first pass, and it is usually removed by model fitting, as previously described (e.g. gamma-variate fitting). However, it has been recently suggested that there is potentially very useful information on the recirculation phase (Kassner et al. 2000; Jackson et al. 2002). One of the problems with this part of *C(t)* is the lack of an accurate kinetic model. Residual effects after the first passage have been described even in cases of an intact BBB (Levin et al. 1995). The cause of these effects are still unclear (Levin et al. 1998), and the model described by Eq. (1) and Eq. (2) is not necessarily valid for the recirculation phase. Therefore, a more heuristic approach was taken by Jackson and co-workers: a relative recirculation (*rR*) parameter was defined to assess the abnormalities in contrast recirculation (Kassner et al. 2000; Jackson et al. 2001, 2002). This parameter was defined as proportional to the area between the idealised first pass (obtained from the gamma-variate curve) and the measured data during the recirculation phase:

$$rR = \frac{\sum_{i=A}^{N}\left(\Delta R_{2,meas}(i) - \Delta R_{2,gamma}(i)\right)}{\Delta R_{2,\max}\cdot(N-A)} \tag{10}$$

where $\Delta R_{2,meas}$ is the measured ΔR_2 value, $\Delta R_{2,gamma}$ is the corresponding value calculated from the gamma-variate fitting[a], $\Delta R_{2,max}$ is the maximum of the $\Delta R_{2,gamma}$, A is the onset time of the recirculation, and N is the end point of the dynamic series. The normalisation to the $\Delta R_{2,max}$ was chosen in an attempt to minimise the expected correlation with CBV[b] (Kassner et al. 2000). Although the biological meaning of *rR* remains to be clarified, it has been suggested that it may provide an index of tumour vessel tortuosity (Kassner et al. 2000; Jackson et al. 2001). Figure 4.8 shows an example of CBV and *rR* maps in a patient with glioblastoma multiforme.

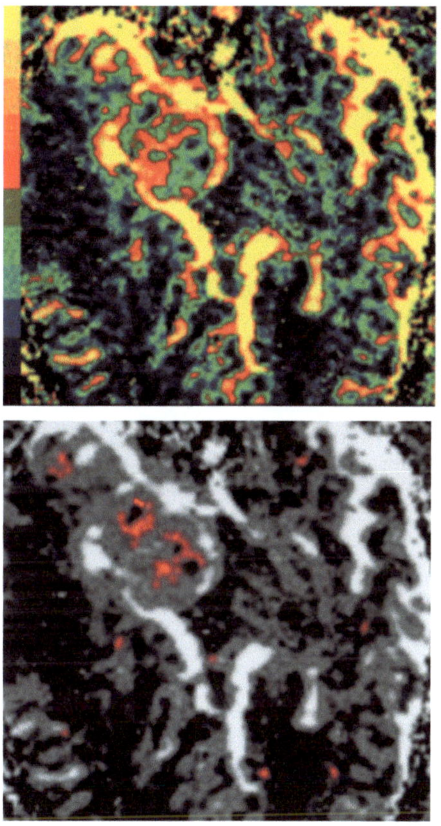

Fig. 4.8. Parametric images of CBV (*left*) and *rR* (*right*) in a patient with glioblastoma multiforme. The CBV map is in colour to aid interpretation. The colour scale is non-linear: *black*, ≤2%; *blue*, 2%–5%; *green*, 5%–10%; *red*, 10%–13%; *orange*, 30%–80%; and *yellow*, ≥80%. The image demonstrates large peripheral vessels (*yellow*) and a central area of high CBV within the tumour core (*red*) and other central areas of poor perfusion or necrosis (*black*). The red areas in the *rR* map indicate pixels with values <0.46, which are seen in fewer than 2% of normal tissue. Single pixel areas of elevated *rR* have been filtered out, and the pixel clusters have been filtered using a 0.5 pixel gaussian filter. The underlying grey scale image shows the CBV map. This image demonstrates areas of elevated *rR*, which occur principally in the centre of the tumour, adjacent to areas of necrosis and away from the central area of elevated CBV. [Image kindly provided by Dr. A. Jackson, and previously published in Jackson et al. (2002)]

4.3 MRI Sequence Type

As discussed throughout the chapter, bolus tracking MRI can be performed using different types of MR sequences. The selection of the optimal sequence depends on a compromise between the different (sometimes competing) factors, as well as the particular application, and the tissue under study. This section discusses some of the main issues associated with the selection of the MR sequence for bolus tracking MRI, and their effects on quantification of perfusion

4.3.1 Gradient-Echo vs. Spin-Echo

As discussed in Sect. 4.1, bolus-tracking MRI can be performed using either gradient-echo (through the T_2^* effect) or spin-echo (through the T_2 effect). Although both MR sequences are sensitive to the susceptibility gradients induced by the contrast agent, it has been shown that gradient-echo is sensitive to the total vasculature, while spin-echo is sensitive primarily to the microvasculature (Weisskoff et al. 1994b; Boxerman et al. 1995). Therefore, the question arises as to the best sequence type for the investigation of perfusion in oncology. Although it was originally thought that the microvascular sensitivity of spin-echo sequences was favourable (since microvascular density has been used as the standard marker to predict tumour angiogenesis), recent studies have suggested that gradient-echo approaches are better (Donahue et al. 2000; Sugahara et al. 2001). This may not be totally unexpected since large, tortuous vessels are often present in tumours. Sugahara et al. (2001), using successive gradient-echo and spin-echo sequences (acquired in random order), detected significant differences between high-grade and low-grade gliomas only with the gradient-echo technique. Similarly, Donahue et al. (2000), using a sequence to simultaneously acquire gradient-echo and spin-echo images, found correlation between CBV and tumour grade only for the CBV calculated from gradient-echo images. The simultaneous acquisition of gradient-echo and spin-echo images also allows the calcu-

[a] As previously mentioned, the gamma-variate function is not always an accurate model for the first pass of the tracer. Since the quantification of *rR* relies on the identification of the first pass by using a gamma-variate fit, this might lead to potential errors (Kassner et al. 2000).

[b] Since part of the recirculation phase represents true contrast recirculation, the area under C(t) in the recirculation phase is expected to correlate with CBV [see Eq. (3)].

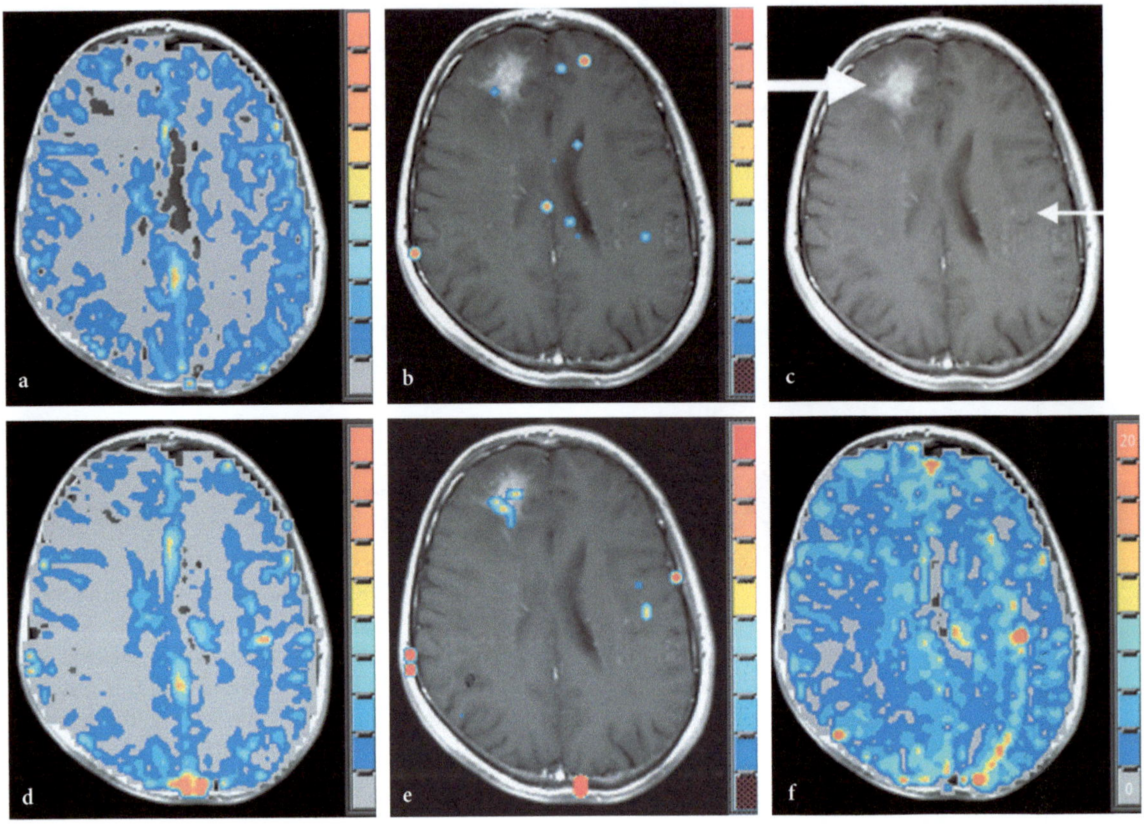

Fig. 4.9a–f. a Example of post-contrast CBV, K_2, and ratio maps from a patient with a malignant lymphoma (*large arrow*) and venous angioma (*small arrow*). While the post contrast T_1-weighted image (**a**) demonstrates signal enhancement in the lymphoma region, neither the gradient-echo CBV (**b**), spin-echo CBV (**e**), or ratio maps (**d**) show increases relative to surrounding brain. However, permeability factor (K_2) increases are apparent on the gradient-echo K_2 map (**c**) but not on the spin-echo K_2 map (**f**). The venous angioma does demonstrate increases on the gradient-echo CBV and ratio maps, a finding consistent with a larger blood volume and increased average vessel diameter in this region. [Image kindly provided by Dr. K. Donahue-Schmainda, and previously published in Donahue et al. (2000)]

lation of the maps of the ratio "$\Delta R_2^*/\Delta R_2$" (see Fig. 4.9), which has been suggested as a potential marker of average vessel diameter due to the different sensitivity to vessel size of the images (Dennie et al. 1998; Donahue et al. 2000). This parameter was found to be strongly correlated to brain tumour grade (Donahue et al. 2000).

4.3.2 Spatial Coverage vs. Time Resolution

Since a proper characterisation of the transient signal drop is required in order to accurately quantify DSC-MRI data, the time resolution is limited by the transit time of the bolus through the tissue. Typically, a repetition time (TR) of 2 s or less is necessary to adequately sample the fast changes during the first pass. Therefore, the spatial coverage will be limited by the maximum number of slices that can be acquired within this TR. The use of EPI sequences typically allows 10–15 slices to be acquired during this time. However, as mentioned in Sect. 4.2, EPI suffers from large image distortions in areas of interface between different susceptibility properties (e.g. tissue–air interfaces), and it is not suitable for imaging in many organs. Therefore, other imaging modalities (e.g. segmented EPI, FLASH, etc.) are sometimes used. In these cases, the 1.5–2 s TR typically allows only a couple of slices to be acquired. In many cases (see for example Rempp et al. 1994; Schreiber et al. 1998; Vonken et al. 1999), one of these slices is positioned through a major artery to estimate the AIF, leaving only one slice for measuring perfusion in the tissue of interest. In summary, the selection of the spatial coverage and time resolution must usually be done as a compromise, according to the particular application and tissue under study.

4.3.3
2D vs. 3D Sequences

The use of 3D sequences has been reported in studies of bolus tracking (see for example van Gelderen et al. 2000 and Kassner et al. 2000). They allow increased coverage, although typically at the expense of an increased TR, or a decreased spatial resolution. The increased coverage is necessary in cases where the location of the abnormality is not known *a priori*, or when the information from multiple areas is required. However, the time of acquisition is not well defined for each image in 3D MRI (cf. the time of acquisition of each slice in multi-slice 2D MRI), and this effect will introduce a smoothing of the time changes during the first pass (due to the signal intensity changes during the MR volume acquisition). If the time resolution is of the order of several seconds, this can lead to an erroneous characterisation of the passage of the bolus. Therefore, 2D sequences are usually preferable, since this effect generally can be neglected.

4.3.4
Single-Echo vs. Dual-Echo Sequences

As discussed in Sect. 42.2, the use of dual-echo sequences allows the quantification of changes in R_2^* without assumption regarding the T_1 behaviour (Miyati et al. 1997; Barbier et al. 1999). This method provides a simple way to remove the relaxivity effects. However, the sampling of two echoes introduces considerable demands on the sampling time, reducing the maximum number of slices that can be acquired (and therefore, coverage) in a given TR. Since the use of a pre-enhancement approach has been shown to be very effective in many situations (Kassner et al. 2000), single-echo sequences with a pre-dose of contrast may be advantageous when spatial coverage is important.

4.4
Conclusion

This chapter has described the more common approaches to quantify bolus tracking data. These ranged from the simple quantification of summary parameters (e.g. TTP calculated directly from the signal intensity time course), to CBF quantification using indicator dilution theory for intravascular tracers, to quantification involving more complex models including contrast extravasation. The analysis of bolus tracking data has therefore a great potential for investigation, diagnosis, and patient management in oncology.

References

Axel L (1980) Cerebral blood flow determination by rapid-sequence computed tomography. Radiology 137:676–686

Axel L (1995) Methods using blood pool tracers, part II. In: Le Bihan D (ed) Diffusion and perfusion magnetic resonance imaging. Raven, New York, pp 205–211

Barbier EL, den Boer JA, Peters AR, Rozeboom Ar, Sau J, Bonmartin A (1999) A model of the dual effect of gadopentetate dimeglumine on dynamic brain MR images. J Magn Reson Imaging 10:242:253

Barbier EL, Lamalle L, Décorps M (2001) Methodology of brain perfusion imaging. J Magn Reson Imaging 13:496–520

Bowtell R, Schmitt F (1998) Echo-planar imaging hardware. In: Schmitt F, Stehling MK, Turner R (eds) Echo-planar imaging. Theory, technique and application. Springer, Berlin Heidelberg New York, pp 31–64

Boxerman JL, Hamberg LM, Rosen BR, Weisskoff RM (1995) MR contrast due to intravascular magnetic-susceptibility perturbations. Magn Reson Med 34:555–566

Calamante F, Thomas DL, Pell GS, Wiersma J, Turner (1999) Measuring cerebral blood flow using magnetic resonance techniques. J Cereb Blood Flow Metab 19:701–735

Calamante F, Gadian DG, Connelly A (2000) Delay and dispersion effects in dynamic susceptibility contrast MRI: simulations using Singular Value Decomposition. Magn Reson Med 44:466–473

Calamante F, Ganesan V, Kirkham FJ, Jan W, Chong WK, Gadian DG, A Connelly (2001) MR perfusion imaging in moyamoya syndrome. Potential implications for clinical evaluation of occlusive cerebrovascular disease. Stroke 32:2810–2816

Calamante F, Gadian DG, Connelly A (2002) Quantification of perfusion using bolus tracking MRI in stroke. Assumptions, limitations, and potential implications for clinical use. Stroke 33:1146–1151

Dennie J, Mandeville JB, Boxerman JL, Packard SD, Rosen BR (1998) NMR imaging of changes in vascular morphology due to tumor angiogenesis. Magn Reson Med 40:793–799

Donahue KM, Krouwer HGJ, Rand SD, Pathak AP, Marszalkowski CS, Censky SC, Prost RW (2000) Utility of simultaneously acquired gradient-echo and spin-echo cerebral blood volume and morphology maps in brain tumor patients. Magn Reson Med 43:845–853

Farrar TC, Becker ED (1971) Pulse and Fourier transform NMR. Introduction to theory and methods. Academic, New York, pp 46–65

Fiehler J, von Bezold M, Kucinski T, Knab R, Eckert B, Wittkugel O, Zeumer H, Röther J (2002) Cerebral blood flow predicts lesion growth in acute stoke patients. Stroke 33:2421–2425

Fischer H, Ladebeck R (1998) Echo-planar imaging image artifacts. In: Schmitt F, Stehling MK, Turner R (eds) Echo-planar imaging. Theory, technique and application. Springer, Berlin Heidelberg New York, pp 179–200

Gillis P, Koenig SH (1987) Transverse relaxation of solvent protons induced by magnetized spheres: application to ferritin erythrocytes and magnetite. Magn Reson Med 5:323–345

Grandin CB, Duprez TP, Smith Am, Mataigne F, Peeters A, Oppenheim C, Cosnard G (2001) Usefulness of magnetic resonance-derived quantitative measurements of cerebral blood flow and volume in prediction of infarct growth in hyperacute stroke. Stroke 32:1147–1153

Gückel FJ, Brix G, Schmiedek P, Piepgras A, Becker G, Kopke J, Gross H, Georgi M (1996) Cerebrovascular reserve capacity in patients with occlusive cerebrovascular disease: assessment with dynamic susceptibility contrast-enhanced MR imaging and the acetazolamide stimulation test. Radiology 201:405–412

Haase A (1990) Snapshot FLASH-MRI. Applications to T1, T2, and chemical-shift imaging. Magn Reson Med 13:77–89

Heiland S, Benner T, Debus J, Rempp K, Reith W, Sartor K (1999) Simultaneous assessment of cerebral hemodynamics and contrast agent uptake in lesions with disrupted blood-brain-barrier. Magn Reson Imaging 17:21–27

Hillis AE, Wityk RJ, Tuffiash E, Beacuchamp NJ, Jacobs MA, Barker PB, Selnes OA (2001) Hypoperfusion of Wenicke's area predicts severity deficit in acute stroke. Ann Neurol 50:561–566

Jackson A, Kassner A, Zhu XP, Li KL (2001) Reproducibility of T2* blood volume and vascular tortuosity maps in cerebral gliomas. J Magn Reson Imaging 14:510–516

Jackson A, Kassner A, Williams DA, Reid H, Zhu XP, Li KL (2002) Abnormalities in the recirculation phase of contrast agent bolus passage in cerebral gliomas: comparison with relative blood volume and tumor grade. AJNR Am J Neuroradiol 23:7–14

Johnson KM, Tao JZT, Kennan RP, Gore JC (2000) Intravascular susceptibility agent effects on tissue transverse relaxation rates in vivo. Magn Reson Med 44:909–914

Kassner A, Annesley DJ, Zhu XP, Li KL, Kamaly-Asl ID, Watson Y, Jackson A (2000) Abnormalities of the contrast re-circulation phase in cerebral tumors demonstrated using dynamic susceptibility contrast-enhanced imaging: a possible marker of vascular tortuosity. J Magn Reson Imaging 11:103–113

Kiselev VG (2001) On the theoretical basis of perfusion measurements by dynamic susceptibility contrast MRI. Magn Reson Med 46:1113–1122

Kluytmans M, van der Grond J, Viergever MA (1998) Gray matter and white matter perfusion imaging in patients with severe carotid artery lesions. Radiology 209:675–682

Levin JM, Kaufman MJ, Ross MJ, Mendelson JH, Maas LC, Cohen M, Renshaw PF (1995) Sequential dynamic susceptibility contrast MR experiments in human brain: residual contrast agent effect, steady state, and hemodynamic perturbation. Magn Reson Med 34:655–663

Levin JM, Wald LL, Kaufman MJ, Ross MJ, Maas LC, Renshaw PF (1998) T1 effects in sequential dynamic susceptibility contrast experiments. J Magn Reson 130:292–295

Lin W, Celik A, Derdeyn C, An H, Lee Y, Videen T, Østergaard L, Powers WJ (2001) Quantitative measurements of cerebral blood flow in patients with unilateral carotid artery occlusion: a PET and MR study. J Magn Reson Imaging 14:659–667

Miyati T, Banno T, Mase M, Kasai H, Shundo H, Imazawa M, Ohba S (1997) Dual dynamic contrast-enhanced MR imaging. J Magn Reson Imaging 7:230–235

Neumann-Haefelin T, Wittsack H-J, Wenserski F, Siebler M, Seitz RJ, Mödder U, Freund H-J (1999) Diffusion- and perfusion-weighted MRI. The DWI/PWI mismatch region in acute stroke. Stroke 30:1591–1597

Østergaard L, Weisskoff RM, Chesler DA, Gyldensted C, Rosen BR (1996a) High resolution measurement of cerebral blood flow using intravascular tracer bolus passages, part I. Mathematical approach and statistical analysis. Magn Reson Med 36:715–725

Østergaard L, Sorensen AG, Kwong KK, Weisskoff RM, Gyldensted C, Rosen BR (1996b) High resolution measurement of cerebral blood flow using intravascular tracer bolus passages, part II. Experimental comparison and preliminary results. Magn Reson Med 36:726–736

Østergaard L, Johannsen P, Poulsen PH, Vestergaard-Poulsen P, Asboe H, Gee AD, Hansen SB, Cold GE, Gjedde A, Gyldensted C (1998) Cerebral blood flow measurements by magnetic resonance imaging bolus tracking: comparison with [O-15] H2O positron emission tomography in humans. J Cereb Blood Flow Metab 18:935–940

Østergaard L, Chesler DA, Weisskoff RM, Sorensen AG, Rosen BR (1999) Modeling cerebral blood flow and flow heterogeneity from magnetic resonance residue data. J Cereb Blood Flow Metab 19:690–699

Press WH, Teukolsky SA, Vetterling WT, Flannery BT (1992) Numerical recipes in C. The art of scientific computing. Cambridge University Press, Cambridge

Perthen JE, Calamante F, Gadian DG, Connelly A (2002) Is quantification of bolus tracking MRI reliable without deconvolution? Magn Reson Med 47:61–67

Rempp KA, Brix G, Wenz F, Becker CR, Guckel F, Lorenz WJ (1994) Quantification of regional cerebral blood flow and volume with dynamic susceptibility contrast-enhanced MR imaging. Radiology 193:637–641

Røhl L, Ostergaard L, Simonsen CZ, Vestergaard-Poulsen P, Andersen G, Sakoh M, Le Bihan D, Gyldented C (2001) Viability thresholds of ischemic penumbra of hyperacute stroke defined by perfusion-weighted MRI and apparent diffusion coefficient. Stroke 32:1140–1146

Rosen BR, Belliveau JW, Vevea JM, Brady TJ (1990) Perfusion imaging with NMR contrast agents. Magn Reson Med 14:249–265

Schlaug G, Benfield A, Baird AE, Siewert B, Lövblad KO, Parker RA, Edelman RR, Warach S (1999) The ischemic penumbra. operationally defined by diffusion and perfusion MRI. Neurology 53:1528–1537

Schreiber WG, Gückel F, Stritzke P, Schmiedek P, Schwartz A, Brix G (1998) Cerebral blood flow and cerebrovascular reserve capacity: estimation by dynamic magnetic resonance imaging. J Cereb Blood Flow Metab 18:1143–1156

Simonsen CZ, Østergaard L, Vestergaard-Poulsen P, Røhl L, Bjørnerud A, Glydensted C (1999) CBF and CBV measurements by USPIO bolus tracking: reproducibility and comparison with Gd-based values. J Magn Reson Imaging 9:342–347

Smith AM, Grandin CB, Duprez T, Mataigne F, Cosnar G (2000) Whole brain quantitative CBF and CBV measurements using MRI bolus tracking: comparison of methodologies. Magn Reson Med 43:559–654

Sorensen AG (2001) What is the meaning of quantitative CBF? AJNR Am J Neuroradiol 22:235–236

Sorensen AG, Reimer P (2000) Cerebral MR perfusion imaging. Principles and current applications. Thieme, Stuttgart, pp 16–20

Sorensen AG, Copen WA, Ostergaard L, Buonanno FS, Gonzalez RG, Rordorf G, Rosen BR, Schwamm LH, Weisskoff RM, Koroshetz WJ (1999) Hyperacute stroke: simultaneous measurement of relative cerebral blood volume, relative cerebral blood flow, and mean transit time. Radiology 210:519–527

Stehling MK, Turner R, Masfield P (1991) Echo-planar imaging: magnetic resonance imaging in a fraction of a second. Science 254:43–50

Stewart GN (1894) Researches on the circulation time in organs and on the influences which affect it, part I–III. J Physiol 15:1–89

Sugahara T, Korogi Y, Kochi M, Ushio Y, Takahashi M (2001) Perfusion-sensitive MR imaging of gliomas: comparison between gradient-echo and spin-echo echo-planar imaging techniques. AJNR Am J Neuroradiol 22:1306–1315

Thompson HK, Starmer F, Whalen RE, McIntosh HD (1964) Indicator transit time considered as a gamma variate. Circ Res 14:502–515

Van Gelderen P, Grandin C, Petrella JR, Moonen CTW (2000) Rapid three-dimensional MR imaging method for tracking a bolus of contrast agent through the brain. Radiology 216:603–608

Van Osch MJP, Vonken EPA, Bakker CJG, Viergever MA (2001) Correcting partial volume artifacts of the arterial input function in quantitative cerebral perfusion MRI. Magn Reson Med 45:477–485

Villringer A, Rosen BR, Belliveau JW, Ackerman JL, Lauffer RB, Buxton RB, Chao YS, Wedeen VJ, Brady TJ (1988) Dynamic imaging with lanthanide chelates in normal brain: contrast due to magnetic-susceptibility effects. Magn Reson Med 6:164–174

Vonken EPA, van Osch MJP, Baker CJG, Viergever MA (1999) Measurement of cerebral perfusion with dual-echo multi-slice quantitative dynamic susceptibility contrast MRI. J Magn Reson Imaging 10:109–117

Vonken EPA, van Osch MJP, Baker CJG, Viergever MA (2000) Simultaneous qualitative cerebral perfusion and Gd-DTPA extravasation measurements with dual-echo dynamic susceptibility contrast MRI. Magn Reson Med 43:820–827

Weisskoff RM, Chesler D, Boxerman JL, Rosen BR (1993) Pitfalls in MR measurement of tissue blood flow with intravascular tracers: which mean transit-time? Magn Reson Med 29:553–559

Weisskoff RM, Boxerman JL, Sorensen AG, Kulke SM, Campbell TA, Rosen BR (1994a) Simultaneous blood volume and permeability mapping using a single Gd-based contrast injection. Proceedings of the 2nd annual meeting of SMRM, San Francisco, p 279

Weisskoff RM, Zuo CS, Boxerman JL, Rosen BR (1994b) Microscopic susceptibility variation and transverse relaxation. Theory and experiment. Magn Reson Med 31:601–610

Wirestam R, Ryding E, Lindgren A, Geijer B, Holtas S, Stahlberg F (2000) Absolute cerebral blood flow measured by dynamic susceptibility contrast MRI: a direct comparison with Xe-133 SPECT. MAGMA 11:96–103

Wu O, Koroshetz WJ, Ostergaard L, Buonanno FS, Copen WA, Gonzalez RG, Rordorf G, Rosen BR, Schwamm LH, Weisskoff RM, Sorensen AG (2001) Predicting tissue outcome in acute human cerebral ischemia combined diffusion- and perfusion-weighted MR imaging. Stroke 32:933–942

Zierler KL (1965) Equations for measuring blood flow by external monitoring of radioisotopes. Circ Res 16:309–321

5 Measuring Contrast Agent Concentration in T_1-Weighted Dynamic Contrast-Enhanced MRI

David L. Buckley and Geoffrey J. M. Parker

CONTENTS

5.1 Introduction

For many years imaging studies have been performed in the nuclear medicine departments of hospitals in which trace amounts of a radioactive isotope are administered to patients with the aim of measuring the subsequent distribution of this isotope in the body, often using a gamma camera. Despite the complications in precisely quantifying such studies, the images obtained provide clear maps of tracer distribution, largely due to the fact that there is no natural background signal to swamp the tracer signal (that is, the underlying tissue does not produce a signal). These images provide functional information about tracer kinetics but provide limited anatomical information. The principle of monitoring tracer kinetics was subsequently extended following the introduction of iodine-based contrast media for use in X-ray computed tomography (CT) examinations. As the name implies, contrast medium is administered to enhance the inherent contrast of the CT image. The signals manifest in such studies represent the combination of the inherent X-ray attenuation due to the tissue plus the additional attenuation due to the iodine distributed in that tissue. These images are said to be contrast enhanced and often provide useful anatomic information. However, the attenuation characteristics of tissue and iodine are purely additive, thus a simple subtraction of an image obtained before the administration of contrast medium from a post-contrast image provides a direct map of iodine distribution (Terada et al. 1992) much like that obtained in the nuclear medicine department. The significant benefit of using CT in this setting is the advantage of also providing anatomical information.

The principle used for CT has subsequently been applied to MR imaging. MR contrast agents (see Chaps. 2 and 3), typically manufactured to include a paramagnetic ion such as gadolinium, may be administered as a means of enhancing contrast in conventional anatomical imaging. Similarly, they can provide functional information when the contrast agent is employed as a tracer and its distribution in the body is assessed, often by means of dynamic imaging. However, there are significant differences between the mode of action of MR contrast agents and those used in X-ray imaging. Fundamentally, while the iodine in X-ray contrast media directly attenuates the X-ray beam, the paramagnetic ion acts upon the surrounding water. An MR signal is not observed from the contrast medium; we rather observe its effect indirectly. In fact the MR contrast agent can be said to catalyse proton relaxation. For the purpose of illustration, this process can be thought of as transient chemical bonding between water protons and the paramagnetic ion. The degree of relaxation enhancement, the relaxivity of the ion, can be interpreted as being related to the number of such bonds that can form and the time scale over which they occur (Engelstad and Wolf 1988). This relaxivity effect can be harnessed,

D. L. Buckley, PhD; G. J. M. Parker, PhD
Imaging Science and Biomedical Engineering, University of Manchester, Stopford Building, Oxford Road, Manchester, M13 9PT, UK

through shortening of relaxation times in tissue, to estimate contrast agent concentrations in vivo and is discussed in greater detail below.

In addition to this catalysis effect the water protons are also influenced by the bulk magnetic susceptibility (BMS) shift. Local magnetic field variations induced by an inhomogeneous distribution of paramagnetic ions produce a relatively long range heterogeneous magnetic field which shortens T_2 and T_2*. This effect can dominate the relaxivity effect when superparamagnetic or ferromagnetic substances (ions with some degree of magnetic ordering) are employed or when compartmentalisation of the contrast agent occurs (SPRINGER 1994). This effect can be harnessed for dynamic susceptibility contrast MRI and is discussed in further detail in Chap. 4. Confusion often arises when contrast agent studies using susceptibility contrast MRI and conventional T_1-weighted (relaxivity enhanced) MRI are discussed. Both techniques measure the influence of the same MR contrast agent through its affect on relaxation times. However, the susceptibility effect is long range in nature and results from compartmentalisation of the contrast agent. Thus a small amount of contrast agent restricted to the vascular spaces in the brain (occupying only a few percent of the brain's volume) can produce a long range effect (stretching way beyond the vessel walls) that dominates any relaxivity effect seen in the water of the vascular spaces. Conversely, if the blood-brain barrier is compromised and the contrast agent leaves the blood vessels, the BMS effect is significantly reduced but short-range relaxivity effects are extended into the interstitial spaces. Thus it is through the appropriate choice of imaging sequence that these two effects may be differentiated. A T_2*-weighted sequence is sensitive to BMS effects and is chosen to monitor contrast agent behind an intact blood-brain barrier. A T_1-weighted sequence may be an inappropriate choice for such a situation since the short-range T_1 effects influence such a small proportion of the overall tissue signal that only very small signal changes result from contrast agent administration. However, once the blood--brain barrier is compromised and contrast agent enters the large water pool of the interstitial spaces these T_1 effects become much larger. For the same reasons T_1-weighted acquisitions are generally favoured for monitoring the distribution of MR contrast agents in tumours at all anatomical locations, as it is generally the case that tumours do not possess a mechanism for preventing contrast agent leakage from the blood pool.

5.2 Contrast Agent Relaxivity

Provided that the BMS shift is negligible, the relationship between relaxation rate ($1/T_1$ and $1/T_2$) and contrast agent concentration can be predicted by the Solomon-Bloembergen equations (GOWLAND et al. 1992):

$$\frac{1}{T_1} = \frac{1}{T_{10}} + r_1[Gd] \tag{5.1}$$

$$\frac{1}{T_2} = \frac{1}{T_{20}} + r_2[Gd] \tag{5.2}$$

where r_1 and r_2 are the spin-lattice and spin-spin relaxivity constants respectively and T_{10} and T_{20} are the spin-lattice and spin-spin relaxation times respectively in the absence of contrast material. These relationships have both been confirmed in vitro (ROSEN et al. 1990; DONAHUE et al. 1994; JUDD et al. 1995) and for T_1 in vivo (WEDEKING et al. 1992) across a range of concentrations. These expressions allow theoretical predictions to be made about the influence of a contrast agent, such as Gd-DTPA, on signal intensity. Relaxivity is dependent upon field strength and the chemical structure of the contrast agent (SPRINGER 1994). While it is normally assumed that the physico-chemical nature of the tissue has little affect upon contrast agent relaxivity, there is strong evidence that compartmentalisation of the agent affects tissue water relaxation (DONAHUE et al. 1994, 1996) (see Sect. 5.5). Furthermore, debate surrounds the issue of whether different tissues may or may not have different relaxivities (DONAHUE et al. 1994; STANISZ and HENKELMAN 2000). This has profound implications for the assessment of contrast agent concentration in vivo since for absolute concentration measurements it may be necessary to know the relaxivity of any tissue being studied, which is a non-trivial consideration.

5.3 Measurement of Contrast Agent Concentration In Vivo

To monitor the kinetic behaviour of a contrast agent in vivo it is necessary to link the changes in concentration to changes seen in MR images. Since most agents are known to alter both the spin-lattice and spin-spin relaxation rates of tissues it should be possible to infer their distribution by observing the influence on the MR signal. Since the concentration of Gd ions is known to be directly proportional to the change in $1/T_1$ (Eq. 5.1), a series of measurements of the T_1 of a tissue as a contrast agent distributes within it could, in principle, be used to monitor the changes in contrast agent concentration. Since the changes in T_1 will also alter the signal intensity of a T_1-weighted imaging sequence, it should also be possible to monitor contrast agent concentration using signal intensity. It is preferable that two further criteria are met. The rate of measurement of contrast agent concentration should be sufficient to monitor the most rapid changes occurring within the tissue. Secondly, the relationship between concentration of contrast agent and the measuring function should be monotonic and minimally affected by small changes in the imaging parameters. In this way the concentration changes can be monitored rapidly and the experimental unknowns are kept to a minimum.

5.3.1 Signal Intensity Change

The choice of pulse sequence to monitor contrast agent kinetics must satisfy the above criteria but also provide acceptable spatial resolution and tissue coverage. As T_2-weighted sequences tend to take more time to collect and the effect of contrast agent on signal intensity is negative [i.e. signal decreases with increasing contrast agent concentration hence reducing the signal to noise ratio (SNR) of the experiment], T_1-weighted sequences offer many advantages. Previously groups have used spin echo sequences to monitor contrast agent accumulation (Larsson et al. 1990; Tofts and Kermode 1991). Larsson et al. (1990) demonstrated a linear relationship between signal intensity and contrast agent concentration when the TR was reduced to 500 ms or less, though their temporal resolution for a single slice was only 68 s. More recently groups have used faster sequences including EPI (Gowland et al. 1992) and turboFLASH (Boetes et al. 1994) to monitor contrast agent accumulation with a temporal resolution of a few seconds. The relationship between the contrast agent concentration and the relative increase in signal intensity can be derived from the Bloch equations (Haase et al. 1986) for any imaging sequence. The signal intensity obtained from the commonly used gradient echo sequence with spoiling of the transverse magnetisation (FLASH) is described below:

$$S = g.\rho.\frac{\sin(\alpha).\left(1-\exp\left(-\frac{TR}{T_1}\right)\right)}{\left(1-\cos(\alpha).\exp\left(-\frac{TR}{T_1}\right)\right)}.\exp\left(-\frac{TE}{T_2*}\right) \quad (5.3)$$

where ϱ is the proton density, α is the flip angle, and g is a constant determined by system receiver and image reconstruction settings. If we assume that Gd ions have no effect on ϱ and that the TE is so short as to be able to neglect the influence of T_2* (or more importantly changes in T_2* during the time series), then the Gd ions influence signal intensity via their effect on T_1 alone. As α approaches 90∞ and TR/T_1 becomes small the relationship between signal intensity and $1/T_1$ is approximately linear (as with a spin echo acquisition):

$$S \approx \frac{g.\rho.TR}{T_1} \quad (5.4)$$

This relationship remains approximately valid across a range of values for TR/T_1 and α. The constant of proportionality is a function of TR, g, ϱ, and, as the flip angle decreases, α. The difficulty in comparing this constant between studies is the sensitive nature of g. The loading of the coil, receiver settings at the MR console and image reconstruction parameters alter the intensity of the signal in the image. Hence, it is necessary to relate the signal intensity to an internal standard. Other groups have used samples with known characteristics (Shames et al. 1993; Vallee et al. 2003) at a fixed location within the field-of-view or the signal from fat located close to the region of interest (Heywang et al. 1989). Placing a sample within the field-of-view in a clinical imaging study is often problematic, complicating patient positioning. The signal from fat (for example in a breast imaging study) is often very variable (Pedevilla et al. 1995) and the values obtained for contrast agent concentration are therefore not reproducible. An alternative is to relate the signal intensity post-contrast to that pre-contrast. This has the advantage of maintaining the position of the standard in relation to the enhancing structure and requires no prior positioning. However, the use of pre-contrast signal intensity also introduces the pre-contrast T_1 of the structure into the analysis. If we assume signal intensity is proportional to $1/T_1$ (Eq. 5.4), then:

$$\frac{S_{Gd}-S_0}{r.\rho.TR} \approx \left(\frac{1}{T_{1Gd}}-\frac{1}{T_{10}}\right) = r_1[Gd] \quad (5.5)$$

where S_0 and S_{Gd}, and T_{10} and T_{1Gd} are the signal intensities and spin-lattice relaxation times before and following administration of contrast agent respectively and r_1 is the relaxivity of contrast agent. Dividing by the pre-contrast signal we obtain:

$$\frac{S_{Gd}-S_0}{S_0} \approx r_1 T_{10}[Gd] \quad (5.6)$$

Consequently the relative increase of signal intensity following administration of contrast agent is related to both the spin-lattice relaxivity of contrast agent and the pre-contrast T_1 of the tissue. This difficulty is highlighted in Fig. 1, which shows data from a study of patients with tumours of the breast (Mussurakis et al. 1997). The enhancement measured using relative changes in signal (Fig. 5.1a) suggests that benign lesions enhance more than malignant lesions. In fact, the benign lesions tend to have a longer native T_1 and actually show less uptake of contrast agent (Fig. 5.1b).

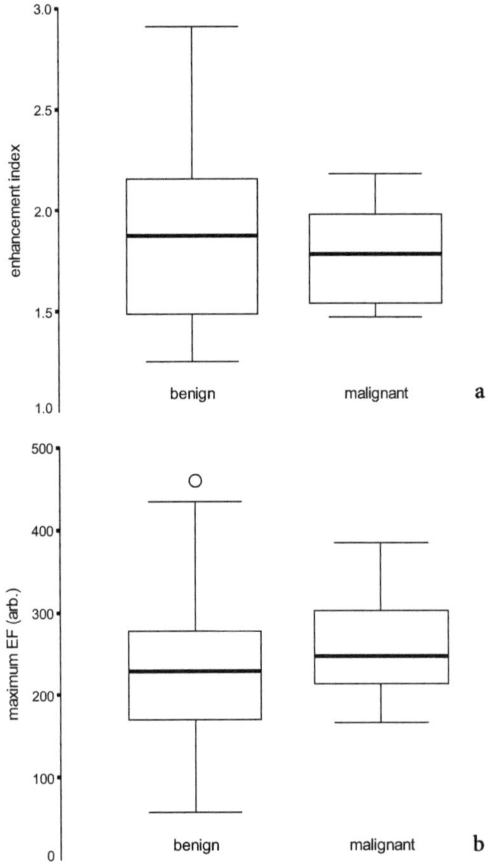

Fig. 5.1a,b. Box and whisker plots showing the distribution of measures of maximum enhancement seen in a study of 58 breast tumours (Mussurakis et al. 1997). The use of relative signal increase as a measure of contrast agent uptake indicates that benign lesions enhance more significantly (**a**). Conversely, the assessment of changes in 1/T1 [by the method of Hittmair et al. (1994)] as a measure of contrast agent uptake indicates that malignant lesions actually take up more contrast agent (**b**)

5.3.2 T_1 Measurement

The linear relationship between signal intensity and contrast agent concentration is only approximately true over a limited range of contrast agent concentration. The linear relationship between $1/T_1$ and contrast agent concentration has been shown to hold over a much wider range of Gd concentrations by a number of groups (Rosen et al. 1990; Donahue et al. 1994; Judd et al. 1995). Unfortunately, the measurement of T_1 in vivo is a non-trivial problem and, of particular significance, accurate measures of T_1 are often time consuming to obtain. A considerable body of literature has developed on this subject and an exhaustive description of the existing techniques is not attempted here, merely an overview of the more common approaches.

The methods of measuring T_1 using MR images fall broadly into two categories:

1. Inversion/saturation recovery prepared imaging sequences, and
2. Variable saturation techniques.

Each technique may be accomplished using a number of imaging sequences including; spin echo, EPI, or gradient-echo imaging with the appropriate additional pulses and subsequent processing algorithms.

5.3.2.1 Inversion or Saturation Recovery Techniques

An inversion pulse (Bluml et al. 1993) or a series of saturation pulses (Parker et al. 2000) provides T_1-weighted preparation of the signal dependent upon the subsequent delay prior to acquisition of a normal imaging sequence. Using a series of images, each obtained with a different delay, the T_1 of the sample may be estimated (BLUML et al. 1993). While this is perhaps the most precise method of obtaining an estimate of T_1, and the precision increases with the number of delay times used, it can be very time consuming. For example, an inversion recovery spin echo image with 128 phase encoding steps collected at four different delay (TI) times between 50 and 950 ms (maximum delay <TR) requires an imaging time of 1.0×128×4=8 min 32 s. In fact, this sequence would only be suitable for estimates of T_1 values up to around 700 ms. Precise estimates of longer T_1 values requires the use of a longer maximum TI and consequently, TR. The time required for such measurements can be reduced significantly by the use of snapshot-FLASH (BLUML et al. 1993) or EPI-based approaches (GOWLAND and MANSFIELD 1993) to sample an entire image at each TI, although these methods can be degraded by severe point spread function artefacts (PARKER et al. 2000). Further time savings can be made using Look-Locker techniques (FREEMAN et al. 1994), albeit with limitation on the number of slices that may be acquired per unit time.

5.3.2.2 Variable Saturation Techniques

T_1 may be estimated using the ratio of two spin echo images collected with different TRs, but again imaging times can become prohibitive. A similar approach is to use gradient echo images with variable flip angles (FRAM et al. 1987). The signal intensity obtained using a FLASH sequence has been described above (Eq. 5.3). Rearranging this equation yields:

$$Y = \exp\left(-\frac{TR}{T_1}\right).X - g.\rho.\left(1-\exp\left(-\frac{TR}{T_1}\right)\right).\exp\left(-\frac{TE}{T_2*}\right) \quad (5.7)$$

where $Y = \dfrac{S_\alpha}{\sin(\alpha)}$ and $X = \dfrac{S_\alpha}{\tan(\alpha)}$. Hence a plot of Y against X for a range of flip angles will result in a straight line, and T_1 may be calculated from the slope. WANG et al. (1987) have described the optimal sequence parameters for minimisation of the error in the calculated T_1 when only two flip angles are used. With a given T_1 two flip angles can be chosen which provide a greater precision in T_1 estimate than a comparable spin echo pair. It may, however, be difficult to choose an appropriate pair of flip angles if the sample contains an unknown or large range of T_1 values. Here the use of a number of equally spaced flip angles may be employed. However, in this case it may be advantageous to use a more computationally intensive non-linear fit using the original FLASH equation (Eq. 5.3) if the precision is to be improved (WANG et al. 1987).

In the field of dynamic contrast-enhanced MRI it is common to measure T_1 using variable flip angle gradient echo acquisitions, usually while keeping TR constant. Such techniques require only relatively short acquisition times, which allows good temporal resolution, and may be used in multi-slice (HITTMAIR et al. 1994) or 3D volume modes to provide tissue coverage (BROOKES et al. 1999) (ZHU et al. 2000). A simple protocol for a quantitative DCE-MRI study utilises a single heavily proton density-weighted (PD-weighted) acquisition, acquired prior to contrast agent administration, followed by numerous T_1-weighted acquisitions over time (see for example PARKER et al. 1997; EVELHOCH 1999; LI et al. 2000). This could be achieved by using a low flip angle for the PD-weighted acquisition, and a higher flip angle for each of the dynamic T_1-weighted acquisitions, whilst keeping TR short to maintain temporal resolution. Note that the PD-weighted acquisition is obtained only once; T_1 is always estimated by comparing the signal intensity of the T_1-weighted acquisitions (before or after contrast agent administration) with this single PD-weighted acquisition. Such a strategy allows T_1 to be estimated rapidly throughout the time course of signal enhancement.

5.3.2.3 Factors Affecting Measurement Accuracy

The accuracy of T_1 measurement may be compromised by a number of factors: machine non-linearities (in the main field, gradients, or radio-frequency (RF) amplifier), which are not accounted for in the calculation, unpredicted sample artefacts (e.g. susceptibility artefacts caused by ferromagnetic objects), sequence dependent errors, partial volume effects or flow and motion. These factors may affect all of the techniques available for T_1 measurement, but certain sequence dependent errors are of particular relevance for the variable saturation techniques described.

Slice imperfections. Slice selective RF-pulses used to excite the imaging slice are never perfectly rectangular and therefore the sample receives a range of different flip angles through the slice. These imperfections (usually manifest in "peaking" at the slice edges) are often magnified in the estimate of T_1 resulting in loss of accuracy. These errors may be corrected via careful calibration (Parker et al. 1997; Brookes et al. 1999) or via modification of the calculation procedure if the true pulse profile is known (Parker et al. 2001). IR techniques employing a non-selective inversion pulse tend to remain largely immune to these problems as the T_1 estimate is principally determined by the TI time and not the read-out sequence. Even in the case of a slice selective IR measurement the effect of an imperfect inversion slice profile can effectively be factored out as a contributing factor to inversion inefficiency.

RF power. Calculations of T_1 rely on a predicted signal behaviour following excitation pulses with controlled flip angles. Often the pulse transmitted may not achieve the expected amplitude due to RF-transmitter coil inhomogeneities or improper calibration of RF amplitude. This will clearly cause a problem in the multiple flip angle technique of T_1 estimation since there will be a tendency to underestimate T_1 if the RF power is too low. These problems may be particularly evident when coils with a non-linear response are used for RF transmission. However, once again, if the RF profile of a given coil is known it is possible to correct for spatially varying RF transmission fields (Parker et al. 2001).

A number of approaches have been developed to minimise the errors associated with slice imperfections (Parker et al. 2000, 2001) or RF power miscalibration (Cron et al. 1999) requiring additional preparation or calculation. If T_1 is to be measured in the clinical setting, then it is desirable that the measurement technique has a minor effect upon the normal imaging protocol. The multiple flip angle approach of Wang et al. (1987), though limited in precision, provides a rapid, and easily implemented methodology and has been employed in 3D mode where slice imperfections are minimised, particularly in the central sections of the 3D block (Brookes et al. 1999; Zhu et al. 2000).

5.4 Dynamic Contrast-Enhanced Imaging

Though T_1 measurements may be made in a matter of seconds (Freeman et al. 1994; Tong and Prato 1994), temporal or spatial constraints usually preclude their use for tracking the passage of a contrast agent bolus (especially for 3D measurements). As discussed above, a common experimental approach is to measure T_1 before contrast agent administration then image the tissue rapidly during uptake using a fast T_1-weighted sequence (Brookes et al. 1999). The pre-contrast T_1 measure provides estimates of T_{10} and the lumped constant, $g.\varrho.\exp(-TE/T_2^*)$ (Eq. 5.3). Thus, by substitution, T_1 following contrast agent administration may be estimated directly from signal intensity measurements (Zhu et al. 2000). To reduce any error associated with RF miscalibration a final, bookend, T_1 measurement may also be made (Cron et al. 1999). Selection of an imaging methodology begins with a series of basic choices such as an appropriate RF coil, imaging plane and sequence to avoid issues such as flow artefacts and field of view aliasing. Subsequently, the choice of T_1-weighted sequence for bolus tracking must fulfil a long list of both generic and study-specific criteria and a wide range of methods have been used in the field (see many of the other chapters in this book). High in the list of generic criteria are: temporal resolution, T_1 sensitivity and dynamic range, spatial coverage, and resolution; each of these competes against noise for the limited MR signal. Perhaps the principal decision to be made when selecting a T_1-weighted sequence for bolus tracking is whether or not an arterial input function (AIF) will be measured (see Chap. 6).

It has been shown that accurate characterisation of the AIF requires a temporal resolution in the order of a second (Henderson et al. 1998). Furthermore, following bolus injection of a typical clinical dose of contrast agent the T_1 of the blood may decrease by more than an order of magnitude (Fritz-Hansen et al. 1996). Monitoring such large changes in relaxation rate requires an imaging sequence with a good dynamic range (Fig. 52). Competing directly with this requirement is the need to monitor much smaller changes in T_1 at the level of the tissue. The location of the AIF, in relation to the tissue of interest, and the extent of that tissue dictates the requirements for spatial coverage. Finally, it is rare to identify a local feeding artery to provide an AIF but the closer the AIF is to the true tissue arterial supply the more accurate the subsequent modelling (Calamante et al. 2000). However, the spatial resolution of the images places a minimum diameter on the artery to be imaged. It is often the case that measurement of the AIF proves to be impossible or inappropriate. Typically this may be due to difficulties in choosing an appropriate artery (e.g. in studies of breast cancer) or the necessity to use an imaging sequence that lacks the necessary temporal resolution or fails to saturate incoming arterial water and thereby makes AIF estimation impossible (Fritz-Hansen et al. 1996). Most MR studies to date have been performed without measurement of

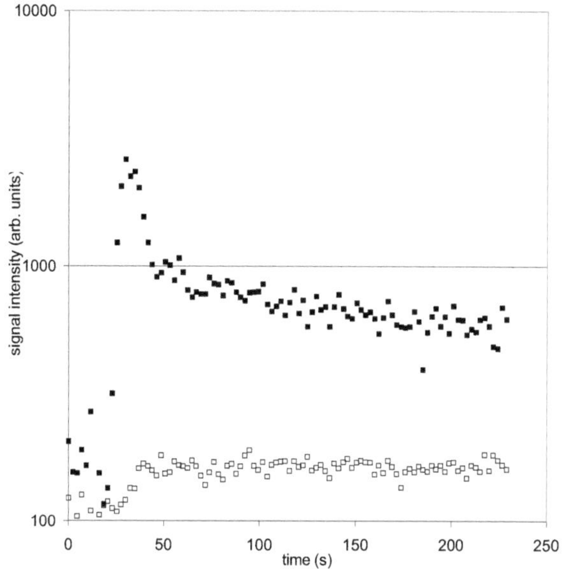

Fig. 5.2. To measure both the signal in the artery and tissue an imaging sequence selected for a bolus tracking experiment requires a large dynamic range. In this example mean values of the raw signal intensity from regions of interest placed in the external iliac artery (*filled squares*) and in muscle (internal obturator; *open squares*) are shown on a semi-logarithmic plot. Note the order of magnitude difference in the level of the signals obtained from each region

the AIF with either a population averaged AIF used in the subsequent data analysis or the AIF has been neglected altogether (see Chap. 6). In such studies the constraints on the required temporal resolution are relaxed and the imaging sequence used may be more T_1 sensitive with a reduced dynamic range. This provides greater opportunity for improving the spatial coverage, resolution or signal to noise ratio of the images acquired.

Finally, with the images acquired and the signal changes converted to changes in contrast agent concentration, quantitative analysis may proceed. Measurement of contrast agent concentration changes in the artery (providing the AIF) are typically made using user-defined regions or automated vessel identification procedures (Rijpkema et al. 2001; Parker et al. 2003). However, the selection of appropriate tissue for analysis remains the subject of considerable debate. Traditionally, a region of interest is defined in, for example, tumour tissue. Once analysed, the characteristics of this region are estimated but are these characteristics, particularly in a heterogeneous cancer, representative of the tumour as a whole? Much research has gone into comparing the results of whole tumour region definition versus the selection of small, highly enhancing, sub-regions or semi-automated region selection (Mussurakis et al. 1997) (Mussurakis et al. 1998). Furthermore, images can be analysed on a per-pixel basis providing tissue characterisation at significantly improved spatial resolutions. Nevertheless, how are the parameters estimated in each pixel combined to provide simple representative values for the tissue as a whole? Mean or median values may not describe the range of characteristics observed. Histogram analysis of the measured uptake parameters present one route for describing the heterogeneity observed (Hayes et al. 2002; Checkley et al. 2003) but a consensus on the best approach for describing heterogeneous contrast enhancement remains the subject of continuing research.

5.5 Water Exchange

In the discussion above, and indeed in many studies employing dynamic contrast-enhanced MRI in oncology, it has been assumed that tissues contain a single, homogenous, water population with well described MR properties that undergoes simple changes when subjected to a contrast agent. However, experiments from the early days of biomedical MR suggested that the situation is more complex than this (Hazlewood et al. 1974). Water is found in a number of different environments in a biological tissue. The largest population is typically found inside cells. For example, water in the brain's grey matter may be coarsely divided into three populations: intracellular water making up around 79% of the total, interstitial water making up around 18% and intravascular water contributing the final 3%. Given the differing physiological environments of these spaces, it is not unreasonable to assume that the MR properties of water in the spaces will also be different (Hazlewood et al. 1974). As such it might be expected that T_1 and T_2 relaxation times (not to mention diffusion and magnetisation transfer coefficients) in the majority of tissues would have multiple components. This is seldom the case in practice, a finding that results from the rapid movement of water between tissue compartments, so called water exchange (Donahue et al. 1997). A water molecule moving from the cell, through the cell membrane and into the interstitial space during, for example, a T_1 measurement will contribute a relaxation behaviour that represents an average of the intrinsic T_1 of the cell and the intrinsic T_1 of the interstitial space weighted by the time it spent in each of those spaces during the measurement. When many millions of water molecules contribute to the MR measurement a pattern or bulk property emerges. The more rapid the motion of the molecules and the smaller the difference between the relaxation times of the two spaces, the closer the relaxation pattern approaches a single average value (Donahue et al. 1994). At this limit, water exchange is said to be "fast". Conversely, as the motion of the water molecules slows down and the difference in relaxation times of the two spaces increases, the pattern of relaxation behaviour approaches that of two distinct populations. At the limit of no exchange between the two spaces during a relaxation measurement, water exchange is said to be "slow". Confusion with this nomenclature often arises since the terms fast and slow do not refer directly to the speed with which the water molecule moves between the two spaces, but to the ratio of this motion to the difference in relaxation rates of the two spaces (Barsky et al. 1997). Two spaces with identical relaxation rates will always be described as being in fast exchange however slowly the water molecules diffuse between them, as there can be no distinction made between their relaxation properties. Conversely, two spaces with orders of magnitude difference in their respective relaxation rates will remain in a slow exchange regime even if water moves very rapidly between them, as their relaxation properties will always be resolved. Moreover, a tissue said to have water undergoing fast exchange may switch to slow exchange without any modification to the speed at which the molecules move. This transformation may be initiated by a simple increase in the difference between the relaxation rates of the two compartments. For example, the addition of a contrast agent to the interstitial space does not change the motion of water molecules; it simply increases the intrinsic relaxation rate of the interstitial space.

The measurement of the tissue concentration of contrast agent is inextricably linked to the rate of water exchange between tissue compartments. Each of the approaches described in the chapter thus far assumes that water exchange is fast and that the tissue has a single, well defined, T_1. In 1994, Donahue et al. performed a series of important experiments to determine the influence of water exchange on the measurement of contrast agent concentrations. They concluded that interstitial-intracellular water exchange was sufficiently rapid (between 8 and 27 Hz) that it was reasonable to assume fast exchange for clinical doses of contrast agent. These findings have been confirmed in a further study, like the Donahue study, performed on isolated perfused hearts (Judd et al. 1999). Nevertheless, both groups stressed the significance of contrast agent dose when considering the influence of water exchange. If the concentration of contrast agent in the interstitial space were to reach much higher levels, then the effects of slow exchange would be felt. These considerations have driven a series of studies by Landis et al. (1999, 2000) and Yankeelov et al. (2003) in which the interstitial-intracellular (transcytolemmal) water exchange process has been examined in detail. Indeed, Yankeelov et al. (2003) have introduced a methodology, BOLERO, for analysing contrast agent kinetics for systems departing the fast exchange limit. Central to their work is the suggestion that transcytolemmal water exchange departs the fast limit, in many tissues, at very low (sub-clinical) doses of contrast agent (Landis et al. 1999). This is at odds with the findings of Donahue et al. (1994) and Judd et al. (1999) and remains an area of continued debate. Critical to these studies is the accurate measurement of water residence times (average time that a water molecule resides in a compartment = 1/exchange rate). Such measurements have been made over a number of years using contrast agents and diffusion measurements (Pirkle et al. 1979; Pfeuffer et al. 1998; Quirk et al. 2003), but no agreement has yet surfaced on the order of magnitude of these residence times.

Less controversial is the issue of intravascular–interstitial water exchange. Donahue et al. (1994) estimated exchange rates in the isolated perfused heart with an upper limit of 7 Hz while Judd et al. (1999) measured an intravascular-interstitial exchange rate of 3 Hz. Given the large difference in relaxation rates of the two spaces immediately following contrast agent administration (due to the fact that the agent will not have had time to pass into the interstitial spaces, whilst being present in high concentration in the intravascular space), these values explain earlier observations of significant departures from the fast exchange limit (Judd et al. 1995). This finding may have a serious practical implication for dynamic contrast-enhanced MRI studies in oncology, even though to date little data has been published on the issue of limited intravascular-interstitial water exchange and its affect on DCE-MRI. It is clear that experiments using intravascular contrast agents are much more sensitive to restricted exchange effects than those employing interstitial agents (Judd et al. 1999). For interstitial agents the degree of first pass extraction plays a major role in determining the magnitude of the effect. If contrast agent enters the interstitial space quickly then the effect of slow water exchange

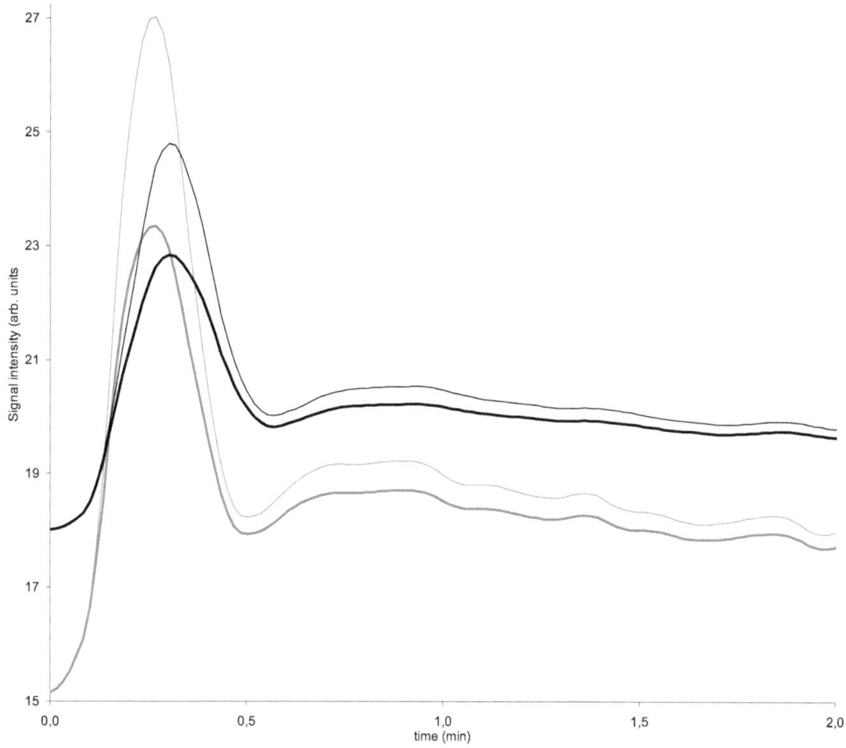

Fig. 5.3. Simulated signal-time curves for normal grey matter (*grey lines*) and white matter (*black lines*). The curves are representative of data obtained in a spoiled gradient echo acquisition (TR, 4.3 ms, flip 35°) when intravascular-interstitial water exchange is either in the fast limit (*faint lines*) or in the slow limit (*bold lines*)

is short-lived. Larsson et al. described such effects in a study of perfusion of the heart and brain (Larsson et al. 2001). With an intact blood-brain barrier agents such as Gd-DTPA behave as intravascular contrast agent. As such, during the first pass of a bolus of Gd-DTPA through the brain intravascular-interstitial exchange approaches the slow limit (Larsson et al. 2001) and contrast agent concentrations (and thereby perfusion) are underestimated. On the other hand first pass extraction of Gd-DTPA in the heart is significant (> 30%) and the slow water exchange effect quickly decreases. In this case the early phase of contrast agent uptake is only underestimated very slightly (Larsson et al. 2001). With the assumption that first pass extraction is ~50%, not unreasonable for tumours, Larsson et al. (2001) conclude that water exchange will have minimal effect on the determination of K^{trans} for typical clinical doses of contrast agent. Further simulations of the signals obtained from the brain and brain tumours (Buckley 2002) support these findings. The initial peak seen in a signal-time plot during the first pass of a contrast agent bolus is flattened by the effect of slow intravascular-interstitial water exchange (Fig. 5.3). If the contrast agent remains intravascular (as seen in normal grey and white matter with an intact blood--brain barrier) this flattening will lead to underestimates in both perfusion and blood volume. Moreover, the mismatch between the first pass peak and the subsequent equilibrium phase (in which the systems returns to the fast exchange regime) may be misinterpreted as contrast agent leakage (Buckley 2002). These effects are negated somewhat in tumour tissue where there is significant first-pass extraction of the contrast agent. However, estimates of blood volume and, to as lesser extent, separate estimates of perfusion and microvascular permeability-surface area product are compromised to some degree as their measurement depends upon very rapid data acquisition in the early phases of enhancement (Buckley 2002). At least two methods for controlling these effects have been proposed. Yankeelov et al. (2003) recommend an approach in which interstitial-intracellular water exchange is explicitly modelled and estimated. Though elegant in concept the additional burden of data analysis has certain limitations (Yankeelov et al. 2003), and these methods have not to date been applied in the consideration of intravascular-interstitial water exchange. Another approach, proposed by Donahue et al., is to minimise the exchange rate dependence of the measurements made. These exchange-minimisation techniques require the use of short inversion time magnetisation prepared sequences or short TR, high flip angle spoiled gradient echo acquisitions (Donahue et al. 1996). These imaging sequences suffer from the drawback of limited SNR, but they produce signals that are largely insensitive to changes or differences in water exchange effects. Finally, experimental design can play a significant role in the influence of water exchange on contrast agent measurements. Consideration must be given to the use of lower contrast agent doses, infusions rather than bolus injections or smaller, rapidly extracted, agents. These considerations will conflict with many other requirements for DCE-MRI acquisitions and a compromise must be reached for each given study.

5.6 Conclusions

Despite the complications in relating MR signals directly to the tissue concentration of typical contrast agents, quantitative MR measurements made in vivo have found an increasing number of applications. Technical developments have allowed for rapid estimates of T_1 and improving temporal resolution in bolus tracking experiments. The straightforward measurement of arterial input functions remains high on the list of requirements for the field to progress; this still remains an area of limited exploitation. The issue of water exchange has mixed implications. For practical applications in oncology the departure from fast exchange provides an experimental confound and may limit the specificity of DCE-MRI unless effectively addressed. Conversely, the opportunities opened up by the possibility of measuring water exchange in vivo provides an exciting new area of research. No doubt developments along both lines will provide much interaction and new directions for future research.

References

Barsky D, Putz B, Schulten K (1997) Theory of heterogeneous relaxation in compartmentalized tissues. Magn Reson Med 37:666-675

Bluml S, Schad LR, Stepanow B, Lorenz WJ (1993) Spin-lattice relaxation-time measurement by means of a turboFLASH technique. Magn Reson Med, 30:289-295

Boetes C, Barentsz JO, Mus RD, van der Sluis RF, van Erning LJ, Hendriks JH, Holland R, Ruys SH (1994) MR characterization of suspicious breast lesions with a gadolinium-enhanced turboFLASH subtraction technique. Radiology 193:777-781

Brookes JA, Redpath TW, Gilbert FJ, Murray AD, Staff RT (1999) Accuracy of T1 measurement in dynamic contrast-enhanced breast MRI using two- and three-dimensional variable flip angle fast low-angle shot. J Magn Reson Imaging 9:163-171

Buckley DL (2002) Uncertainty in the analysis of tracer kinetics using dynamic contrast-enhanced T1-weighted MRI. Magn Reson Med 47:601-606

Calamante F, Gadian DG, Connelly A (2000) Delay and dispersion effects in dynamic susceptibility contrast MRI: simulations using singular value decomposition. Magn Reson Med 44:464-473

Checkley D, Tessier JJ, Wedge SR, Dukes M, Kendrew J, Curry B, Middleton B, Waterton JC (2003) Dynamic contrast-enhanced MRI of vascular changes induced by the VEGF-signalling inhibitor ZD4190 in human tumour xenografts. Magn Reson Imaging 21:475-482

Cron GO, Santyr G, Kelcz F (1999) Accurate and rapid quantitative dynamic contrast-enhanced breast MR imaging using spoiled gradient-recalled echoes and bookend T-1 measurements. Magn Reson Med 42:746-753

Donahue KM, Burstein D, Manning WJ, Gray ML (1994) Studies of Gd-DTPA relaxivity and proton-exchange rates in tissue. Magn Reson Med 32:66-76

Donahue KM, Weisskoff RM, Chesler DA, Kwong KK, Bogdanov AA Jr, Mandeville JB, Rosen BR (1996) Improving MR quantification of regional blood volume with intravascular T1 contrast agents: accuracy, precision, and water exchange. Magn Reson Med 36:858-867

Donahue KM, Weisskoff RM, Burstein D (1997) Water diffusion and exchange as they influence contrast enhancement. J Magn Reson Imaging 7:102-110

Engelstad BL, Wolf GL (1988) Contrast agents. In: Stark DD, Bradley WG Jr (eds) Magnetic resonance imaging. Mosby, St Louis, pp 161-181

Evelhoch JL (1999) Key factors in the acquisition of contrast kinetic data for oncology. J Magn Reson Imaging 10:254-259

Fram EK, Herfkens RJ, Johnson GA, Glover GH, Karis JP, Shimakawa A, Perkins TG, Pelc NJ (1987) Rapid calculation of T1 using variable flip angle gradient refocused imaging. Magn Reson Imaging 5:201-208

Freeman A, Gowland P, Jellineck D, Wilcock D, Firth J, Worthington B, Mansfield P, Ratcliffe G (1994) Gd uptake in brain tumours: application of fast T1 mapping using LL_EPI (abstract). Proceedings of the Society of Magnetic Resonance, San Francisco, CA p 863

Fritz-Hansen T, Rostrup E, Larsson HB, Sondergaard L, Ring P, Henriksen O (1996) Measurement of the arterial concentration of Gd-DTPA using MRI: a step toward quantitative perfusion imaging. Magn Reson Med 36:225-231

Gowland P, Mansfield P (1993) Accurate measurement of T(1) in-vivo in less-than 3 seconds using echo-planar imaging. Magn Reson Med 30:351-354

Gowland P, Mansfield P, Bullock P, Stehling M, Worthington B, Firth J (1992) Dynamic studies of gadolinium uptake in brain tumors using inversion-recovery echo-planar imaging. Magn Reson Med 26:241-258

Haase A, Frahm J, Matthaei D, Hanicke W, Merboldt KD (1986) FLASH imaging - rapid NMR imaging using low flip-angle pulses. J Magn Reson 67:258-266

Hayes C, Padhani AR, Leach MO (2002) Assessing changes in tumour vascular function using dynamic contrast-enhanced magnetic resonance imaging. NMR Biomed 15:154-163

Hazlewood CF, Chang DC, Nichols BL, Woessner DE (1974)

Nuclear magnetic resonance transverse relaxation times of water protons in skeletal muscle. Biophys J 14:583-606

Henderson E, Rutt BK, Lee TY (1998) Temporal sampling requirements for the tracer kinetics modeling of breast disease. Magn Reson Imaging 16:1057-1073

Heywang SH, Wolf A, Pruss E, Hilbertz T, Eiermann W, Permanetter W (1989) MR imaging of the breast with Gd-DTPA: use and limitations. Radiology 171:95-103

Hittmair K, Gomiscek G, Langenberger K, Recht M, Imhof H, Kramer J (1994) Method for the quantitative assessment of contrast agent uptake in dynamic contrast-enhanced MRI. Magn Reson Med 31:567-571

Judd RM, Atalay MK, Rottman GA, Zerhouni EA (1995) Effects of myocardial water exchange on T1 enhancement during bolus administration of MR contrast agents. Magn Reson Med 33:215-223

Judd RM, Reeder SB, May-Newman K (1999) Effects of water exchange on the measurement of myocardial perfusion using paramagnetic contrast agents. Magn Reson Med 41:334-342

Landis CS, Li X, Telang FW, Molina PE, Palyka I, Vetek G, Springer CS (1999) Equilibrium transcytolemmal water-exchange kinetics in skeletal muscle in vivo. Magn Reson Med 42:467-478

Landis CS, Li X, Telang FW, Coderre JA, Micca PL, Rooney WD, Latour LL, Vetek G, Palyka I, Springer CS (2000) Determination of the MRI contrast agent concentration time course in vivo following bolus injection: effect of equilibrium transcytolemmal water exchange. Magn Reson Med 44:563-574

Larsson HBW, Stubgaard M, Frederiksen JL, Jensen M, Henriksen O, Paulson OB (1990) Quantitation of blood-brain barrier defect by magnetic resonance imaging and gadolinium-DTPA in patients with multiple sclerosis and brain tumors. Magn Reson Med 16:117-131

Larsson HBW, Rosenbaum S, Fritz-Hansen T (2001) Quantification of the effect of water exchange in dynamic contrast MRI perfusion measurements in the brain and heart. Magn Reson Med 46:272-281

Li KL, Zhu XP, Waterton J, Jackson A (2000) Improved 3D quantitative mapping of blood volume and endothelial permeability in brain tumors. J Magn Reson Imaging 12:347-357

Mussurakis S, Buckley DL, Drew PJ, Fox JN, Carleton PJ, Turnbull LW, Horsman A (1997) Dynamic MR imaging of the breast combined with analysis of contrast agent kinetics in the differentiation of primary breast tumours. Clin Radiol 52:516-526

Mussurakis S, Gibbs P, Horsman A (1998) Peripheral enhancement and spatial contrast uptake heterogeneity of primary breast tumours: quantitative assessment with dynamic MRI. J Comput Assist Tomogr 22:35-46

Parker GJ, Suckling J, Tanner SF, Padhani AR, Revell PB, Husband JE, Leach MO (1997) Probing tumor microvascularity by measurement, analysis and display of contrast agent uptake kinetics. J Magn Reson Imaging 7:564-574

Parker GJ, Baustert I, Tanner SF, Leach MO (2000) Improving image quality and T-1 measurements using saturation recovery turboFLASH with an approximate K-space normalisation filter. Magn Reson Imaging 18:157-167

Parker GJ, Barker GJ, Tofts PS (2001) Accurate multislice gradient echo T(1) measurement in the presence of non-ideal RF pulse shape and RF field nonuniformity. Magn Reson Med 45:838-845

Parker GJ, Jackson A, Waterton JC, Buckley DL (2003) Automated arterial input function extraction for T1-weighted DCE-MRI (abstract). Proc ISMRM 11th annual meeting, Toronto, Canada, p 1264

Pedevilla M, Stollberger R, Schmidt F, Wach P, Ebner F (1995) Comparison of various methods used for breast tumor characterization in Gd-DTPA enhanced MR imaging (abstract). Proc Society of Magnetic Resonance, Nice, France, p 1600

Pfeuffer J, Flogel U, Dreher W, Leibfritz D (1998) Restricted diffusion and exchange of intracellular water: theoretical modelling and diffusion time dependence of H-1 NMR measurements on perfused glial cells. NMR Biomed 11:19-31

Pirkle JL, Ashley DL, Goldstein JH (1979) Pulse nuclear magnetic resonance measurements of water exchange across the erythrocyte membrane employing a low Mn concentration. Biophys J 25:389-406

Quirk JD, Bretthorst GL, Duong TQ, Snyder AZ, Springer CS, Ackerman JJH, Neil JJ (2003) Equilibrium water exchange between the intra- and extracellular spaces of mammalian brain. Magn Reson Med 50:493-499

Rijpkema M, Kaanders J, Joosten FBM, van der Kogel AJ, Heerschap A (2001) Method for quantitative mapping of dynamic MRI contrast agent uptake in human tumors. J Magn Reson Imaging 14:457-463

Rosen BR, Belliveau JW, Vevea JM, Brady TJ (1990) Perfusion imaging with NMR contrast agents. Magn Reson Med 14:249-265

Shames DM, Kuwatsuru R, Vexler V, Muhler A, Brasch RC (1993) Measurement of capillary-permeability to macromolecules by dynamic magnetic resonance imaging: a quantitative noninvasive technique. Magn Reson Med 29:616-622

Springer CS Jr (1994) Physiochemical principles influencing magnetopharmaceuticals. In: Gillies RJ (ed) NMR in physiology and biomedicine. Academic Press, San Diego, pp 75-100

Stanisz GJ, Henkelman RM (2000) Gd-DTPA relaxivity depends on macromolecular content. Magn Reson Med 44:665-667

Terada T, Nambu K, Hyotani G, Miyamoto K, Tsuura M, Nakamura Y, Nishiguchi T, Itakura T, Hayashi S, Komai N (1992) A method for quantitative measurement of cerebral vascular- permeability using X-ray CT and iodinated contrast-medium. Neuroradiology 34:290-296

Tofts PS, Kermode AG (1991) Measurement of the blood-brain barrier permeability and leakage space using dynamic MR imaging. 1. Fundamental concepts. Magn Reson Med 17:357-367

Tong CY, Prato FS (1994) A novel fast T1-mapping method. J Magn Reson Imaging 4:701-708

Vallee JP, Ivancevic M, Lazeyras F, Kasuboski L, Chatelain P, Righetti A, Didier D (2003) Use of high flip angle in T1-prepared FAST sequences for myocardial perfusion quantification. Eur Radiol 13:507-514

Wang HZ, Riederer SJ, Lee JN (1987) Optimizing the precision in T1 relaxation estimation using limited flip angles. Magn Reson Med 5:399-416

Wedeking P, Sotak CH, Telser J, Kumar K, Chang CA, Tweedle MF (1992) Quantitative dependence of MR signal intensity on tissue concentration of Gd(HP-DO3A) in the nephrectomized rat. Magn Reson Imaging 10:97-108

Yankeelov TE, Rooney WD, Li X, Springer CS (2003) Variation of the relaxographic „shutter-speed" for transcytolemmal water exchange affects the CR bolus-tracking curve shape. Magn Reson Med 50:1151-1169

Zhu XP, Li KL, Kamaly-Asl ID, Checkley DR, Tessier JJ, Waterton JC, Jackson A (2000) Quantification of endothelial permeability, leakage space, and blood volume in brain tumors using combined T1 and T2* contrast-enhanced dynamic MR imaging. J Magn Reson Imaging 11:575-585

6 Tracer Kinetic Modelling for T_1-Weighted DCE-MRI

Geoffrey J. M. Parker and David L. Buckley

CONTENTS

6.1 Introduction

The use of exogenous extracellular contrast media has lead to marked improvements in the sensitivity of detection and delineation of tumours using MRI. Additionally, it has been demonstrated that the time course of contrast agent accumulation may be used to extract quantitative information regarding the functional integrity of tumour microvasculature. This is potentially of great use, as it has been established that tumour growth is heavily dependent on the recruitment of a vascular supply and that the characteristics of the resultant tumour microvasculature may differ from those of normal tissues (Folkman 1990, 1995).

Tumour blood supply is the target of a number of putative anti-cancer treatments that aim to impede tumour growth by either stopping the process of angiogenesis or by destroying existing tumour vasculature. Such treatments are expected to be cytostatic when applied in isolation, and as such are unlikely to produce immediate reductions in tumour size; arresting tumour growth is a more likely initial indication of a treatment effect (see Chap. 16). In this scenario, established radiological measures of response to treatment, based around measurements of tumour size reduction are redundant, especially when assessing short-term drug effects. DCE-MRI methods that are sensitive to the blood supply of tumours, and that are able to quantify features such as blood flow and capillary wall permeability, are likely to be of use in this situation, as these features match the targets of the therapies.

In this chapter, the basics of DCE-MRI quantitative tracer kinetic analyses will be reviewed from the most basic analyses to more challenging modelling approaches. The inter-relationships between perfusion and contrast agent extravasation in tumours will be emphasized.

G. J. M. Parker, PhD; D. L Buckley, PhD
Imaging Science and Biomedical Engineering, University of Manchester, Stopford Building, Oxford Road, Manchester, M13 9PT, UK

6.2 Model-Free Quantification

Perhaps the simplest method for providing reliable and reproducible information that has a bearing on the kinetics of contrast agent accumulation is an integration of the concentration of the agent observed in the tissue of interest over time [the initial area under curve (IAUC), unit mmol] (Evelhoch 1999). The value of *IAUC* obtained in a tumour is dependent on the period over which integration is performed; typically this will be from the time point representing contrast agent administration or arrival in the tissue to a time t, typically 60–120 s after the start time. $IAUC_t$ may then be defined as

$$IAUC_t = \int_0^t [CA](t')dt' \quad (1)$$

where [CA](t') represents the concentration of contrast agent measured in the tissue at time *t*'. Figure 6.1 shows an example of IAUC60 calculated in a liver metastasis, and clearly indicates spatial

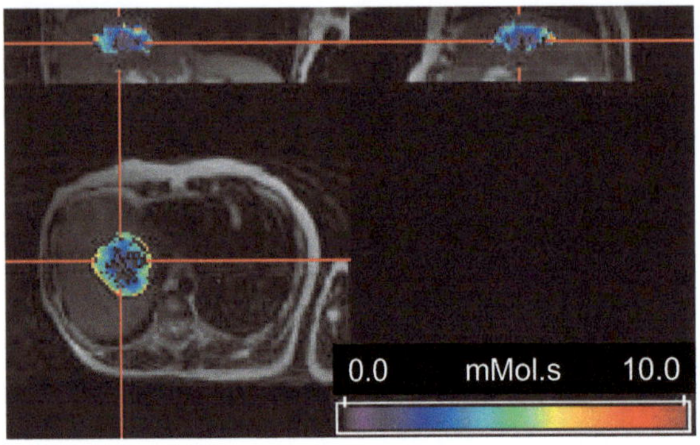

Fig. 6.1. Multiplanar map of $IAUC_{60}$ defined in a liver metastasis (orthogonal coronal sagittal and axial views). Note ring pattern indicating a higher degree of contrast agent arrival in the tumour periphery than the tumour core, which shows near zero *IAUC*, indicating a severely hypoperfused area. Red crosshairs indicate point of intersection of three orthogonal planes

heterogeneity in the vascular characteristics of this tumour.

Other approaches to quantification of the general shape of the contrast agent concentration time course include measurement of the peak concentration, gradient of concentration increase, time of contrast agent arrival, and time to maximum enhancement. Whilst some of these parameters are perhaps simpler still to estimate than *IAUC*, they are in general more prone to the effects of data noise, and are therefore considered to be less reproducible than *IAUC*.

Most model-free measurements of contrast agent delivery to tissues are dependent upon the arterial input function (AIF) to the tumour of interest, which may be difficult to measure. This has led to attempts to normalise against reference measurements obtained from normal tissues. For example, *IAUC* measurements in tumours have been made using a reference *IAUC* obtained from muscle (Evelhoch 1999; Evelhoch et al. 2002). However, in situations where the effects of intervention are to be monitored using DCE-MRI, care must be taken to ensure that the normalisation tissue is unaffected by the procedure (e.g. due to the action of a blood flow altering pharmaceutical or the spatially extended effects of a radiotherapy field). An alternative method to minimise the influence of variable AIFs is to use a power injector or other highly reproducible contrast agent administration protocol (Parker et al. 2003b).

Model-free quantification methods aim to provide indices related to contrast agent accumulation that are easy to derive and relatively reproducible. The major disadvantage in using such parameters is the difficulty in their interpretation. For example, whilst *IAUC* could be defined as 'a measure of the amount of contrast agent delivered to and retained within the tumour within the stated time period', such a definition is imprecise, and does not tell us anything specifically about potentially useful tissue properties, such as blood flow and microvascular endothelial permeability. In general (although the relative contributions are dependent upon the integration time period) *IAUC* reflects contrast agent kinetics determined by a combination of blood flow, blood volume, endothelial permeability, and the extracellular extravascular space volume. Similarly, measures such as the gradient of contrast agent uptake, and the peak concentration during the time course are influenced by each of these parameters in generally intractable proportions.

6.3 Tissue Compartmentalisation and Functional Parameters

To allow us to attempt to quantify the observed contrast agent kinetics in terms of physiologically meaningful parameters we first need to define the elements of the tumour or tissue structure and the functional processes that affect the distribution of the tracer (the contrast agent). It is customary to represent tissue as comprising three or four compartments, each of which is a bulk tissue characteristic (that is, we are unable to observe these compartments at their natural microscopic scale, but we can observe their aggregate effects at the image voxel scale or in a region of interest). These compartments are the vascular plasma space, the extracellular extravascular space (EES), and the intracellular space (Tofts et al. 1999) (Fig. 6.2). A fourth tissue component forms a catch-all for all the other microscopic tissue components, such as membranes, fibrous tissues, etc. All clinically utilised MRI contrast agents, and most experimental agents, do not pass into the intracellular space of the tissue, due to their size, inertness, and non-lipophilicity, making the intracellular space un-probable using DCE-MRI; for this reason, the intracellular and 'other' volumes are usually lumped together as a loosely defined 'intracellular' space. These three compartments may be expressed either in absolute terms (units ml per ml tissue or ml per g tissue if

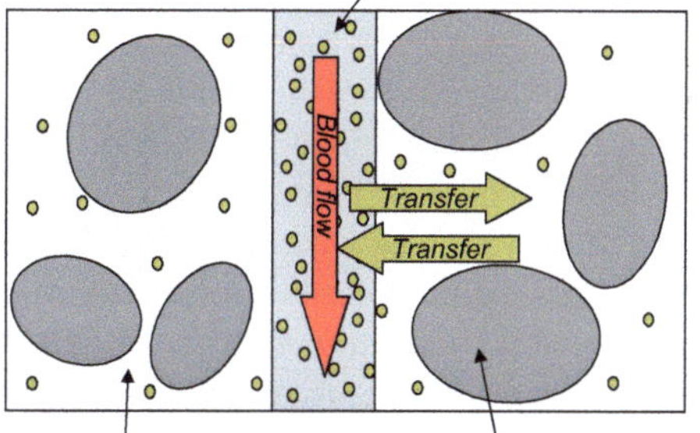

Fig. 6.2. Major compartments and functional variables involved in the distribution of a contrast agent tracer (small circles) in tissue (from PARKER and PADHANI 2003)

tissue density is known) or as fractions of tissue volume. In the latter case,

$$v_e + v_p + v_i = 1 \;, \qquad v_p = (1 - \mathrm{Hct})\, v_b \tag{2}$$

where v_e is the fractional EES, v_p is the fraction occupied by blood plasma, vi is the fraction occupied by the 'intracellular' space, v_b is the fraction occupied by whole blood, and Hct is the haematocrit (typically about 0.4).

The functional parameters, or delivery mechanisms, that influence contrast agent distribution in the intravascular space and the EES are usually assumed to be restricted to blood flow F [units ml blood (g tissue)$^{-1}$ min^{-1}] and the endothelial permeability-surface area product PS (units ml g^{-1} min^{-1}), which describes how 'leaky' a capillary wall is (Fig. 6.2). In reality there are additional functional parameters that may affect contrast agent distribution, including the rate of lymphatic drainage, the rate or degree of contrast agent mixing within compartments, and the degree of tracer diffusion in regions without an effective blood supply (e.g. areas of necrosis or oedema). Whilst some modelling approaches attempt to consider some of these factors explicitly, they are in general viewed as confounding factors in the kinetic analysis of DCE-MRI data, and a simplified model of diffusive transport is generally assumed.

6.4 Compartmental Models

6.4.1 The Rate Equation

The kinetic modelling of contrast agent distribution has a basis in the simple rate equation describing diffusive flux across a semi-permeable membrane (KETY 1951; RENKIN 1959; CRONE 1963). The diffusive transport of a dissolved substance across a semi-permeable membrane is determined by the difference in concentration of that substance between the two compartments that it separates and the freedom with which the membrane allows molecules to diffuse from one side to the other. The amount of substance per unit time that diffuses through the membrane, or diffusive flux Φ_d, may be defined as:

$$\Phi_d = PA(C_1 - C_2) \tag{3}$$

where P is the trans-membrane permeability coefficient (units m min-1), A is the cross-sectional area of the membrane through which transport is occurring (m^2), and C_1 and C_2 are the concentrations of permeant on each side of the membrane (mmol). The process outlined by Eq. 3 governs the transport of lipophilic molecules across intact membranes. Small polar molecules, such as water, are also subject to such diffusive transport across membranes, but to a lesser extent than lipid-soluble molecules. A second transport mechanism is important in this case; through 'pores' or channels in the membrane. The rate of transport due to this effect is influenced by pressure differences across the membrane or partial pressure differences induced by an osmotic potential.

The system that is being probed using contrast agents in vivo is far more complex than a single membrane, and the simple rate equation of Eq. 3 is overly simplistic. In addition to multiple possible transport mechanisms the model of a single membrane is obviously naive (as noted above, contrast agent molecules do not readily diffuse through intact lipid bilayer membranes), and the true situation that we wish to probe, the tissue capillary wall, is one of multiple cell membranes, basement membranes, and gaps of varying sizes. Fortunately, we can still utilise Eq. 3 with the aid of two basic assumptions (SHA'AFI 1981). Firstly, we assume that multiple transport mechanisms via a given obstacle (e.g. diffusive transport and osmotic transport across a single membrane) may be treated as a system of n diffusive permeabilities in parallel, leading to a total permeability for a given obstacle of π:

$$\pi = \sum_{i}^{n} P_i \tag{4}$$

Secondly, we assume that multiple obstacles to transport (e.g. overlapping membranes) may be treated as N permeabilities in series:

$$\frac{1}{P_T} = \sum_{i}^{N} \frac{1}{\pi_i} \tag{5}$$

We can therefore relate the total flux of a molecular species Φ_T to the overall permeability:

$$\Phi_T = P_T S(C_1 - C_2), \tag{6}$$

where S is the *effective* surface area of, in our case, the capillary wall. As well as providing a means for combining the effects of more than one transport mechanism occurring via a set of obstacles at the capillary scale, this result also allows us to define permeability to a first order as a bulk, or volume averaged, quantity. This therefore allows us to make use of Eq. 6 in data analysis of imaging voxels or volumes of interest that are large relative to the scale of the phenomenon of interest, as long as we are aware that our estimates of PS and other parameters are always bulk parameters. It should also always be remembered that (for the purposes of this chapter, at least) all transport processes are being modelled as diffusive transport.

6.4.2 Modelling Using the Rate Equation

Low molecular weight contrast media available for clinical MRI studies, such as gadopentetate dimeglumine (Gadolinium diethylenetriaminepentaacetic acid or Gd-DTPA, molecular weight approximately 500 Daltons (Mitchell 1997)) and medium weight experimental molecules (see Chap. 2), transfer between the blood pool and the extravascular extracellular space (EES) of the tissue or tumour at a rate determined by the blood flow to the tissue, the permeability of the blood vessel walls, and the surface area of the perfusing vessels. The contrast agent does not cross cell membranes and its volume of distribution is therefore effectively the EES (Fig. 6.2). On T_1-weighted images, T_1 relaxation time shortening caused by the contrast agent in the interstitial space is the dominant mechanism of enhancement seen, although a variable contribution from contrast agent in the blood vessels, especially in highly vascular tumours, may also be observed. The early phase of contrast enhancement (often referred to as the first pass in bolus injection studies) involves the arrival of the contrast medium in the tissue of interest via the arterial supply, and lasts a number of cardiac cycles. In the presence of well-perfused and leaky vessels contrast medium immediately begins to pass into the EES, in a manner similar to that predicted by Eq. 6. Over a period lasting many hours, the contrast agent is removed from the blood stream by renal excretion, and it therefore washes out of the EES, again in a manner similar to that predicted by Eq. 6.

Equation 6 may be modified to account for contrast agent kinetic parameters that are more familiar to those working with DCE-MRI. The rate of accumulation and wash-out of an extracellular contrast medium in the EES, under the assumption that contrast agent is well-mixed in the vascular plasma space, v_p, and in the EES, can be described by a modified general rate equation [see for example Kety (1951)]:

$$v_e \frac{dC_e(t)}{dt} = K^{\text{trans}}(C_p(t) - C_e(t)), \tag{7}$$

where C_e is the concentration of agent in v_e, C_p is the concentration of agent in v_p, and K^{trans} is the volume transfer constant between v_p and v_e (Tofts et al. 1999). If the delivery of contrast medium to the tissue is ample (meaning that the rate of extraction of contrast agent via the leaky capillary wall is small compared with the rate of replenishment via perfusion), K^{trans} is equal to the product of the capillary wall permeability and capillary wall surface area per unit volume, $PS\rho$. PS is the permeability surface area product per unit mass of tissue (ml g^{-1} min^{-1}), and ρ is the tissue density (g ml^{-1}). However, if the delivery of the contrast medium to a tissue is insufficient, blood perfusion will be the dominant factor determining contrast agent kinetics, and K^{trans} approximates Fρ(1-Hct), where F is blood flow [units ml blood (g tissue)$^{-1}$ min^{-1}]. Thus in regions with a poor blood supply, low transfer coefficient values may be observed despite high intrinsic vessel permeability (Su et al. 1994; Degani et al. 1997). Consideration of a general mixed perfusion- and permeability-limited regime leads to K^{trans} being equal to EFρ(1-Hct) (Tofts et al. 1999), where E is the extraction fraction of the tracer (i.e. the fraction of tracer that is extracted from v_p into ve in a single capillary transit) (CRONE 1963):

$$E = 1 - \exp\left(-\frac{PS}{F(1-\text{Hct})}\right). \tag{8}$$

The relationships described above form the basis of the models used to describe contrast agent kinetics by a number of researchers, and the conventions for the names and symbols used are now generally accepted (TOFTS et al. 1999). In normal tissues, the vascular volume is a small fraction of the total tissue volume (approximately 5%, although it can be considerably higher in some tissues), and it is sometimes assumed (largely as a matter of convenience) that the tracer concentration in the tissue as a whole, C_t, is not influenced to a large degree by the concentration in the vessels (i.e. that $C_t \approx v_e C_e$) (TOFTS and KERMODE 1991). Whilst this assumption is acceptable in abnormalities with no large increase in blood volume that are situated in tissues with a relatively low normal blood volume, it is invalid in many contexts, especially as blood volume can increase markedly in tumours. Models of additional sophistication are required to allow these cases to be described, and a number of investigators [see for example PARKER (1997); TOFTS (1997); TOFTS et al. (1999); DALDRUP et al. (1998); FRITZ-HANSEN et al. (1998); HENDERSON et al. (1998)] have attempted to incorporate the effects of a significant vascular signal. Perhaps the most straightforward approach is to extend Eq. 7 to include the concentration of contrast agent in the blood plasma, giving $C_t = v_p C_p + v_e C_e$. Using this relationship and by rearrangement of Eq. 7 we then have

$$C_t(t) = v_p C_p(t) + K^{\text{trans}} \int_0^t C_p(t') \exp\left(\frac{-K^{\text{trans}}(t-t')}{v_e}\right) dt', \tag{9}$$

which may be re-expressed as

$$C_t(t) = v_p C_p(t) + C_p(t) \otimes H(t)' \tag{10}$$

where H(t) is the impulse response (or residue) function,

$$C_t(t) = v_p C_p(t) + K^{\text{trans}} \int_0^t C_p(t') dt', \tag{11}$$

and $\otimes$ represents the convolution operation.

It has been noted that the special case of analysis only of the first passage of a contrast agent bolus is amenable to a simplified form of kinetic model (TOFTS 1997; LI et al. 2000; VONKEN et al. 2000). It is assumed that during the first passage of a bolus of contrast agent through a tissue (a period that lasts up to approximately 1 min), that the flux of contrast agent from the vascular compartment into the interstitial space, ve, will not be enough to make the concentration of contrast agent in ve significant when compared with the extremely large concentrations seen in the vascular space during this period (that is $C_p(t) >> C_e(t)$). Under these conditions, it can be shown that (PATLAK et al. 1983; LI et al. 2000):

$$C_t(t) = v_p C_p(t) + K^{\text{trans}} \int_0^t C_p(t') dt'. \tag{12}$$

This model variant is therefore attractive in situations where a limited period of acquisition is either necessary or desirable. This may include the study of restless patients or studies in regions that may benefit from the use of a breath-hold (for example lung or liver tumours) (JACKSON et al. 2002). Note that this convenience is achieved at the expense of determining the parameter v_e, and that for situations where the ratio K^{trans}/v_e (Eq. 9) is large the assumptions underlying this approach break down.

6.4.3 Model Variants and the Influence of Acquisition Method

The study of contrast agent kinetics using MRI has seen a range of compartmental modelling approaches suggested [see for example LARSSON et al. (1990); BRIX et al. (1991); TOFTS and KERMODE 1991; LI et al. 2000)]. Most of these have been shown to be theoretically compatible, after some manipulation, with the general compartmental contrast agent kinetic equation given in Eq. 9 (TOFTS 1997; TOFTS et al. 1999). However, there have historically often been significant differences in the assumptions used in the application of each model. The determination of an arterial input function (AIF) to define $C_p(t)$ is an example where differences frequently occur. The early model proposed by TOFTS and KERMODE (1991) used an assumed AIF of biexponential form drawn from literature regarding elimination of Gd-DTPA in the normal population (WEINMANN et al. 1984). The model originally proposed by BRIX et al. (1991) attempted to define an exponentially decaying AIF on a patient-by-patient basis by including this as a free fitting parameter. The approach proposed by Larsson et al. (1990) utilised an AIF measured from blood samples drawn from the brachial artery at intervals of 15 s during the DCE-MRI data acquisition. Often such differences may be traced to the fact that the development of many analysis methods has proceeded in tandem with a specific data acquisition programme, and the modelling assumptions frequently reflect limitations imposed by the data. Care must therefore be taken in applying these methods in settings other than those originally intended and in comparing apparently compatible results from different studies using different models and/or data acquisitions.

Whilst quantification of contrast agent kinetics in absolute physical terms apparently promises inter-study comparability, in practice this is rarely the case, although intra-study and, to a lesser degree, same-model comparisons are more likely to be valid. For example, a transfer coefficient K derived by a particular model will rarely be directly comparable to K derived using another. There is also a wide range of data-imposed limitations that can lead to differences, including temporal resolution (Henderson et al. 1998), T_1 contrast dynamic range (Roberts 1997) and spatial resolution (this can lead to differences in partial volume averaging, which may be a significant issue in very heterogeneous tumours). Therefore, results derived with the same model but under different image acquisition conditions may not be comparable. Perhaps the most significant data-related influences on model output are the ability to quantify T_1 (Tofts et al. 1995; Roberts 1997; Evelhoch 1999) and the ability to measure the arterial input function non-invasively for each patient (Tofts et al. 1995; Parker 1997; Parker et al. 1996). Although these measurements are today more practical than previously, due to sequence and hardware improvements, they are still technically challenging to perform accurately and reproducibly. For this reason many workers make an assumption of a predictable linear relationship between contrast agent concentration and signal intensity change and may assume an input function for all individuals. Often such assumptions are a pragmatic answer to the need for kinetic analysis in a clinical environment when the technical resources to allow a more sophisticated acquisition are lacking. Modelling under these conditions is likely to produce less accurate kinetic parameters than may be obtained under ideal measurement conditions, often with a bias in the results obtained. Such assumptions may be particularly dangerous in a study that is wishing to assess treatment-related changes in kinetic parameters, as a treatment or disease progression that could alter tumour T_1 (for instance by inducing oedema) or affect the AIF (for instance by altering kidney function or heart output rate) could cause apparent changes in kinetic parameters that are in fact due to changes in tissue composition or unrelated physiological state changes. Whilst observations of apparent changes in kinetic parameters in these conditions may well indicate an effect due to intervention they will not be specific to the microvascular parameters that are the aim of the study, leading to severe difficulties in data interpretation.

6.4.4
The Importance of the Arterial Input Function

Some early attempts to model contrast agent kinetics did not use an explicitly measured AIF. Rather a population-averaged AIF was used, which lacked the detail of a true AIF, but provided a convenient solution to a potentially difficult measurement problem. However, if it is possible to measure an AIF there may be benefits for the accuracy of derived kinetic parameters. AIF measurement is feasible given a suitable image acquisition protocol [see for example Fritz-Hansen et al. (1998); Li et al. (2000)], and can be shown to be reproducible – Fig. 6.3 shows AIF measurements repeated in six glioma patients (Parker et al. 2003b). Whilst it is clear that reproducible measurement of the AIF is possible for a given patient, it is also apparent that the exact shape may vary between patients, as it is a function of injection timing and dose, heart output rate, distribution of the contrast agent about the body and kidney function.

Figure 6.4 shows model fits using Eq. 9 to the contrast agent concentration time course observed in a lung tumour (Parker et al. 2003a). It is clear that when a standard AIF, as used by Tofts and Kermode (1991), and originally described in Weinmann et al. (1984), is employed the fitted function fails to describe the time course adequately. In particular the concentration changes immediately after contrast agent arrival in the tissue are poorly modelled when an obvious 'first pass' peak is observed. The net result of these poor fits is that the various parameters in Eq. 9 are inaccurately estimated (Parker et al. 1996; Parker 1997). However, the fit of Eq. 9 to the data is improved when an explicitly measured

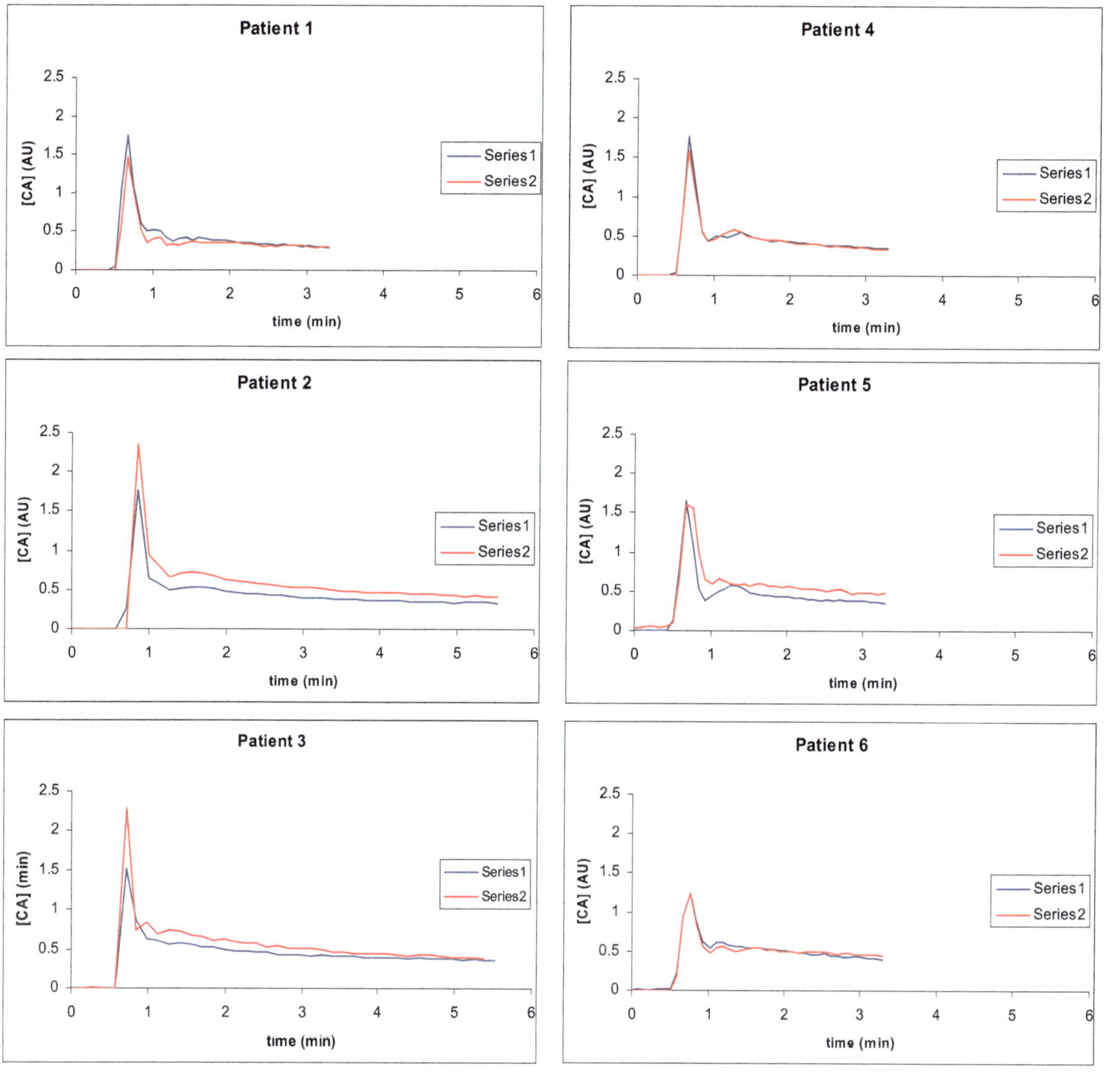

Fig. 6.3. Arterial input functions measured in the middle cerebral artery of six glioma patients. Patients were scanned on average 1.5 days apart (series 1 and series 2). Manual injection used for all cases, with an automated arterial input function definition algorithm (PARKER et al. 2003b)

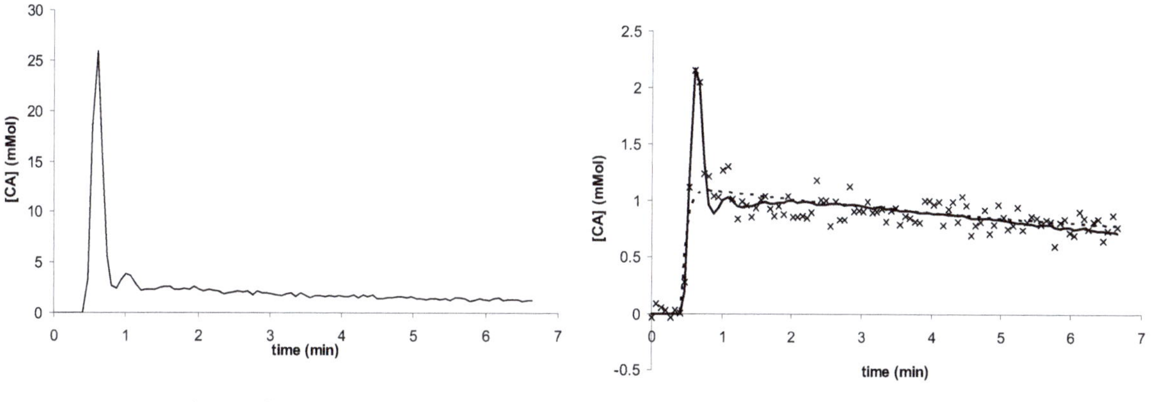

Fig. 6.4. Top, arterial input function measured in the aorta. *Right*, tissue response curve measured in a region of interest placed in a lung tumour (*crosses*) with fits using Eq. 9. *Dashed line*, fit with assumed AIF. *Solid line*, fit with AIF as measured in the *left* graph

AIF is utilised, and the derived parameter estimates are more likely to be more accurate.

6.5 Modelling Blood Flow

Equation 9 provides a straightforward model to allow the estimation of the transfer constant, K^{trans}, the vascular plasma volume, v_p, and the interstitial space, v_e, from a series of concentration values obtained over time. However, as was noted earlier, these approaches rely on the assumption that contrast agent is instantaneously and well-mixed in each of the compartments it occupies (this is, in fact, inherent to the definition of a compartmental model). Additionally, there is no term allowing independent determination of blood plasma flow rates (as opposed to blood plasma volume).

Two possible approaches exist to determine tumour blood flow. Methods based on the indicator dilution theory have been applied in tumours using first pass T_2*-weighted data acquisitions [see Chap. 4 for a detailed description of these dynamic susceptibility contrast (DSC-MRI) approaches]. The same modelling approaches may be applied to T1-weighted time series, if the data acquisition is rapid enough, although to date this possibility has only been exploited in normal tissues (Hatabu et al. 1999; Ohno et al. 2004). The indicator dilution theory in its standard form assumes that contrast agent does not leave the blood pool during its first passage through tissue. This assumption is likely to be an oversimplification in many tumours when using small molecular weight contrast agents (see Chaps. 2 and 3), leading to possible errors in blood flow estimates, and the missed opportunity to estimate contrast agent leakage characteristics. However, in situations where blood volume is likely to be low, and where extravasation is low, DSC-MRI methods are likely to provide more sensitive indications of blood volume than T_1-weighted methods.

Recent work (St. Lawrence and Lee 1998; Henderson et al. 2000; Koh et al. 2001; Buckley 2002), using well-established concepts (Kety 1951; Renkin 1959; Crone 1963; Zierler 1963; Johnson and Wilson 1966), has attempted to extend the modelling processes described in Sect. 6.4 to account additionally for blood flow, thus providing a comprehensive assessment of bulk microvascular characteristics. These workers have shown that the tissue homogeneity model introduced in (Johnson and Wilson 1966) may be adapted to the dynamic contrast-enhanced MRI experiment via an adiabatic approximation (St. Lawrence and Lee 1998; Henderson et al. 2000). The tissue homogeneity model divides tissue into a vascular plasma volume, v_p, and the EES, v_e, in the same way as the previously discussed models, but differs in that it defines the tracer concentration within v_p as a function of both time and distance along the length of the capillary, while v_e is assumed still to be a well-mixed compartment. The adiabatic approximation to the tissue homogeneity model (AATH) invokes the additional assumption that the rate of change in Ce is low relative to that of C_p, an approximation that makes application of the model practically possible.

The AATH model requires the following residue functions, separating the time course of contrast agent arrival into a 'vascular' phase ($t < T_c$, where T_c is the transit time through a capillary), and a 'parenchymal tissue' phase ($t > T_c$) (St. Lawrence and Lee 1998; Koh et al. 2001):

$$\begin{aligned} H(t) &= F\rho(1-\text{Hct}) && (0 \le t < T_c \\ H(t) &= EF\rho(1-\text{Hct})\exp\left(-\frac{EF\rho(1-\text{Hct})}{v_e}(t-T_c)\right) && (t \ge T_c)\ . \end{aligned} \tag{13}$$

As may be seen immediately, H(t) at t ≥ Tc in this model is equivalent to that defined previously for the compartmental model of diffusive transport (Eq. 11) under the general mixed perfusion- and permeability-limited regime (when K^{trans} = EFϱ(1-Hct)). The presence of the separate residue function for $0 \le t < T_c$ is what allows the estimation of blood plasma flow as an independent variable when using this model. Note that for consistency with Eqs. 1–12, the form of Eq. 13 is slightly different to that in (St. Lawrence and Lee 1998). The tissue homogeneity approach models the tissue contrast agent concentration as:

$$C_t(t) = F\rho(1-\text{Hct})\int_0^{\tau} C_p(t-t')dt' + EF\rho(1-\text{Hct})\int_{\tau}^{t} C_p(t')\exp\left(\frac{-EF\rho(1-\text{Hct})(t-\tau-t')}{v_e}\right)dt' \tag{14}$$

where τ is the mean capillary transit time within a voxel (typically a few seconds). τ may be expressed as v_p/(Fϱ(1-Hct)), thus reducing Eq. 12 to Eq. 9 if τ is small and by substituting K^{trans} for EFϱ(1-Hct). It is therefore clear that the AATH model is compatible with the compartmental approaches to examining the kinetics of contrast agent accumulation in tissues, whilst providing the possibility for extracting

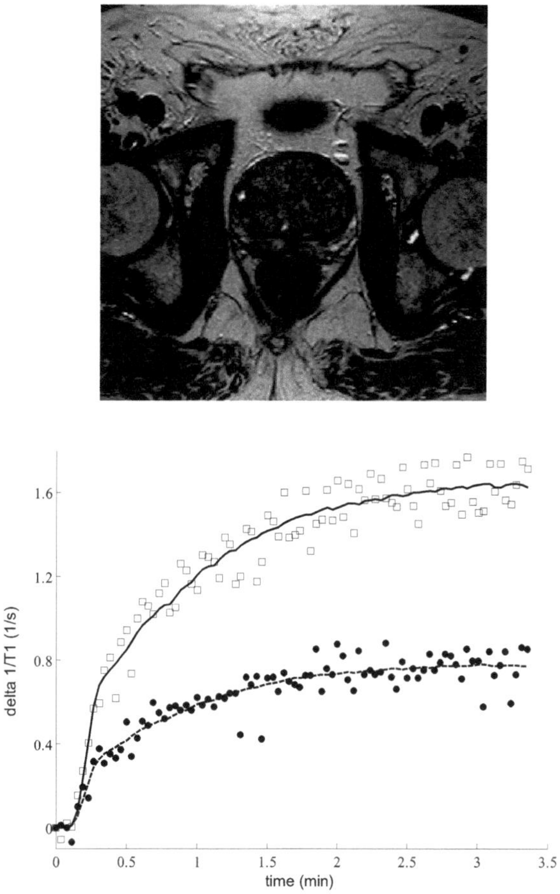

Fig. 6.5. *Top*, axial T_2-weighted turbo spin echo image showing the prostate with tumour in the posterior area of the peripheral zone. *Bottom*, graph showing enhancement over time in tumour (*open squares*) and normal prostate (*filled circles*) regions. The lines represent the fit of the AATH model (St. Lawrence and Lee 1998) to the data. Parameter estimates were obtained for tumour: *F*, 59.3 ml/100 ml/min; V_b, 1.0 ml/100 ml; *PS*, 19.0 ml/100 ml/min; V_e, 40.0 ml/100 ml; normal: F, 23.39 ml/100 ml/min; V_b, 0.4 ml/100 ml/min; *PS*, 10.3 ml/100 ml/min; V_e, 18.4 ml/100 ml (a tissue density ϱ of 1 g/ml is assumed)

additional information. Figure 6.5 gives examples of fits of this model to time courses extracted from ROIs placed in normal prostate and prostate tumour. Recent observations in a group of 22 men show that it is possible to differentiate tumour from normal peripheral zone prostate based on the derived values of blood flow and EES, whilst blood volume and capillary permeability surface area product (*PS*) do not appear to be different (Buckley et al. 2003; Buckley et al. in press). It is clear that the use of a sophisticated modelling approach such as this is viable in a practical setting and that it can provide specific functional information, allowing a rich characterisation of tumour microvasculature.

6.6 Sources of Error

As mentioned in Sect. 6.4 of this chapter, major sources of error may include the common assumption of a straightforward signal change - contrast agent concentration relationship and of a standard arterial input function. Other factors that can significantly affect both the accuracy and the precision of kinetic modelling results include the blood haematocrit, contrast agent relaxivity, errors in T_1 measurement, image artefact, in particular those caused by motion, model over-simplification and spatial undersampling of data. Leaving aside errors introduced during to data acquisition, the major remaining modelling-related sources of error are therefore the haematocrit, model over-simplification, and spatial undersampling.

The blood haematocrit (Hct) plays a significant role in the measurement of kinetic parameters, as MRI contrast agents occupy only the plasma compartment of the blood. It is therefore the concentration of contrast agent in the blood plasma (C_p), rather than the whole blood, that drives diffusive transport across endothelial walls (Eq. 7). Any arterial input function measurement therefore needs to convert the measured blood concentration C_b to a measure of C_p via the relationship $C_p = C_b/(1\text{-Hct})$. Most DCE-MRI studies have not included an explicit measurement of Hct, and a value of approximately 0.4 is generally assumed, or Hct is ignored entirely. This lack of Hct measurement will undoubtedly lead to errors, as Hct may vary in patients with advanced cancer. An additional frequently-overlooked consideration is the likely difference in Hct between large vessels (where the AIF measurement is performed) and the microvasculature (where the modelling of contrast agent transfer is performed). The packing of red blood cells in the capillary bed is less dense, leading to a smaller value of Hct. In normal capillary beds Hct is approximately 0.7 of that in large vessels; however, in the poorly-formed and variably perfused vessels within a tumour there is scope for this factor to vary considerably.

As discussed in Sect. 6.4.2, Hct is factored into the definition of a number of key modelling parameters. K^{trans} is equal to EFϱ(1-Hct) (Tofts et al. 1999), and it also occurs in a number of other modelling steps (see for example Eqs. 8, 13, 14). Errors in Hct will consequently also affect the modelling process at this stage.

All modelling of a complex biological system requires significant simplification of reality to generate an analysis method that is robust but that provides useful summary functional parameters. Modelling methods therefore provide an approximation of true tissue status, and the simpler the model the greater the degree of approximation, which can lead to systematic errors in the physiological parameters being assessed and a weakened relationship between the true physiological parameter and its modelling analogue. This is due to the simplified model 'compensating' for parameters that have been omitted by distorting the fitted parameters to include effects that are caused by an omitted factor. Examples of this are the lack of a blood pool component and the lack of a measured AIF in modelling analysis, which has been shown to lead to artefactually high estimated K^{trans} values and an incorrect assignment of high K^{trans} in blood vessels (Fig. 6.6) (Parker 1997; Parker et al. 1996) . These findings are in agreement with simulation studies that also suggest that more complex models also reduce bias in parameter estimates of v_e and v_p (Buckley 2002). However, typically the loss of accuracy caused by the use of simpler models is accompanied by an increase in precision as fitting processes become more stable (as fewer fitted parameters are being extracted from the data) and noise levels are reduced (as acquisition times often become longer). It is therefore not to be presupposed that precisely measured, but difficult-to-interpret parameters are of more or less use in understanding, assessing, and following disease process than the more comprehensive (but noisier) complex model parameters.

DCE-MRI analysis should ideally take into account the heterogeneity of tumour vascular characteristics. As may be appreciated in Figs. 6.1 and 6.6, tumours in all regions of the body may exhibit large internal variation in their microvascular characteristics, reflecting variation in microvessel density, VEGF expression, and areas of avascularity or necrosis. Many DCE-MRI studies utilize user-defined regions of interest (ROIs), yielding graphical outputs with good signal-to-noise ratio (Figs. 6.4 and 6.5), but which lack spatial resolution and are prone to partial volume averaging errors. Additionally, the placement of ROIs can have profound effects on the outcome of analysis (Liney et al. 1999), and tell you nothing of the heterogeneity of tumour characteristics, which may in itself be a useful diagnostic/prognostic feature. Another approach to the analysis problem is to utilise parametric mapping (Figs. 6.1 and 6.6). This type of display has a number of advantages including the appreciation of heterogeneity of enhancement and the removal of the need for selective placement of user-defined regions of interest. An important advantage is being able to spatially match tumour vascular characteristics such as blood volume, blood flow, K^{trans}, and v_e. The risk of missing important diagnostic information and of creating ROIs that contain more than one tissue type is reduced. However, voxel mapping

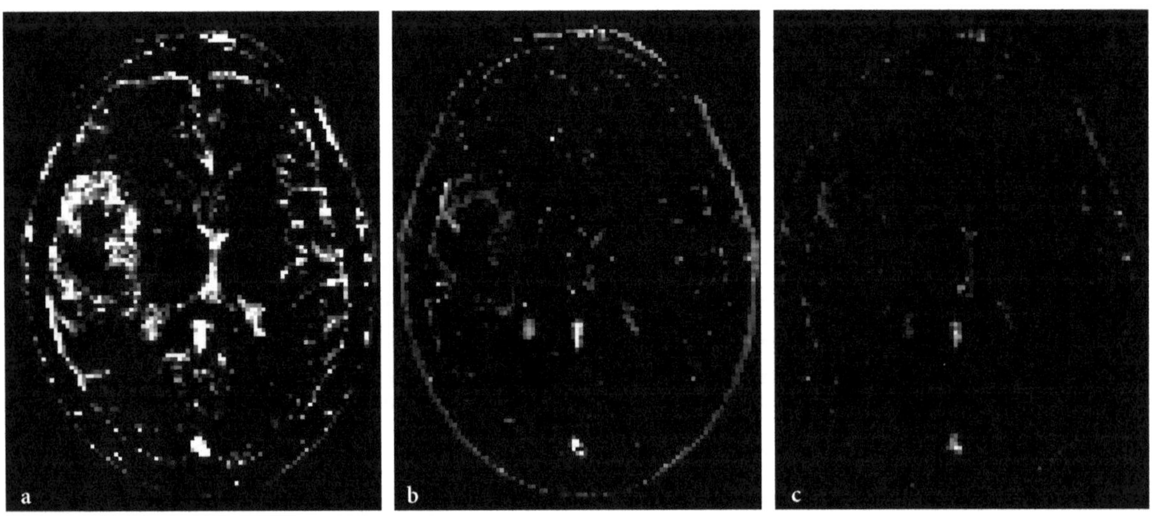

Fig. 6.6. a,b K^{trans} maps in an axial brain slice incorporating a high-grade glioma. **a** Result of fitting Eq. 9 without a blood pool contribution and with an assumed AIF. **b** Result of fitting Eq. 9 with a blood pool contribution and an AIF measured in the middle cerebral artery. Note the assignment of high K^{trans} in sulcal vessels when using the simpler model (**a**), which should show theoretically zero K^{trans}. The more complex model (**b**) shows much lower values in these vessels. **c** v_p Map generated during the more complex model fitting for (**b**). Note that many areas of apparently high vessel K^{trans} in (**a**) are now correctly identified as high blood volume areas

has a poorer signal-to-noise ratio than ROI analysis and it can be difficult to compare model outputs between two tumours or during a programme of treatment in a meaningful way (see Chap. 16). Recently, histogram analysis has been used to quantify the heterogeneity of tumours for comparative and longitudinal studies, for monitoring the effects of treatment, and to show the regression or development of angiogenic hot spots (MAYR et al. 2000; HAYES et al. 2002). Simple frequency distributions can be plotted and descriptive statistics can be used to quantify the variability therein.

6.7 Summary

A set of well-established modelling procedures is available for T_1-weighted DCE-MRI data that can provide useful estimates of microvascular functional parameters. The simpler approaches have now been available for over 10 years, and their limitations are well recognised. Notwithstanding these limitations, they have been shown to be of use in the characterisation of tumours and in the monitoring of treatment. More complex modelling methods are now emerging, which, when coupled with the best current data acquisition methods, can provide a still wider range of physiological information. These developments will allow more specific and more informative studies of tumour microvascular function.

Acknowledgements

We are grateful to Judith Harrer for the provision of the images used for Fig. 6.6.

References

Brix G, Semmler W, Port R, Schad LR, Layer G, Lorenz WJ (1991) Pharmacokinetic parameters in CNS Gd-DTPA enhanced MR imaging. J Comput Assist Tomogr 15:621–628

Buckley DL (2002) Uncertainty in the analysis of tracer kinetics using dynamic contrast-enhanced T1-weighted MRI. Magnetic Resonance in Medicine 47(3):601–606

Buckley DL, Roberts C, Parker GJ, Logue JP, Hutchinson CH (2003) In vivo determination of the microvascular characteristics of prostate cancer using dynamic contrast-enhanced MRI. 11th Meeting of the International Society for Magnetic Resonance in Medicine:461

Crone C (1963) The permeability of capillaries in various organs as determined by use of the 'indicator diffusion' method. Acta Physiol Scand 58:292–305

Daldrup HE, Shames DM, Husseini W, Wendland MF, Okuhata Y, Brasch RC (1998) Quantification of the extraction fraction for gadopentetate across breast cancer capillaries. Magn Reson Med 40:537–543

Degani H, Gusis V, Weinstein D, Fields S, Strano S (1997) Mapping pathophysiological features of breast tumors by MRI at high spatial resolution. Nat Med 3:780–782

Evelhoch JL (1999) Key factors in the acquisition of contrast kinetic data for oncology. J Magn Reson Imaging 10:254–259

Evelhoch JL, LoRusso P, DelProposto Z, Stark K, Latif Z, Morton P, Waterton J, Wheeler C, Barge A (2002) Dynamic contrast-enhanced MRI evaluation of the effects of ZD6126 on tumour vasculature in a phase I clinical trial. Proceedings of the Annual Meeting of the ISMRM:2095

Folkman J (1990) What is the evidence that tumors are angiogenesis dependent?. J Natl Cancer Inst 82:4–6

Folkman J (1995) Angiogenesis in cancer, vascular, rheumatoid and other disease. Nat Med 1:27–31

Fritz-Hansen T, Rostrup E, Søndergaard L, Ring PB, Amtorp O, Larsson HB (1998) Capillary transfer constant of Gd-DTPA in the myocardium at rest and during vasodilation assessed by MRI. Magn Reson Med 40:922–929

Hatabu H, Tadamura E, Levin DL, Chen Q, Li W, Kim D, Prasad PV, Edelman RR (1999) Quantitative assessment of pulmonary perfusion with dynamic contrast-enhanced MRI. Magn Reson Med 42:1033–1038

Hayes C, Padhani AR, Leach MO (2002) Assessing changes in tumour vascular function using dynamic contrast-enhanced magnetic resonance imaging. NMR Biomed 15:154–163

Henderson E, Rutt BK, Lee TY (1998) Temporal sampling requirements for the tracer kinetics modeling of breast disease. Magn Reson Imaging 16:1057–1073

Henderson E, Sykes J, Drost D, Weinmann HJ, Rutt BK, Lee TY (2000) Simultaneous MRI measurement of blood flow, blood volume, and capillary permeability in mammary tumors using two different contrast agents. J Magn Reson Imaging 12:991–1003

Jackson A, Haroon H, Zhu XP, Li KL, Thacker NA, Jayson G (2002) Breath-hold perfusion and permeability mapping of hepatic malignancies using magnetic resonance imaging and a first-pass leakage profile model. NMR Biomed 15:164–173

Johnson JA, Wilson TA (1966) A model for capillary exchange. Am J Physiol 210:1299–1303

Kety SS (1951) The theory and applications of the exchange of inert gas at the lungs and tissues. Pharmacol Rev 3:1–41

Koh TS, Zeman V, Darko J, Lee TY, Milosevic MF, Haider M, Warde P, Yeung IW (2001) The inclusion of capillary distribution in the adiabatic tissue homogeneity model of blood flow. Phys Med Biol 46:1519–1538

Larsson HBW, Stubgaard M, Fredricksen JL, Jensen M, Henriksen O, Paulson OB (1990) Quantitation of blood-brain barrier defect by magnetic resonance imaging and gadolinium-DTPA in patients with multiple sclerosis and brain lesions. Magnetic Resonance in Medicine 16:117–131

Li KL, Zhu XP, Waterton J, Jackson A (2000) Improved 3D quantitative mapping of blood volume and endothelial permeability in brain tumors. J Magn Reson Imaging 12:347–357

Liney GP, Gibbs P, Hayes C, Leach MO, Turnbull LW (1999) Dynamic contrast-enhanced MRI in the differentiation of breast tumors: user-defined versus semi-automated region-of-interest analysis. J Magn Reson Imaging 10(6):945–949

Mattiello, J and Evelhoch JL (1991) Relative volume-average murine tumor blood flow measurement via deuterium nuclear magnetic resonance spectroscopy. Magn Reson Med 18(2):320–334

Mayr NA, Yuh WT, Arnholt JC, Ehrhardt JC, Sorosky JI, Magnotta VA, Berbaum KS, Zhen W, Paulino AC, Oberley LW, Sood AK, Buatti JM (2000) Pixel analysis of MR perfusion imaging in predicting radiation therapy outcome in cervical cancer. J Magn Reson Imaging 12:1027–1033

Mitchell DG (1997) MR imaging contrast agents--what's in a name?. J Magn Reson Imaging 7:1–4

Ohno Y, Hatabu H, Higashimo T, Takenaka D, Watanabe H, Nishimura Y, Yoshimura M, Sugimura K (2004) Dynamic perfusion MRI versus perfusion scintigraphy: prediction of postoperative lung function in patients with lung cancer. American Journal of Roentgenology 182:73–78

Parker GJM (1997) Monitoring contrast agent kinetics using dynamic MRI: Quantitative and qualitative analysis. Institute of Cancer Research

Parker GJM, Tanner SF, Leach MO (1996) Pitfalls in the measurement of tissue permeability over short time-scales using multi-compartment models with a low temporal resolution blood input function. 4th Meeting of the International Society for Magnetic Resonance in Medicine:1582

Parker GJ, Clark D, Watson Y, Buckley DL, Berrisford C, Anderson H, Jackson A, Waterton JC (2003) T1-weighted DCE-MRI applied to lung tumours: Pre-processing and modelling. 11th Meeting of the International Society for Magnetic Resonance in Medicine:1255

Parker GJ, Jackson A, Waterton JC, Buckley DL (2003) Automated arterial input function extraction for T1-weighted DCE-MRI. 11th Meeting of the International Society for Magnetic Resonance in Medicine:1264

Patlak CS, Blasberg RG, Fenstermacher JD (1983) Graphical evaluation of blood-to-brain transfer constants from multiple-time uptake data. J Cereb Blood Flow Metab 3:1–7

Renkin EM (1959) Transport of potassium-42 from blood to tissue in isolated mammalian skeletal muscles. Am J Physiol 197:1205–1210

Roberts TP (1997) Physiologic measurements by contrast-enhanced MR imaging: expectations and limitations. J Magn Reson Imaging 7:82–90

Sha'afi RI (1981) Permeability for water and other polar molecules. Membrane Transport:29–60

St Lawrence KS, Lee T-Y (1998) An adiabatic approximation to the tissue homogeneity model for water exchange in the brain: 1. Theoretical derivation. J Cereb Blood Flow Metab 18:1365–1377

Su MY, Jao JC, Nalcioglu O (1994) Measurement of vascular volume fraction and blood-tissue permeability constants with a pharmacokinetic model: studies in rat muscle tumors with dynamic Gd-DTPA enhanced MRI. Magn Reson Med 32:714–724

Tofts PS (1997) Modeling tracer kinetics in dynamic Gd-DTPA MR imaging. J Magn Reson Imaging 7:91–101

Tofts PS, Kermode AG (1991) Measurement of the blood-brain barrier permeability and leakage space using dynamic MR imaging. 1. Fundamental concepts. Magn Reson Med 17:357–367

Tofts PS, Berkowitz B, Schnall MD (1995) Quantitative analysis of dynamic Gd-DTPA enhancement in breast tumors using a permeability model. Magn Reson Med 33:564–568

Tofts PS, Brix G, Buckley DL, Evelhoch JL, Henderson E, Knopp MV, Larsson HBW, Lee T-Y, Mayr NA, Parker GJM, Port RE, Taylor J, Weisskoff RM (1999) Estimating kinetic parameters from dynamic contrast-enhanced T(1)-weighted MRI of a diffusable tracer: standardized quantities and symbols. J Magn Reson Imaging 10(3):223–232

Vonken EP, van Osch MJ, Bakker CJ, Viergever MA (2000) Simultaneous quantitative cerebral perfusion and Gd-DTPA extravasation measurement with dual-echo dynamic susceptibility contrast MRI. Magn Reson Med 43:820–827

Weinmann H-J, Laniado M, Mützel W (1984) Pharmacokinetics of GdDTPA/dimeglumine after intravenous injection into healthy volunteers. Physiol Chem Phys Med NMR 16:167–172

Zierler KL (1963) Theory of use of indicators to measure blood flow and extracellular volume and calculation of transcapillary movement of tracers. Circulation Research 7:464–471

Imaging Techniques

7 Imaging Techniques for Dynamic Susceptibility Contrast-Enhanced MRI

Michael Pedersen, Peter van Gelderen, and Chrit T. W. Moonen

CONTENTS

7.1 Introduction

Dynamic susceptibility contrast-enhanced MRI (DSC-MRI) is based on the detection of the first passage of a bolus of non-diffusible paramagnetic contrast media (Axel 1980). The DSC-MRI approach offers a window on tissue perfusion (Villringer et al. 1988) and is currently widely used in the clinic in the case of suspected stroke. Specific pulse sequences have proven to be sensitive to tissue perfusion and blood volume, permitting the functional aspects of tumor neovascularity to be assessed in vivo in a non-invasive and repeatable way.

The contrast in such MR images is largely based on local magnetic field inhomogeneities generated in extravascular space by the passage of the contrast material through the vasculature. A transient decrease of gradient echo (GE) signal due to the resulting decrease in T2* is obtained which can be converted into curves of contrast agent concentration versus time for each voxel (Warach et al. 1996; Sorensen et al. 1996; Rosen et al. 1990; Edelman et al. 1990; Guckel et al. 1996; Moseley et al. 1991). Integration of such curves lead to regional cerebral blood volume (rCBV) and, upon deconvolution with the arterial input curve, to blood mean transit time (MTT), and regional cerebral blood flow (rCBF) (Boxerman et al. 1995; Fisel et al. 1991; Perman et al. 1992; Weisskoff et al. 1993). DSC-MRI can also be performed by spin echo (SE) T2-weighted pulse sequences, although these are not frequently used in daily clinical practice. The behavior of T2 and T2* DSC-MRI is different when a paramagnetic contrast agent passes through the cerebral vascular system, but both are, to some extent, capable of measuring important hemodynamic parameters. However, direct measurements of rCBF, rCBV, and MTT alone provide limited insight into the underlying physiology of the brain tumor. Various physiological pharmacokinetic models have been proposed to describe dynamic contrast agent enhancement, often described in physiological terms such as vascular permeability and surface area, and extracellular volume. These models are often attributed to situations when the blood–brain barrier is compromised, resulting in extravasation of the contrast agent through the leaky capillaries. By modeling data obtained with T1-weighted sequences, this extravasation of contrast agent is used as a measure of the extravascular uptake.

In this chapter, we discuss different imaging techniques for T2 and T2* dynamic contrast-enhanced MRI (DCE-MRI), in particular as it is related to spa-

M. Pedersen, PhD
Imagerie Moléculaire et Fonctionnelle: de la Physiologie à la Thérapie, ERT CNRS/ Universite Victor Segalen Bordeaux 2, 146 Rue Leo Saignat, Case 117, 33076 Bordeaux Cedex, France *and* MR Research Center, Institute of Experimental Clinical Research, Aarhus University Hospital, Skejby, 8200 Aarhus N, Denmark
P. van Gelderen, PhD
Laboratory for Molecular and Functional Imaging, NINDS, NIH, Bethesda, MD 20892, USA
C. T. W. Moonen, PhD
Imagerie Moléculaire et Fonctionnelle: de la Physiologie à la Thérapie, ERT CNRS/ Universite Victor Segalen Bordeaux 2, 146 Rue Leo Saignat, Case 117, 33076 Bordeaux, France

tial and temporal resolution, and the advantages and limitations with regard to feasibility of perfusion related parameters. In addition, we describe the fundamental mechanisms of T1 DCE-MRI in order to present available techniques based on simultaneous detection of changes in magnetic susceptibility and relaxation rate. We review current existing techniques and discuss potential improvements and recently developed acquisition strategies. Special attention is devoted to the DSC-MRI mechanisms in the brain, since most clinically relevant DSC-MRI studies are done with the purpose of obtaining insight in cerebral perfusion.

7.2 DSC-MRI with Gadolinium-Based Contrast Agents in the Brain

For the time being, five gadolinium-based contrast agents are approved for clinical use: gadopentetate dimeglumine (Magnevist, Berlex Lab, Berlin, Germany) gadoteridol (ProHance, Bracco s.p.A, Milan, Italy), gadodiamide (Omniscan, Nycomed Amersham Health, Oslo, Norway), gadoversetamide (OptiMark, Mallinckrodt, St Louis, Missouri) and gadolinium-DOTA (Dotarem, Guerbet, Paris, France). The first three agents all share similar MR properties with regard to the subject of this paper, and since these three are by far the most used in daily clinical practice, only mechanisms governed by these agents are discussed here.

The brain provides a unique environment for contrast agent use. By virtue of the blood–brain barrier, small solutes such as gadolinium-based contrast agents are retained intravascularly, in contrast to the remainder of the body where such small agents are known as extracellular fluid markers and rapidly equilibrate between intravascular and extravascular compartments. Thus, the physiological environment of the brain compared with the rest of the body gives these contrast agents markedly different behavior and offers considerably different applications, analysis opportunities, and physiological properties (Roberts et al. 2000).

Due to its high sensitivity for brain water, MRI is generally sensitive for detecting brain abnormalities. For example, in cases of cerebral infarction, brain tumors or infections may easily be recognized as a breakdown of the blood–brain barrier.

7.3 T2*-Weighted Perfusion

The passage of gadolinium induces differences in local magnetic susceptibility between vessels and the surrounding tissue. Although the fraction of the vascular space is in the range of 4%–5% of the total tissue blood volume, this compartmentalization of contrast agent leads to a disruption of the local magnetic field homogeneity extending beyond the immediate confines of the vascular compartment, within which the contrast agent is retained. Thus, both intra- and extravascular spins undergo a reduction in T2* that leads to a significant transient signal loss, even in normal white matter with a standard dose of 0.1 kg/mmol body weight. In the blood, the signal dependency of the contrast agent concentration is given by the signal equation of the spoiled GE sequence,

$$S_{blood}(c) \propto \frac{1-E_1(c)}{1-\cos\alpha \cdot E_1(c)} E_2(c) \cdot \exp[-TE \cdot c/k] \tag{1}$$

with $E_1(c)$=exp[-TR/T1(c)] and $E_2(c)$=exp[-TE/T2*(c)], where TE is the echo time, TR is the repetition time, c denotes the contrast agent concentration and á the flip angle. The constant k depends on the characteristics of the contrast agent (through the specific relaxivity constant), magnet strength and the pulse sequence parameters. The first two factors in this equation describe the change in signal intensity due to the shortening of the relaxation times T1 and T2, whereas the third factor reflects the long-range susceptibility effect caused by the contrast agent. Because of the intravascularity of the contrast agent, the signal intensity in the tissue is only dependent on the susceptibility effect (Edelman et al. 1990)

$$S_{tissue}(c) \propto \exp[-TE \cdot c/k] \tag{2}$$

It has been shown, both theoretically and experimentally, that the apparent transverse relaxation rate, R2*, in a volume of interest (VOI) depends on the concentration of the contrast agent within the VOI. Consequently, changes in the transverse relaxation rate relative to a pre-contrast period can be determined from the GE signal–time curve using the relationship,

$$c_{tissue}(t) = k\Delta R2^*(t) = -\frac{k}{TE} \ln \frac{S(t)}{S0} \tag{3}$$

where S(t) is the signal intensity in the VOI at time t, and S0 is the mean signal intensity in the VOI for the images acquired before the contrast agent arrives in the VOI. During the first pass bolus of contrast agent, the signal intensity will change according to the microvascular concentration. By use of the dynamic GE approach, absolute quantitative values of rCBV and rCBF have been derived from the indicator dilution theory by assuming that the constant k in Eq. 1 is equal to the constant k in Eq. 2 (Rempp et al. 1994). Later studies, however, revealed that as vascular structures in brain tissue have a complex network, a signal decrease during a bolus passage of contrast agent in capillary influences signal intensity in other capillaries, and the constants in Eqs. 1 and 2 must differ (Koshimoto et al. 1999).

Equation 1 is, however, only valid in the stationary limit in which inflow effects can be neglected. In larger vessels with a high blood flow perpendicular to the image plane, it is assumed that saturated spins in blood vessels are completely replaced by fully relaxed spins, and the T1-dependent factor in Eq. 1 can be approximated by unity (Aumann et al. 2003). As seen from Eq. 1, inappropriate values of TE, TR, and flip angle may lead to unwanted T1-weighting or an insufficient signal-to-noise ratio. Changing the acquisition procedure from a standard GE pulse sequence to a faster scheme such as echo planar imaging (EPI) will also change the weighting due to different TE, TR, and flip angle. Long TE amplifies the signal intensity change, but substantial lengthening of the TE worsens diamagnetic susceptibility artifacts. The use of longer TE in the echo planar technique involves larger signal changes than the standard spoiled GE method. More important, calculated hemodynamic parameters have been shown to differ between these two methods. For example, the rCBV in the deep gray versus white matter structures based on the EPI method was 28% larger as compared with the standard spoiled GE sequence (Speck et al. 1999), which has been suggested to be the result of the different parameters used.

The effect of T1 enhancement is rarely negligible. The subsequent extravasation of contrast agent into the tissue will not only cause a biasing T1-based signal enhancement, but will also give rise to extra T2* shortening in the tissue (Vonken et al. 2000). Thus, a small pre-dose of contrast agent can be given prior to the DSC-MRI experiment to reduce systematic effects from changes in tissue T1 (Simonsen et al. 2000). Such procedures are essential in DSC-MRI of tumors, particularly in malignant types (Aronen et al. 1994). There is another way in which tissue T1 may be affected, although minor in the normal brain. Water exchange between the intravascular and the extravascular space induces a change in the extravascular T1 (Donahue et al. 1997). The residence time of water in capillaries has been considered to be in the order of 500 ms, meaning that water exchange is slow with respect to MR measurement times, resulting in a limited shortening of tissue T1 (Eichling et al. 1974).

The sensitivity to T1-based signal enhancement can be accounted for by the use of a dual-echo acquisition GE technique. This sequence permits the interleaved acquisition of T2*-weighted images from two different parallel sections with different echo times (Perman et al. 1992). Nevertheless, the magnitude of the T2* shortening in the tissue due to the contrast agent extravasation is unpredictable beforehand in the dual-echo acquisition method. More sophisticated techniques have, nevertheless, been invented, dealing with both T1 enhancement and T2* shortening (see Sect. 7.8).

7.4 T2-Weighted Perfusion

SE-based DSC-MRI exhibits some distinct physical and physiological differences as compared with the GE method. Simulation studies as well as in vivo experiments have demonstrated that SE-based measurements are less sensitive to microvasculature than GE-based techniques. In computer simulations, the sensitivity of a GE-based method is relatively independent of the size and distribution of vessels and capillaries within a given voxel, whereas SE-based methods show a maximum sensitivity for vessels between 5 and 10 ìm; the size of capillaries in the human brain (Boxerman et al. 1995). Although the selective sensitivity to microvascularization of SE methods is advantageous, DSC-MRI is often performed by the GE method due to better coverage of the brain (given the same repetition time, it is possible to obtain more slices during a single acquisition). Furthermore, due to the susceptibility contrast sensitivity of GE sequences for all dimensions, the contrast dose used in the GE experiment is typically only half of that used for SE (Simonsen et al. 2000)

On the other hand, since the SE sequence has the inherent advantage of being less sensitive to differences in magnetic susceptibility at air–tissue interfaces, the method offers better anatomic precision and detail in clinical studies. Parametric analysis using

SE-based EPI techniques are predominantly based on the susceptibility effects rather than changes in the T2 relaxation rate. This is in fact a consequence of the relatively long sampling period of the EPI acquisition technique. The susceptibility contrast arising from compartmentalization of the contrast agent is therefore used to determine relative tissue and relative arterial concentration levels according to Eq. 3, allowing subsequent calculations of rCBV, rCBF, and MTT in the same manner as described in Chap. 4.2.1 (Fig. 7.1).

Importantly, several studies have indicated that the SE- and GE- (both EPI) based methods are comparable with regard to the relevant parameter for in vivo studies (SPECK et al. 2000; WEISSKOFF et al. 1994). In fact, the SE method shows smaller standard deviations than the GE method for small regions that are adjacent to or contain large vessels (SPECK et al. 2000). However, great care must be taken when comparing parametric information based on SE DSC-MRI with those obtained by GE DSC-MRI in tumors. It has for example been shown that the tumor rCBV obtained with the GE method was significantly higher than the SE method in the high-grade gliomas, but not in the low-grade gliomas. Simulations of the microscopic susceptibility variation in SE and GE imaging have subsequently demonstrated that the SE method produces a relaxivity peak for capillary or smaller sized vessels, and that SE-based DSC-MRI demonstrated a capillary blood volume that is more physiologically interesting than total blood volume obtained from conventional GE-based DSC-MRI (WEISSKOFF et al. 1994).

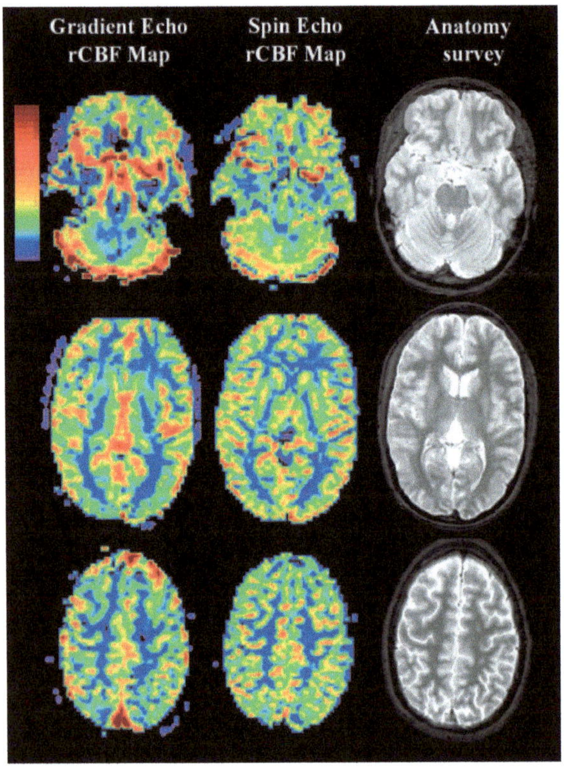

Fig. 7.1. rCBF maps calculated from the gradient-echo, the spin-echo, and the corresponding anatomic inversion recovery images. It was found that the appearance of large vessels was markedly reduced in the spin-echo images, resulting primarily in a representation of capillary perfusion. [Reprinted with permission from SPECK et al. (2000)]

Another method to integrate the SE technique into DSC-MRI of the brain is the half-Fourier acquisition single-shot SE (HASTE), enabling heavy T2-weighting during its imaging interval of approximately 1–2 s (MIYAZAKI et al. 1996). The HASTE method has successfully demonstrated its availability to measure cerebral microvascular hemodynamics (KOSHIMOTO et al. 1999). In addition, because of HASTE's low susceptibility to artifact, image degradation was effectively eliminated, but the major drawback of the HASTE method is its limitation to obtain more than one or a few slices as compared with the whole brain coverage of the EPI methods.

In general, SE-based DSC-MRI is promising with respect to the detection of the cerebral intravascular effects. Unfortunately, SE methods are only sensitive for T2* changes in and around capillaries, and not around larger size venules and arteries (BOXERMAN et al. 1995). As a consequence, signal changes during passage of the bolus are lower, and SE techniques are therefore hardly used in the clinic for bolus tracking. Nevertheless, this may change with the use of very high field magnet strengths.

7.5 Temporal and Spatial Resolution: Current Limitations

In order to determine the arterial input curve, a temporal resolution of 1–2 s is required, although this constraint may differ, depending on the contrast agent dosage, volume, and injection rate. Such constraints make the EPI technique the first choice. However, the long acquisition window in EPI, as compared to T2* during the bolus passage, leads to significant signal decay during the echo train and may therefore pose problems for DSC-MRI. For example, in order to have optimal sensitivity to the T2* changes induced by the contrast agent, all image data should be acquired at a relatively long echo time, and for optimal contrast in

GE MRI, the TE should be equal to T2*. Due to the range of echo times at which single-shot EPI acquires data, the different parts of the echo train cannot all be optimally sensitive to the bolus effect. Secondly, the actual spatial resolution of the EPI images depends, among other parameters, on the T2* which changes dramatically during the bolus passage, particularly in gray matter and in the vicinity of vessels. Thus, the resolution of single-shot EPI will typically be degraded during the passage of the bolus, leading to partial volume effects. Thirdly, in EPI off-resonance effects lead to distortion, which is proportional to the length of the acquisition period. The bolus itself results in a changing distortion as the presence of the paramagnetic material in large vessels leads to significant field shifts in the neighboring tissue.

As a consequence of these restrictions, many approaches for improving temporal resolution over conventional Fourier transform have been reported. One such dynamic imaging approach is the reduced field of view (rFOV) method, which improves the temporal resolution without compromising the spatial resolution. This technique relies on the assumption that the region of change is localized to a small portion of the overall field of view. Another way to improve temporal resolution while retaining Fourier sampling is the keyhole technique, where only the central region of the Fourier domain is acquired. The spatial resolution of the dynamic data is governed by the size of the central Fourier domain sampled in the keyhole since approximately 90% of the signal power in an image is retained within the keyhole segment (D'Arcy et al. 2002). However, the basic prerequisite of keyhole imaging demands that the change in the dynamic series is of a nature that can be sufficiently covered by the central part of the Fourier domain. Thus, the expected change has to be of low spatial frequency content, meaning it should be of broad nature without any important fine details or sharp edges (Oesterle et al. 2000). Based on this assumption, the correct width of the keyhole segment should be determined before the dynamic experiment; for example by manually inspecting the Fourier domain data. The dangers of inappropriate positioning of the keyhole window are especially prominent in T2*-weighted DSC-MRI (Oesterle et al. 2000). Therefore, despite the advantage of improved temporal resolution, neither the rFOV nor the keyhole technique has widely been adopted for clinical use. Likewise, non-Cartesian Fourier sampling can be performed very fast in conjunction with a strong T2*-weighting, and has in fact been proven feasible for DSC-MRI studies in the brain (Hou et al. 1999; Lia et al. 2000). However, the non-trivial reconstruction problems associated with the regridding, filtering, and interpolation procedures have so far prevented these experimental sequences from being applied into the clinical environment.

Sufficient temporal and spatial resolution in DSC-MRI is often obtained at the expense of signal-to-noise and contrast-to-noise. In the conventional spoiled GE sequence, these parameters are inefficient in that all transverse magnetization is destroyed before the next encoding step. In the steady-state free precession (SSFP) balanced GE sequences, often called TrueFISP or FIESTA, however, the transverse magnetization is refocused after data acquisition, resulting in an addition of the refocused transverse magnetization of the next phase encoding step. At very short repetition times, SSFP offers higher signal-to-noise ratios than the spoiled GE sequence. The drawback of this sequence is that artifact-free images can be obtained only if the two echo signals overlap precisely, which seems technically close to impossible to achieve, or if they are sufficiently separated in time to allow sampling of only one of the signals (Zur et al. 1988). In previous studies, SSFP has been used with inversion or saturation preparation schemes to improve the T1 weighting. The feasibility for dynamic myocardial perfusion studies has been evaluated by Schreiber et al. (2002), demonstrating a substantial improvement of image quality and spatial resolution and was suggested as being well suited for first pass perfusion. Observed artifacts together with the need for high-power gradients systems capable of rapid gradient switching may, however, limit its practical use.

7.6 An Echo-Shifting Approach for T2*-Weighted DSC-MRI

The T2* decay during the acquisition results in a loss of EPI image resolution (Farzaneh et al. 1992). This effect is stronger for a long acquisition window and a (relatively) short T2*. Extension of the echo train beyond about the T2* value will not increase the resolution any further, in spite of the extra data points acquired as the signal has already decayed. This limits the useful acquisition window to the short T2* values occurring during passage of the bolus. Furthermore, a disadvantage of single shot EPI techniques is the relatively long duration of the data acquisition window, where the entire Fourier domain needs to be acquired within a single repetition time. When motion or macroscopic susceptibility problems exist, the severity of artifacts is unfortunately related to the repetition time. One method to alleviate these problems is the use of a segmented EPI method with an acquisition window (echo train) that is significantly shorter than the shortest T2* value during bolus passage in order to maintain image resolution throughout the experiment. A high temporal resolution can be maintained by shifting the acquisition into a subsequent TR period. This class of GE echo-shifted MRI results in TE longer than TR by using pulsed gradients flanking the radiofrequency excitation to delay the signal beyond a subsequent excitation, and the resulting PRESTO (principles of echo-shifting with a train of observations) sequence is most conveniently realized as a 3D method, acquiring one complete volume every 1–2 s (Liu et al. 1993)

The effects of the T2* decay during the acquisition on the image can be analyzed using the point spread function (PSF). The PSF shows how signal originating from a single point source is distributed over the actual image, and determines the actual resolution of the image. The wider the PSF, the coarser is the true resolution. Fig. 7.2 demonstrates the PSF for an acquisition period of 50 ms which is typical for a single-shot EPI method as compared to the 10 ms acquisition period used in the PRESTO method, before (long T2*) and during passage of the bolus (shorter T2*). The results demonstrate the significant loss in resolution in EPI during the bolus passage and the much smaller effect in PRESTO.

The measured frequency shifts close to larger vessels can lead to significant distortions in EPI acquisitions. Limiting the acquisition window as in PRESTO reduces the artifacts from these bolus-induced frequency changes and the blurring effects of the short T2* during bolus passage. This results in an improved image quality, actual resolution, and ultimately more accurate rCBV and delay maps from PRESTO acquisitions (Flacke et al. 2000)

The advantages of the PRESTO sequence are threefold: it allows a reduction of macroscopic susceptibility artifacts, a constant spatial resolution during the passage of the bolus, as well as high temporal resolution (Duyn et al. 1996). In addition, an important advantage is that PRESTO has a more optimal sensitivity and speed, because the gradient echo train is delayed beyond the next excitation pulse (Moonen et al. 1992; Liu et al. 1993; van Gelderen et al. 2000). 3D susceptibility-based perfusion maps have been dem-

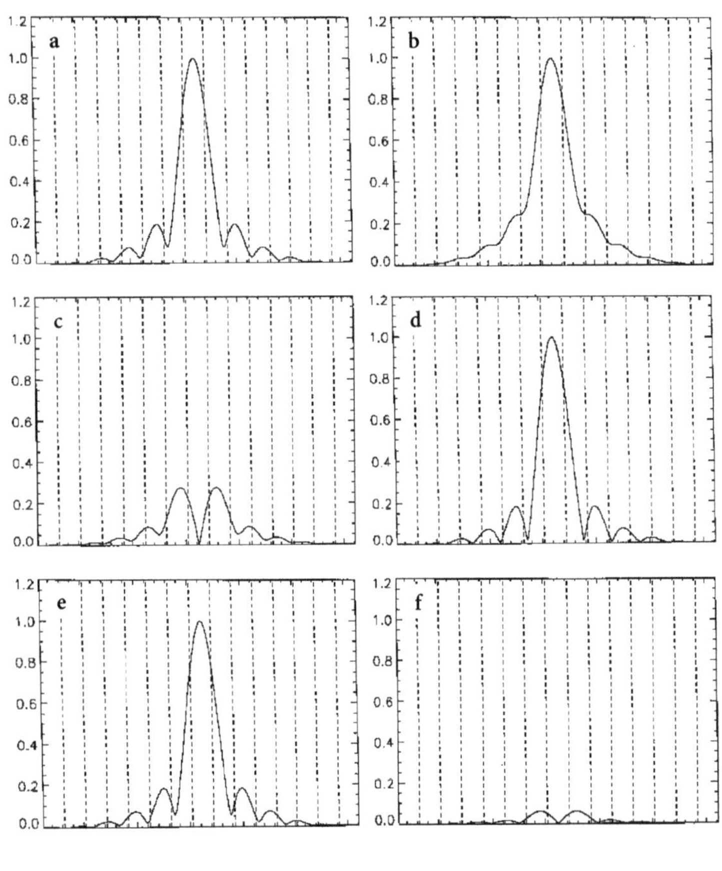

Fig. 7.2a–f. The point spread function (PSF) for the EPI (*left*) and the PRESTO (*right*) sequence, before (*top row*) and during the bolus passage (*middle row*) including their difference (*bottom row*), showing the effect of the transient T2* decay due to passage of the bolus on the actual resolution. The horizontal and vertical axes represent voxel number, and signal intensity in arbitrary units, respectively. **a,c,e** A 50-ms single-shot EPI acquisition. **b,d,f** A 10-ms PRESTO window. **a,b** 70-ms T2*; c,d 25-ms T2*. **e,f** The absolute value of the difference of the PSFs due to the T2* change upon the bolus passage [(**a**) minus (**c**) for EPI, and (**b**) minus (**d**) for PRESTO, respectively]. The *dotted lines* indicate the image voxel size as calculated from field-of-view divided by the number of phase encode steps. The relative broadening of the PSF can be expressed as the induced change in area of the PSF within the nominal dimensions of a voxel relative to the total area. For the simulations, changes of 18.7% and 1.8% were found for EPI and PRESTO, respectively. The total intensity of the difference is 44% of the intensity of the PSF without bolus for EPI and 9% for PRESTO. These measurements provide an indication of to what extent signal is displaced during bolus passage for the two methods [see also van Gelderen et al. (2000)]

onstrated in the assessment of acute stroke patients and provided information on perfusion parameters from the entire brain (Fig. 7.3) (Flacke et al. 2000). Although the method is promising, a comparison of obtained rCBV and rCBF needs to be validated against a conventional EPI technique.

7.7 T1-Weighted Perfusion

It is important to emphasize the fundamental differences between susceptibility enhanced and relaxation enhanced data. Besides T2- or T2*-based DSC-MRI, several important parameters can be assessed by T1-weighted DCE-MRI, where changes in signal intensity indicate changes in the relaxation rate. We refer to Chap. 5 for a more in depth understanding of the mechanisms that underlie the T1-weighted methods. However, since DSC-MRI techniques exist that combine both strategies (Sect. 7.8), a short description of the technical possibilities and theories behind T1-weighted perfusion follows for completeness.

Absolute quantification of the tissue concentration of contrast agent can usually be calculated by the linear relationship to the T1 relaxation rate and the specific relaxivity (r1),

$$\frac{1}{T1(t)} = \frac{1}{T1(0)} + r1 \cdot c(t) \tag{4}$$

where T1(0) is the bulk relaxation time in the tissue without contrast agent. In principle, this means

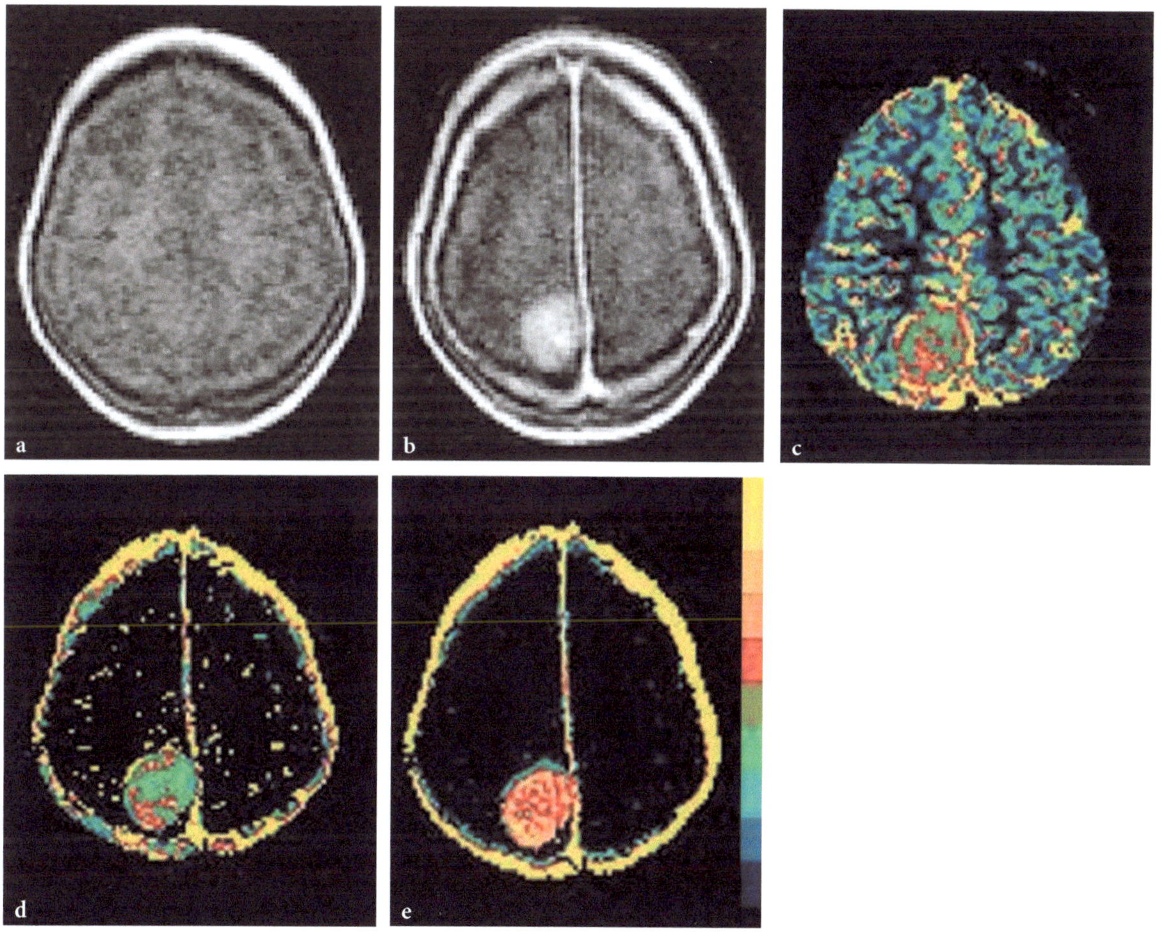

Fig. 7.3a–e. A combination of T1-weighted and T2*-weighted dynamic contrast enhanced acquisitions was used in the imaging protocol for DCE-MRI studied in a patient with a posterior falcine meningioma (Zhu et al. 2000). The pulse sequence consisted of a series of three-dimensional radiofrequency spoiled T1-weighted gradient echo acquisitions. Calculated maps are (**a**) and (**b**): Pre- and postcontrast images from T1-weighted dynamic series. There is diffuse enhancement of leptomeninges on the postcontrast T1-weighted images. **c** Parametric map of rCBV from T2*-weighted data. **d** Map of extravascular uptake from T1-weighted data. **e** Map of leakage space from T1-weighted data. [Reprinted with permission from Zhu et al. (2000)]

that a precontrast measurement of T1 should be performed before injection the contrast agent. One important factor contributing to erroneous measurements of T1(0) in blood is flow. The coherent movement of flowing fluid can alter T1 of the signal arising from spins therein (LEE et al. 1999). Placing the central-encoding lines within the low blood flow window can, to some extent, minimize the problems of the inflow effect (ZHENG et al. 1999). Triggering difficulties, saturation effects, or problems related to the mathematical conversion of signal intensity into relaxation rates may further complicate the measurement of the T1(0). Another problem may arise when the signal saturates less than what is predicted theoretically, originating from the assumption that the slice profile exhibits a rectangular slice profile. In reality, the excitation flip angle will vary across the slice, concomitantly implying an incorrect measurement of T1(t), since the flip angle is included in the theoretical signal equation. Very fast T1 mapping is one possibility to eliminate this problem (McKENZIE et al. 1999; FREEMAN et al. 1998), but their practical applications in DCE-MRI studies are yet to be investigated. The value of T1(0) for blood is therefore difficult to measure in vivo, since the experimenter may not stop the blood flow. Consequently, the T1 of the arterial blood has in many studies not actually been measured, but has been assumed from previously measured T1 values in humans (NOSEWORTHY et al. 1999) or has been assumed to be the same as the value measured ex vivo (STROUSE et al. 1996).

Since tumor contrast kinetics is often sampled over several minutes, the potential for patient motion is substantial. T1-weighted DCE-MRI is therefore vulnerable to partial volume errors (also present in the T2 and T2* DSC-MRI), both in AIF(t) and c(t), and since tumors are known to be heterogeneously distributed, the result may be a heterogeneous enhancement. From simulation studies, it has been demonstrated that the partial volume effect gives rise to a non-linear relationship of 1/T1-1/T1(0) with respect to the contrast agent concentration (VAN OSCH et al. 2001). Another approach to the problem could be to separate the signal from a voxel into contributions from large vessels and capillaries by independent component analysis (CARROLL et al. 2002). By this method, a pure large-vessel signal can be obtained from a voxel, with no partial volume effect.

The signal as a function of TE, TR and flip angle can usually be calculated on the basis of the Ernst formula (ZHU et al. 2000). Different pulse sequences result in different mathematical expressions, and the T1 weighting may therefore strongly depend on the pulse sequence parameters applied. Moderate T1 weighting can easily be performed using standard spoiled GE techniques, either as the standard or as the EPI mode. However, pre-pulses such as saturation or inversion pulses allow a significantly heavier T1 weighting. FRITZ-HANSEN and coworkers (1996), for example, demonstrated that the arterial input function could be accurately measured using an inversion-prepared spoiled GE sequence. HENDERSON et al. (2000) used a similar approach with the use of a saturation prepared spoiled GE sequence, which was related to T1 through the complex equation,

$$S_n \propto \left[(1-e^{-TI/TR)}(e^{-TR/T1}\cos\alpha)^{n-1} + (1-e^{-TR/T1)} \frac{1-(e^{-TR/T1}\cos\alpha)^n}{1-e^{-TR/T1}\cos\alpha} \right] \quad (5)$$

where TI is the time from the saturation pulse to the first excitation pulse in the GE sequence, and n is the number of phase-encoding lines before acquisition of the center of the Fourier space. Based on this methodology, it was demonstrated that simultaneous measurements of the tissue blood flow and the permeability surface product could be accurately performed within minutes after administration of contrast agent. Sequences equipped with prepulses, especially inversion prepared pulses, suffer from their longer scan times, and a compromise must be found between sufficient temporal resolution, multiple slices and the extent of T1 weighting.

As seen from Eq. 4, the mathematical conversion of signal intensities into molar units depends on the specific relaxivity (r1) of the gadolinium complex in the tissue of interest; the ability of a contrast agent to enhance proton relaxation is defined as the relaxivity. This conversion is occasionally based on the assumption that the relaxivity is the same in plasma and tissue. Ex vivo measurements of the relaxivity in the myocardium of frogs (DONAHUE et al. 1994) and in vivo measurements of the relaxivity in the rat kidney (PEDERSEN et al. 2000), however, suggest that there is a distinct difference between plasma and tissue. To our knowledge, relaxivities of brain tumors have not been fully investigated, although steps have been taken to quantify such cell cultures ex vivo (SCHMALBROCK et al. 2001).

Correspondingly, accurate determination of c(t) based on measured signal intensities and precontrast T1 is not trivial. Alternatively, conversions of signal intensities into quantitative measures have been performed from empirical in vitro calibrations (LOMBARDI et al. 1997). Although it is not strictly correct to do so, such methods will often allow a representative determination of the tissue contrast concentration time curve and the qualitative arterial input function, but one must be aware of the limitations in such a process, and usually the same calibration profile cannot be used if pulse sequence parameters are different from the in vivo experiment.

The extravascular uptake occurs more slowly than compared with changes in AIF(t). The temporal resolution is therefore determined by AIF(t) more than c(t). The temporal requirements for sampling AIF(t) have been considered in detail by Henderson et al. (2000). They concluded that an accurate representation of AIF(t), requires 1 s of temporal resolution. The temporal requirements are furthermore closely linked to the pharmacokinetic model used. It should be emphasized that different pharmacokinetic models exist to handle T1-weighted DCE-MRI data (Tofts 1997; Fenstermacher et al. 1981; Parker and Tofts 1999; St Lawrence and Lee 1998), allowing calculations of tissue blood flow, permeability surface area, mean capillary transit time, fractional volume of the extracellular extravascular space, fractional blood volume, extraction ratio, and more. These parameters are of great interest as indicators of tumor angiogenesis, and essential component of growth of tumors. On the other hand, since more parameters are modeled, it is likely that there is a reduction in the precision with which each parameter can be measured. The subsequent need for high signal-to-noise ratio may therefore be more than realistically obtainable, and the alleviation of this problem may be going from parametric pixel-by-pixel analyses to region-of-interest analyses (Henderson et al. 2000). Another way is to reduce the number of model parameters, such as in the flow-limited models or in the permeability surface-limited models, where it is supposed that flow or extravascular uptake is the dominating factor. Characterization of the transcapillary transport of contrast agent in brain tumors have been extensively investigated by Brix and coworkers (1991) and by Gowland and coworkers (1992), to whom we refer for more information about T1-weighted DCE-MRI modeling.

7.8 Techniques Combining T2*, T2, or T1

As earlier depicted, the absence of an intact BBB implies that the contrast agent does not remain intravascular. Even when only considering the first pass, the extravascular compartment will show contrast uptake (Heiland et al. 1999). To incorporate this effect in the measurement of quantitative hemodynamic parameters, such as rCBV, rCBF, and MTT, combined techniques may be useful, involving simultaneous measurements of the changes in the T1, T2, or T2* relaxation rate.

To be insensitive to T1 enhancement, a radiofrequency-spoiled gradient echo sequence is preferred with dual echo acquisition. The change in the T2* relaxation rate, ÄR2*(t), can then be readily calculated from (Vonken et al. 2000),

$$\Delta R2^*(t) = \frac{\ln[S1(t)/S2(t)] - \ln[S1_0/S2_0]}{TE2 - TE1} \qquad (6)$$

where S1(t) and S2(t) correspond to the signal intensities at the first echo time (TE1) and the second echo time (TE2), respectively, and $S1_0$ and $S2_0$ are the signal intensities before contrast agent is administered. Through extrapolation of the signal back to TE=0, a signal intensity is obtained that is influenced by T1 changes only, with no T2* dependence. The advantages are clear: no additional MRI examination time is necessary for obtaining dynamic T1-weighted information and the pixel-by-pixel computational load is low. This separation of the intravascular hemodynamic response and the extravascular component leads to the corrected tissue response function for perfusion quantification, as well as the extravascular uptake rate. Feasibility studies in a glioblastoma multiforme have led to simultaneous maps of rCBV, rCBF, MTT, and the extraction fraction (Vonken et al. 2000). An alternative dynamic multi-T1–T2*-weighted sequence has been presented by Tian et al. (1999) for myocardial perfusion. The interleaved T1–T2* imaging sequence consisted of one preparatory 90° pulse and a GE imaging sequence with a dynamically variable echo time. With a bolus injection of contrast agent, the maximum changes in T2* signal intensity occurred significantly earlier than the changes in T1 signal, and secondly, it was found that the maximum change in T1 signal intensity during the first pass of contrast agent was significantly greater in a reperfused infarcted region than in normal regions. In brain tumors, Li et al. (2000) have proposed a similar strategy for simultaneous mapping of endothelial permeability and blood volume from the same single bolus data (Fig. 7.4), where the concentration versus time curves were considered as a superposition of an intravascular and a slow and a fast exchange term, whereas modeling was based on a modified version of the mapping methods devised by Tofts and Kermode (1991), giving a necessary scan duration

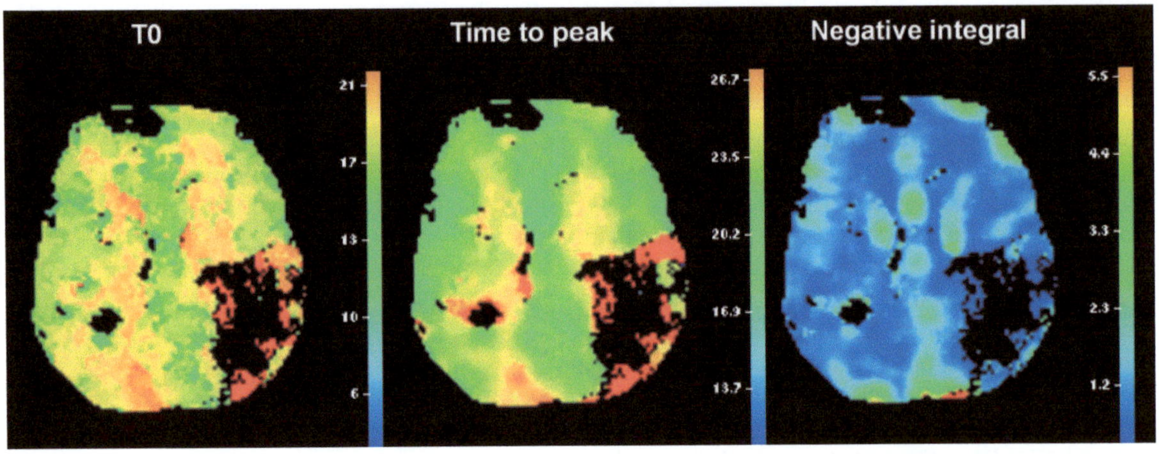

Fig. 7.4. DSC-MRI performed by a 3D PRESTO sequence in a stroke patient. The images show examples of parametric images during the passage of a bolus of contrast agent for one slice out of 32 slices, with each 3D volume acquired with 2 s time resolution. Gd-DTPA (range 0.1–0.2 mmol/kg) was injected manually in the antecubital vein in a short bolus over 4 s. Duration of the study was 37 s. Each TR period (16.5 ms) contained 13 gradient echoes with alternating polarity of the readout gradient. The average TE was 24 ms, and the difference in TE between first and last echo was 10 ms. Nominal spatial resolution was maintained at 3.75 mm isotropic. To shift the echo train into the following acquisition period, additional gradients were required. The duration of these gradients was 1–2 ms for the first and second additional gradient pulse of each TR with amplitudes of 10–20 mT/m for x,y and z gradient (physical axis). Before each echo train, a zero order navigator echo was acquired allowing a correction for translation in the direction of the additional gradients. Radiofrequency spoiling must be used to avoid contributions of stimulated and spin echoes. A 95-mm thick axial slab was selected covering more than 90% of the brain. Fourier domain was split in 13 equal sections in the second dimension, each of which was encoded by one of the 13 gradient echoes, in a linear order. The effective sampling frequency was 100 kHz, and a flip angle of 9° (Ernst angle) was used

of less than 1 min. Simultaneous T2 and T2* weighting is not commonly used, but it can be accomplished by using an asymmetric SE EPI sequence, where the echo center is displaced, resulting in a mixture of T2 and T2* weighting. The degree of asymmetry can be adjusted to trade off sensitivity against susceptibility to artifacts (Stables et al. 1998).

7.9 Parallel Imaging Techniques

The fact that perfusion is often only studied in a few slices of the brain is occasionally a significant limitation of the study. Future increments in gradient strength and reductions in the gradient switching time may allow the time required for a given matrix to be reduced, albeit with reduced signal-to-noise for the former due to increased sampling bandwidth. Even with today's hardware performance EPI-based DSC-MRI occasionally suffers from inappropriate image quality. Therefore, a considerable amount of work has been devoted to the development of more effective acquisition strategies.

SMASH (simultaneous acquisition of spatial harmonics) (Sodickson and Manning 1997) and the subsequently developed SENSE (sensitivity encoding) (Pruessmann et al. 1999) are related methods that are based on the independent spatial information of multiple receiver coils. The data acquisition for both techniques is identical to that of conventional imaging, except that the FOV is decreased to values that are smaller than the object. An independent data set (a sensitivity map) must be acquired for each receiver coil. As compared to conventional rectilinear Fourier space coverage with n phase encoding lines, Fourier space coverage for SENSE and SMASH is thus limited. The reduction of FOV leads to aliasing (backfolding) of the image upon a conventional Fourier transform in the phase encoding direction. SMASH and SENSE are special processing methods in which this aliasing is avoided based on the known spatial sensitivities of the independent receiver coils. The difference between SMASH and SENSE is that SMASH works in the Fourier space domain, whereas SENSE works in the object domain. The disadvantages of SMASH and SENSE are that new, specially designed, receiver coils should be used, and that, in general, sensitivity is no longer spatially homogeneous.

Advances in image reconstruction algorithms, coil sensitivity calibration procedures, and coil array designs have improved the practical robustness of such parallel MRI techniques. Recently, it has been demonstrated that the SENSE technique can be combined with PRESTO in fMRI measurements of activating areas in the human brain (Golay et al. 2000).

7.10 Conclusion s and Future Directions

DSC-MRI has long been adapted into daily clinical use with the rapid 2D or 3D GE EPI being the most predominant method due to its sensitivity to contrast agent in both capillaries and larger vessels (Weisskoff et al. 1994), whereas the SE method is more rarely used since it is sensitive only to contrast agent within the capillaries. However, single-shot, GE EPI techniques for bolus tracking of the brain suffer from a substantial loss of resolution at the maximum of the bolus associated with the field inhomogeneities due to the distribution of paramagnetic contrast agent. The 3D PRESTO method offers several advantages over EPI for bolus tracking such as: (1) nearly identical image resolution throughout the passage of the bolus, (2) improved image quality due to relatively short echo train, (3) near optimum TE for all Fourier space lines. A similar result may be obtained using SENSE (or SMASH) techniques to shorten EPI echo train length. However, this will require a specially designed head coil with multiple, independent, receivers.

The main advantage of T1-weighted perfusion is that both tissue blood flow and permeability surface can be measured simultaneously. One important element in the T1-weighted DCE-MRI is the underlying model from which the calculated parameters are derived. Many assumptions about the tissue system are made with respect to the nature of contrast agent kinetic modeling, and these assumptions may affect the accuracy of the parameter calculations. In practice, a compromise must be struck between reality and the precision of the measured data (signal-to-noise ratio, temporal resolution, etc.). The different models proposed by leading experts in this field have unfortunately been presented with a variety of quantities, meaning that comparisons between different groups are almost impossible. Consequently, a standardization of quantities associated with analysis of T1-weighted DCE-MRI has been proposed in the review by Tofts et al. (1999), as part of building up a common language for the estimation and description of physiologic quantities that determines the dynamic behavior of diffusible and non-diffusible contrast agents. Future applications of DCE-MRI may introduce additional physiological parameters. Furthermore, techniques like texture analysis may allow discrimination between tumors by applying common first-order and second-order statistics. Until now, such methods have been restricted to fine discrimination between white matter, cortical gray matter, and cerebrospinal fluid (Kjær et al. 1995). Nevertheless, the use of DCE-MRI to estimate perfusion in pathological conditions, such as different types of brain tumors, should not be used without careful consideration of the underlying vascular nature of the tissue of interest.

The interpretation of parametric data from T2, T2*, and T1 DCE-MRI may depend on whether the signal versus time curves are quantified into true absolute units or can only be measured on a relative basis. Heuristic enhancement parameters such as the rate of enhancement, the time to peak, etc., may have some relation to physiologic parameters although they also depend on the particular MRI pulse sequence parameters. Although these semi-quantitative parameters derived from DCE-MRI do not allow for pharmacokinetic modeling, a significant correlation of these parameters and treatment outcome of cancer patients has been established (Mayr et al. 1998). However, the lack of clear physiologic significance can make their interpretation difficult. Methods for fast, reliable and precise conversion of the signal versus times curves of blood and tissues have not yet been fully settled, and future strategies in this area would be very welcome.

Technical developments in hardware and post-processing software are likely to continue, thus enabling more rapid, higher spatial resolution, multi-slice, and simultaneous T1 and T2* techniques. Refinements in pharmacokinetic modeling analysis will allow an improved understanding of contrast enhancement, and such techniques are likely to be incorporated into patient management. In cancer patients, DCE-MRI studies are expected to play an important role in the monitoring of the efficiency of anti-angiogenics. Such treatments may be directed at already formed vessels or against the angiogenic process itself. DCE-MRI is ideally suited to monitor such treatment in vivo.

Acknowledgements
We express our gratitude to those who provided data and figures: Fig. 7.1, Dr. Oliver Speck, Department of Diagnostic Radiology, Medical Physics Section, Uni-

versity of Freiburg Medical Center, Freiburg, Germany; Fig. 7.3, Alan Jackson, Division of Imaging Science and Biomedical Engineering, Stopford Medical School, University of Manchester, Manchester, United Kingdom; Fig. 7.4, Christoph Manka, Department of Radiology, University of Bonn, Germany.

References

Aronen HJ, Gazit IE, Louis DN, Buchbinder BR, Pardo FS, Weisskoff RM, Harsh GR, Cosgrove GR, Halpern EF, Hochberg FH (1994) Cerebral blood volume maps of gliomas: comparison with tumor grade and histologic findings. Radiology 191:41–51

Aumann S, Schoenberg SO, Just A, Briley-Saebo K, Bjørnerud A, Bock M, Brix G (2003) Quantification of renal perfusion using an intravascular contrast agent (part 1): results in a canine model. Magn Reson Med 49:276–287

Axel L (1980) Cerebral blood flow determination by rapid-sequence computed tomography. Radiology 137:679–686

Boxerman JL, Hamberg LM, Rosen BR, Weisskoff RM (1995) MR contrast due to intravascular magnetic susceptibility perturbations. Magn Reson Med 34:555–566

Brix G, Semmler W, Port R, Schad LR, Layer G, Lorenz WJ (1991) Pharmacokinetic parameters in CNS Gd-DTPA enhanced MR imaging. J Comput Assist Tomogr 15:621–628

Carroll TJ, Haughton VM, Rowley HA, Cordes D (2002) Confounding effect of large vessels on MR perfusion images analyzed with independent component analysis. AJNR Am J Neuroradiol 23:1007–1012

D'Arcy JA, Collins DJ, Rowland IJ, Padhani AR, Leach MO (2002) Applications of sliding window reconstruction with cartesian sampling for dynamic contrast enhanced MRI. NMR Biomed 15:174–183

Donahue KM, Burstein D, Manning WJ, Gray ML (1994) Studies of Gd-DTPA relaxivity and proton exchange rates in tissue. Magn Reson Med 32:66–76

Donahue KM, Weisskoff RM, Burstein D (1997) Water diffusion and exchange as they influence contrast enhancement. Cerebral blood volume maps of gliomas: comparison with tumor grade and histologic findings. J Magn Reson Imaging 7:102–110

Duyn JH, Yang Y, Frank JA, Mattay VS, Hou L (1996) Functional magnetic resonance neuroimaging data acquisition techniques. Neuroimage 4:S76–S83

Edelman RR, Mattle HP, Atkinson DJ, Hill T, Finn JP, Mayman C, Ronthal M, Hoogewoud HM, Kleefield J (1990) Cerebral blood flow: assessmen with dynamic contrast-enhanced T2*-weighted MR imaging at 1.5T. Radiology 176:211–220

Eichling JO, Raichle ME, Grubb RL Jr, Ter-Pogossian MM (1974) Evidence of the limitations of water as a freely diffusible tracer in brain of the rhesus monkey. Circ Res 35:358–364

Farzaneh F, Riedrer SJ, Pelc NJ (1992) Analysis of T2 limitations and off-resonance effects on spatial resolution and artifacts in Echo-Planar imaging. Magn Reson Med 14:123–139

Fenstermacher JD, Blasberg RG, Patlak CS (1981) Methods for Quantifying the transport of drugs across brain barrier systems. Pharmacol Ther 14:217–248

Fisel CR, Ackerman JL, Buxton RB, Garrido L, Belliveau JW, Rosen BR, Brady TJ (1991) MR contrast due to microscopically heterogeneous magnetic susceptibility: numerical simulations and applications to cerebral physiology. Magn Reson Med 17:336–347

Flacke S, Urbach H, Folkers PJ, Keller E, van den Brink JS, Traber F, Block W, Gieseke J, Schild HH (2000) Ultra-fast three-dimensional MR perfusion imaging of the entire brain in acute stroke assessment. J Magn Reson Imaging 11:250–259

Freeman AJ, Gowland PA, Mansfield P (1998) Optimization of the ultrafast Look-Locker echo-planar imaging T1 mapping sequence. Magn Reson Med 16:765–772

Fritz-Hansen T, Rostrup E, Larsson HB, Søndergaard L, Ring P, Henriksen O (1996) Measurement of the arterial concentration of Gd-DTPA using MRI: a step toward quantitative perfusion imaging. Magn Reson Med 36:225–231

Golay X, Pruessmann KP, Weiger M, Crelier GR, Folkers PJ, Kollias SS, Boesiger P (2000) PRESTO-SENSE: an ultrafast whole-brain fMRI technique. Magn Reson Med 43:779–786

Gowland P, Mansfield P, Bullock P, Stehling M, Worthington B, Firth J (1992) Dynamic studies of gadolinium uptake in brain tumors using inversion-recovery echo-planar imaging. Magn Reson Med 26:241–258

Guckel FJ, Brix G, Schmiedek P, Piepgras Z, Becker G, Kopke J, Gross H, Georgi M (1996) Cerebraovascular reserve capacity in patients with occlusive cerebrovascular disease: assessment with dynamic susceptibility contrast-enhanced MR imaging and the acetozalomide stimulation test. Radiology 201:405–412

Heiland S, Benner T, Debus J, Rempp K, Reith W, Sartor K (1999) Simultaneous assessment of cerebral hemodynamics and contrast agent uptake in lesions with disrupted blood-brain-barrier. Magn Reson Imaging 17:21–27

Henderson E, Sykes J, Drost D, Weinmann HJ, Rutt BK, Lee TY (2000) Simultaneous MRI measurement of blood flow, blood volume, and capillary permeability in mammary tumors using two different contrast agents. J Magn Reson Imaging 12:991–1003

Hou L, Yang Y, Mattay VS, Frank JA, Duyn JH (1999) Optimization of fast acquisition methods for whole-brain relative cerebral blood volume (rCBV) mapping with susceptibility contrast agents. J Magn Reson Imaging 9:233–239

Kjær L, Ring P, Thomsen C, Henriksen O (1995) Texture analysis in quantitative MR imaging. Tissue characterisation of normal brain and intracranial tumours at 1.5 T. Acta Radiol 36:127–135

Koshimoto Y, Yamada H, Kimura H, Maeda M, Tsuchida C, Kawamura Y, Ishii Y (1999) Quantitative analysis of cerebral microvascular hemodynamics with T2-weighted dynamic MR imaging. J Magn Reson Imaging 9:462–467

Lee JH, Li X, Sammi MK, Springer CS Jr (1999) Using flow relaxography to elucidate flow relaxivity. J Magn Reson 136:102–113

Li KL, Zhu XP, Waterton J, Jackson A (2000) Improved 3D quantitative mapping of blood volume and endothelial permeability in brain tumors. J Magn Reson Imaging 12:347–357

Lia TQ, Guang Chen Z, Østergaard L, Hindmarsh T, Moseley ME (2000) Quantification of cerebral blood flow by bolus tracking and artery spin tagging methods. Magn Reson Imaging 18:503–512

Liu G, Sobering G, Duyn J, Moonen CT (1993) A functional MRI technique combining principles of Echo-Shifting with a train of observations (PRESTO). Magn Reson Med 30:764–768

Lombardi M, Jones RA, Westby J, Kvaerness J, Torheim G, Michelassi C, L'Abbate A, Rinck PA (1997) MRI for the evaluation of regional myocardial perfusion in an experimental animal model. J Magn Reson Imaging 7:987–995

Mayr NA, Yuh WT, Zheng J, Ehrhardt JC, Magnotta VA, Sorosky JI, Pelsang RE, Oberley LW, Hussey DH (1998) Prediction of tumor control in patients with cervical cancer: analysis of combined volume and dynamic enhancement pattern by MR imaging. AJR 170:177–182

McKenzie CA, Pereira RS, Prato FS, Chen Z, Drost DJ (1999) Improved contrast agent bolus tracking using T1 FARM. Magn Reson Med 41:429–435

Miyazaki T, Yamashita Y, Tsuchigame T, Yamamoto H, Urata J, Takahashi M (1996) MR cholangiopancreatography using HASTE (half-Fourier acquisition single-shot turbo spin-echo) sequences. AJR 166:1297–1303

Moonen CT, Liu G, van Gelderen P, Sobering G (1992) A fast gradient-recalled MRI technique with increased sensitivity to dynamic susceptibility effects. Magn Reson Med 26:184–189

Moseley ME, Vexler Z, Asgari HS, Mintorovitch J, Derugin N, Rocklage S, Kucharczyk J (1991) Comparison of Gd- and Dy-chelates for T2* contrast-enhanced imaging. Magn Reson Med 22:259–264

Noseworthy MD, Kim JK, Stainsby JA, Stanisz GJ, Wright GA (1999) Tracking oxygen effects on MR signal in blood and skeletal muscle during hyperoxia exposure. J Magn Reson Imaging 9:814–820

Oesterle C, Strohschein R, Kohler M, Schnell M, Hennig J (2000) Benefits and pitfalls of keyhole imaging, especially in first-pass perfusion studies. J Magn Reson Imaging 11:312–323

Parker GJ, Tofts PS (1999) Pharmacokinetic analysis of neoplasms using contrast-enhanced dynamic magnetic resonance imaging. Top Magn Reson Imaging 10:130–142

Pedersen M, Mørkenborg J, Jensen FT, Stødkilde-Jørgensen H, Djurhuus JC, Frøkiær J (2000) In vivo measurements of relaxivities in the rat kidney cortex. J Magn Reson Imaging 12:289–296

Perman WH, Gado MH, Larson KB, Perlmutter JS (1992) Simultaneous MR acquisition of arterial and and brain signal-time curves. Magn Reson Med 28:74–83

Pruessmann KP, Weiger M, Scheidegger MB, Boesiger P (1999) SENSE: sensitivity encoding for fast MRI. Magn Reson Med 42:952–962

Rempp KA, Brix G, Wenz F, Becker CR, Guckel F, Lorenz WJ (1994) Quantification of regional cerebral blood flow and volume with dynamic susceptibility contrast-enhanced MR imaging. Radiology 193:637–641

Roberts TPL, Chuang N, Roberts HC (2000) Neuroimaging: do we really need new contrast agents for MRI? Eur J Radiol 34:166–178

Rosen BR, Belliveau JW, Vevea JM, Brady TJ (1990) Contrast agents and cerebral hemodynamics. Magn Reson Med 14:249–265

Schmalbrock P, Hines JV, Lee SM, Ammar GM, Kwok EW (2001) T1 measurements in cell cultures: a new tool for characterizing contrast agents at 1.5T. J Magn Reson Imaging 14:636–648

Schreiber WG, Schmitt M, Kalden P, Mohrs OK, Kreitner KF, Thelen M (2002) Dynamic contrast-enhanced myocardial perfusion imaging using saturation-prepared TrueFISP. J Magn Reson Imaging 16:641–652

Simonsen CZ, Østergaard L, Smith DF, Vestergaard-Poulsen P, Gyldensted C (2000) Comparison of gradient- and spin-echo imaging: CBF, CBV, and MTT measurements by bolus tracking. J Magn Reson Imaging 12:411–416

Sodickson DK, Manning WJ (1997) Simultaneous acquisition of spatial harmonics (SMASH): ultra-fast with radio-frequency coil arrays. Magn Reson Med 38:591–603

Sorensen AG, Buonanno FS, Gonzalez RG, Schwamm LH, Lev MH, Huang-Hellinger FR, Reese TG, Weisskoff RM, Davis TL, Suwanwela N, Can U, Moreira JA, Copen WA, Look RB, Finklestein SP, Rosen BR, Koroshetz WJ (1996) Hyperacute stroke: evaluation with combined multi-section diffusion weighted and haemodynamically weighted echo planar MR imaging. Radiology 199:391–401

Speck O, Chang L, Itti L, Itti E, Ernst T (1999) Comparison of static and dynamic MRI techniques for the measurement of regional cerebral blood volume. Magn Reson Med 41:1264–1268

Speck O, Chang L, DeSilva NM, Ernst T (2000) Perfusion MRI of the human brain with dynamic susceptibility contrast: gradient-echo versus spin-echo techniques. J Magn Reson Imaging 12:381–387

St Lawrence KS, Lee TY (1998) An adiabatic approximation to the tissue homogeneity model for water exchange in the brain. I. Theoretical derivation. J Cereb Blood Flow Metab 18:1365–1377

Stables LA, Kennan RP, Gore JC (1998) Asymmetric spin-echo imaging of magnetically inhomogeneous systems: theory, experiment, and numerical studies. Magn Reson Med 40:432–442

Strouse PJ, Prince MR, Chenevert TL (1996) Effect of the rate of gadopentetate dimeglumine administration on abdominal vascular and soft-tissue MR imaging enhancement patterns. Radiology 201:809–816

Tian G, Shen JF, Dai G, Sun J, Xiang B, Luo Z, Somorjai R, Deslauriers R (1999) An interleaved T1-T2* imaging sequence for assessing myocardial injury. J Cardiovasc Magn Reson 1:145–151

Tofts PS (1997) Modeling tracer kinetics in dynamic Gd-DTPA MR imaging. J Magn Reson Imaging 7:91–101

Tofts PS, Kermode AG (1991) Measurement of the blood-brain barrier permeability and leakage space using dynamic MR imaging. 1. Fundamental concepts. Magn Reson Med 17:357–367

Tofts PS, Brix G, Buckley DL, Evelhoch JL, Henderson E, Knopp MV, Larsson HB, Lee TY, Mayr NA, Parker GJ, Port RE, Taylor J, Weisskoff RM (1999) Estimating kinetic parameters from dynamic contrast-enhanced T(1)-weighted MRI of a diffusable tracer: standardized quantities and symbols. J Magn Reson Imaging 10:223–232

Van Gelderen P, Grandin C, Petrella JR, Moonen CTW (2000) Rapid three-dimensional MR imaging method for tracking a bolus of contrast agent through the brain. Radiology 216:603–608

Van Osch MJ, Vonken EJ, Bakker CJ, Viergever MA (2001) Correcting partial volume artifacts of the arterial input function in quantitative cerebral perfusion MRI. Magn Reson Med 45:477–485

Villringer A, Rosen BR, Belliveau JW, Ackerman JL, Lauffer RB, Buxton RB, Chao YS, Wedeen VJ, Brady TJ (1988) Dynamic

Imaging with Lanthanide Chelates in normal brain: contrast due to magnetic susceptibility effects. Magn Reson Med 6:164–174

Vonken EJ, van Osch MJ, Bakker CJ, Viergever MA (2000) Measurement of cerebral perfusion with dual-echo multi-slice quantitative dynamic susceptibility contrast MRI. Magn Reson Med 43:820–827

Warach S, Dashe JF, Edelman RR (1996) Clinical outcome in ischemic stroke predicted by early diffusion-weighted and perfusion magnetic resonance imaging: a preliminary analysis. J Cerebral Blood Flow Metabol 16:3–59

Weisskoff RM, Chesler D, Boxerman JL, Rosen BR (1993) Pitfalls in MR measurement of tissue blood flow with intravascular tracers: which mean transit time? MR contrast due to intravascular magnetic susceptibility perturbations. Magn Reson Med 29:553–559

Weisskoff RM, Zuo CS, Boxerman JL, Rosen BR (1994) Microscopic susceptibility variation and transverse relaxation: theory and experiment. Magn Reson Med 31:601–610

Zheng J, Venkatesan R, Haacke EM, Cavagna FM, Finn PJ, Li D (1999) Accuracy of T1 measurements at high temporal resolution: feasibility of dynamic measurement of blood T1 after contrast administration. J Magn Reson Imaging 10:576–581

Zhu XP, Li KL, Kamaly-Asl ID, Checkley DR, Tessier JJ, Waterton JC, Jackson A (2000) Quantification of endothelial permeability, leakage space, and blood volume in brain tumors using combined T1 and T2* contrast-enhanced dynamic MR imaging. J Magn Reson Imaging 11:575–585

Zur Y, Stokar S, Bendel P (1988) An analysis of fast imaging sequences with steady-state transverse magnetization refocusing. Magn Reson Med 6:175–193

8 Consensus Recommendations for Acquisition of Dynamic Contrasted-Enhanced MRI Data in Oncology

Jeffrey L. Evelhoch

CONTENTS

8.1 Introduction

As demonstrated by many of the chapters in this book, dynamic contrast-enhanced magnetic resonance imaging (DCE-MRI), using clinically available, diffusible, extracellular contrast agents [e.g., gadopentetate dimeglumine (Mitchell 1997)], has emerged over the past decade as a promising method for cancer diagnosis, staging, response assessment and evaluation of biological activity in early clinical trials of novel drugs targeting the tumor vasculature. Remarkably, these positive results have been obtained despite considerable variation in both the methods of data acquisition (e.g., pulse sequences, acquisition parameters, temporal resolution, spatial resolution, and coverage) and analysis (e.g., visual inspection, parametric analysis, pharmacokinetic, or physiologic modeling). This suggests there are substantial physiologic / pharmacokinetic differences (e.g., between benign and malignant, or between non-responsive and responsive tumors) underlying these observations that are evident independent of the methods used for acquisition and analysis of the DCE-MRI data. Clearly, there is a promising future for use of DCE-MRI as both a clinical research tool and in routine clinical practice. However, there are several issues that should be addressed to expedite the realization of this promise in both routine clinical practice and drug development.

In order to better define and address these issues, two consensus workshops were held by the Cancer Imaging Program of the United States National Cancer Institute on October 28 and 29, 1999 and November 13, 2000 to consider this problem (Evelhoch et al. 2000; see also http://www3.cancer.gov/bip/DCEMRIrpt.htm). This chapter summarizes the consensus opinions of the attendees of those workshops[1] (see Sects. 8.5 and 8.6) and provides examples of recent research that may help to address some of these critical issues.

8.2 General Recommendations

A fundamental issue impeding realization of the potential of DCE-MRI in oncology is the need to compare and/or integrate results across multiple institutions. This is due to the variety of methods used for data acquisition and analysis that have resulted from the lack of an established consensus regarding how best to acquire and/or analyze DCE-MRI data. Moreover, evaluation of the relative merits of the various data analysis methods is difficult, if not impossible, when disparate acquisition methods are used. Consequently, the clinical relevance of the information provided by different analysis approaches is difficult to assess, further hindering the establishment of a consensus on analysis methods.

J. L. Evelhoch, PhD
Director, Structural Imaging, World Wide Clinical Technology, Pfizer Global Research and Development, 2800 Plymouth Road, Mail Stop: 50-M129, Ann Arbor, MI 48105, USA

[1] These recommendations are an independent statement of the workshop participants and do not represent a policy statement of the United States National Institutes of Health or the United States Federal Government.

Ultimately, for DCE-MRI to be useful for routine oncologic applications in a clinical setting, including clinical trials of new cancer therapies, it should be straightforward to implement and provide a 3D representation of relevant quantitative information, voxel-wise quantitative results, and a statistical summary of quantitative results. Regardless of the method used for data analysis, some of the factors that could impact the information derived from DCE-MRI data are intra- or inter-patient variation in either the initial T1 or the blood contrast agent concentration as a function of time (i.e., 'arterial input function'), tumor heterogeneity, patient motion during the study (Evelhoch 1999), and signal artifacts, particularly in the baseline image (Dale et al. 2003). A recent paper from Rijpkema and co-workers (2001), which showed that use of rapidly and highly-enhancing pixels to define a 'vascular' input function in each subject substantially improves the test-retest reproducibility of DCE-MRI measurements, illustrates the importance of using an individual input function for analysis of DCE-MRI data (see also Chap. 6). Although the importance of each of these factors for clinical research tools and routine clinical practice may differ depending on the specific application, it would be useful to establish baseline requirements for general research and/or clinical studies. This would facilitate both integration of data from different institutions and comparison of various approaches for analysis of the kinetic data.

Optimal properties for DCE-MRI methods include the following. The entire tumor should be imaged (3D measurements are preferred because single slice measurements may be prone to sampling error) with the best possible spatial and temporal resolution (trade-offs among coverage, spatial and temporal resolution depend on the specific application), while specification of the field of view (FOV) should be flexible. The results should be independent of system (e.g., site, manufacturer and field strength) and contrast dose, free from spurious correlations, motion artifacts, and errors due to tissue properties, include an estimate of error, and highly reproducible (although accurate assessment of the underlying pathophysiology should not be sacrificed for reproducibility).

Requirements for temporal resolution may vary over the course of a contrast-enhanced examination, with resolution on the order of seconds targeted for the initial contrast enhancement phase and becoming progressively longer to tens of seconds during later phases, which may occur minutes later. Thus, 'adaptive imaging' methods that allow dynamic trade-off between temporal and spatial resolution throughout the time course are preferable (see schematic representation of this concept in Fig. 8.1). Krishnan and Chenevert (2004) recently published an intuitive formalism applicable to DCE-MRI for a set of targeted/anticipated dynamic events as well as spatial features that should facilitate consideration of the trade-off between spatial and temporal resolution. Many meritorious technical directions are possible for attempting to get high 3D spatial and good temporal resolution. Examples of methods allowing arbitrary selection from among several combinations of temporal/spatial resolutions during post-processing were recently published by Song et al. (2001) and d'Arcy et al. (2002). However, they have yet to be used widely for DCE-MRI studies.

In order to have adequate patient power (see Chap. 15) clinical studies using a specific DCE-MRI technique will likely require the collaboration of oncologic investigators from multiple institutions. Recruitment of such oncologic collaborators may likely involve sites that focus primarily on clinical MRI rather than those with extensive technical research capability. Furthermore, not all collaborators will necessarily be located near each other. Most commercial MRI manufacturers provide to their academic collaborators some level of capability to develop acquisition techniques of their own. In general, this requires a technical knowledge of the MRI machine at the academic site which is much deeper than that required for routine clinical operation. Consequently, translation of new methods developed at one institution to other institutions for clinical use requires personnel dedicated to sup-

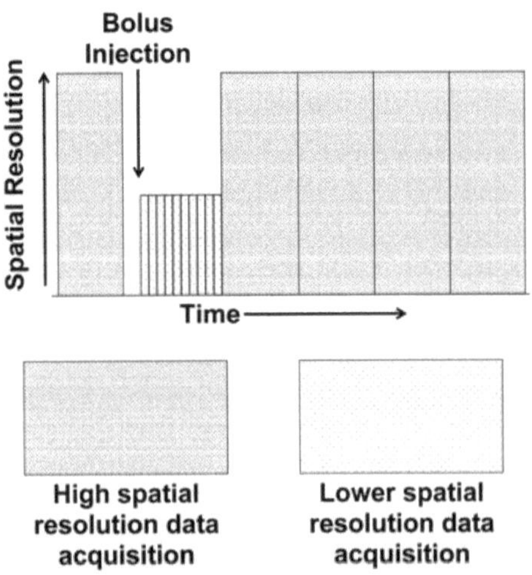

Fig. 8.1. Recommended 'adaptive' imaging approach

porting implementation at sites with considerably less technical knowledge of the MRI machine. In the short term, given the difficulty in translating new methods to multiple sites, it may be best to specify minimum requirements, rather than more sophisticated sequences, to facilitate cooperation from more clinically oriented sites.

8.3 Specific Recommendations

- When feasible, entire tumor volume should be included in field of view.
- Prior to injection, T1 should be measured using the same resolution and field of view used for acquisition of dynamic data, if possible (see Chap. 5). This should be facilitated by the use of methods such as Look-Locker (Karlsson and Nordell 2000) and DESPOT1 (Deoni et al. 2003).
- A power injector should be used for contrast agent injection to minimize variation and a saline flush should always be used.
- The input function (preferably arterial) should be measured with no in-flow effect.
- In general, a minimum of 5- to 30-s temporal resolution (or the fastest sampling possible consistent with spatial resolution requirements) should be used for the first 90–150 s after bolus injection. It should be noted that poorer temporal resolution would not support the use of sophisticated data analysis methods (Henderson et al. 2000; Jackson et al. 2002), but may be unavoidable if high spatial resolution is critical. Higher spatial resolution images could be acquired out to 10 min with 1- to 4-min temporal resolution. In some cases, sufficient spatial resolution may be achieved with higher temporal resolution and there is no need to change spatial resolution.

8.4 Discussion

The rationale behind the recommended strategy is to record potentially rapid signal changes as they occur (albeit at reduced spatial resolution), then transition to high spatial resolution imaging as the intensity changes become less abrupt. Use of the recommended 'adaptive imaging' approach would facilitate both comparisons among different groups and evaluation of the merits of the various approaches available for data analysis. The Workshop Attendees recognize that all interested investigators may not be able to implement the recommended 'adaptive imaging' approach immediately. This is primarily for two reasons: (1) 'adaptive imaging' and input function sampling requires very flexible control of clinical MR scanners to permit dynamic switching between high spatial and temporal resolution; and (2) methods optimizing the number of pixels imaged per unit time may not be available on all systems. Increased cooperation among investigators, manufacturers and the relevant government agencies should facilitate more widespread use of the recommendations.

As mentioned previously, approaches providing dynamic trade-offs between high spatial resolution adequate to fully characterize the tumor morphology and high temporal resolution adequate to characterize contrast pharmacokinetics have already been introduced. However, numerous other approaches, including those routinely applied to contrast-enhanced magnetic resonance angiography (CE-MRA; Mazaheri et al. 2002) may be valuable if adapted to DCE-MRI. In addition, new ways for providing improved spatial resolution per unit time could be applied to DCE-MRI [e.g. multiple coil techniques such as sensitivity encoding, SENSE (Tsao et al. 2003), and simultaneous acquisition of spatial harmonics, SMASH (Sodickson and McKenzie 2001), and effective utilization of high capability gradient technology]. In any case, it is critical to verify that the fidelity of the quantitative contrast agent concentration is maintained throughout the observation period.

The use of acquisition protocols meeting at least the minimum requirements recommended herein should facilitate the establishment of a database that could be used to evaluate data analysis approaches. In this light, it is important to note that such protocols would support the recommendations made by the Pharmacodynamic/Pharmacokinetic Technologies Advisory Committee (PTAC) of the Cancer Research United Kingdom for use of DCE-MRI in clinical trials of drugs targeting the tumor vasculature (Leach et al. 2003). The Cancer Research UK PTAC recommended Ktrans (Tofts et al. 1999) and/or Initial Area Under the (concentration-time) Curve (IAUC) (Mattiello and Evelhoch 1991) should be the primary endpoint for clinical trials of anti-angiogenic or anti-vascular drugs. Both Ktrans and IAUC require calculation of instantaneous tumor contrast agent concentration, based on the change in relaxivity due to contrast uptake. Since that calculation requires a measurement of tumor T1 immediately prior to contrast uptake and the arterial input function is required for Ktrans and can be used as a reference for IAUC (Evelhoch 1999; Redman et al. 2003), the recommended acquisition protocols would allow determination of the recommended primary endpoints.

8.5 Participants (Affiliation at Time of Workshop) of the Dynamic Contrast-Enhanced Magnetic Resonance Imaging Workshop

Truman Brown (Fox Chase Cancer Center, Philadelphia, PA)
Thomas Chenevert (University of Michigan, Ann Arbor, MI)
Laurence Clarke (National Cancer Institute, Bethesda, MD)
Bruce Daniel (Stanford University, Palo Alto, CA)
Hadassa Degani (Weizmann Institute, Rehovot, Israel)
Jeffrey Evelhoch - Chair (Wayne State University, Detroit, MI)
Nola Hylton (University of California, San Francisco, CA)
Michael Knopp (German Cancer Research Center, Heidelberg, Germany)
Jason Koutcher (Memorial Sloan Kettering, New York, NY)
Ting-Yim Lee (University of Western Ontario, London, Ontario, Canada)
Nina Mayr (University of Iowa, Iowa City, IA)
Daniel Sullivan (National Cancer Institute, Bethesda, MD)
June Taylor (St. Jude Children's Research Hospital, Memphis, TN)
Paul Tofts (University College London, London, UK)
Robert Weisskoff (EPIX Medical, Cambridge, MA)

8.6 Participants (Affiliation at Time of Workshop) of the Future Technical Needs in Contrast-Enhanced MRI of Cancer Workshop

Houston Baker (National Cancer Institute, Bethesda, MD)
Thomas Chenevert (University of Michigan, Ann Arbor, MI)
Laurence Clarke (National Cancer Institute, Bethesda, MD)
W. Thomas Dixon (General Electric Medical Systems, Schenectady, NY)
William Edelstein (General Electric Medical Systems, Schenectady, NY)
Jeffrey Evelhoch - Co-Chair (Wayne State University, Detroit, MI)
Robert Herfkens (Stanford University, Palo Alto, CA)
Alan Jackson (University of Manchester, Manchester, UK)
Andrea Kassner (Philips Medical Systems, Hammersmith, London, UK)
Larry Kasuboski (Marconi Medical Systems, Cleveland, OH)
Leon Kaufman (Toshiba America Medical Systems, Tustin, CA)
Elaine Keeler (Marconi Medical Systems, Cleveland, OH)
Michael Knopp (Ohio State University, Columbus, OH)
Charles Mistretta (University of Wisconsin, Madison, WI)
James Pipe (Barrow Neurological Institute, Phoenix, AZ)
Stephen Riederer - Co-Chair (Mayo Clinic, Rochester, MN)
Mitchell Schnall (University of Pennsylvania, Philadelphia, PA)
Edward Staab (National Cancer Institute, Bethesda, MD)
David Thomasson (Siemens Medical Solutions, Malvern, PA)

References

Dale BM, Jesberger JA, Lewin JS, Hillenbrand CM, Duerk JL (2003) Determining and optimizing the precision of quantitative measurements of perfusion from dynamic contrast enhanced MRI. J Magn Reson Imaging 18:575–584

d'Arcy JA, Collins DJ, Rowland IJ, Padhani AR, Leach MO (2002) Applications of sliding window reconstruction with Cartesian sampling for dynamic contrast enhanced MRI. NMR Biomed 15:174–183

Deoni SC, Rutt BK, Peters TM (2003) Rapid combined T1 and T2 mapping using gradient recalled acquisition in the steady state. Magn Reson Med 49:515–526

Evelhoch JL (1999) Key factors in the acquisition and analysis of contrast uptake data for oncology. J Magn Reson Imaging 10:254–259

Evelhoch JL, Brown T, Chenevert T, Clarke L, Daniel B, Degani H, Hylton N, Knopp M, Koutcher J, Lee T-Y, Mayr N, Sullivan D, Taylor J, Tofts P, Weisskoff R (2000) Recommendation for acquisition of dynamic contrasted-enhanced MRI data in oncology. Proc 8th Mtg Int Soc Magn Reson Med, Denver, CO

Henderson E, Sykes J, Drost D, Weinmann H-J, Rutt BK, Lee T-Y (2000) Simultaneous MRI measurement of blood flow, blood volume, and capillary permeability in mammary tumors using two different contrast agents. J Magn Reson Imaging 12:991–1003

Jackson A, Haroon H, Zhu XP, LI KL, Thacker NA, Jayson G (2002) Breath-hold perfusion and permeability mapping of hepatic malignancies using magnetic resonance imaging and a first-pass leakage profile model. NMR Biomed 15:164–173

Karlsson M, Nordell B (2000) Analysis of the Look-Locker T1 mapping sequence in dynamic contrast uptake studies: simulation and in vivo validation. Magn Reson Imaging 18:947–954

Knopp MV (1999) Estimating kinetic parameters from dynamic contrast-enhanced T1-weighted MRI of a diffusable tracer – standardized quantities and symbols. J Magn Reson Imaging 10:223-232

Knopp MV, Maxwell RJ, McIntyre D, Padhani A, Price P, Rathbone R, Rustin G, Tofts P, Tozer GM, Vennart W, Waterton JC, William SR, Workman P (2003) Assessment of anti-angiogenic and anti-vascular therapeutics using magnetic resonance imaging: recommendations for appropriate methodology for clinical trials. Proc Am Assoc Cancer Res

Krishnan S, Chenevert TL (2004) Spatio-temporal bandwidth-based acquisition for dynamic contrast-enhanced magnetic resonance imaging. J Magn Reson Imaging 20:129–137

Leach MO, Brindle KM, Evelhoch JL, Griffiths JR, Horsman MR, Jackson A, Jayson G, Judson IR, Tofts PS, Port R, Brix G, Larsson HBW, Shames DM, Parker GJ, Weisskoff RM, Evelhoch JL, Taylor JS, Mattiello J, Evelhoch JL (1991) Relative volume-average tumor blood flow measurement via deuterium nuclear magnetic resonance spectroscopy. Magn Reson Med 18:320-334

Mazaheri Y, Carroll TJ, Du J, Block WF, Fain SB, Hany TF, BDL Aagaard, Strother CM, Mistretta CA, Grist TM (2002) J Magn Reson Imaging 15:291–301

Mitchell DG (1997) MR imaging contrast agents – what's in a name? J Magn Reson Imaging 7:1–4

Redman BG, Esper P, Pan Q, Dunn RL, Hussain HK, Chenevert T, Brewer GJ, Merajver SD (2003) Phase II trial of tetrathiomolybdate in patients with advanced kidney cancer. Clin Cancer Res 9:1666-1672

Rijpkema M, Kaanders JHAM, Joosten FBM, van der Kogel AJ, Heerschap A (2001) Method for quantitative mapping of dynamic MRI contrast agent uptake in human tumors. J Magn Reson Imaging 14:457–463

Sodickson DK, McKenzie CA (2001) A generalized approach to parallel magnetic resonance imaging, Med Phys 28:1629-43

Song HK, Dougherty L, Schnall MD (2001) Simultaneous acquisition of multiple resolution images for dynamic contrast enhanced imaging of the breast. Magn Reson Med 46:503–509

Tsao J, Boesiger P, Pruessmann KP (2003) k-t BLAST and k-t SENSE: dynamic MRI with high frame rate exploiting spatiotemporal correlations. Magn Reson Med 50:1031–1042

Clinical Applications

9 Dynamic Contrast-Enhanced MRI in Cerebral Tumours

Xiao Ping Zhu, Kah Loh Li, and Alan Jackson

CONTENTS

X. P. Zhu, MD PhD
MR Unit and Department of Radiology, VA Medical Center, University of California, San Francisco, 4150 Clement Street, San Francisco, CA 94121 USA
K. L. Li, PhD
Department of Radiology, University of California, San Francisco, 4150 Clement Street, San Francisco, CA 94121 USA
A. Jackson, MBChB (Hons), PhD, FRCP, FRCR
Professor, Imaging Science and Biomedical Engineering, The Medical School, University of Manchester, Stopford Building, Oxford Road, Manchester, M13 9PT, UK

9.1 Introduction

The growth of new blood vessels, a process known as angiogenesis, is essential for tumour growth and development. Angiogenesis is regulated by stimulatory and inhibitory cytokines released from tumour cells, endothelial cells and macrophages (Assimakopoulou et al. 1997; Costello 1994; Lund et al. 1998; Luthert et al. 1986). Numerous studies have identified the role of the endothelial mitogen vascular endothelial growth factor (VEGF), which plays a critical role in stimulating angiogenesis in many cerebral tumours including gliomas and meningiomas. The production of VEGF and of other angiogenesis promoters is stimulated by regional hypoxia and hypoglycaemia as the tumour outgrows its existing blood supply (Damert et al. 1997a,b). Regional production of VEGF increases local capillary endothelial permeability, promotes endothelial mitosis and new vessel formation and has direct effects on local tissues to promote in-growth of the new vessels into the interstitial stroma. The level of expression of VEGF varies with tumour type and grade (Amoroso et al. 1997; Dietzmann et al. 1997) and the greatest expression is seen in cerebral gliomas where levels are directly related to tumour grade (Amoroso et al. 1997; Pietsch et al. 1997; Plate and Risau 1995). A similar relationship between VEGF and tumour behaviour is seen in meningiomas where high levels of VEGF expression have been described in vascular aggressive tumours and in tumours associated with peri-tumoral oedema (Bharara et al. 1996; Goldman et al. 1997; Provias et al. 1997). These observations have led to the development of new therapeutic agents designed to inhibit angiogenesis with the intention of restricting tumour growth and dissemination (Lund et al. 1998; Martiny-Baron and Marme 1995; Stratmann et al. 1997). The introduction of this class of therapeutic agent requires the development of imaging strategies to demonstrate and quantify the presence of angiogenic activity and to provide surrogate biological markers of anti-angiogenic activity of angiogenic inhibition strategies.

A striking histological feature of increasing grade and progression of many tumour types is an increase in the density of tumoral neovascularization which can be measured by estimation of microvascular density (MVD) (Aronen et al. 2000; Brem et al. 1972; Maxwell et al. 1991). This relationship between tumour behaviour and microvascular density on histological studies led several groups to investigate the measurement of fractional cerebral blood volume (CBV), using MR-based methods as a potential surrogate marker for MVD. Estimates of CBV can be derived from "bolus-tracking techniques" which use rapid dynamic MRI acquisitions to monitor the passage of standard contrast media through a capillary bed following venous injection (Rosen et al. 1989). In gliomas, cerebral blood volume derived in this way appears to correlate closely with both tumour grade and prognosis (Aronen et al. 1995). Other workers have attempted to improve the specificity of imaging techniques by identifying parameters which are sensitive to the marked variation in vessel size (Dennie et al. 1998; Donahue et al. 1999) or the resulting disturbances in blood flow, associated with angiogenesis (Kassner et al. 2000). These imaging parameters, describing vascular density, size variation and flow disturbance provide direct quantifiable indicators of the amount and structure of new blood vessels which have developed as a tumour has grown. An alternative approach to imaging angiogenesis relies on the direct effects of VEGF and other vasoactive cytokines which produce rapid and significant increases in local capillary endothelial permeability. Teleologically this permeability promoting effect can be envisaged as promoting easier transfer of nutrients and waste products and has been directly implicated in the passage of tumour cells into the blood stream and therefore with metastatic potential. Many groups have described methods based on dynamic MRI to quantify the permeability of vascular endothelia (Amoroso et al. 1997; Brasch et al. 1997; Dvorak et al. 1995; Jensen 1998). This can be achieved by the application of established pharmacokinetic models of contrast distribution (Brix et al. 1991; Larsson et al. 1990; Shames et al. 1993; Tofts et al. 1995; Tofts and Kermode 1991) to data acquired from dynamic MR relaxivity-based studies. In practice, the simultaneous acquisition of endothelial permeability, blood volume and vascular morphology data is desirable and several groups have proposed methods to combine these measurements in the same subject (Bhujwalla et al. 1999; Donahue et al. 1999; Henderson et al. 1999; Weisskoff et al. 1994a; Zhu et al. 1999, 2000c).

Dynamic contrast-enhanced MR imaging represents an important transition in radiology from a purely anatomy-based discipline to one in which physiologic patterns and tissue microstructure also can be investigated (Cha 2002; Hobbs et al. 2002). Dynamic contrast-enhanced MR imaging-derived parametric maps offer a new dimension to imaging in which both anatomic structures and spatial derivatives of specific descriptive physiological variables can be depicted with numerical accuracy and excellent spatial resolution. In vivo measurements of regional microvascular structure offer tremendous research and clinical opportunities (Cha 2002). In addition, calculation of true absolute values for physiological variables such as endothelial permeability surface area product (PS) and fractional tumour blood volume should be independent of patient and acquisition related variables, such as receive gain, contrast dosage, patient cardiovascular status etc.

9.2 Dynamic Susceptibility Contrast-enhanced MRI (DSCE-MRI)

Heterogeneous distributions of high and low magnetic susceptibility within voxels of an MR image can have a pronounced effect on the signal intensity due to the modification of local proton resonance, resulting in susceptibility induced T2 and T2* shortening (Wismer et al. 1988). Paramagnetic chelates, such as gadolinium DTPA alter the magnetic susceptibility of tissue and these effects monitored during the passage of a contrast bolus form the basis of T2-weighted and T2*-weighted contrast dynamic MRI (Rosen et al. 1989). The methodological considerations associated with data collection and analysis for dynamic susceptibility contrast-enhanced MRI (DSCE-MRI) have been discussed at length in previous chapters (see Chaps. 1, 4 and 7). In brief, a rapid dynamic collection with a temporal resolution of 2 s or less is performed using a heavily T2- or T2*-weighted acquisition. An intravenous bolus of paramagnetic contrast media is administered during the acquisition following a reasonable number of pre-injection baseline data sets and data collection is continued for 1–2 min. This data demonstrates the signal change due to the passage of the initial contrast bolus through the vascular bed and due to the early recirculation of contrast through the brain.

The analysis of the data requires transformation of signal change data into contrast concentration data which is a simple non-linear transformation (Rosen et al. 1990, 1991; Villringer et al. 1988). Contrast concentration time course data can then be used to extract parametric variables the most important of which in tumour studies is the relative cerebral blood volume (rCBV) (Aronen et al. 1994, 1995). A number of analysis techniques have been described and a common feature of many is the use of a gamma variate fitting procedure to define the shape and position of the first pass bolus (Rosen et al. 1990, 1991). The need for this step arises from the signal changes which occur after passage of the contrast bolus as contrast recirculates into the cerebral circulation from the periphery. As a result of this recirculation the contrast concentration fails to return to pre-enhancement levels. The use of a curve fitting technique allows us to estimate what the first pass bolus would have looked like if no recirculation had occurred and to calculate parametric images from the curve fit results, which are an idealised and smoothed representation of contrast concentration changes occurring due to the primary bolus passage.

9.2.1 Application of DSCE-MRI in Enhancing Tissues

The standard analysis techniques for DSCE-MRI contrast bolus studies assume that the signal change observed results entirely from contrast within the blood vessels. However, leakage of contrast into the interstitial space will cause additional signal changes, principally by relaxivity mechanisms. The effects of relaxivity and susceptibility contrast mechanisms on the observed signal change are contradictory and successful application of DSCE-MRI techniques in enhancing tissues requires that competing relaxivity effects be removed. In theory, the use of a true intra-vascular contrast media, or a contrast with negligible relaxivity effects, such as iron oxide, would allow pure susceptibility measurements. In fact, the restrictions of standard contrast media force us to use methods to separate as far as possible the susceptibility and relaxivity effects. The use of techniques with reduced T1 sensitivity such as low flip angle gradient echo-based sequences has become a common technique (Maeda et al. 1994; Aronen et al. 1994, 1995; Kassner et al. 1999). This technique effectively removes relaxivity effects although some workers have still observed residual effects in rapidly enhancing tumours (Maeda et al. 1993). The major problem with this method is the loss of signal to noise ratio produced by the reduction in flip angle (Kassner et al. 1999). However, this can be partially compensated by increased contrast doses (Aronen et al. 1994). Another approach to reducing T1 sensitivity is to use a dual echo technique in which the T1-weighted first echo is used to correct the predominantly T2-weighted second echo (Miyati et al. 1997). This method is a simple and effective way to remove relaxivity effects. Unfortunately, the requirement for two echoes places considerable demands on the sampling time and inevitably restricts the number of samples, and therefore slices which can be obtained. In addition, the use of a calculated parameter of this type derived from single pixels in noisy data sets produces a mathematical coupling effect on background noise which adversely affects the signal-to-noise ratio (Kassner et al. 1999). The use of pre-enhancement techniques is based on the signal changes that occur in response to changing concentrations of contrast media. The change in signal intensity resulting from T1 shortening is bi-exponential so that for any given sequence, there is a plateau phase during which signal intensity remains relatively constant. The position and length of this plateau phase will vary with the sequence. The effect of this response curve is that pre-enhancement of tumours will reduce the relaxivity based signal intensity responses to subsequent contrast doses (Fig. 9.1). The major problem with this approach is that the efficiency of the technique is dependent on the interstitial contrast concentration at the time of the bolus passage. Since tumours show differing contrast diffusion rates (Bullock et al. 1991; Gowland et al. 1992) this concentration cannot be accurately predicted although it can be measured (Kassner et al. 1999). Elimination of T1 shine through requires an interstitial contrast concentration (Gd DTPA-BMA) greater that 0.4 mmol/l and this figure will differ slightly depending on the relaxivity of the contrast (Kassner et al. 1999). Another problem with pre-enhancement techniques is the residual contrast effects seen in sequential dynamic susceptibility contrast experiments (Levin et al. 1995). The use of sequential injections demonstrates a change in the signal response curve so that it no longer conforms to a gamma variate pattern and demonstrates elevation of $\Delta R2$ during the re-circulation phase. The cause of these effects is unclear although, they do not appear to result from relaxivity changes (Kuhl

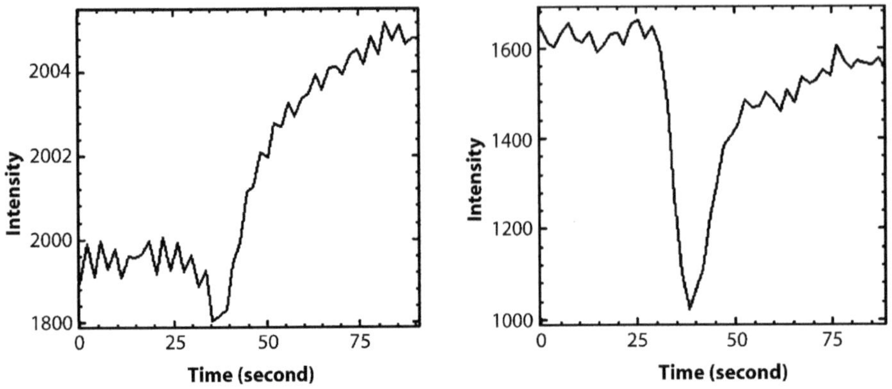

Fig. 9.1. Signal intensity time course curves from a T2*-weighted dynamic acquisition. The data is from a region of interest over a large meningioma. The *left-hand* curve shows an initial signal drop due to susceptibility effects followed by a rapid rise due to residual relaxivity effects. The *right-hand* curve shows the same region of interest following a pre-enhancement with contrast administered approximately 5 min prior to the dynamic experiment. The relaxivity effects have been removed and there is no residual T1 "shine through"

et al. 1997; LEVIN et al. 1995). The effect of these changes is that analysis of the perfusion data using standard gamma variate fitting techniques becomes less accurate and may result in significant levels of fitting error (KUHL et al. 1997).

Each of these methods has specific advantages and disadvantages. The choice of technique must be made by considering the requirements of the individual study. Dual echo methods will reliably eliminate relaxivity effects but suffer from poor signal to noise ratio and limited sampling volume. Low flip angle methods provide reliable elimination of relaxivity effects as long as TR times are adequate. These methods are fast but also suffer from poor signal to noise ratios in normal tissue. Pre-enhancement techniques allow the use of sequences which are sensitive to relaxivity changes and consequently provide good signal to noise ratio. However, the effect of pre-enhancement on T1 shine through will be highly dependent on tumour contrast concentration which will vary from tumour to tumour. In addition, pre-enhancement causes residual bolus effects, which affect the dynamics of signal change during subsequent bolus passage in normal tissue.

9.2.2 Parametric Images from DSCE-MRI

A number of physiological parametric maps can be calculated from DSCE-MRI data. Parametric images of relative blood volume (rCBV) can be derived from the area under the contrast concentration time course curve either by direct calculation from the fitted gamma variate curve or by numerical integration based on contrast concentration time course data. Parametric images of bolus arrival time such as time of arrival (T0) or time to peak concentration (TTP) can also be accurately derived (CASE et al. 1987; FISEL et al. 1991), however accurate calculation of regional blood flow is difficult due to effects such as bolus dispersion and arrival time delays (CALAMANTE et al. 2000) (see Chap. X). In tumours accurate estimation of regional blood flow is further complicated by the difficulty of identifying an appropriate arterial input function and the presence of large high flow vessels in some aggressive tumours, which invalidate the assumptions underlying the standard perfusion analysis techniques.

Another interesting parameter provides an estimate of abnormality during the recirculation phase of the contrast passage. This "relative recirculation" (rR) parameter quantifies any abnormal elevation of the contrast concentration in the period immediately after the passage of the contrast bolus (KASSNER et al. 1999, 2000, 2001). In theory the rR will be increased by local vascular factors such as absolute flow rate, flow rate heterogeneity and therefore by local perfusion pressure.

Another useful parametric map represents the scaled fitting error (SFE) of the curve fitting process used in the analysis (Fig. 9.2). At first sight this is counterintuitive but it is important to realise that where the analysis technique involves fitting gamma variant curve and it is the curve fit data which is then used to calculate the physiological parameters. In areas where the original data is noisy or does not

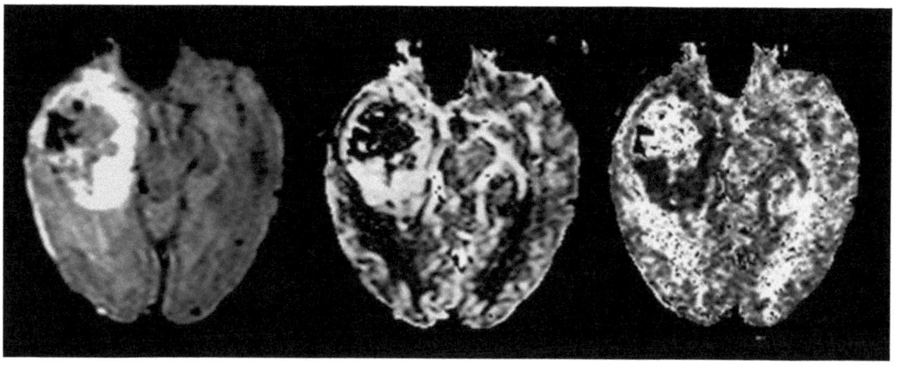

Fig. 9.2. Images from a patient with a large right-sided grade four glioma. The *left-hand* image is a post-contrast T1-weighted image from the dynamic data series showing typical enhancement within the rim of the tumour. The central image shows a map of rCBV which demonstrates both the enhancing portion of the tumour, cerebral blood vessels and the raised values in grey matter compared to white. The *right-hand* image is a map of scaled fitting error (SFE) which shows low fitting error is in the area of enhancing tumour, blood vessels and grey matter but higher fitting error is in white matter and in the central non-enhancing portion of the tumour

conform well to the gamma variant shape a curve fit will be produced which is inaccurate. The effect of this is that estimates of physiological parameters such as rCBV will have different confidence levels in each voxel. Comparisons of CBV and other parameters between areas become meaningless if the measurements in one of these areas simply represent artefacts due to the application of curve fitting technique to noisy data. The SFE map is a parametric image of the level of confidence which can be given to individual pixels in the associated parametric map (Jackson et al. 2000; Zhu et al. 1997, 2000b). The SFE map has the advantage of providing both visual and quantitative comparison of measurement accuracy between anatomical regions within an image which allows subjective identification of areas with low fitting accuracy. More importantly, individual pixel values can be used to exclude parametric measurements from voxels with high errors prior to numerical or statistical analysis. The SFE map is also particularly useful as a quality control tool allowing immediate identification of variations in data quality.

Figure. 9.3a,b shows rCBV and SFE maps from a patient with a supra-tentorial meningioma. White matter surrounding the tumour has a low blood volume and high SFE. The tumour blood volume is higher than that of either grey or white matter. SFE in the tumour is similar to that in grey matter except some focal areas (arrow). In these areas both tumour blood volume and SFE are much higher than in grey matter or in the bulk of the tumour. The $\Delta R2^*(t)$ from a voxel in normal grey matter and from one of these areas of elevated SFE fitted with gamma variate functions are shown in Fig. 9.3c and d, respectively. The $\Delta R2^*(t)$ from the area of high SFE shows marked broadening of the temporal response to contrast medium passage with poor fitting of the later part of the gamma variate function. In this case the SFE map allows identification of poor quality data due to low perfusion in the peri-tumoral oedema and also identifies areas of poor conformance to the gamma variant shape due to flow disturbances within the tumour despite high signal-to-noise ratio in the original data.

9.2.3
Effect of Acquisition Protocol on Calculated Values of CBV

DSCE-MRI data may be collected using either T2-weighted spin-echo or T2*-weighted gradient-echo techniques. T2-weighted spin-echo imaging techniques are relatively more sensitive to the small vessels than T2*-weighted gradient-echo imaging techniques which have equivalent sensitivity to both small and large vessels (Fisel et al. 1991; Weisskoff et al. 1994c). This aspect of susceptibility physics has been demonstrated using simulation studies (Boxerman et al. 1995). In some applications the use of T2-weighted spin-echo imaging is considered a significant advantage since it allows the collection of signal change arising predominantly in the capillary beds. However, tumour angiogenesis is a complex process, and the existence of larger neovascular structures (such as feeding arteries and draining veins) associated with malignant tumours

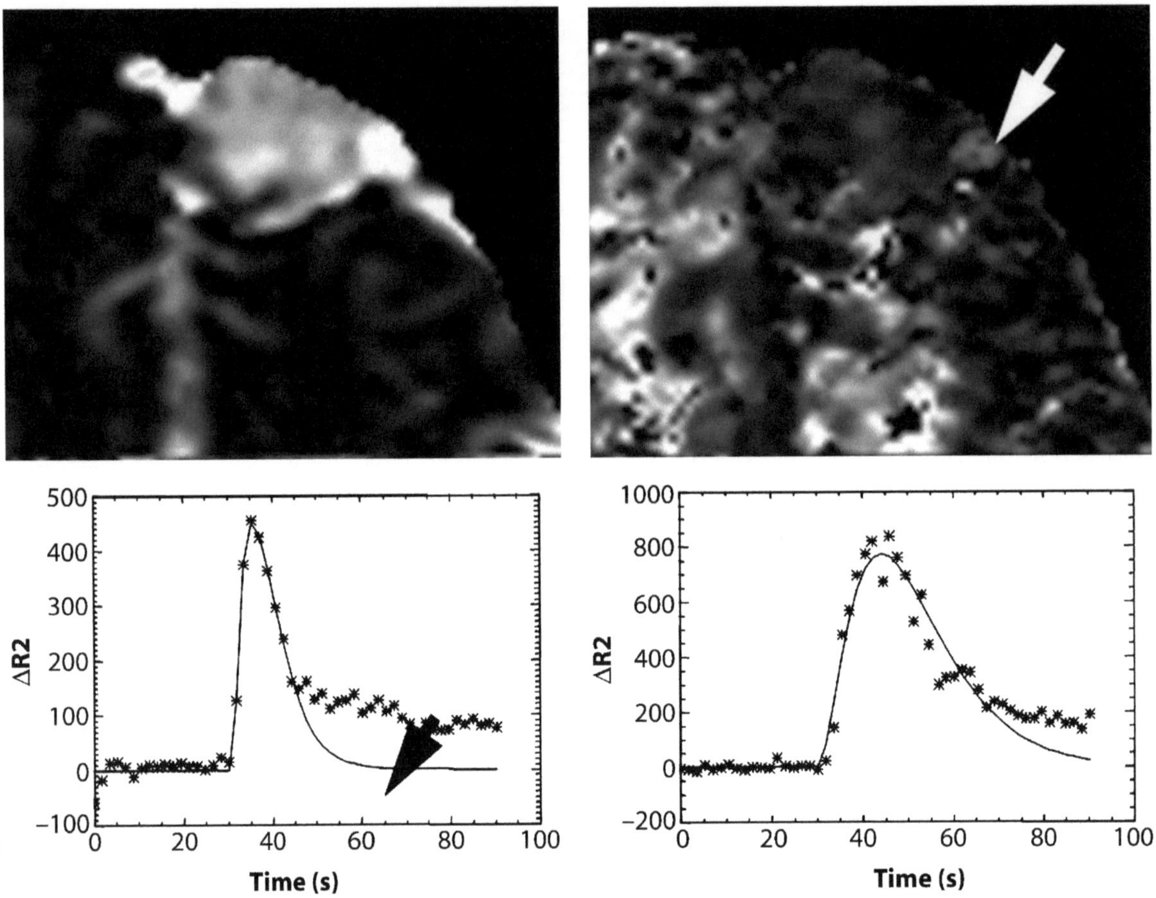

Fig. 9.3a-d. a,b rCBV (**a**) and SFE (**b**) maps from a patient with a supra-tentorial meningioma. Pre-enhancement has been used to eliminate residual T1 effects. rCBV values in tumour are higher than in normal grey and white matter. SFE values in the majority of the tumour are similar to grey matter, but a focal bright area of increased SFE is seen (*arrow*). **c,d** $\Delta R2^*(t)$ curves from a voxel in grey matter and a voxel from the areas of increased SFE in the tumour. The $\Delta R2^*(t)$ curve in tumour has a high SNR (=28.9) and much higher SFE (=0.21) compared to grey matter (SNR=12.9, SFE=0.02). The increase in SFE in the tumour can be seen to result from broadening of the underlying first pass curve

is common (Reinhold and Endrich 1986; Song et al. 1984). Many tumours also have abnormally large capillary vessels which may be two to three times those of normal tissues (Dewhirst et al. 1989). The gradient-echo echo-planar technique (GE-EPI) provides hemodynamic maps by representing the effects of total blood volume from capillaries to large vessels and weights all vessels approximately equally (Boxerman et al. 1995). In addition, this technique, because of its greater sensitivity to the susceptibility effect, can provide an rCBV map with the use of a low-dose contrast medium (Boxerman et al. 1995). Therefore, GE-EPI imaging may be expected to be suitable for evaluating brain tumours, especially those with large vessels (Sugahara et al. 1998). Sugahara et al. (2001) compared GE-EPI and SE-EPI for differentiating between low and high-grade gliomas. Six patients with low-grade gliomas and 19 patients with high-grade gliomas underwent two DSCE-MRI examinations, one produced by a GE- and the other by an SE-EPI technique. Maximum rCBV ratios normalized to the rCBV of contralateral white matter were calculated for comparison. Maximum rCBV ratios of high-grade gliomas obtained with the GE-EPI technique (mean, 5.0±2.9) were significantly higher than those obtained with the SE-EPI technique (mean, 2.9±2.3) (p=0.02). Maximum rCBV ratios of low-grade gliomas obtained with the GE-EPI technique (mean, 1.2±0.7) were almost equal to those obtained with the SE-EPI technique (mean, 1.2±0.6, p=0.66). These authors concluded that the GE-EPI technique is more useful for differentiating low grade from high-grade gliomas than the SE-EPI.

9.2.4
Clinical Applications of DSCE-MRI in Cerebral Tumours

9.2.4.1
Glioma

Despite a large number of studies describing quantitative imaging of angiogenesis in brain tumours none of these techniques has yet passed into routine clinical use. There is however considerable evidence that these quantitative techniques can provide valuable clinical data concerning tumour type, tumour grade and therapeutic response. Tables 9.1 and 9.2 show maximum and mean literature values for rCBV derived from DSCE-MRI in brain tumours. In gliomas tumour capillary blood volumes measured by dynamic susceptibility contrast-enhanced MR imaging have been shown to correlate with and predict tumour grade (Aronen et al. 1994, 1995, 2000; Knopp et al. 1999). More importantly rCBV maps identify areas of malignant transformation or tumour dedifferentiation allowing more accurate targeting of stereotaxic biopsies and therefore more accurate estimation of maximal tumour grade (Knopp et al. 1999). Histological comparisons show close relationships between rCBV values within tumours and histological features indicative of tumour aggression including mitotic activity and vascularity (Aronen et al. 1995; Sugahara et al. 1998). Direct comparison between rCBV mapping and other indicators of malignancy such as fluorodeoxyglucose positron emission tomography shows close agreement between local rCBV values and glucose uptake and moderate significant correlation between maximal glucose uptake and rCBV (n=21; r=0.572; p=0.023) (Aronen et al. 2000). Similar comparisons with thallium-201 SPECT show greater sensitivity to glioma grade (Lam et al. 2001) and also higher sensitivity for demonstrating early tumour recurrence after therapy. In addition to potential uses in identifying grade and histological heterogeneity it has also been shown that susceptibility based methods may be helpful in differentiating between primary gliomas and solitary cerebral metastasis on the basis of difference in peri-tumoral rCBV measurements (Cha et al. 2000, 2002). In metastatic tumours,

Table 9.1. Comparison of literature values of maximum tumour relative cerebral blood volume

Authors	Gliomas				Lymphomas	Methods
	I	II or "low grade"	III or AG	IV or GB		
Sugahara et al. 1998		1.26±0.55 (n=4)	5.84±1.82 (n=10)[a]	7.32±4.39 (n=12)		GE EPI
			1.53±0.75 (n=4)[b]			
Knopp et al. 1999		1.44±0.68 (n=3)	6.53±2.67 (n=5)	4.72±2.76 (n=21)		GE EPI
Lam et al. 2001	2.306±1.135 (n=6)	4.263±3.552 (n=3)	11.308±6.863 (n=4)	8.268±2.020 (n=6)		GE EPI
Cha 2001		2.27 (n=1)	10.92±7.42 (n=3)	9.24±8.29 (n=3)	2.11±.53 (n=4)	GE EPI
Cha 2002		1.86±0.77 (n=13)	4.03±2.43 (n=19)	5.5±4.5 (n=51)	1.44±0.67 (n=19)	GE EPI
Sugahara 2001		1.2±0.7 (n=6)	5.0±2.9 (n=19)			GE EPI
		1.2±0.6 (n=6)	2.9±2.3 (n=19)			SE EPI
Aronen et al. 2000		1.51±0.94 (n=8)	4.13±2.92 (n=13)			SE EPI

AG, anaplastic gliomas; GB, glioblastoma.
[a] Anaplastic gliomas with enhancement.
[b] Anaplastic gliomas without enhancement.

Table 9.2. Comparison of literature values of overall mean of tumour relative cerebral blood volume

Authors	Gliomas			Meningioma	AN	Metastasis	GC	Methods
	II or "low grade"	III or AG	IV or GB					
DOMINGO 1998				6.01±2.52 (*n*=8)				SE Turbo-Flash
PARDO 1994	3.07±3.81 (*n*=3)	1.03±0.14 (*n*=2)	2.35±2.14 (*n*=3)					SE EPI
ZHU 2000a		0.91±.11 (*n*=2)	2.00±0.23 (*n*=3)	3.10±1.73 (*n*=5)	1.67±1.08 (*n*=5)			GE EPI
JACKSON et al. 2001		2.51±1.09 (III, *n*=7; IV, *n*=4)						GE EPI
LAW 2002		2.87±1.89[a] (*n*=24)				3.05±1.79[a] (*n*=12)		GE EPI
		1.31±0.97[b] (*n*=24)				0.39±0.97[b] (*n*=12)		
YANG 2002							1.02±0.42 (*n*=7)	
WENZ et al. 1996	2.77±1.98 (*n*=19)							GE (SD-Flash)
FUSS et al. 2000	1.9±1.1 (*n*=25)							GE (SD-Flash)
MIYATI et al. 1997		2.21 (*n*=1)	3.08 (*n*=1)	6.34±2.45 (*n*=6)	2.58±1.59 (*n*=8)	4.35±1.17 (*n*=3)		Double echo
UEMATSU et al. 2000				9.88±9.99 (*n*=11)	2.14±1.71 (*n*=11)			Double echo
VONKEN et al. 2000			4.74 (*n*=1)					Double echo

AG, anaplastic gliomas; GB, glioblastoma; AN, acoustic neuroma; GC, gliomatosis cerebri.
[a] Tumoral region.
[b] Immediate peritumoral region.

peri-tumoral oedema represents pure vasogenic oedema caused by increased interstitial water due to microvascular extravasation of plasma fluid and proteins through the inter-endothelial spaces (STRUGAR et al. 1994, 1995). In high grade gliomas the peri-tumoral region represents a variable combination of vasogenic oedema and tumour cells infiltrating along the perivascular space (STRUGAR et al. 1995). JACKSON et al. (2002b) have described a correlation between the skewness of the distribution of rR values in glioma and tumour grade (Fig. 9.4). Unlike rCBV and rCBF measurements the rR parameter is affected only by areas of local ischemia, decreased perfusion pressure or vascular tortuosity. Changes are therefore typically seen at the boundary between well vascularized peripheral growing tissue and central tumour necrosis in high grade gliomas. Abnormalities of rR affected only a small subpopulation of the tumour pixels and presumably represent areas of incipient ischemia where the ischaemic drive for production antigenic cytokines such as VEGF is high. The limitation of abnormalities of rR to small subgroups of pixels within the tumour explains the lack of any relationship between mean values and tumour grade in the presence of a strong correlation between skewness of the distribution and grade.

9.2.4.2 Other Tumour Types

Susceptibility contrast MRI has also been reported to have some clinical benefit in a number of other tumour types. Comparisons of rCBV values between

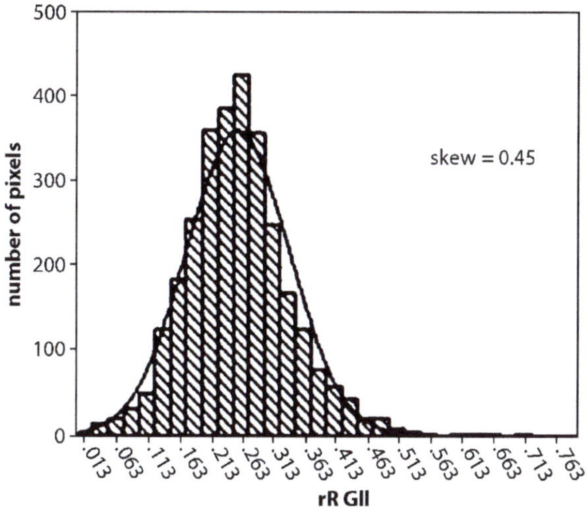

a

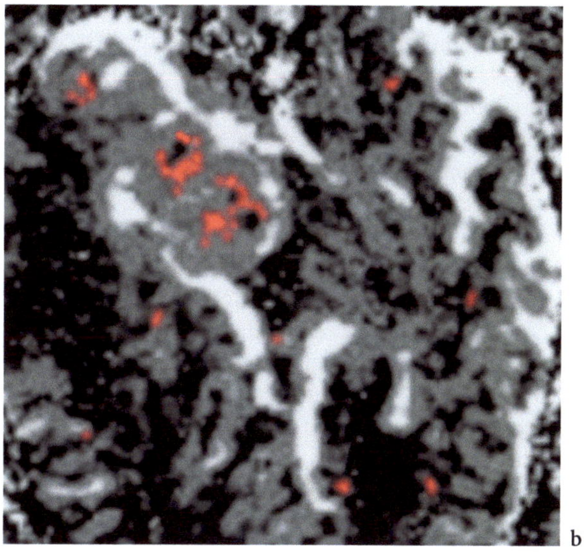
b

Fig. 9.4a,b. a The histograms on the *top left* and *lower left* illustrate the frequency of particular values of rR in a low-grade (*top*) and high-grade (*bottom*) glioma. Note that the low-grade glioma shows close conformity to a normal distribution whereas the high-grade glioma shows a marked skewness of the distribution with a tail to the right. This pattern can be explained by the image on the right side (**b**) which shows a grey scale map of rCBV in a high-grade glioma overlaid with red areas which indicate significant elevation of rR. Only a small number of the total tumour pixels exhibit abnormally high values. Because of this mean values from large regions of interest are not significantly different between different tumour types; however, the shape of the distribution is affected

common extra-axial tumours and typical enhancing inter-axial tumours shows the extra-axial masses to typically have higher values. This may be helpful in differentiating between intra and extra-axial masses where other features are equivocal (Sugahara et al. 1999a). Measurements of rCBV can also help to differentiate between meningiomas, schwannomas and neurinomas, meningiomas having higher values of rCBV than those seen in neurinomas or schwannomas (Maeda et al. 1994; Miyati et al. 1997; Sugahara et al. 1999b; Uematsu et al. 2000). rCBV can also provide supplementary information to differentiate between malignant lymphoma and glioma because the absence of tumour neovascularization of malignant lymphoma leads to low rCBV, in contrast to malignant gliomas (Sugahara et al. 1999b). Cha et al. (2002) in their series of 19 consecutive patients (Cha et al. 2002) with primary cerebral lymphoma found the maximum rCBV ranged from 0.42 to 3.41 (mean, 1.44±0.67) compared to mean rCBV of 5.5±4.5 in 51 patient with glioblastoma multiforme ($p<0.01$) (Frazzini et al. 1999). Cha et al. (2002) have also identified DSCE-MRI as a useful diagnostic tool in differentiating tumefactive demyelinating lesions (TDL) from intracranial neoplasms. The rCBV values of TDLs ranged from 0.22 to 1.79 ($n=12$), with a mean of 0.88±0.46 (SD) compared to rCBV values of 1.55–19.20 ($n=11$), with a mean of 6.47±6.52 in intracranial neoplasms ($p=0.009$).

9.3 Dynamic Relaxivity Contrast-enhanced MRI (DRCE-MRI)

DRCE-MRI has significant advantages over DSCE-MRI in tumour applications since it is possible to calculate physiological parameters related to contrast leakage including endothelial permeability surface area product, the size of the EES and the size of the intravascular space (rCBV) from a single data acquisition. Unfortunately estimates of blood flow using DRCE-MRI remain difficult and relatively inaccurate. The acquisition strategies for T1-weighted dynamic data have been discussed in detail in Chap. 5. Briefly, T1-weighted multislice or volume acquisitions are acquired during injection of intravenous bolus of paramagnetic contrast. Assuming that the baseline T1 of the tissues can be measured, the relaxivity effects of the contrast observed in the data can be used to measure concentration time course curves from blood vessels and pixels of interest. The analysis techniques for this type of data range from simple observational measurements based on signal intensity changes to more complex pharmacokinetic analyses based on calculated contrast concentration maps many of which require estimation or measurement of a vascular input function. The additional complexity of the pharmacokinetic analyses is intended to produce parametric measurements which are independent of scanner and acquisition variables and which truly reflect important physiological features. A detailed overview of the analysis methods available for the interpretation of DRCE-MRI data is presented in Chap. 6.

As early as at the beginning of 1990s investigators from three different groups almost simultaneously proposed the use of T1W dynamic MRI to measure endothelial permeability of pathological capillary beds (Brix et al. 1991; Larsson et al. 1990; Tofts and Kermode 1991). Each of these groups based their technique around a slightly different pharmacokinetic model which described the leakage of contrast from the blood into the extravascular extracellular space (EES) and the pattern of contrast loss into other tissues and by renal excretion. These models were developed before the descriptions of T2* based techniques and before bolus passage analysis methods were described. Only 1 year later, Dean et al. (1992) showed that high temporal resolution T1 dynamic data can also be used to calculate other indices of cerebral hemodynamics, particularly cerebral blood volume (Dean et al. 1992) based on the first passage of the contrast bolus through the vascular bed. Over the next several years workers continued to analyse DRCE-MRI data to produce measurements of rCBV (Bruening et al. 1996; Hacklander et al. 1996b, 1997; Kwong et al. 1995) whilst other groups used similar data to estimate the bulk transfer coefficient K^{trans} as an indicator of endothelial permeability. By the end of the last century several groups started to extend analysis techniques for DRCE-MRI to provide simultaneous mapping of both rCBV and K^{trans}.

9.3.1
Data Analysis: Simple Descriptive Metrics

The extraction of appropriate parameters to describe the microvasculature from DRCE-MRI data is extremely complex and many analysis approaches are available. The choice of analysis approach will depend on the expected changes in the tissue and the features of the microvasculature that are of particular interest. The simplest approach relies on quantification of the signal changes observed which avoids the complexity of calculating contrast concentrations from the baseline data and does not require estimation or measurement of a vascular input function. A number of such metrics have been described and most characterise the shape of the signal time course curve based on the maximal amplitude of the signal and the time taken to reach this value. Other metrics compare the signal intensity achieved in any given period of time (Bullock et al. 1991; Fujii et al. 1992; Nagele et al. 1993; Worthington et al. 1991). Typical metrics include T90 (Stack et al. 1990) which is the time taken to reach 90% of the maximal enhancement value and the maximal intensity change per time interval ratio (MITR) (Flickinger et al. 1993) which measures the maximal rate of enhancement. These simple metrics suffer from problems which have limited their application and led many authors to adopt the more complex pharmacokinetic analysis techniques. Firstly, metrics based entirely on signal change characteristics will reflect both intra- and extra-vascular contrast concentrations and separation of signal change effects due to blood flow and contrast leakage is impossible. More worryingly these measurements will be unpredictably affected by variations in scanning protocol, including those that may change from scan to scan such as receiver gain. Despite these limitations, simple measurements based on signal change alone can be diagnostically useful and have found application in a number of clinical applications (Knopp et al. 1999, 2001).

9.3.2
Pharmacokinetic Analysis

As discussed in Chaps. 1 and 6 pharmacokinetic analyses of T1-weighted DRCE-MRI data have a number of theoretical advantages (Tofts et al. 1999). The use of pharmacokinetic models leads to the derivation of parameters which are independent of the scanning acquisition protocol or any features associated with it. In theory, such parameters should reflect only tissue characteristics supporting the use of these measurements in multicentre studies employing varying image acquisition protocols and equipment (Padhani and Husband 2001) (see also Chap. 16). In practice pharmacokinetic analysis is complex and the choice of pharmacokinetic model controls the range of parameters that can be extracted. Each of the pharmacokinetic analysis approaches uses curve fitting techniques to characterise the arterial input function (AIF) and the tissue contrast concentration curve. These two functions are then used to derive the parameters which control the relationship between AIF and tissue contrast content. The simplest of the pharmacokinetic models, such as that described by TOFTS and KERMODE (1991), use a single arterial input function and time course data describing contrast concentration from individual voxels to calculate the size of the EES (v_e) and the bulk transfer coefficient K^{trans} (Fig. 9.5). The transfer constant K^{trans} is simply a mathematical function which describes the relationship between the AIF and contrast concentration changes occurring in the voxel. The measurements of K^{trans} will be affected by blood flow, blood volume, endothelial permeability and endothelial surface area. Changes in any of these variables can produce observable changes in K^{trans} and the specific contribution of the individual components cannot be identified. This simple model is also based on an invalid assumption that the signal changes within the measurement voxels will result entirely from extravasated contrast medium within the EES. This gives rise to significant errors in voxels which contain intra vascular contrast where measured values of K^{trans} will be artificially elevated. Despite these shortcomings the model described by TOFTS and KERMODE (1991) is widely used (PARKER et al. 1997). Nonetheless, many workers have attempted to refine the pharmacokinetic analysis to provide more accurate estimates of individual microvascular parameters, particularly permeability surface area product and fractional plasma volume (v_p). One reason for the refinement of analysis techniques is that these techniques are widely used in drug development and discovery and, particularly, for the study of new anti antiangiogenic therapies (see Chap. 16). As we have described the angiogenic cytokine VEGF has a specific action in promoting endothelial permeability so that measurements of endothelial permeability, uncontaminated by other factors, are highly desirable (GOSSMANN et al. 2000; VAN DER SANDEN et al. 2000; WEISSLEDER and MAHMOOD 2001). The basic pharmacokinetic model described by the TOFTS and KERMODE can be modified to specifically model the signal contribution produced by contrast medium within the plasma (TOFTS et al. 1999). This reduces errors due to the so-called pseudo permeability effect where intravascular contrast gives rise to falsely elevated values of K^{trans}. The K^{trans} values from this modified model will differ significantly from those obtained with the classic model and will more accurately reflect changes in permeability surface area product although they will still be dependent on adequate blood flow to the tissue to support contrast leakage. It is important to realise that the exact meaning of the K^{trans} variable depends on the method of analysis used. More complex pharmacokinetic models such as those described by ST. LAWRENCE and LEE (1998) allow direct estimation of local tissue blood flow (F) in addition to v_e, v_b and permeability surface area product (P.S). It seems clear that models such as these are more desirable than simpler approaches to the analysis; however, separate identification of extra fitting parameters requires more accurate and reliable curve fitting and is associated with increased variability and susceptibility to noise (Fig. 9.6). The choice of analysis techniques is therefore not straightforward and must be made based on the likely quality of the data to be obtained and the specific biological question to be answered.

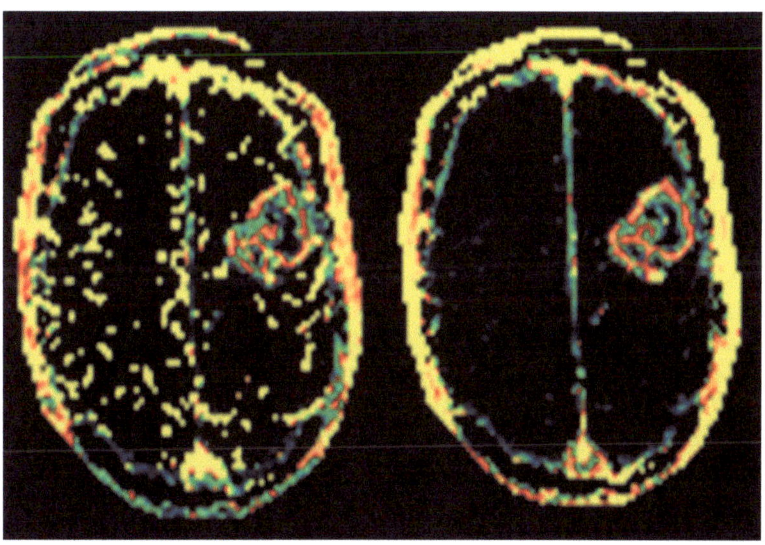

Fig. 9.5. Parametric images of K^{trans} and v_e in a patient with a left-sided high-grade glioma generated using a simple pharmacokinetic model. Note the elevated values of K^{trans} within the enhancing rim of the glioma and also in the region of large blood vessels. These areas of pseudo-permeability reflect an error in the model which fails to account for the presence of large amounts of intravascular contrast

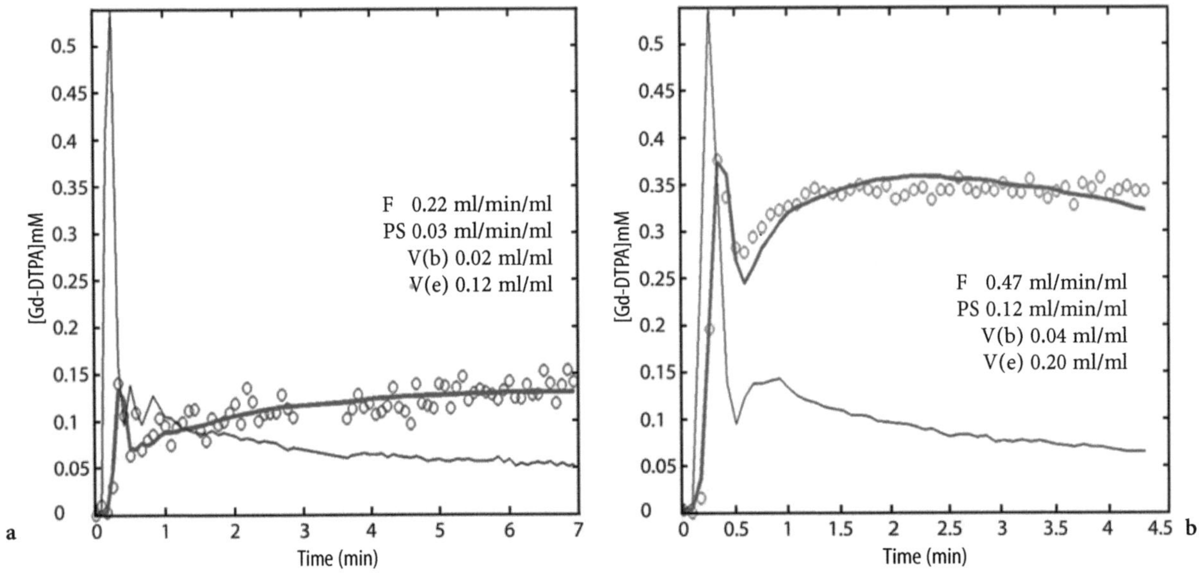

Fig. 9.6. Curve fitting analysis using the model described by St. Lawrence and Lee (1998) to produce estimates of flow (F), endothelial permeability surface area product (PS), vascular distribution volume ($V_{(b)}$) and extra-cellular extra-vascular space ($v_{(e)}$). The curve fits are performed on data from regions of interest in order to improve signal-to-noise ratio. The underlying data points are indicated by *circles*. The arterial input function used in the analysis is shown in grey. The results showed clear differences in all parameters between a low-grade glioma (*left*) and a high-grade glioma (*right*).

9.3.3 Clinical Applications of DRCE-MRI in Cerebral Tumours

9.3.3.1 Gliomas

Simplistic metrics based on estimations of the time taken to reach maximal enhancement or some proportion of maximal enhancement was shown to relate to tumour type as early as 1992 (Fujii et al. 1992; Nagele et al. 1993). Studies comparing different tumours found significant differences in the early enhancement pattern, which were believed to relate both to tumour vascularity and to the rates of contrast leakage into the interstitial space. Later studies concentrated on the use of derived physiological parameters using pharmacokinetic analysis (Andersen and Jensen 1998) in an attempt to separate variations in enhancement pattern resulting from local variations in blood volume from those resulting from variations in blood flow and endothelial permeability. In an early study of this type Andersen and Jensen (1998) examine differences in tumour capillary permeability, K^{trans} and extra-cellular distribution volume (v_e) in human intracranial tumours based on a two compartment pharmacokinetic model (Ohno et al. 1978; Paulson and Hertz 1983; Larsson et al. 1990; Tofts and Kermode 1991). Results from 17 brain tumour patients (seven glioblastoma, four metastasis, and six meningioma) showed that K^{trans} was significantly higher in meningiomas than in glioblastomas and metastasis (see Tables 9.3-9.5). Studies comparing tumour grade and estimates of K^{trans} show a close correlation in gliomas with higher values seen in grade three and grade four tumours than in low-grade minimally enhancing masses (Haroon et al. 2002b; Roberts et al. 2000, 2001). These studies also showed strong correlation between K^{trans} and mitotic indices derived from histological samples and that the correlation between K^{trans} and grade was higher than the correlation between mitotic indices and fractional blood volume measurements and grade suggesting high sensitivity for aggressive behaviour in glial tumours.

9.3.3.2 Other Tumour Types

Zhu et al. (2000c) calculated K^{trans} and v_e maps in patients with meningioma, glioma and acoustic neuroma. K^{trans} in acoustic neuromas was found to be consistently lower than those observed in meningioma and glioma however an observation of interest was

Table 9.3. Comparison of literature values of overall mean of transfer constant, K^{trans} (min^{-1})

Authors	Gliomas			Meningiomas	AN	Lymphoma	Metastasis	Methods
	Grade II	Grade III or AG	Grade IV or GB					
Andersen and Jensen 1998			.039±.026 (*n*=7)	.143±.109 (*n*=6)			.072±.042 (*n*=4)	a
Zhu et al. 2000b		.085±.021 (*n*=2)	.220±.044 (*n*=3)	.278±.195 (*n*=5)	.228±.078 (*n*=5)			Tofts
Roberts et al. 2000	.019±.017 (*n*=8)	.046±.029 (*n*=7)	.094±.038 (*n*=6)					In house
Johnson et al. 2002			.084±.096 (*n*=7)	.378±.192 (*n*=7)		.036±.012 (*n*=6)		b
Haroon et al. 2002b	<0.01 (*n*=1)	.034±.006 (*n*=2)	.044±.020 (*n*=8)					FPLP

GB, glioblastoma; AG, anaplastic glioma; AN, acoustic neuroma.
[a] A generalised plasma-interstitial two-compartment kinetic model ignoring the direct intravascular tracer contribution.
[b] A generalised plasma-interstitial two-compartment kinetic model including the direct intravascular tracer contribution fitted to first pass data.

Table 9.4. Comparison of literature values of overall mean of fractional plasma volume, v_p (%)

Authors	Gliomas			Meningioma	Lymphoma	Methods
	Grade II	Grade III	Grade IV or GB			
Roberts et al. 2000	1.88±1.64 (*n*=8)	3.20±1.96 (*n*=7)	4.13±2.54 (*n*=6)			In house
Ludemann et al. 2000	1.13±0.19 (*n*=5)	1.69±0.38 (*n*=6)	3.13±1.26 (*n*=12)			In house
Johnson et al. 2002			2.84±1.72 (*n*=7)	5.35±3.92 (*n*=7)	1.42±0.53 (*n*=6)	a
Haroon et al. 2002a	<1.0 (*n*=1)	3.51±.5 (*n*=2)	4.93±.84 (*n*=8)			FPLP

[a] A generalised plasma-interstitial two-compartment kinetic model including the direct intravascular tracer contribution fitted to first pass data.

Table 9.5. Comparison of literature values of overall mean of leakage space, v_e (%)

Authors	Gliomas		Meningiomas	AN	Metastasis	Methods
	Grade III	Grade IV or GB				
Andersen and Jensen 1998		19.1±13.5 (*n*=7)	29.5±17.4 (*n*=6)		16.0±6.9 (*n*=4)	a
Zhu et al. 2000b	25.0±1.4 (*n*=2)	19.3±3.8 (*n*=3)	33.8±8.5 (*n*=5)	53.4±10.4 (*n*=5)		Tofts

AN, acoustic neuroma.
[a] A generalised plasma-interstitial two-compartment kinetic model ignoring the direct intravascular tracer contribution.

that measurements of v_e (the size of the extra-cellular extra-vascular space) were higher in meningioma (34%±7%) than in glioma (22%±4%) and were consistently highest in the acoustic neuroma (53%±9%, $p<0.001$) (see Tables 9.3-9.5). These results agree with those of previous investigators (LONG 1973) who demonstrating very large extra-cellular spaces in schwannomas using fluorescence and electron microscopy. Several workers have shown relationship between the severity of peri-tumoral or oedema in meningioma, VEGF expression and measure values of K^{trans} (BITZER et al. 1998).

9.3.3.3 Therapeutic Monitoring

Several workers have documented radiation induced changes in normal brain and tumours including gliomas and meningioma (FUSS et al. 2000; WENZ et al. 1996). These studies have shown short-term increases and medium to long-term decreases in K^{trans} in response to radiation therapy in both normal brain and tumour. More importantly some studies have shown that DRCE-MRI data can differentiate between patients who show subsequent local tumour control and those who do not (HAWIGHORST et al. 1998). Other workers have also suggested that dynamic contrast-enhanced techniques may be useful in differentiating between tumour recurrence, characterised by high rCBV and K^{trans}, and radiation necrosis characterised by low values (ALAVI et al. 1988; BOXERMAN et al. 1995; COHEN and WEISSKOFF 1991; FRAHM et al. 1986; MAEDA et al. 1993; REINHOLD and ENDRICH 1986; ROSEN et al. 1993; SONG et al. 1984). However, radiation necrosis represents a heterogeneous process with features resembling inflammation. Immature vessels may grow into previously necrotic areas (GOBBEL et al. 1992), and viable tumour cells may still be found in the areas with decreased blood volume. Therefore, when this technique is used to monitor irradiated areas, the risk of overlooking foci of tumour recurrence cannot be excluded and further studies are necessary (SUGAHARA et al. 1999b).

One of the main proposed roles for DRCE-MRI is to detect and quantify responses to novel antiangiogenic therapies. Measurements of K^{trans} and regional blood volume have been used successfully in a number of therapeutic studies in peripheral tumours . However, due to anxieties that antiangiogenic therapy may cause cerebral haemorrhage no systematic trials of specific antiangiogenic therapies in cerebral tumours have been conducted. In animal models (GOSSMANN et al. 2002b) the effects of a neutralising anti-vascular endothelial growth factor (anti-VEGF) antibody on tumour microvascular permeability have been evaluated and demonstrated significant inhibition of tumour microvascular permeability (6.1±3.6 ml/min per 100 cc tissue), compared to the control, saline-treated tumours (28.6±8.6 ml/min per 100 cc tissue) together with significant suppression of tumour growth ($p<0.05$). Similar decreases have been seen in human trials of the monoclonal anti-VEGF antibody HumV 883 in abdominal and pelvis epithelial carcinomas (see Chap. 16).

9.4 Are K^{trans}, v_p and v_e Method-Independent?

One of the major justifications for the use of complex pharmacokinetic analyses for DRCE-MRI is the assumption that correctly calculated physiological parameters should be reproducible independent of scanner type, acquisition protocol or variation in other patient related physiological parameters. This is important if these techniques are to be used for multicentre studies or as routine diagnostic tests. This section presents and compares the findings of a number of independent groups applying pharmacokinetic analyses to DRCE-MRI in brain tumours. Tables 9.3-9.5 list literature values of overall mean values for tumour K^{trans}, fractional plasma volume (v_p), v_e, and bulk efflux constant (K_{ep}), reported by investigators from different centres. Direct comparison between the data listed in the tables must be made cautiously, keeping in mind the influence that the variation in both the methods of data acquisition (particularly the temporal resolution) and pharmacokinetic modelling might give to the derived quantities. Despite these variations in the acquisition and analysis methods some common trends can still be observed from these studies. In Table 9.3, for example, the range of the mean values of K^{trans} in gliomas were consistently reported (approximately from 0.004 to 0.100 min^{-1}) by three centres (ANDERSEN and JENSEN 1998; ROBERTS et al. 2000; HAROON et al. 2002a,b).

Table 9.4 shows that the mean values of v_p in grade III and grade IV gliomas reported by ROBERTS et al. (2000) (3.20%±1.96%, and 4.13%±2.54%, respectively) are very close to those calculated by HAROON et al. (3.51%±0.5% and 4.93±0.84, respectively) despite the use of quite different methodological approaches (HAROON et al. 2002b).

Table 9.5 shows a comparison of estimated v_e values in two studies. Zhu et al. (2000c) 15 patients with brain tumours (five gliomas, five meningiomas, and five acoustic neuromas). Mean values of v_e were significantly greater in acoustic schwannomas (53%±9%) than in meningiomas (34%±7%) or gliomas (22%±4%). Mean values of V_e in meningioma were significantly greater than those of gliomas. Andersen and Jensen (1998) studied 17 brain tumour patients (seven glioblastomas, six meningioma, and four metastasis). They reported that there were no significant differences between pre-treatment v_e among the different tumour types. However, both the mean v_e of the six meningioma patients (29.5%±17.4%) and the mean v_e of the seven glioblastoma patients (19.1%±13.5 %) were close to the results of Zhu et al.'s (2000c) study.

Table 9.6 compares calculated values for K_{ep} between studies. Hawighorst et al. (1997) studied 20 patients with intracranial meningiomas and reported that all meningiomas showed a high K_{ep} (median, 5.7 min^{-1}; range, 1.9–23.0 min^{-1}). In fact, all groups listed in Table 9.6, which studied meningiomas (Hawighorst et al. 1997; Zhu et al. 2000c; Andersen and Jensen 1998; Gowland et al. 1992), observed a relatively high average value and broad distribution range in K_{ep} for this tumour. Acoustic neuroma showed low K_{ep} in both studies which examined them.

These observations suggest remarkable coherence in the results of groups studying cerebral tumours using DRCE-MRI in combination with pharmacokinetic analyses. Although there are differences in absolute values general trends between studies degree reasonably well. These agreements are all the more remarkable considering a wide variation in acquisition protocol and analysis technique used by the various groups.

9.5 Comparison of rCBV Obtained by T1 and T2* Methods

Contrast-enhanced MR imaging has spawned two distinct methods, based on differing philosophies, for the measurement of CBV in brain tumours. Most commonly used are the first-pass techniques, based on the magnetic susceptibility contrast phenomenon and the resulting changes in T2* (Aronen et al. 1995; Maeda et al. 1993; Rosen et al. 1991; Siegal et al. 1997). T2* methods, albeit studied widely, have certain disadvantages. Without a reference blood signal, only relative CBV values can be obtained. Even with more sophisticated analyses, including arterial input function estimation (Ostergaard et al. 1996, 1998; Perman et al. 1992), CBF determinations remain approximate. The anatomic coverage is typically limited unless echo planar collection techniques are available and these will increase the severity of susceptibility based distortions. In most cases whatever

Table 9.6. Comparison of literature values of overall mean of rate constant, K_{ep} (min^{-1})

Authors	Gliomas		Meningiomas	AN	Metastasis	Methods
	Grade III	Grade IV or GB				
Larsson et al. 1990		0.074; .113 (*n*=1)[a]				Larsson
Brix et al. 1991		0.734; 2.59 (*n*=1)[a]				Brix
Hawighorst et al. 1997			6.4±8.7 (*n*=18)			Brix
Gowland et al. 1992		1.55±1.46 (*n*=3)	2.34±3.22 (*n*=2)	0.94 (*n*=1)	1.76 (*n*=1)	b
Andersen and Jensen 1998		0.21±.07 (*n*=7)	0.54±.40 (*n*=6)		0.41±.14 (*n*=4)	b
Zhu et al. 2000c	.405±.049 (*n*=2)	1.51±.48 (*n*=3)	0.898±.593 (*n*=5)	0.434±0.101 (*n*=5)		Tofts

[a] Two representative ROIs.
[b] A generalised plasma-interstitial two-compartment kinetic model ignoring the direct intravascular tracer contribution.
AN, acoustic neuroma.

sequences are employed there is some degree of sensitivity to susceptibility artefacts near large vessels or osseous structures, which, for example, make it difficult to assess infratentorial tumours (Siegal et al. 1997). T1-based methods do not suffer from these disadvantages but have their own inherent problems caused mainly by the additive effects of intravascular and extra vascular contrast media on the observed signal changes. Despite this, several techniques have been developed to estimate rCBV measurements from DRCE-MRI (Dean et al. 1992; Li et al. 2003; Schwarzbauer et al. 1993).

Bruening et al. (1996) compared T1-rCBV maps with T2-rCBV maps in 19 patients with suspected intra-axial brain tumours. For the T1-rCBV maps, a low-dose bolus of contrast material was given during T1-weighted interleaved SE-EPI imaging. This was followed by a second injection during dynamic T2-weighted imaging for generation of the T2-rCBV maps. Among patients with low-grade lesions (n=9), T1-based and T2-based rCBV maps showed comparably low rCBV in seven subjects. In the other two patients, with confirmed tumour dedifferentiation, elevation of rCBV values was seen on maps obtained with both techniques. In patients with the high-grade lesions exhibiting conventional contrast enhancement, lesions had higher estimated values on T1-rCBV maps than on the T2-rCBV maps. Despite this the T1-rCBV maps showed comparable ability to distinguish between low and high grade tumours.

Hacklander et al. (1996a,b, 1997) investigated the application of T1-rCBV measurements in healthy subjects, and patients with gliomas. A mean CBV value of 4.1±1.1 vol% averaged over the entire brain area was found in the normal subjects with the T1 method, which was in agreement with those obtained by nuclear medicine techniques. The value obtained with the T2* method was 2.6±1.1 vol%. Similar underestimations of the rCBV values were also found using the T2* method when evaluating regions of interest in tumour patients.

9.6 Advanced Parametric Imaging Techniques

Standard approaches to DSCE-MRI and DRCE-MRI analysis described above all suffer from method specific limitations related to the number of physiological variables they are able to derive. This is essentially a trade off between the complexity of the mathematical pharmacokinetic model used to fit to the observed data and the stability of the curve fitting routines used to derive values for individual variables. In general terms the more free variables used in a model (i.e. PS, v_e, v_p, F) the more specific the physiological information but the less robust the performance of the fitting algorithm. This has led to widespread use of relatively simple pharmacokinetic modelling approaches in the interest of reproducibility and sensitivity at the expense of physiological specificity. For example, DRCE-MRI data is commonly analysed using a simple model similar to that described by Tofts and Kermode (1991), which will provide K^{trans} values that are affected by blood flow and blood volume as well as PS product, in preference to more specific models such as that described by St. Lawrence and Lee (1998). A further important restriction of many of these pharmacokinetic analyses is that they are commonly designed to quantify the vasculature (i.e. rCBV) or contrast leakage (i.e. K^{trans}) but seldom address both processes. Several groups have described alternate approaches to data analysis designed to measure both perfusion and extravasation (Griebel et al. 1997; Kovar et al. 1997, 1998; Su et al. 1998; Ludemann et al. 2000). This section will briefly describe techniques for simultaneous blood volume and vessel permeability mapping from first pass dynamic MRI data (T2*-weighted, T1-weighted, and double-echo dynamic perfusion-weighted MRI).

9.6.1 Simultaneous Mapping of rCBV and K^{trans} from DSCE-MRI

The first attempt to use dynamic MRI to separate effect of contrast medium leakage from the presence of extravascular contrast medium was described by Weisskoff et al. (1994b). They modelled the combined T1 and T2 relaxivity effects of Gd compounds on T2*W MR signals to allow simultaneous measurement of rCBV volume and permeability. The observed signal, $\Delta R2_{obs}$, is a weighted linear combination of the T2-weighted intravascular effects and the T1-weighted extravascular effects of a gadolinium-based contrast agent, as illustrated in the equation

$$\Delta R2_{obs}(t) = K_1 \Delta R2_{ave}(t) - K_2 \int_0^t \Delta R2_{ave}(t')dt'$$

where $\Delta R2$ is assumed to be proportional to local blood volume. K_1 represents the weighting of the intravascular component, and K_2 represents the weighting of the extravascular component. $\Delta R2_{ave}$ is computed by averaging $\Delta R2_{obs}$ for all nonenhancing pixels in the brain, defined by signal enhancement no greater than one standard deviation above baseline. For each pixel in the brain, K_1 and K_2 are determined by a least-squares fit to the equation, which allows determination of permeability weighting (K_2).

OSTERGAARD et al. (1999) applied this technique to simultaneously measure blood flow, blood volume and blood–brain barrier permeability following dexamethasone treatment in a group of patients with brain tumours. PROVENZALE et al. (2002) compared permeability measurements in high-grade and low-grade glial neoplasms using a T2*-weighted method. Their results showed that permeability values for high-grade tumours obtained using a T2*-weighted method were significantly greater than those for low-grade tumours and are consistent with previous studies reporting results using T1-weighted techniques.

Unfortunately the accuracy of decomposition of intra- and extra-vascular contrast agent (CA) contributions from T2*W dynamic data are related to extraction rate of CA into the tumour tissue. In the presence of large extraction fractions, the authors expect the model to be less appropriate (WEISSKOFF et al. 1994b). The presence of significant "T1 shine through" represents the worst case (KASSNER et al. 2000), where parts of a tumour with highly permeable capillary beds shows significant T1-related enhancement during the first passage of the CA bolus. WEISSKOFF et al. (1994b) suggested that in such cases the use of other techniques to minimise relaxivity enhancement would be required in order to more aggressively correct for CA leakage. Possible approaches might include the use of a larger dose of CA pre-load or of macromolecular intravascular agents to reduce first pass extraction (BRASCH et al. 1997; WEISSKOFF et al. 1994b).

9.6.2 Simultaneous Acquisition of T1- and T2*-Weighted Data Using Double-Echo Techniques

HIETSCHOLD et al. (1993), KLENGEL et al. (1994) and later MIYATI et al. (1997) proposed the use of a gradient echo pulse sequence with two echoes for collection of DCE-MRI data to allow simultaneous acquisition of T2*-weighted and T1-weighted data. T2* is then derived from the two echoes and used to remove the transverse decay effects from the first echo. Such an acquisition yields well separated T2* and T1 variations that form a stronger and more complete basis for physiologic analysis. Removal of T1 shortening effects which occur with conventional DSCE-MRI reduces artifactual underestimation of tumour blood volume (UEMATSU et al. 2001). Several groups (LI et al. 1998; BARBIER et al. 1999; HEILAND et al. 1999; UEMATSU et al. 2000; VONKEN et al. 2000) have proposed analysis methods for simultaneous quantification of vascularity and permeability of tissues using double-echo acquisitions.

In tumours, analysis of DSCE-MRI data is further complicated by the direct T2* effects of extravasated contrast (ARONEN et al. 1994; HEILAND et al. 1999). Extravasation of contrast into the tissue does not only cause a biasing T1-based signal enhancement, but will also give rise to extra T2*-shortening in the tissue. In spite of the use of a dual-echo acquisition, which is insensitive to T1-changes, the magnitude of the T2*-effect due to extravasation is unpredictable. A technique that corrects for possible contrast extravasation has been described (VONKEN et al. 2000). The technique is based on a two-compartment kinetic model which separates the effects of contrast in the active circulation from that trapped within the tumour (either by extravasation or delayed flow). Results of simulation experiments with different degrees of contrast extravasation are also presented (VONKEN et al. 2000).

9.6.3 Simultaneous Mapping of rCBV and K^{trans} from DRCE-MRI

Simultaneous mapping of rCBV and K^{trans} from DRCE-MRI is complicated by the synergistic effects of intravascular and extravascular contrast agent on measured signal intensity. LI et al. (2000b) (LI and JACKSON 2003) have described a novel method which uses an initial data decomposition step to derive separate concentration time course data for intravascular and extravascular contrast media. This data decomposition is performed using prior knowledge of the shape of the intravascular contrast concentration time course curve, which is assumed to be identical to that observed in large vessels and the predicted shape of the contrast concentration time course curve in the extravascular extracellular space, known as the leakage profile. The shape of the leakage profile can be shown to be the integral of the intravascular concentration time course curve and the ratio between this integral and the observed data represents the K^{trans}, which is known as K_{fp} (first pass) for this technique. The estimation of the leakage profile is based on an assumption that, since the data collection period is very short, there is no significant contrast in the EES through the measurement period. This assumption breaks down where extraction fractions are high and leads to systematic underestimation of high values of K^{trans} (Li and Jackson 2003).

This approach has a number of significant advantages over conventional techniques. The generation of both intravascular and extravascular contrast time course data allows separate analysis for CBV and K^{trans}. This removes the tendency of simplistic analyses to overestimate K^{trans} in the region of large blood vessels so that pixel rejection rates due to inappropriately high K^{trans} estimations are reduced from as high as 50% in gliomas to zero. The separation of the curve fitting analyses for intravascular and extravascular data reduces the number of free parameters and makes the technique more robust to images with poor signal-to-noise ratio (Li and Jackson 2003).

The technique is described in more detail in Chap. 6. Briefly it is based on a tumour leakage profile (LP), which could be defined in a two-compartment kinetic model:

$$C(t) = K^{trans} \cdot LP + C_v(t)$$

where $C_v(t)$ denoted the intravascular component of C(t); and

$$LP = \int^{t} C_p(t') \cdot e^{-\frac{K^{trans}}{v_e}(t-t')} dt'$$

If it is assumed that the backflow of the CA from extravascular space to the blood space during the first passage of CA was negligible, LP reduces to:

$$LP = \int_0^t C_p(t')dt'$$

LP is calculated from the time dependent plasma-contrast concentration function, $C_p(t)$, in 3D T1W dynamic studies. Dynamic MRI was performed using 3D T1W radio-frequency (rf) spoiled gradient-echo images (T1W-GRE) of large volume (acquisition matrix size = 128×128×25) and high time resolution ($\Delta t < 5.1$ seconds). Substantial improvements have been made in the implementation of this approach since its first description (Jackson et al. 2002a; Li et al. 2003).

The performance of the method was evaluated by comparing results to those obtained from more conventional methods in patients with primary brain neoplasms. The technique produced maps of K^{trans} that appeared to be free of any contribution from intra-vascular contrast agent. Maps of v_p showed close correlation with maps of blood volume calculated from independently acquired dynamic susceptibility weighted MRI examinations with no evidence of residual permeability effects (Haroon et al. 2002a; Li et al. 2000b, 2003; Zhu et al. 2000a). The novel feature of the first pass leakage profile (FPLP) method is that it uses only data collected during the first passage of the contrast bolus through the target tissue so that data acquisition is extremely fast compared to conventional method for measuring permeability (Brix et al. 1991; Larsson et al. 1990; Tofts and Kermode 1991).

Monte Carlo simulations have been performed to assess the accuracy and precision of the FPLP method (Li and Zhu 2002). These show that FPLP method produces accurate measurements of fractional plasma volume and of transfer constant where the leakage rate is not high ($K^{trans} < \sim 0.2$ min^{-1}); Measurements of K^{trans} and v_p were highly reproducible and were less affected by low SNR than conventional curve fitting approaches. However, FPLP will underestimate K^{trans} when the backflow of the CA from extravascular space to the plasma during the first passage of CA is not negligible. A combined use of the FPLP and conventional curve fitting methods has shown the potential to overcome this problem and provide optimal accuracy and precision in quantification of K^{trans} and v_p (Li and Zhu 2002).

Figure 9.7 shows images from the central slice of the tumour volume from a patient with a glioma. The T1-CBV map calculated using conventional methods (Bruening et al. 1996; Hacklander et al. 1996b) has mixed contribution from both perfusion and contrast leakage (Fig. 9.7a). The geometrical distribution of high values (yellow and red) conforms poorly to that seen on either T2*-CBV (Fig. 9.7e) or K^{trans} maps calculated using a modified version of the standard Tofts' analysis (Zhu et al. 2000c) (Fig. 9.7c). There is close concordance between maps of $T1\text{-}CBV_{corrected}$ (Fig. 9.7d) and T2*-CBV (Fig. 9.7e). The distribution of high permeability values on K_{fp} maps (Fig. 9.7b) is similar to that on K^{trans} maps (c) but K^{trans} maps again show higher levels of noise and considerable residual contributions from first pass effects in vessels. Figure 9.7f compares the histograms of K_{fp} and K^{trans} from the whole tumour. The tumour region of interest was manually drawn and summed from 12 slices. There is a drop in the number of pixels seen on K^{trans} (denoted as k in Fig 9.7f) maps at low levels (<0.1 min^{-1}). This is due to the high rate of fitting failures when the tri-exponential model is applied to the data points with low C(t). The K^{trans} histogram is also skewed at the high end with a long thin tail due to the inclusion of mea-

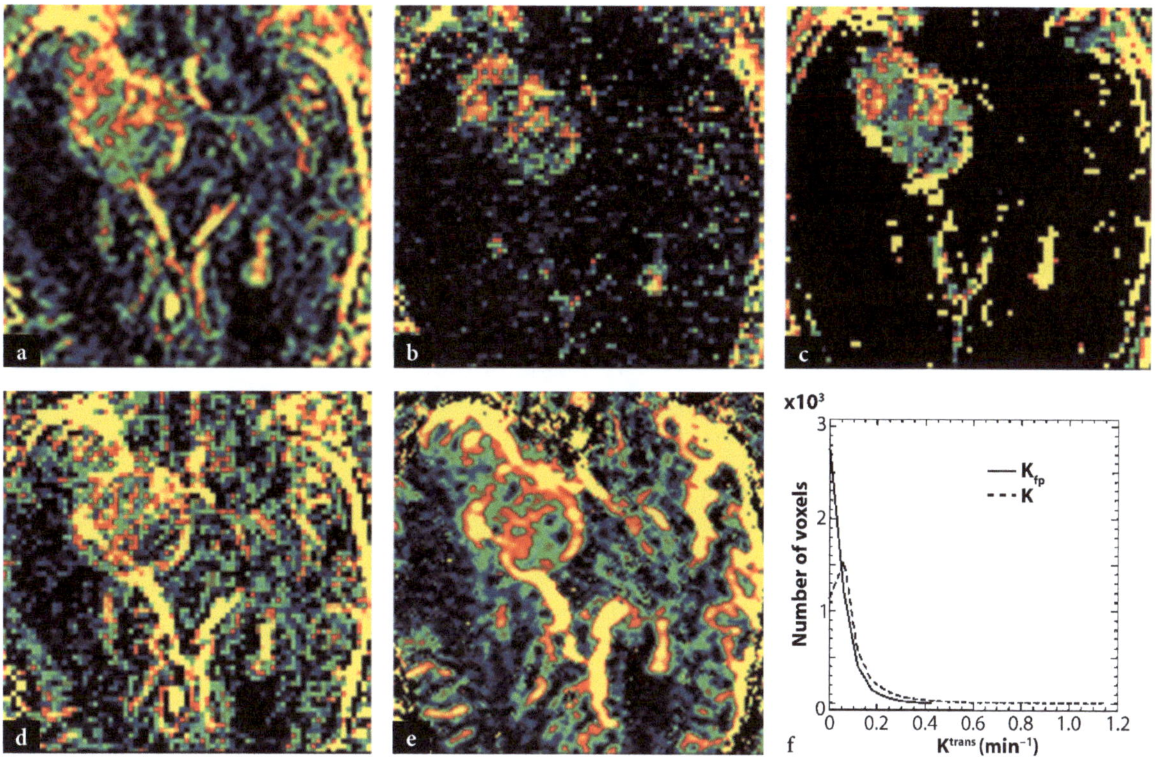

Fig. 9.7a–f. Colour-rendered parametric maps of T1-rCBV (**a**), K_{fp} (calculated using the FPLP model) (**b**), K^{trans} [calculated using a modified version of the standard Tofts' analysis (Zhu et al. (2000c)] (**c**), 3D T1-$CBV_{corrected}$ (**d**), and T2*-rCBV (**e**) from a patient with a glioma. The distribution of "hot" (*yellow* and *red*) areas in the tumour on the uncorrected T1-rCBV map does not agree with either the k or T2*-rCBV map. In contrast, the distribution of high values on the K_{fp} map is similar to that in the K^{trans} map but with less noise. 3D T1-$rCBV_{corrected}$ agrees closely with the T2*-CBV map. The histograms of permeability surface product values of whole tumour of the patient, calculated using the new (*solid line*) and conventional methods (*dashed line*), are compared in (**f**)

surements from voxels with large vessel components. The two histograms correspond reasonably well where k values lie in the mid range (0.1–0.4 min^{-1}).

9.7 Clinical Applications of Advanced Parametric Analysis Techniques

A small number of studies have recently appeared describing results from combined studies of both tumour blood volume and vessel leakage.

9.7.1 Combined T1W and T2W Image Acquisition

Zhu et al. (2000c) used sequential T1W DRCE-MRI and T2*-weighted DSCE-MRI to produce estimates of K^{trans} and rCBV. The pre-enhancement method worked well for all cases in this study whether contrast leakage was large or small. Production of parametric maps of rCBV, K^{trans} and v_e calculated from T1W and T2W dynamic MRI of 15 patients with brain tumours (five glioma, five meningioma, five acoustic neuroma) (Zhu et al. 2000c) allowed comparison of parameters on a pixel by pixel basis. This comparison demonstrated strong correlation between rCBV and K^{trans} in 11 of 15 patients. However, decoupling between pixel-wise rCBV and K^{trans} was found in four patients who had lesions with moderate K^{trans} and v_e elevation but no increase of rCBV. Figure 9.8 shows an example of one of these cases with low rCBV. The rCBV map in this case with meningioma demonstrates a heterogeneous tumour with low values between normal grey and white matter. Both K^{trans} and v_e maps delineate the tumour clearly against the background of non-enhancing brain tissues.

Apparently, in some tumours areas of high contrast leakage are not associated with increases of tumour blood volume. Such decoupling between permeability and blood flow may be of immediate significance, not only indicating inefficient blood supply (Jain and

Gerlowski 1984), but also reflecting the difference of time scales involved in the different angiogenic processes. Anti-angiogenic treatment (VEGF inhibition) has been shown to reduce K^{trans} in a period of hours whilst CBV remains unaltered (Amoroso et al. 1997; Hawighorst et al. 1998; Jensen 1998; Lund et al. 1998), reflecting the continuous modulation of VEGF activity according to the metabolic demands in tumours (Hawighorst et al. 1998). It may be postulated therefore that a loss of co-location of K^{trans} and rCBV associated with low tumour blood volume described here will be one initial marker of successful inhibition of angiogenic drive. Evidence that this type of de-coupling does occur is also seen in the case illustrated in Fig. 9.9, where a meningioma is associated with extensive increases in K^{trans} in the

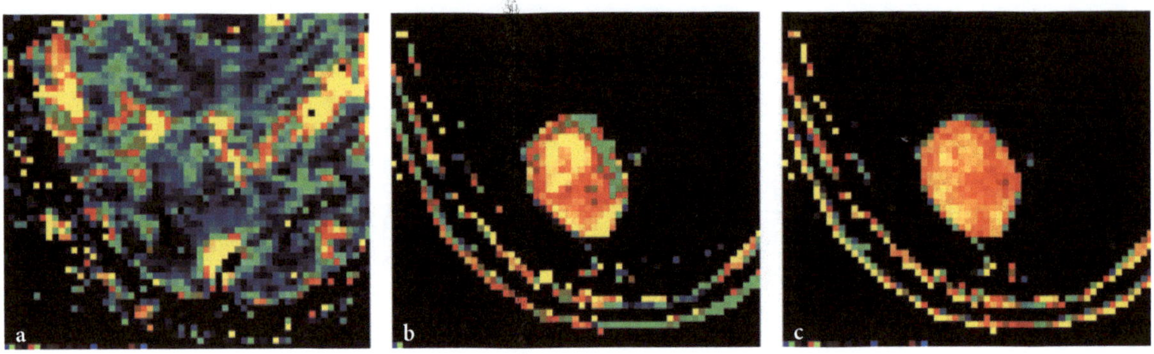

Fig. 9.8a–c. Images from a patient (case 10) with a tentorial mengioma. **a** Parametric map of rCBV from T2*W data. **b** Map of K^{trans} from T1W data. **c** Map of v_e from T1W data. The tumour shows very low values of rCBV below those of normal grey matter but is conspicuous on both K^{trans} and v_e maps. Spatial distribution of K^{trans} is weakly correlated with v_e but not at all with rCBV, with $R(k,v_e)$ =0.22, R(k,CBV) = 0.10, respectively

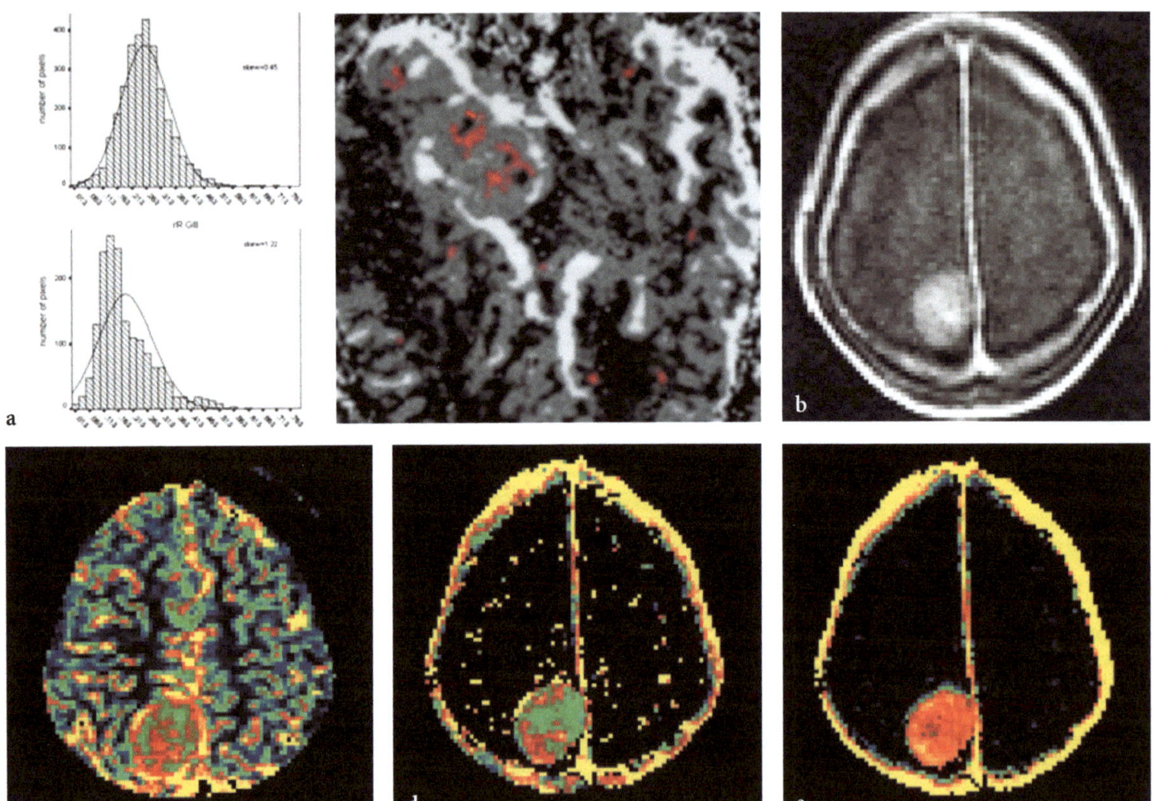

Fig. 9.9a–e. Images from a patient with a posterior falcine meningioma. **a,b** Pre- and post-contrast image from T1W dynamic series. There is diffuse enhancement of leptomeninges on the post-contrast T1W images. **c** Parametric map of rCBV from T2*W data. **d** Map of K^{trans} from T1W data. **e** Map of v_e from T1W data. The tumour shows relatively homogeneous distribution of all parameters. The calvarial meninges show marked increase in K^{trans} and v_e whilst the rCBV map shows no apparent abnormality

tumour and in apparently normal meninges with normal values of rCBV. These observations suggest that biological decoupling of permeability effects and changes in local blood volume do occur and that independent measures of K^{trans} and rCBV may provide different, complementary biological information about the angiogenic process.

9.7.2 Simultaneous Mapping of Tumour Blood Volume and Permeability

Li et al. (2000b) assessed simultaneously calculated tumour blood volume and permeability transfer coefficients of 11 patients with brain tumours using T1W dynamic MRI analysed using the FPLP model described above. In patients with high grade glioma, voxels with high values of rCBV were significantly more common than in low grade gliomas and benign tumours. Haroon et al. (2002a,b) used the same method for calculation of tumour blood volume, T1-CBV, and permeability from 3D T1W GE dynamic MRI (Zhu et al. 2000c). K^{trans} (fp) was compared to the permeability-surface area product by two compartmental pharmacokinetic models by Tofts and Kermode, K^{trans} (TK). The results show the predicted overestimation of K^{trans} (TK) using conventional techniques (Haroon et al. 2002a). Mean T1-CBV of glioma grade III was significantly lower than grade VI glioma. Mean K^{trans} (fp) of grade III glioma was also lower than grade VI glioma but did not reach significance (Haroon et al. 2002b).

Gossman et al. (2002) calculated endothelial transfer coefficient, K^{trans} and fractional plasma volume v_p based on method by Daldrup et al. (1998) in 25 patients with brain tumours. They found that high grade tumours showed a marked intra-tumoral heterogeneity, as reflected in the histologic samples and in corresponding permeability values and that this heterogeneity had a strong correlation to tumour grade. In all patients, permeability values depicted the most aggressive area within the tumour mass. In contrast, v_p values showed no correlation to tumour grade (Gossman et al. 2002). Johnson et al. (2002) simultaneously calculated blood volume and K^{trans} in patients with meningiomas (n=7), glioblastomas (n=7) and lymphomas (n=6) by first pass pharmacokinetic modelling. There was good discrimination between the three different types of brain tumours. Ludemann et al. (2002) calculated rCBV, interstitial volume, cellular volume and K^{trans} in a group of patients gliomas (n=39), meningiomas (n=6) and metastasis (n=8) using a three compartmental model. Grading of gliomas with an accuracy of 73% was achieved based on the vascular volume maps alone. Furthermore, meningiomas and gliomas differed in the voxel patterns plotted in interstitial vs. vascular volume diagrams. Additionally, gliomas showed an increased K^{trans} compared to other two tumour types (Ludemann et al. 2002).

9.8 Reproducibility of Tumour Blood Volume and Transfer Coefficient Measurements

The development of anti-angiogenic therapies for tumours has led to a demand for imaging-based surrogate markers of the angiogenic process. The utility of such markers is highly dependent on their test-retest reproducibility. The use of these techniques in brain tumours can be expected to be associated with good reproducibility compared to applications in other anatomical areas since normal brain has no significant permeability and the area is not subject to physiological motion. To date only a small handful of publications have addressed the reproducibility of these parametric imaging techniques in the brain (Galbraith et al. 2002; Jackson et al. 2001, 2003; Li et al. 2000a, 2003; Padhani et al. 2002).

A reproducibility study on tumour rCBV and vascular tortuosity as estimated by measurement of rR based on T2*W DSCE MRI was conducted in 11 patients with glioma who were scanned on two occasions 36–56 h apart (Jackson et al. 2001). The observed reliability estimates were used to calculate 95% confidence limits for detection of differences between groups and for changes in individual cases. The results showed that measurement of rCBV in consecutive studies was statistically capable of reliable detecting changes in excess of 15% for between group studies and 25% for individual patients. Measurement of vascular tortuosity using rR was less reproducible but was able to confidently identify changes in excess of 30% in group studies and 35% in individuals.

Li et al. (2000a) assessed reproducibility of K^{trans} and v_e measurements derived from dynamic T1-weighted data for nine patients with cerebral glioma. Patients were scanned on two occasions 36–56 h apart. Data were analysed using the conventional multi-compartmental model (Tofts and Kermode 1991) with improvements designed to improve reproducibility (Zhu et al. 2000c). Mean values of K^{trans} and the v_e for the whole tumour volume were calculated for each patient and for each observation. Measures of the test-retest CoV were used to assess reproducibility. The differences between the two scans were 7.1±4.9% in mean K^{trans}, and 8.7±2.5% in mean v_e. These values compare very favourably with similar values are calculated for non-cerebral tumours by other workers (Padhani et al. 2002).

The FPLP method described above has shown the highest test-retest reproducibility in the brain to date (Li et al. 2003). Reproducibility was assessed in patients with cerebral glioma (n=5) by examining K^{trans} and v_p on two separate DRCE-MRI scans performed at a 2-day interval (Fig. 9.10). The CoV of mean values ranged from 0.02%–6.34% for K^{trans} and from 0.7%–5.34% for v_p. The CoV of 97.5 percentile values ranged from 0.07%–6.54% for K^{trans} and 2.06%–4.79% for v_p and there were strong intra-class correlation between mean values and 97.5 percentiles of K^{trans} measured on day 0 and day 2 with intra-class correlation coefficients rI>0.997 and rI>0.984, respectively. There were also strong intra-class correlation between mean values and 97.5 percentiles of v_p measured on day 0 and day 2 with intra-class correlation coefficients rI>0.981 and rI>0.979, respectively. The results showed that the FPLP method is highly reproducible.

These results suggest that the reproducibility of parametric imaging data in brain tumours is more than adequate to support longitudinal studies and that studies in brain tumours benefit specifically from a lack physiological motion.

9.9 Conclusions

The studies that we have reviewed show that dynamic contrast-enhanced MRI techniques are able to provide valuable and often unique information in a wide variety of cerebral tumours. It is clear that these techniques are relatively free from the problems of physiological motion and susceptibility artefacts that are

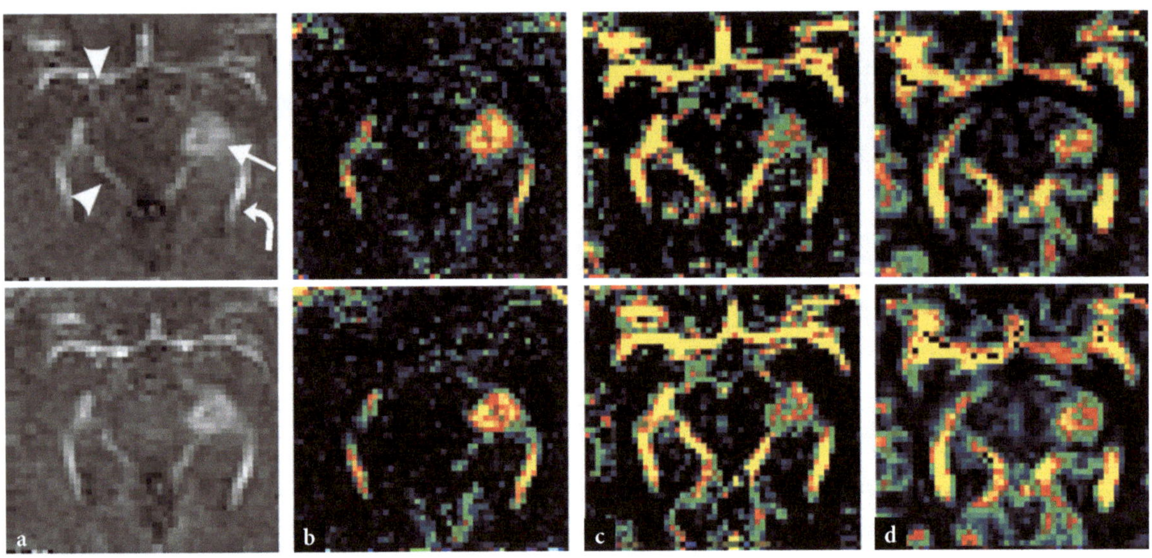

Fig. 9.10a–d. Trans-axial images from a patient with an anaplastic astrocytoma, acquired on day 0 (*upper row*) and day 2 (*lower row*). **a** T1W GRE images in a contrast-enhanced dynamic series 41 s after CA bolus injection, approximately at the beginning of first recirculation of the CA bolus to the sagittal sinus. **b** Colour-rendered K_{fp} maps (*dark blue* represents K_{fp} = 0–0.05 min^{-1}; green, 0.05–0.10 min^{-1}; *red*, 0.10–0.17 min^{-1}; *yellow*, 0.17–0.25 min^{-1}). **c** $\mathrm{rCBV}^{T1}_{corrected}$ maps (*dark blue*, 0–0.06; *green*, 0.06–0.08; *red*, 0.08–0.12; *yellow*, 0.12–1.0). $\mathrm{rCBV}^{T1}_{corrected}=\sum C(t)/\sum C_p(t)$, where $C_p(t)$ was measured from superior sagittal sinus. **d** T2-CBV maps (*dark blue*, 0–1.9 A.U.; *green*, 1.9–2.3 A.U.; *red*, 2.3–4.1 A.U.; *yellow*, 4.1–6.5 A.U. A.U. is a unit of normalised $\Delta R2^*$. T2-CBV = $\sum \Delta R2^*(t)/\sum \Delta R2^*_{normal\ brain}(t)$, where $\Delta R2^*_{normal\ brain}(t)$ was measured from the hemisphere contralateral to the tumour (excluding CSF). There is marked enhancement of the astrocytoma (**a**, *arrow*), blood vessels (*arrowheads*) and choroid plexus. The tumour and choroid plexus are best distinguished in the permeability maps (**b**) due to high K_{fp} values. Since normal brain tissue with intact brain-blood barrier is not permeable to Gadodiamide, both normal brain tissue and blood vessels have low K_{fp} values. Large blood vessels are clearly depicted on the blood volume maps (**c**). On the $\mathrm{T1\text{-}CBV}_{correlated}$ maps, the grade III glioma shows heterogeneous increases of blood volume (*yellow*, *red* and *green* representing tumour blood volume ranging from 0.06 to 0.14). Maps of $\mathrm{T1\text{-}CBV}_{corrected}$ (**c**) and T2-CBV (**d**) show close agreement in appearance although blood vessels are slightly narrower on $\mathrm{T1\text{-}CBV}_{corrected}$ maps

seen in other body areas and this has led many workers to develop advanced analysis techniques for use in the head. The principal physiological parameters that appear to be of value are the transfer coefficient K^{trans} and the regional relative blood volume, rCBV. These parameters provide useful information about tumour diagnosis and grade, tumour tissue heterogeneity and therapeutic response. Unfortunately, probably due to the complexity of the analysis techniques, these methods have not yet penetrated into routine clinical practice. This is likely to change in the near future with increasing computer power and better understanding of the analysis methodologies required.

References

Alavi JB, Alavi A et al (1988) Positron emission tomography in patients with glioma. A predictor of prognosis. Cancer 62:1074–1078

Amoroso A, Del Porto F et al (1997) Vascular endothelial growth factor: a key mediator of neoangiogenesis. A review. Eur Rev Med Pharmacol Sci 1:17–25

Andersen C, Jensen FT (1998) Differences in blood-tumour-barrier leakage of human intracranial tumours: quantitative monitoring of vasogenic oedema and its response to glucocorticoid treatment. Acta Neurochir (Wien) 140:919–924

Aronen HJ, Gazit IE et al (1994) Cerebral blood volume maps of gliomas: comparison with tumor grade and histologic findings. Radiology 191:41–51

Aronen HJ, Glass J et al (1995) Echo-planar MR cerebral blood volume mapping of gliomas. Clinical utility. Acta Radiol 36:520–528

Aronen HJ, Pardo FS et al (2000) High microvascular blood volume is associated with high glucose uptake and tumor angiogenesis in human gliomas. Clin Cancer Res 6:2189–2200

Assimakopoulou M, Sotiropoulou-Bonikou G et al (1997) Microvessel density in brain tumors. Anticancer Res 17:4747–4753

Barbier EL, den Boer JA et al (1999) A model of the dual effect of gadopentetate dimeglumine on dynamic brain MR images. J Magn Reson Imaging 10:242–253

Bharara S, Goldman CK et al (1996) Vascular endothelial growth factor expression correlates with cerebral edema in meningiomas (meeting abstract). Proc Annu Meet Am Assoc Cancer Res 37:A399

Bhujwalla ZM, Artemov D et al (1999) Comparison of vascular volume and permeability for tumors derived from metastatic human breast cancer cells with and without the metastasis suppressor gene nm23. Proceedings of the 7th scientific meeting of the International Society of Magnetic Resonance, Philadelphia, p 146

Bitzer M, Opitz H et al (1998) Angiogenesis and brain oedema in intracranial meningiomas: influence of vascular endothelial growth factor. Acta Neurochir (Wien) 140:333–340

Boxerman JL, Hamberg LM et al (1995) MR contrast due to intravascular magnetic susceptibility perturbations. Magn Reson Med 34:555–566

Brasch R, Pham C et al (1997) Assessing tumor angiogenesis using macromolecular MR imaging contrast media. J Magn Reson Imaging 7:68–74

Brem S, Cotran R et al (1972) Tumour angiogenesis a quantitative method for histological grading. J Natl Cancer Inst 28:347–356

Brix G, Semmler W et al (1991) Pharmacokinetic parameters in CNS Gd-DTPA enhanced MR imaging. J Comput Assist Tomogr 15:621–628

Bruening R, Kwong KK et al (1996) Echo-planar MR determination of relative cerebral blood volume in human brain tumors: T1 versus T2 weighting. AJNR Am J Neuroradiol 17:831–840

Bullock PR, Mansfield P et al (1991) Dynamic imaging of contrast enhancement in brain tumors. Magn Reson Med 19:293–298

Calamante F, Gadian DG et al (2000) Delay and dispersion effects in dynamic susceptibility contrast MRI: simulations using singular value decomposition. Magn Reson Med 44:466–473

Case TA, Durney CH et al (1987) A mathematical model of diamagnetic line broadening in lung tissue and similar heterogeneous systems. J Magn Reson 73:304–314

Cha S (2002) Relative recirculation: what does it mean? AJNR Am J Neuroradiol 23:1–2

Cha S, Law M et al (2000) Peritumoral region: differentiation between primary high-grade neoplasma and solitary metastasis using dynamic contrast-enhanced T2*-weighted echo-planar perfusion MR imaging. Proceeding of the 38th annual meeting of the American Society of Neuroradiology. American Society of Neuroradiology, Atlanta, p 22

Cha S, Pierce S et al. (2001) Dynamic contrast-enhanced T2*-weighted MR imaging of tumefactive demyelinating lesions. AJNR Am J Neuroradiol 22:1109-1116

Cha S, Knopp EA et al (2002) Intracranial mass lesions: dynamic contrast-enhanced susceptibility-weighted echo-planar perfusion MR imaging. Radiology 223:11–29

Cohen MS, Weisskoff RM (1991) Ultra-fast imaging. Magn Reson Imaging 9:1–37

Costello PC (1994) Human cerebral microvascular endothelial involvement in neovascularization of malignant glial tumors. Diss Abstr Int [B] 54:5543

Daldrup H, Shames DM et al (1998) Correlation of dynamic contrast-enhanced MR imaging with histologic tumor grade: comparison of macromolecular and small-molecular contrast media. AJR Am J Roentgenol 171:941–949

Damert A, Ikeda E et al (1997a) Activator-protein-1 binding potentiates the hypoxia-induciblefactor-1-mediated hypoxia-induced transcriptional activation of vascular-endothelial growth factor expression in C6 glioma cells. Biochem J 327:419–423

Damert A, Machein M et al (1997b) Up-regulation of vascular endothelial growth factor expression in a rat glioma is conferred by two distinct hypoxia-driven mechanisms. Cancer Res 57:3860–3864

Dean BL, Lee C et al (1992) Cerebral hemodynamics and cerebral blood volume: MR assessment using gadolinium contrast agents and T1-weighted Turbo-FLASH imaging. AJNR Am J Neuroradiol 13:39–48

Dennie J, Mandeville JB et al (1998) NMR imaging of changes

in vascular morphology due to tumor angiogenesis. Magn Reson Med 40:793–799

Dewhirst MW, Tso CY et al (1989) Morphologic and hemodynamic comparison of tumor and healing normal tissue microvasculature. Int J Radiat Oncol Biol Phys 17:91–99

Dietzmann K, von Bossanyi P et al (1997) Immunohistochemical detection of vascular growth factors in angiomatous and atypical meningiomas, as well as hemangiopericytomas. Pathol Res Pract 193:503–510

Domingo Z, Rowe G et al. (1998) Role of ischaemia in the genesis of oedema surrounding meningiomas assessed using magnetic resonance imaging and spectroscopy. Br.J Neurosurg. 12:414–418

Donahue K, Pathak A et al (1999) Utility of acquiring vascular blood volume, permeability and morphology information from dynamic susceptibility contrast agent studies in patients with brain tumors. Proceedings of the 7th scientific meeting of the International Society of Magnetic Resonance in Medicine, Philadelphia, p 149

Dvorak HF, Brown LF et al (1995) Vascular permeability factor/vascular endothelial growth factor, microvascular hyperpermeability, and angiogenesis. Am J Pathol 146:1029–1039

Fisel CR, Ackerman JL et al (1991) MR contrast due to microscopically heterogeneous magnetic susceptibility: numerical simulations and applications to cerebral physiology. Magn Reson Med 17:336–347

Flickinger FW, Allison JD et al (1993) Differentiation of benign from malignant breast masses by time-intensity evaluation of contrast enhanced MRI. Magn Reson.Imaging 11:617–620

Frahm J, Haase A et al (1986) Rapid three-dimensional MR imaging using the FLASH technique. J Comput Assist Tomogr 10:363–368

Frazzini VI, Cha S et al (1999) Dynamic contrast enhanced T2*-weighted echo-planar perfusion MR imaging of primary CNS lymphoma and glioblastoma multiforme. Proceedings of the 37th annual meeting of the American Society of Neuroradiology. American Society of Neuroradiology, San Diego, p 185

Fujii K, Fujita N et al (1992) Neuromas and meningiomas: evaluation of early enhancement with dynamic MR imaging. AJNR Am J Neuroradiol 13:1215–1220

Fuss M, Wenz F et al (2000) Radiation-induced regional cerebral blood volume (rCBV) changes in normal brain and low-grade astrocytomas: quantification and time and dose-dependent occurrence. Int J Radiat Oncol Biol Phys 48:53–58

Galbraith SM, Lodge MA et al (2002) Reproducibility of dynamic contrast-enhanced MRI in human muscle and tumours: comparison of quantitative and semi-quantitative analysis. NMR Biomed 15:132–142

Gobbel GT, Seilhan TM et al (1992) Cerebrovascular response after interstitial irradiation. Radiat Res 130:236–240

Goldman CK, Bharara S et al (1997) Brain edema in meningiomas is associated with increased vascular endothelial growth factor expression. Neurosurgery 40:1269–1277

Gossmann A, Helbich TH et al (2000) Magnetic resonance imaging in an experimental model of human ovarian cancer demonstrating altered microvascular permeability after inhibition of vascular endothelial growth factor. Am J Obstet Gynecol 183:956–963

Gossman A, Bangard C et al (2002a) Quantitative MRI estimates of microvascular permeability in human brain tumors: detection of reginal heterogeneityh and correlation with histological grade. Proc 10th Intern Magn Reson Med Hawaii

Gossmann A, Helbich TH et al (2002b) Dynamic contrast-enhanced magnetic resonance imaging as a surrogate marker of tumor response to anti-angiogenic therapy in a xenograft model of glioblastoma multiforme. J Magn Reson Imaging 15:233–240

Gowland P, Mansfield P et al (1992) Dynamic studies of gadolinium uptake in brain tumors using inversion- recovery echo-planar imaging. Magn Reson Med 26:241–258

Griebel J, Mayr NA et al (1997) Assessment of tumor microcirculation: a new role of dynamic contrast MR imaging. J Magn Reson Imaging 7:111–119

Hacklander T, Hofer M et al (1996a) Cerebral blood volume maps with dynamic contrast-enhanced T1-weighted FLASH imaging: normal values and preliminary clinical results. J Comput Assist Tomogr 20:532–539

Hacklander T, Reichenbach JR et al (1996b) Measurement of cerebral blood volume via the relaxing effect of low-dose gadopentetate dimeglumine during bolus transit. AJNR Am J Neuroradiol 17:821–830

Hacklander T, Reichenbach JR et al (1997) Comparison of cerebral blood volume measurements using the T1 and T2* methods in normal human brains and brain tumors. J Comput Assist Tomogr 21:857–866

Haroon HA, Buckley DL et al (2002a) A comparison of Ktran measurements in gliomas obtaqined with conventional and first pass model. Proc 10th Intern Magn Reson Med Hawaii, p 663

Haroon HA, Patankar TA et al (2002b) Relationship between vasculoar endothelial permeability and histological grade in human gliomas using a novel first pass model. Proc 10th Intern Magn Reson Med Hawaii, p 2113

Hawighorst H, Engenhart R et al (1997) Intracranial meningeomas: time- and dose-dependent effects of irradiation on tumor microcirculation monitored by dynamic MR imaging. Magn Reson Imaging 15:423–432

Hawighorst H, Knapstein PG et al (1998) Uterine cervical carcinoma: comparison of standard and pharmacokinetic analysis of time-intensity curve for assessment of tumor angiogenesis and patient survival. Cancer Res 58:3598–3602

Heiland S, Benner T et al (1999) Simultaneous assessment of cerebral hemodynamics and contrast agent uptake in lesions with disrupted blood-brain-barrier. Magn Reson Imaging 17:21–27

Henderson E, Sykes J et al (1999) Measurement of blood flow, blood volume and capillary permeablility in a canine spontaneous breast tumor model using tow differen contrast agents. Proceedings of the 8th sientific meeting of the International Society of Magnetic Resonance in Medicine, Philadelphia, p 148

Hietschold V, Klengel S et al (1993) Simultaneous dynamic measurement of tissue contrast enhancement and perfusion at 0.5 Tesla: method and postprocessing. Proceedings of the 12th scientific meeting of the International Society of Magnetic Resonance in Medicine, New York, p 619

Hobbs SK, Homer RJ et al (2002) Image guided protoemics in human gliomas. Proc 10th Ann Meeting Intern Soc Magn Res Med, Hawaii, p 662

Jackson A, Zhu XP et al (2000) Parametric mapping of scaled

fitting error in dynamic susceptibility contrast enhanced MR perfusion imaging, part II. Clinical application. ISMRM, 8th scientific meeting, Denver, p 619

Jackson A, Kassner A et al (2001) Reproducibility of T2* blood volume and vascular tortuosity maps in cerebral gliomas. J Magn Reson Imaging 14:510–516

Jackson A, Haroon H et al (2002a) Breath-hold perfusion and permeability mapping of hepatic malignancies using magnetic resonance imaging and a first-pass leakage profile model. NMR Biomed 15:164–173

Jackson A, Kassner A et al (2002b) Abnormalities in the recirculation phase of contrast agent bolus passage in cerebral gliomas: comparison with relative blood volume and tumor grade. AJNR Am J Neuroradiol 23:7–14

Jackson A, Jayson GC et al (2003) Reproducibility of quantitative dynamic contrast-enhanced MRI in newly presenting glioma. Br J Radiol 76:153–162

Jain R, Gerlowski L (1984) Extravascular transport in normal and tumour tissues. Crit Rev Onc Haematol 5:115–170

Jensen RL (1998) Growth factor-mediated angiogenesis in the malignant progression of glial tumors: a review. Surg Neurol 49:189–195; discussion 196

Johnson G, Wetzel S et al (2002) Simultaneous measurement of blood volume and vascular transfer contstant by first pass pharmacokinetic modeling. Proc 10th Intern Magn Reson Med Hawaii, p 2123

Kassner A, Annesley D et al (1999) Abnormalities of the contrast re-circulation phase in cerebral tumours demonstrated using dynamic susceptibility contrast-enhanced MR imaging: a possible marker of vascular tortuosity. Proceedings of the 7th scientific meeting of the International Society of Magnetic Resonance in Medicine, Philadelphia, p 151

Kassner A, Annesley DJ et al (2000) Abnormalities of the contrast re-circulation phase in cerebral tumors demonstrated using dynamic susceptibility contrast-enhanced imaging: a possible marker of vascular tortuosity. J Magn Reson Imaging 11:103–113

Kassner A, Zhu XP et al (2001) A marker of vascular tortuosity (relative recirculation) in gliomas: comparison with blood volume and tumor grade. ISMRM, 9th scientific meeting, Glasgow, p 2247

Klengel S, Hietschold V et al (1994) Simultaneous acquisition of dynamic contrast enhancement and perfusion of normal and pathologic human brain at 0.5 Tesla. Proceedings of the SMR, 2nd annual meeting, San Francisco, p 1467

Knopp EA, Cha S et al (1999) Glial neoplasms: dynamic contrast-enhanced T2*-weighted MR imaging. Radiology 211:791–798

Knopp MV, Giesel FL et al (2001) Dynamic contrast-enhanced magnetic resonance imaging in oncology. Top Magn Reson Imaging 12:301–308

Kovar DA, Lewis MZ et al (1997) In vivo imaging of extraction fraction of low molecular weight MR contrast agents and perfusion rate in rodent tumors. Magn Reson Med 38:259–268

Kovar DA, Lewis M et al (1998) A new method for imaging perfusion and contrast extraction fraction: input functions derived from reference tissues. J Magn Reson Imaging 8:1126–1134

Kuhl CK, Bieling H et al (1997) Breast neoplasms: T2* susceptibility-contrast, first-pass perfusion MR imaging. Radiology 202:87–95

Kwong KK, Chesler DA et al (1995) MR perfusion studies with T1-weighted echo planar imaging. Magn Reson Med 34:878–887

Lam WW, Chan KW et al (2001) Pre-operative grading of intracranial glioma. Acta Radiol 42:548–554

Larsson HB, Stubgaard M et al (1990) Quantitation of blood-brain barrier defect by magnetic resonance imaging and gadolinium-DTPA in patients with multiple sclerosis and brain tumors. Magn Reson Med 16:117–131

Law M, Cha S et al. (2002) High-grade gliomas and solitary metastases: differentiation by using perfusion and proton spectroscopic MR imaging.Radiology 222:715-721

Levin JM, Kaufman MJ et al (1995) Sequential dynamic susceptibility contrast MR experiments in human brain: residual contrast agent effect, steady state, and emodynamic pertubation. Magn Reson Med 34:655–663

Li KL, Jackson A (2003) New hybrid technique for accurate and reproducible quantitation of dynamic contrast-enhanced MRI data. Magn Reson Med 50:1286–1295

Li KL, Zhu XP (2002) Quantification of plasma volume and permeability using first pass pharmacokinetic models: an assessment of accuracy and precision by Monte Carlo simulation. Proceedings of the 10th scientific meeting of the International Society for Magnetic Resonance in Medicine, Honolulu

Li KL, Zhu XP et al (2003) Simultaneous mapping of blood volume and endothelial permeability surface area product in gliomas using iterative analysis of first-pass dynamic contrast enhanced MRI data. Br J Radiol 76:39–50

Li KL, Zhu XP et al (2000a) Quantitative dynamic contrast-enhanced MRI in tumors. A reproducible technique in the head? A reproducible technique in the breast? ISMRM, 8th scientific meeting, Denver, p 724

Li KL, Zhu XP et al (2000b) Improved 3D quantitative mapping of blood volume and endothelial permeability in brain tumors. J Magn Reson Imaging 12:347–357

Li TQ, Ostergaard L et al (1998) Simultaneous blood flow, blood volume and permeability mapping using dual-echo spiral imaging: methodology and initial validation with spiral-FAIR. Proceedings of the 6th scientific meeting of the International Society for Magnetic Resonance in Medicine, Sydney, p 1189

Long DM (1973) Vascular ultrastructure in human meningiomas and schwannomas. J Neurosurg 38:409–419

Ludemann L, Hamm B et al (2000) Pharmacokinetic analysis of glioma compartments with dynamic Gd-DTPA-enhanced magnetic resonance imaging. Magn Reson Imaging 18:1201–1214

Ludemann L, Grieger W et al (2002) Pharmacokinetic imaging of brain tumors. Proc 10th Intern Magn Reson Med Hawaii, p 2079

Lund EL, Spang-Thomsen M et al (1998) Tumor angiogenesis – a new therapeutic target in gliomas. Acta Neurol Scand 97:52–62

Luthert PJ, Deane BR et al (1986) The vasculature of experimental brain tumours: angiogenesis, vascular pathology, and permeability studies. In: Walker MD, Thomas DGT (eds) Biology of brain tumour. Nijhoff, Boston, pp 197–202

Maeda M, Itoh S et al (1993) Tumor vascularity in the brain: evaluation with dynamic susceptibility-contrast MR imaging. Radiology 189:233–238

Maeda M, Itoh S et al (1994) Vascularity of meningiomas and neuromas: assessment with dynamic susceptibility-contrasst MRF imaging. Am J Roentgenol 163:181–186

Martiny-Baron G, Marme D (1995) VEGF-mediated tumour angiogenesis: a new target for cancer therapy. Curr Opin Biotechnol 6:675–680

Maxwell M, Naber SP et al (1991) Expression of angiogenic growth factor genes in primary human astrocytomas may contribute to their growth and progression. Cancer Res 51:1345–1351

Miyati T, Banno T et al (1997) Dual dynamic contrast-enhanced MR imaging. J Magn Reson Imaging 7:230–235

Nagele T, Petersen D et al (1993) Dynamic contrast enhancement of intracranial tumors with snapshot-FLASH MR imaging. AJNR Am J Neuroradiol 14:89–98

Ohno K, Pettigrew KD et al (1978) Lower limits of cerebro-vascular permeability to nonelectrolytes in the conscious rat. Am J Physiol 235:H299–307

Ostergaard L, Weisskoff RM et al (1996) High resolution measurement of cerebral blood flow using intravascular tracer bolus passages, part I. Mathematical approach and statistical analysis. Magn Reson Med 36:715–725

Ostergaard L, Smith DF et al (1998) Absolute cerebral blood flow and blood volume measured by magnetic resonance imaging bolus tracking: comparison with positron emission tomography values. J Cereb Blood Flow Metab 18:425–432

Ostergaard L, Hochberg FH et al (1999) Early changes measured by magnetic resonance imaging in cerebral blood flow, blood volume, and blood-brain barrier permeability following dexamethasone treatment in patients with brain tumors. J Neurosurg 90:300–305

Padhani AR, Hayes C et al (2002) Reproducibility of quantitative dynamic MRI of normal human tissues. NMR Biomed 15:143–153

Padhani AR, Husband JE (2001) Dynamic contrast-enhanced MRI studies in oncology with an emphasis on quantification, validation and human studies. Clin Radiol 56:607–620

Pardo FS, Aronen HJ et al. (1994) Functional cerebral imaging in the evaluation and radiotherapeutic treatment planning of patients with malignant glioma. Int J Radiat Oncol Biol Phys 30:663-669

Parker GJ, Suckling J et al (1997) Probing tumor microvascularity by measurement, analysis and display of contrast agent uptake kinetics. J Magn Reson Imaging 7:564–574

Paulson OB, Hertz MM (1983) Tracer kinetics and physiologic modelling. In: Lambrecht RM, Rescigno A (eds) Theory to practice. Lecture notes in biomathematics. Springer, Berlin Heidelberg New York, pp 428–444

Perman WH, Gado MH et al (1992) Simultaneous MR acquisition of arterial and brain signal-time curves. Magn Reson Med 28:74–83

Pietsch T, Valter MM et al (1997) Expression and distribution of vascular endothelial growth factor protein in human brain tumors. Acta Neuropathol (Berl) 93:109–117

Plate KH, Risau W (1995) Angiogenesis in malignant gliomas. Glia 15:339–347

Provenzale JM, Wang GR et al (2002) Comparison of permeability in high-grade and low-grade brain tumors using dynamic susceptibility contrast MR imaging. AJR Am J Roentgenol 178:711–716

Provias J, Claffey K et al (1997) Meningiomas: role of vascular endothelial growth factor/vascular permeability factor in angiogenesis and peritumoral edema. Neurosurgery 40:1016–1026

Reinhold HS, Endrich B (1986) Tumour microcirculation as a target for hyperthermia. Int J Hyperthermia 2:111–137

Roberts HC, Roberts TP et al (2000) Quantitative measurement of microvascular permeability in human brain tumors achieved using dynamic contrast-enhanced MR imaging: correlation with histologic grade. AJNR Am J Neuroradiol 21:891–899

Roberts HC, Roberts TP et al (2001) Correlation of microvascular permeability derived from dynamic contrast-enhanced MR imaging with histologic grade and tumor labeling index: a study in human brain tumors. Acad Radiol 8:384–391

Rosen BR, Belliveau JW et al (1989) Perfusion imaging by nuclear magnetic resonance. Magn Reson Q 5:263–281

Rosen BR, Belliveau JW et al (1990) Perfusion imaging with NMR contrast agents. Magn Reson Med 14:249–265

Rosen BR, Belliveau JW et al (1991) Susceptibility contrast imaging of cerebral blood volume: human experience. Magn Reson Med 22:293–299

Rosen BR, Aronen HJ et al (1993) Advances in clinical neuroimaging: functional MR imaging techniques. Radiographics 13:889–896

Schwarzbauer C, Syha J et al (1993) Quantification of regional blood volumes by rapid T1 mapping. Magn Reson Med 29:709–712

Shames DM, Kuwatsuru R et al (1993) Measurement of capillary permeability to macromolecules by dynamic magnetic resonance imaging: a quantitative noninvasive technique. Magn Reson Med 29:616–622

Siegal T, Rubinstein R et al (1997) Utility of relative cerebral blood volume mapping derived from perfusion magnetic resonance imaging in the routine follow up of brain tumors. J Neurosurg 86:22–27

Song CW, Lokshina A et al (1984) Implication of blood flow in hyperthermic treatment of tumors. IEEE Trans Biomed Eng 31:9–16

Stack JP, Redmond OM et al (1990) Breast disease: tissue characterization with Gd-DTPA enhancement profiles. Radiology 174:491–494

St Lawrence KS, Lee TY (1998) An adiabatic approximation to the tissue homogeneity model for water exchange in the brain. I. Theoretical derivation. J Cereb Blood Flow Metab 18:1365–1377

Stratmann A, Machein MR et al (1997) Anti-angiogenic gene therapy of malignant glioma. Acta Neurochir Suppl (Wien) 68:105–110

Strugar J, Rothbart D et al (1994) Vascular permeability factor in brain metastases: correlation with vasogenic brain edema and tumor angiogenesis. J Neurosurg 81:560–566

Strugar JG, Criscuolo GR et al (1995) Vascular endothelial growth/permeability factor expression in human glioma specimens: correlation with vasogenic brain edema and tumor-associated cysts. J Neurosurg 83:682–689

Su MY, Muhler A et al (1998) Tumor characterization with dynamic contrast-enhanced MRI using MR contrast agents of various molecular weights. Magn Reson Med 39:259–269

Sugahara T, Korogi Y et al (1998) Correlation of MR imaging-determined cerebral blood volume maps with histologic and angiographic determination of vascularity of gliomas. AJR Am J Roentgenol 171:1479–1486

Sugahara T, Korogi Y et al (1999a) Contrast enhancement of intracranial lesions: conventional T1-weighted spin-echo versus fast spin-echo MR imaging techniques. AJNR Am J Neuroradiol 20:1554–1559

Sugahara T, Korogi Y et al (1999b) Value of dynamic susceptibility contrast magnetic resonance imaging in the evaluation of intracranial tumors. Top Magn Reson Imaging 10:114–124

Sugahara T, Korogi Y et al (2001) Perfusion-sensitive MR imaging of gliomas: comparison between gradient-echo and spin-echo echo-planar imaging techniques. AJNR Am J Neuroradiol 22:1306–1315

Tofts PS, Kermode AG (1991) Measurement of the blood-brain barrier permeability and leakage space using dynamic MR imaging. 1. Fundamental concepts. Magn Reson Med 17:357–367

Tofts P, Berkowitz B et al (1995) Quantitative analysis of dynamic Gd-DTPA enhancement in breast tumours using a permeability model. Mag Res Med 33:564–568

Tofts PS, Brix G et al (1999) Estimating kinetic parameters from dynamic contrast-enhanced T(1)-weighted MRI of a diffusable tracer: standardized quantities and symbols. J Magn Reson Imaging 10:223–232

Uematsu H, Maeda M et al (2000) Vascular permeability: quantitative measurement with double-echo dynamic MR imaging - theory and clinical application. Radiology 214:912–917

Uematsu H, Maeda M et al (2001) Blood volume of gliomas determined by double-echo dynamic perfusion-weighted MR imaging: a preliminary study. AJNR Am.J Neuroradiol 22:1915–1919

Van der Sanden BP, Rozijn TH et al (2000) Noninvasive assessment of the functional neovasculature in 9L-glioma growing in rat brain by dynamic 1H magnetic resonance imaging of gadolinium uptake. J Cereb Blood Flow Metab 20:861–870

Villringer A, Rosen BR et al (1988) Dynamic imaging with lanthanide chelates in normal brain: contrast due to magnetic susceptibility effects. Magn Reson Med 6:164–174

Vonken EP, van Osch MJ et al (2000) Simultaneous quantitative cerebral perfusion and Gd-DTPA extravasation measurement with dual-echo dynamic susceptibility contrast MRI. Magn Reson Med 43:820–827

Weisskoff R, Boxerman J et al (1994a) Simultaneous blood volume and permeability mapping using a single Gd-based contrast injection. Proceedings of the 2nd scientific meeting of the International Society of Magnetic Resonance in Medicine, San Francisco, p 279

Weisskoff RM, Boxerman JL et al (1994b) Simultaneous blood volume and permeability mapping using Gd-based contrast injection. Proceedings of the 2nd scientific meeting of the International Society of Magnetic Resonance in Medicine, San Francisco, p 279

Weisskoff RM, Zuo CS et al (1994c) Microscopic susceptibility variation and transverse relaxation: theory and experiment. Magn Reson Med 31:601–610

Weissleder R, Mahmood U (2001) Molecular imaging. Radiology 219:316–317

Wenz F, Rempp K et al (1996) Effect of radiation on blood volume in low-grade astrocytomas and normal brain tissue: quantification with dynamic susceptibility contrast MR imaging. AJR Am J Roentgenol 166:187–193

Wismer GL, Boxton RB et al (1988) Susceptibility induced MR line broadening: application to brain iron mapping. J Comput Assist Tomogr 12:259–265

Worthington BS, Bullock P et al (1991) Clinical experience with contrast enhanced echo-planar imaging of the brain. Magn Reson Med 22:255–258

Yang S, Wetzel S et al. (2002) Dynamic contrast-enhanced T2*-weighted MR imaging of gliomatosis cerebri. AJNR Am.J Neuroradiol 23:350-355

Zhu XP, Li KL et al (1997) Parametric error maps in MR perfusion imaging. British chapter of ISMRM, Manchester, UK

Zhu X, Hawnaur JM et al (1999) Quantification of relative blood volume and endothelium permeability of breast neoplasm using dynamic MR imaging. Proceedings of the 7th scientific meeting of the International Society of Magnetic Resonance in Medicine, Philadelphia, p 1076

Zhu XP, Jackson A et al (2000a) The choroid plexus as an internal reference for quantitative permeability studies in brain tumors. ISMRM, 8th scientific meeting, Denver, 1969

Zhu XP, Laing AD et al (2000b) Parametric mapping of scaled fitting error in dynamic susceptibility contrast enhanced MR perfusion imaging, part I. Statistical analysis. ISMRM, 8th scientific meeting, Denver, p 748

Zhu XP, Li K. L et al (2000c) Quantification of endothelial permeability, leakage space, and blood volume in brain tumors using combined T1 and T2* contrast-enhanced dynamic MR imaging. J Magn Reson Imaging 11:575–585

10 Dynamic Magnetic Resonance Imaging in Breast Cancer

Mei-Lin W. Ah-See and Anwar R. Padhani

CONTENTS

10.1 Introduction

Breast cancer is the most common cancer in women worldwide accounting for 25% of all female malignancies. The incidence of breast cancer is high in Western, developed countries and in England and Wales almost 36,000 new cases were diagnosed in the year 2000 with a mortality of 11,500 in 2002 (www.statistics.gov.uk). In the UK, one in nine women will develop breast cancer at some point in their life. Risk factors for breast cancer include early menarche, nulliparity, increasing age, family history and a previous history of breast cancer.

Radiological imaging of the breast using mammography and ultrasound have well defined roles which are currently central in the diagnosis, assessment of treatment response and follow-up of breast cancer. Indeed, a third of the reduction in mortality rates from breast cancer in England and Wales (a reduction of 20% in 1999 compared with the mid-1980s) has been attributed to the National Health Service mammographic breast screening programme introduced in 1988. Early detection and advances in systemic therapy have also contributed to the overall improvement in disease specific survival.

Both mammography and ultrasound have limitations however. In terms of screening, increased density of the breasts (common in younger patients) significantly reduces the sensitivity of mammography; in one screening series from 80% to 30% (Mandelson et al. 2000). Ultrasound as an adjunct to mammography can reduce the false-negative rate in high risk patients with dense breasts (Crystal et al. 2003), however its routine use in population screening is not recommended (Teh and Wilson 1998). Radiation exposure during screening mammography must also be considered particularly in younger high-risk women who may be carriers of the hereditary breast cancer disposing genes (BRCA-1 or BRCA-2) in whom the implications of defective repair mechanisms for ionising-radiation-induced DNA double-strand breaks is not yet fully known (Nieuwenhuis et al. 2002). In the case of lobular carcinoma of the breast, mammographic changes are often subtle due to the tendency for the lobular carcinoma cells to infiltrate in a "single-file" ("Indian-files") fashion and induce a less prominent

M. W. Ah-See, MD
Paul Strickland Scanner Centre, Mount Vernon Hospital, Rickmansworth Road, Northwood, Middlesex, HA6 2RN, UK
A. R. Padhani, MRCP, FRCR
Consultant Radiologist and Lead in MRI, Paul Strickland Scanner Centre, Mount Vernon Hospital, Rickmansworth Road, Northwood, Middlesex, HA6 2RN, UK

desmoplastic reaction. This results in the cancers being missed or the volume of disease being underestimated. In addition, lobular carcinomas are associated with an increased incidence of multifocality (additional tumour within the same quadrant as the index lesion) and multicentricity (additional tumour within a separate quadrant to the index lesion) which can result in incomplete excision with breast-conserving surgery (Takehara et al. 2004). In contrast, when monitoring response to primary systemic therapy (PST) for patients with more advanced disease, mammography and ultrasound often overestimate the extent of residual disease (Feldman et al. 1986; Helvie et al. 1996; Vinnicombe et al. 1996; Herrada et al. 1997; Fiorentino et al. 2001) and are unable to distinguish between active disease and post-chemotherapy fibrosis. The inability to distinguish fibrosis from active disease also limits the role of mammography for differentiating post-surgical scarring from recurrent tumour.

With the aforementioned limitations in mind, dynamic contrast medium enhanced MRI (DCE-MRI) has emerged as a powerful imaging tool in the management of patients with breast cancer. The success of DCE-MRI of the breast is dependent on its ability to demonstrate intrinsic differences that affect contrast medium behaviour in a variety of tissue types. Early breast MRI studies showed that breast cancers enhanced following contrast-agent administration (Heywang et al. 1989; Kaiser and Zeitler 1989); however, further studies showed that benign breast diseases, such as fibroadenoma, fibrocystic disease, radial scar and mastitis, also enhanced (Heywang et al. 1989; Harms et al. 1993; Orel et al. 1994; Fobben et al. 1995; Stomper et al. 1995). In addition, normal breast tissue can enhance to a variable degree depending on the phase of the menstrual cycle (Kuhl et al. 1997b). Thus, the presence of enhancement alone cannot be used to distinguish malignant from benign disease. Optimal interpretation of DCE-MRI in the breast uses the combination of morphological and functional characteristics to aid in diagnoses.

DCE-MRI studies can be performed to be sensitive to the vascular phase of contrast medium delivery (so-called T_2* DCE-MRI) which reflect on tissue perfusion and blood volume (Padhani and Dzik-Jurasz 2004). Similarly, sequences sensitive to the presence of contrast medium in the extravascular extracellular space (EES) which reflect on microvessel permeability and extracellular leakage space in addition to perfusion can be performed (so-called T_1 DCE-MRI). In this chapter, we discuss the technical aspects of data acquisition and analysis together with the biological basis and pathophysiological correlates of both MRI techniques. We will review all the potential roles of DCE-MRI including breast cancer screening, evaluation of response to primary systemic chemotherapy and diagnosis of recurrent disease (Table 10.1). A brief summary of angiogenesis in breast cancer is also discussed as this forms the biological basis for the differing functional DCE-MRI characteristics of tumours compared with benign lesions and normal breast tissue.

10.2 Breast Cancer and Angiogenesis

In solid tumours, growth beyond a millimetre cannot occur without vascular support (Folkman 1996). Transgenic animal tumour model experiments have shown that progression from an in-situ to invasive cancer is accompanied by the onset of angiogenesis (Rak et al. 1995). There are a number of clinical examples where vascularization has been related to tumour progression (e.g., in the change from breast ductal carcinoma in-situ to invasive cancer (Gilles et al. 1995); Bose et al. 1996). Immunohistochemical techniques show changes consistent with this observation; for example, expression of the endothelial cell-specific tyrosine kinase receptor, Tie-2 (TEK) is increased during the transition from benign to invasive breast cancer (Bernsen et al. 1998).

The most potent pro-angiogenic factor in breast tumours is vascular endothelial growth factor (VEGF), initially termed vascular permeability factor due to its hyperpermeable effect on vessels (Senger et al. 1983). VEGF leads to endothelial cell migration, proliferation and hence neovascularization within tumours (Toi et al. 2001). The vessels so created lack normal hierarchical structures and are disorganised, chaotic and hyperpermeable (Jain 2003). Elevated levels of VEGF within breast tumours are associated with a poorer overall prognosis when measured from the tumour cytosol (Gasparini et al. 1997); Linderholm et al. 2000) or by immunohistochemistry (Toi et al. 1995). Levels of circulating VEGF have also been shown to be elevated in breast cancer patients when compared with normal controls (Adams et al. 2000). In one study, a decline in serum VEGF was shown to correlate with a partial response or stable disease following docetaxel chemotherapy in metastatic disease (Lissoni et al. 2000). In a series of patients receiving PST however, neither baseline nor changes in serum or plasma VEGF measurements during treatment predicted for final response (BURCOMBE et al. 2002). Furthermore, no correlation has been found between circulating VEGF levels and tumour VEGF expression (ADAMS et al. 2000; LANTZSCH et al. 2002) although a correlation has been noted between serum VEGF levels and primary tumour bulk and intratumoral microvessel density (ADAMS et al. 2000).

Intratumoral microvessel density (MVD) is the currently recognised method for assessing the angiogenic status of breast tumours and employs immunohistochemical staining of tumour sections using one of a number of panendothelial antibodies include factor VIII-related antigen, CD34 and CD31 (VERMEULEN et al. 1996). One of the most commonly used techniques for assessing MVD in breast cancers is Chalkley point counting, which has been used to assess areas with the highest vascularity (termed "vascular hotspots") or random fields within the tumour section (FOX et al. 1995).

The majority of studies assessing MVD and outcome in breast cancer have shown that MVD is an independent prognostic factor for both relapse-free and overall survival as well as a predictor of metastases [see HASAN et al. (2002) for a comprehensive review of these studies]. A small number of studies assessing MVD, however, have failed to demonstrate any prognostic value of MVD and breast cancer outcome. Indeed, the College of American Pathologists consensus statement on prognostic factors in breast cancer in 1999, ranked MVD in category III which encompasses "all factors which are not sufficiently studied to demonstrate their prognostic significance" (FITZGIBBONS et al. 2000). Variations in the endothelial antibody employed, counting technique used, region of tumour assessed and defined cut-off for increased vascularity may have contributed to conflicts in study results (FOX 1997). To minimise these problems, the "International Consensus on the methodology and criteria of evaluation of angiogenesis quantification in solid human tumours" was updated in 2002 and aims at standardising the techniques used for MVD quantification as well as other aspects of angiogenesis assessment (VERMEULEN et al. 2002). The latter include endothelial cell proliferation fraction and pericyte coverage index which reflect upon the functional status of the tumour microvasculature. It must be borne in mind, however, that any pathological assessment of angiogenesis can only ever provide a quantitative, non-functional measure of the vascular status of a tumour.

10.3 T_2*-weighted Breast Cancer DCE-MRI

10.3.1 Data Acquisition

Perfusion-weighted images of the breast can be obtained using "bolus-tracking techniques" that monitor the passage of contrast material through a capillary bed as magnetic field (Bo) inhomogeneities are produced which result in a decrease in signal intensity of surrounding tissues (SORENSEN et al. 1997; BARBIER et al. 2001). The decrease in signal intensity of tissues can be observed with susceptibility-weighted sequences. The degree of signal loss observed is dependent on the vascular concentration of the contrast agent and microvessel size (DENNIE et al. 1998) and density. The signal to noise ratio (SNR) of such images can be improved by using high doses of contrast medium (i.e., ≥0.2 mmol/kg body weight) (BRUENING et al. 2000). High specification MR systems are ideally suited to this task allowing multi-slice acquisitions. Such studies are also possible on conventional MRI systems using standard gradient-echo techniques but the need to acquire data every 1–2 s requires that the matrix size is relatively coarse and often limited to 1–2 slices. We use a spoiled gradient-echo fast low angle shot (FLASH) sequence in the breast as a single sagittal slice imaging technique with data acquisitions every 2 s over 2 min (TE=20 ms, TR=30 ms, alpha=10°, 1 slice, field of view 13 × 26cm, matrix size 64 × 128) with 0.2 mmol/kg of contrast medium (gadopentetate dimeglumine; Gd-DTPA) injected at 4 ml/s after 20 s (Fig. 10.1).

10.3.2 Data Quantification and Limitations

Tracer kinetic principles can be used to provide estimates of relative blood volume (rBV), relative blood flow (rBF) and mean transit time (MTT) derived from the first-pass of contrast agent through the microcirculation (ROSEN et al. 1991; SORENSEN et al. 1997; BARBIER et al. 2001) (Fig. 10.2). These variables are related by the central volume theorem equation (BF = BV/MTT). Quantification of T_2*-weighted DCE-MRI and its application for leaky vasculature is discussed in detail elsewhere in this book (see Chap. 4). The most robust parameter which can be extracted reliably from first pass techniques is rBV, which is obtained from the integral of the time

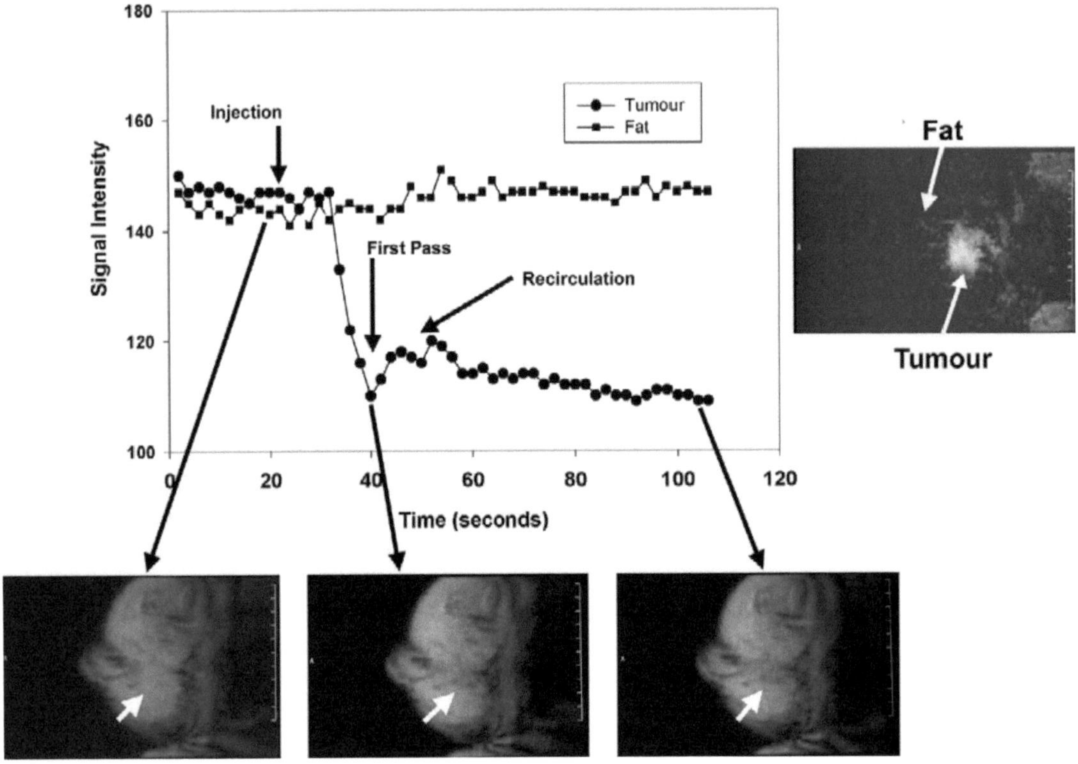

Fig. 10.1 Typical T2*-weighted DCE-MRI study. 30-year old woman with a grade 3 invasive ductal cancer of the right breast. The same patient is illustrated in Figures 10.2, 10.3, 10.6, 10.8 and 10.12. 22 ml of IV contrast Gd-DTPA was given after the 10th data point (*arrow*). First pass T2* susceptibility effects cause darkening of the tumour (*arrows*) (subtle but the effect is better appreciated in the subtraction image (insert)). No darkening of the breast parenchyma is seen. The first pass and recirculation phases are indicated. Signal intensity changes for 2 regions of interest are shown on the subtraction image of the nadir point for ROI corresponding to the tumour.

series data during the first pass of the contrast agent (Ostergaard et al. 1996). This cannot readily be done for the breast because of the loss of compartmentalisation of the contrast medium. Solutions to counter T_1 enhancing effects of gadolinium chelates include pre-dosing with contrast medium to saturate the leakage space and idealised model fitting (Fig. 10.2); the time series data is fitted to a gamma-variate function from which the parameters rBV, rBF and MTT are derived. This is the methodology that we employ using custom analysis software called Magnetic Resonance Imaging Workbench (MRIW) which has been developed at the Institute of Cancer Research, Royal Marsden Hospital, London (D'Arcy et al. 2004). This software is also used to analyse T_1-weighted DCE-MRI data (see Sect. 10.4.4.4). From a practical perspective, it is not always necessary to quantify T_2*-weighted DCE-MRI data to obtain insights of the spatial distribution of tissue perfusion. Simple subtraction images can demonstrate the maximal signal attenuation (also termed rela-

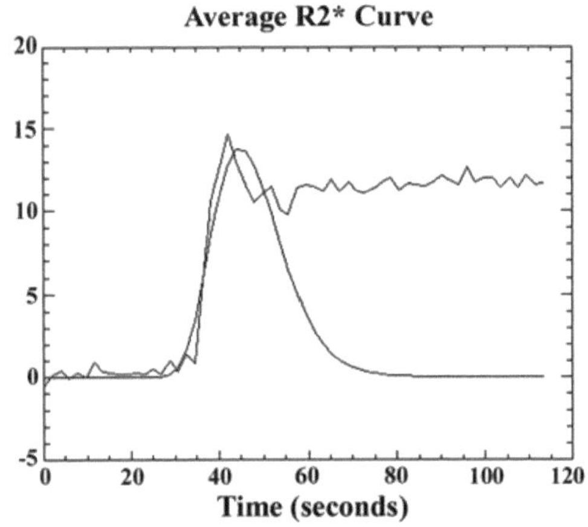

Fig. 10.2 Model fitting of T2*-weighted data. Same patient as illustrated in Figures 10.1, 10.3, 10.6, 10.8 and 10.12. T2* signal intensity data from the tumour ROI in Figure 10.1 are converted into R2* (1/T2*) and then fitted with a gamma variate function. The computed values of rBV, rBF and MTT for this region of interest are 265, 10.6 arbitrary units and 25.1 seconds. Parametric maps representing blood flow kinetics are shown in Figure 10.3.

tive maximum signal drop, rMSD), which has been strongly correlated with relative blood flow and volume in tumours (Liu et al. 2002). Subtraction analysis should only be done, however, if there is a linear relationship between rBV and rBF, that is when MTT is in a narrow range; this is often the case in non-necrotic breast tumours (Fig. 10.3).

10.3.3 Clinical Experience

Both Kuhl et al. (1997a) and Kvistad et al. (1999) have evaluated the value of T_2*W DCE-MRI for characterising breast lesions using visual assessments. Both studies showed strong decreases in signal intensity in malignant tissues whereas susceptibility effects in fibroadenomas were minor. Thus, it was possible to differentiate carcinomas from fibroadenomas with high specificity using T_2*W characteristics despite significant overlap in T_1 enhancement patterns. The pathophysiological explanation probably relates to differences in the microvessel arrangements, density and size between malignant tumours and fibroadenomas. Weind et al. (1998) conducted a histopathological study comparing microvessel distributions in invasive breast cancer and fibroadenomas that shed light on these observations. They showed that there was complete overlap in MVD counts between the breast cancers and the fibroadenomas but the distribution of microvessels was distinctive. Microvessel distribution in the breast carcinomas showed marked regional variations with fewer vessels seen within the centre of tumours compared to the periphery whereas the microvessel distribution in fibroadenomas was

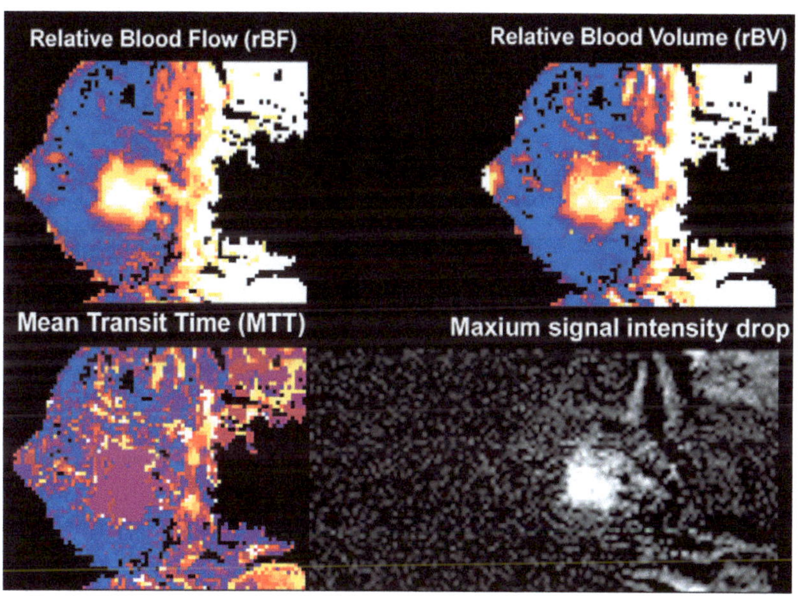

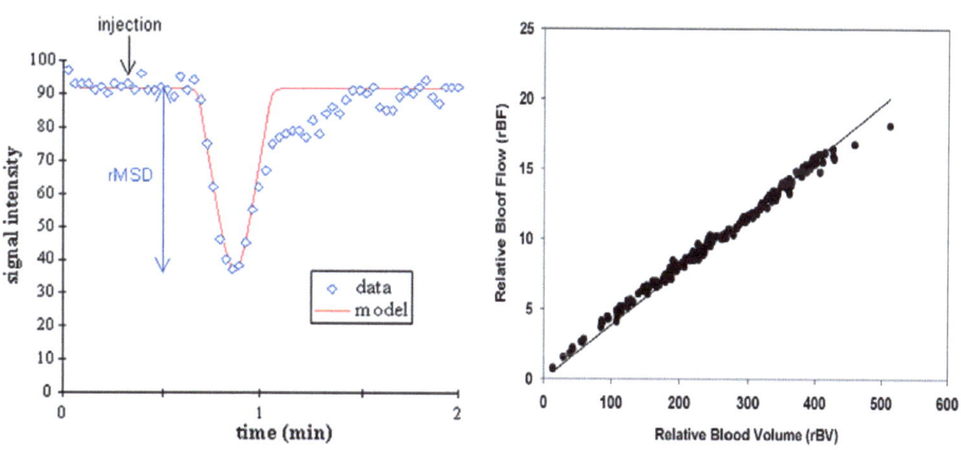

Fig. 10.3. Parametric T2*-weighted DCE-MRI images. This is the same tumour illustrated in Figures 10.1, 10.2, 10.6, 10.8 and 10.12. Parametric images of relative blood volume (rBV), relative blood flow (rBF) and mean transit time (MTT) are derived on a pixel-by-pixel basis using gamma variate fitting as shown in Figure 10.2. Also shown is the T2*-weighted DCE-MRI subtraction image (identical to the Figure 10.1 insert). The graphs show how relative maximum signal drop (rMSD) is calculated and that there is a linear correlation between relative blood volume and flow at a pixel level within the tumour ROI.

more evenly spread within the stroma. Other investigators have also shown microvessel density "hot spots" in breast cancers (Buadu et al. 1996; Jitsuiki et al. 1999). Both Kuhl et al. (1997a) and Kvistad et al. (1999) comment that the T_2* effects occurred in focal areas which may correspond to the "hot spot" MVD counts in tumours. Our experience in the use of quantitative T_2*-weighted DCE-MRI has been in monitoring tumour response to neoadjuvant chemotherapy and is discussed in detail in Sect. 10.4.3.5 below.

10.4 T_1-Weighted Breast Cancer DCE-MRI

A large body of literature has shown that breast cancer enhance earlier and to a greater extent than benign breast diseases on T_1-weighted DCE-MRI. This difference is most marked in the early period (1–3 min) after bolus contrast medium administration (Kaiser and Zeitler 1989; Flickinger et al. 1993; Gilles et al. 1993; Boetes et al. 1994). However, other investigators have demonstrated that there is an overlap in the enhancement rates of benign and malignant lesions (Heywang et al. 1989; Fobben et al. 1995; Stomper et al. 1995). Thus, any kinetic parameter used for tissue characterisation has to take into consideration the relative contrast medium behaviour in different tissues.

Extracellular contrast media readily diffuse from the blood into the EES of tissues at a rate determined by the permeability of the capillaries and their surface area. Shortening of T_1 relaxation rate caused by contrast medium is the mechanism of tissue enhancement. Most DCE-MRI studies employ gradient-echo sequences to monitor the tissue enhancing effects of contrast media. This is because gradient-echo sequences have good contrast medium sensitivity, yield images with high signal to noise ratio and enable data acquisition to be performed rapidly. The degree of signal enhancement seen on T_1-weighted images is dependent on a number of physiological and physical factors. These include tissue perfusion, capillary permeability to contrast agent, volume of the extracellular leakage space, native T_1-relaxation rate of the tissue, contrast agent dose, imaging sequence and parameters utilised and on-machine scaling factors (Roberts 1997). The relative effects of these factors on DCE-MRI are discussed in more detail elsewhere in this book (see Chap. 5 and 6).

10.4.1 Data Acquisition, Analysis and Processing

10.4.1.1 Dosage Considerations

Most modern breast MR examinations are performed with gradient-echo techniques due to their high T_1 sensitivity with a contrast medium dose of 0.1 mmol/kg body weight (b.w.). In general, there is uncertainty about the most effective contrast agent dose for breast MR examinations. One report claimed a higher sensitivity with 0.16 mmol/kg b.w. of gadopentetate dimeglumine (Gd-DTPA) compared with standard dose (0.1 mmol/kg b.w.) (Heywang-Kobrunner et al. 1994). Using 3D gradient-echo sequences, they found that the ability to detect small lesions and to discriminate benign from malignant lesions was improved with higher doses of contrast medium (Fig. D.4). This dose-related effect has been confirmed by a recent study comparing three doses of gadobenate dimeglumine (0.05, 0.1 and 0.2 mmol/kg b.w.) (Knopp et al. 2003). Significant dose-related increases in lesion detection were observed for gadobenate dimeglumine (0.1 and 0.2 mmol/kg were better than 0.05); the sensitivity for detection was comparable for 0.1 and 0.2 mmol/kg and specificity was highest with the 0.1 mmol/kg dose. The higher lesion detection rate with higher doses of contrast medium has led to higher dose contrast medium being adopted when screening for breast cancer (Brown et al. 2000).

10.4.1.2 Pulse Sequence Timing

The optimal timing for breast MRI sequences is dependent on the goal of imaging. This is because high-resolution and short imaging-time requirements represent competing examination strategies on current equipment. A single high sensitivity, high-resolution, fat suppression 3D technique (with voxel sizes less than 1 mm) is sometimes used. This can be performed within approximately 2–4 min which is in the time window where the differential enhancement between malignant and benign lesions is greatest; image interpretation has to be based on morphological characteristics alone. Several investigators have reported on the architectural features that can be used for lesion diagnosis using this technique (Harms et al. 1993; Orel et al. 1994; Stomper et al. 1995; Orel et al. 1997).

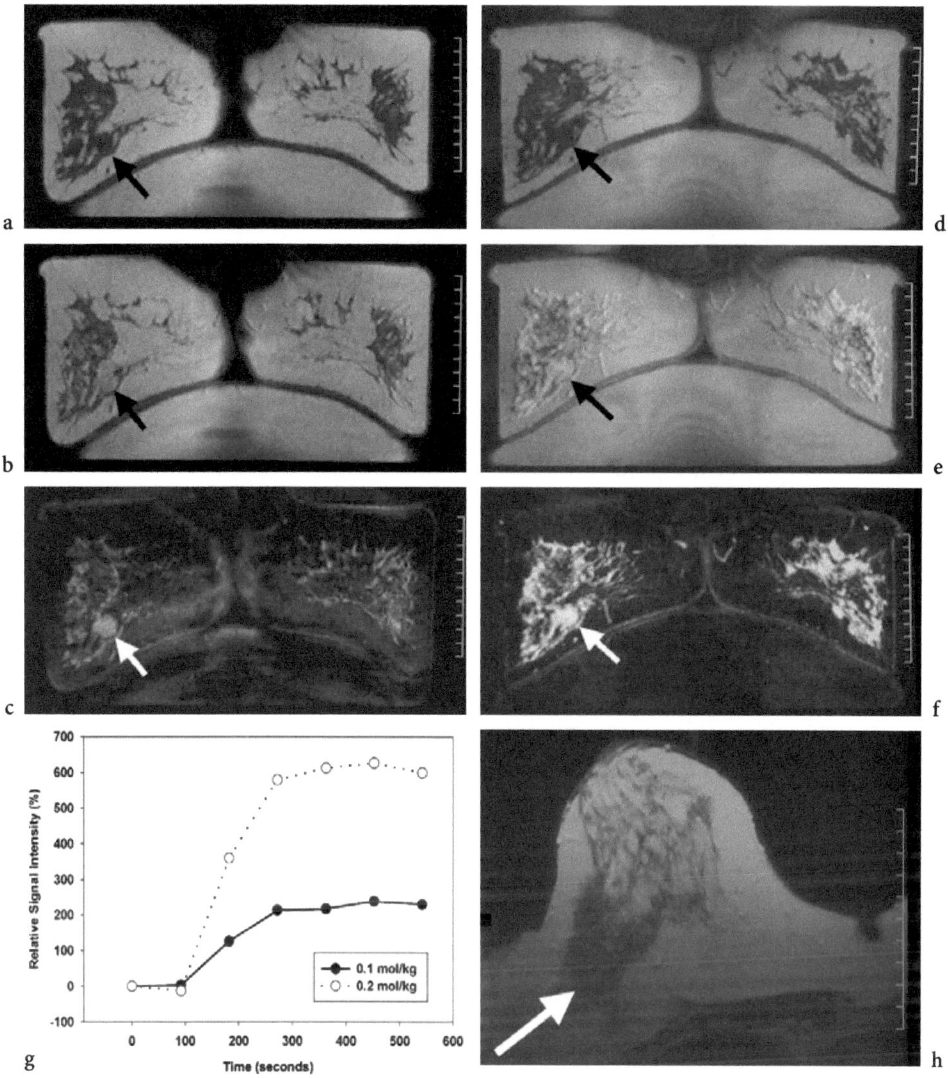

Fig. 10.4a–h. Normal versus double dose of contrast medium. 54 year old woman with infiltrating lobular carcinoma of the breasts. An 8 cm mass was palpable in the right breast. Mammograms were suspicious of tumour in the right breast only. **a–c** and **d–e** represent coronal T1-weighted 3D GRE images before and after 0.1 and 0.2 mmol/kg body weight of Gd-DPTA contrast medium and corresponding subtraction images respectively. The image datasets are acquired 1 month apart. Subjective, a greater intensity of enhancement is seen in the breast parenchyma of both breasts with the higher dose of contrast medium which is diffusely infiltrated with tumour. In a more localised lesion (*arrow*), shows a greater intensity of enhancement with the higher dose of contrast medium which is confirmed on the relative signal intensity (%) time curve shown in image **g**. Image **h** is a T2-weighted spin-echo image of the right breast showing a mass with distortion of the parenchyma in the region of the axillary tail (*arrow*). Pathological examination following bilateral mastectomy showed an 8 cm invasive lobular carcinoma of the right breast and a 1.6 cm tumour on the left of the same histology. We are grateful to Dr. Kausar Raza (Hillingdon Hospital, Hillingdon, UK) for providing this illustration.

If the goal is to maximise specificity, then repeated data acquisitions are recommended. On current MRI systems, higher temporal resolution necessitates reduced spatial resolution, decreased coverage or a combination thereof. The highest specificity MRI breast studies have used 2D dynamic techniques with data acquisition rates of 1–12 s. Higher temporal resolution techniques appear to improve the specificity of examinations because of better characterisation of the signal intensity time curve; one study has suggested that characterisation of breast lesions is optimal at 1–2 s image acquisition (Boetes et al. 1994). It should be recognised that the temporal requirements of fast dynamic imaging limits the spatial resolution and coverage and therefore multi-focal disease may be overlooked and such sequences are not suitable for breast cancer screening. In the latter cases, volume coverage at the slower data acquisition rates is often used.

It is now clear that both kinetic and morphologic information is required to achieve optimal discrimination between benign and malignant disease. Therefore, as a compromise, a dynamic high-resolution 3D technique (slower dynamic 3D technique) is used by many with each time point being acquired every 60–90 s. This approach does not require prior knowledge of lesion location. Both Kuhl et al. (1999) and Liu et al. (1998) have reported on the value of such techniques where an integration of contrast agent kinetics and architecture evaluation is performed. Both groups make the very important point that there must be concordance between the kinetic information and the morphologic features. There are some malignant lesions including invasive ductal and lobular carcinomas or certain ductal carcinoma in-situ lesions, that do not enhance rapidly but in which lesion morphology suggests the presence of malignancy e.g. architectural distortion or mass with spiculated borders. Thus, in general, if the goal is to determine the likelihood that a previously identified lesion is malignant, then the best strategy is to use the minimal number of slices necessary to cover the lesions and obtain a series of acquisitions with high temporal resolution. However, if the goal is to detect multicentric or multi-focal disease, search for a primary tumour or demonstrate the extent of a known cancer, then good spatial resolution and coverage are essential and is best achieved with a 3D MR technique.

10.4.1.3
Post-processing of Data

A breast MRI study can yield many hundreds of images, the analysis of which can be very time consuming. When comparisons have to be made with previous MRI studies the whole endeavour can be made impossible. Many of the assessment challenges can be eased by dedicated workstations and specialist software. Both commercial and academic software is now available with a high degree of sophistication. Tools incorporated into software include simple subtraction, multiplanar reconstruction capability, maximum intensity projections (MIP) and signal-intensity time analysis; these are the minimum tools needed for the analysis of breast MRI. More sophisticated tools that are not widely available include software for deriving contrast agent kinetics with physiological properties, tumour segmentation with volume estimation, image fusion, automated detection and classification of abnormalities (computed assisted diagnosis or CAD) and 3D image registration.

As already noted, both morphological and kinetic analysis needs to be performed to maximise the diagnostic yield of breast MRI. Depending on the scanner type, appropriate weighting to morphological and kinetic characteristics aids in diagnosis as suggested by Brown et al. (2000). It should be noted that diagnostic accuracy of such scoring systems are technique and machine dependent. In practice, subtraction images are performed first. Both early and late subtraction images should be generated. Early subtraction studies are obtained by subtraction of the images obtained at approximately 90–120 s from the pre-contrast images and the late subtraction comprises images obtained after 5 min minus the pre-contrast images. Viewing of the subtraction images will display any significant patient motion; even the slightest movement can generate pseudolesions on subtraction images. If significant motion is present, then the images can be realigned using 3D registration software (Denton et al. 1999) or alternatively, the appropriate individual (non subtracted) images of the dynamic series are reviewed for morphological and kinetic analysis. Both early and late subtraction images should be inspected for the presence of enhancing lesions. When these are identified, morphological assessments should take into account the enhancement pattern. Morphological and kinetic descriptions should follow the MRI lexicon adapted from the American College of Radiology Breast Imaging and Data Reporting System terminology for breast MRI (Ikeda et al. 2001). Readers are referred to Sect. 10.4.3.3 on the usage of these descriptors for breast lesion diagnosis.

Kinetic analysis in everyday practise uses the region of interest (ROI) method. Where possible the early subtraction images are used to determine the position for ROI placement. If enhancement is low or minimal on early subtraction images then the late subtraction dataset can be used. If no enhancement is seen, the baseline data (non-enhanced) is used for ROI placement. When lesion enhancement is homogenous the entire outline of the lesion is used. The outer limit of the lesion should act as a boundary of the ROI to minimise partial volume effects. When lesion enhancement is heterogeneous on the early subtraction images, the maximally enhancing area of the lesion should be used excluding areas of necrosis, artefacts and adjacent blood vessels. The ROI should be constant in position and size for each image in the series under analysis adjusting as required if there is movement. The ROI must be small enough (at least 3–5 pixels) but not too small, because a ROI of 1–2 pixels may include a vessel and is likely to have increased pixel noise. In small lesions it is essential to check that a ROI is placed without partial volume averaging which may occur in the through-plane direction. ROIs should also be placed on background (air surrounding breasts) away from phase encoding artefacts and another ROI should be placed on an area of fat adjacent to or parallel with the enhancing region within the breast if possible. These additional ROIs can help to scale the y-axis of the signal intensity time curves and aid interpretation.

The simplest way of assigning pathological significance to signal intensity time curves is to provide a description of the initial enhancement (1–2 min) followed by an assignment of the late enhancement pattern. The initial enhancement can be described as fast, medium and slow and the late pattern patterns as persistent, plateau and wash-out (patterns I–III) (Fig. 10.5). This scheme has been widely adopted following its initial clinical evaluation by KUHL et al. (1999). They examined 263 breast lesions and found that the distribution of curve types in 101 malignant lesions was type I, 8.9%; type II, 33.6%; and type III, 57.4% and in 165 benign lesions was type I, 83.0%; type II, 11.5%; and type III, 5.5%. The diagnostic efficacy of pattern evaluation of the signal intensity time curves were: sensitivity, 91%; specificity, 83%; and diagnostic accuracy, 86%, which was more informative than enhancement rate alone (sensitivity, 91%; specificity, 37%; and diagnostic accuracy, 58%). It should also be noted that heterogeneity of curve shapes within a lesion is also diagnostically informative. A malignant tumour may well enhance

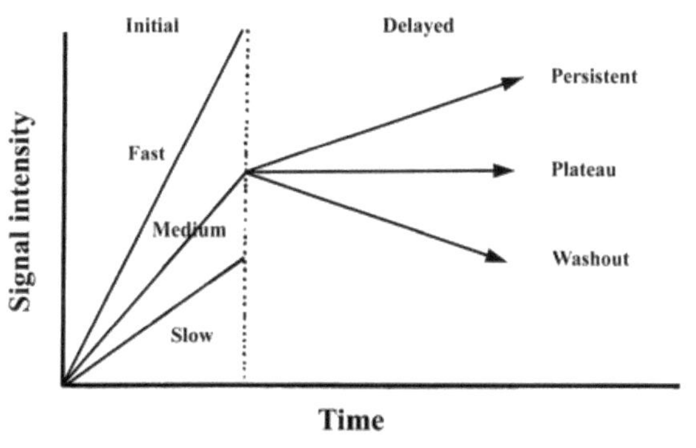

Fig. 10.5. Stylized diagram illustrating classification of signal-intensity curves time in breast MRI. Modified from SCHNALL MD and IKEDA DM (1999) (SCHNALL and IKEDA 1999)

with type I and II shapes but the heterogeneous nature of enhancement and variety of curve shapes in different anatomical areas is strongly suggestive of malignancy (Fig. 10.6, 10.7, 10.9 and 10.10).

10.4.1.4
Data Quantification and Limitations

Signal enhancement seen on a dynamic acquisition of T_1-weighted images can be measured in two ways: by the analysis of signal intensity changes (semi-quantitative) and/or by quantifying contrast agent concentration change using pharmacokinetic modelling techniques. Semi-quantitative parameters describe tissue signal intensity enhancement using a number of descriptors. These parameters include onset time (time from injection or appearance in an artery to the arrival of contrast medium in the tissue (BOETES et al. 1994), gradient of the upslope of enhancement curves, maximum signal intensity and washout gradient. As the rate of enhancement is important for improving the specificity of examinations, parameters that include an additional time element are also used (e.g., maximum intensity time ratio (MITR) (FLICKINGER et al. 1993) and maximum focal enhancement at one minute (KAISER and ZEITLER 1989); GRIBBESTAD et al. 1994). The uptake integral or initial area under the time signal curve (initial AUC) has also been studied (EVELHOCH 1999).

Semi-quantitative parameters have the advantage of being relatively straightforward to calculate but have a number of limitations. These limitations include the fact that they do not accurately reflect contrast medium concentration in the tissue of interest and can be influenced by scanner settings (including gain and scaling factors) which can alter baseline values and therefore can alter calculated semi-quantitative parameter values. Semi-quantitative parameters have a close but complex and undefined link to underlying tissue physiology and contrast agent kinetics. These factors limit the usefulness of semi-quantitative parameters and make between-patients and system comparisons difficult.

Quantitative techniques use pharmacokinetic modelling techniques that are usually applied to tissue contrast agent concentration changes. Quantitative analysis required conversion of the MRI signal intensities to T_1-relaxation rates and then Gd-DTPA concentrations. We use the methods detailed by PARKER et al. (1997, 2000). Concentration-time curves are then mathematically fitted using one of a number of recognised pharmacokinetic models

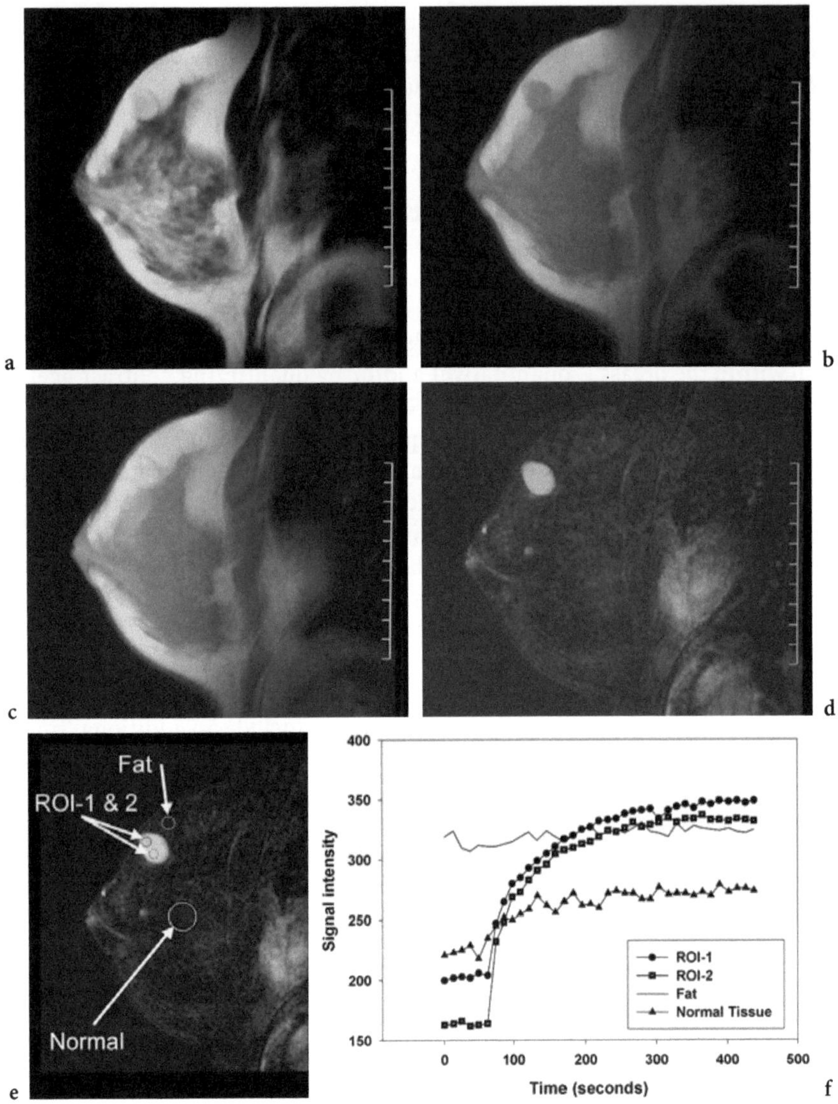

Fig. 10.6a–f. Morphological and kinetic patterns of enhancement of a fibroadenoma. This is the same patient illustrated in Figures 10.1-10.3, 10.8 and 10.12 but of a different slice in the same right breast. (**a**) T2-weighted, (**b**) T1-weighted pre-contrast and (**c**) post contrast enhanced T1-weighted (100 seconds after 0.1 mmol/kg contrast medium) images of a fibroadenoma. (**d**) Subtraction image with regions of interest indicated in (**e**). (**f**) Time signal intensity data from the regions indicated. The fibroadenoma is homogenously enhancing, with clear edges. Invasive ductal cancer shows marked heterogeneous enhancement with irregular margins as illustrated in Figures 10.7 and 10.9.

and quantitative kinetic parameters are derived for tumour regions of interest and on a pixel by pixel basis (Fig. 10.8). For a detailed discussion on pharmacokinetic modelling techniques and their limitations reader are directed to the review by Tofts (1997) and elsewhere in this book (see Chap. 6). Examples of modelling parameters include the volume transfer constant of the contrast agent (K^{trans} – formally called permeability-surface area product per unit volume of tissue), leakage space as a percentage of unit volume of tissue (ve) and the rate constant (kep – also called K21). These standard parameters are related mathematically (kep = K^{trans}/v_e) (Tofts et al. 1999). The details of what physiological processes are depicted by these kinetic parameters are detailed elsewhere in this book (see Chap. 6). To summarise, the transfer constant (K^{trans}) describes the transendothelial transport of low molecular weight contrast medium by diffusion. If the delivery of the contrast medium to a tissue is insufficient (flow-limited situations or where vascular permeability is greater than inflow) then blood perfusion will be the dominant

a b c d

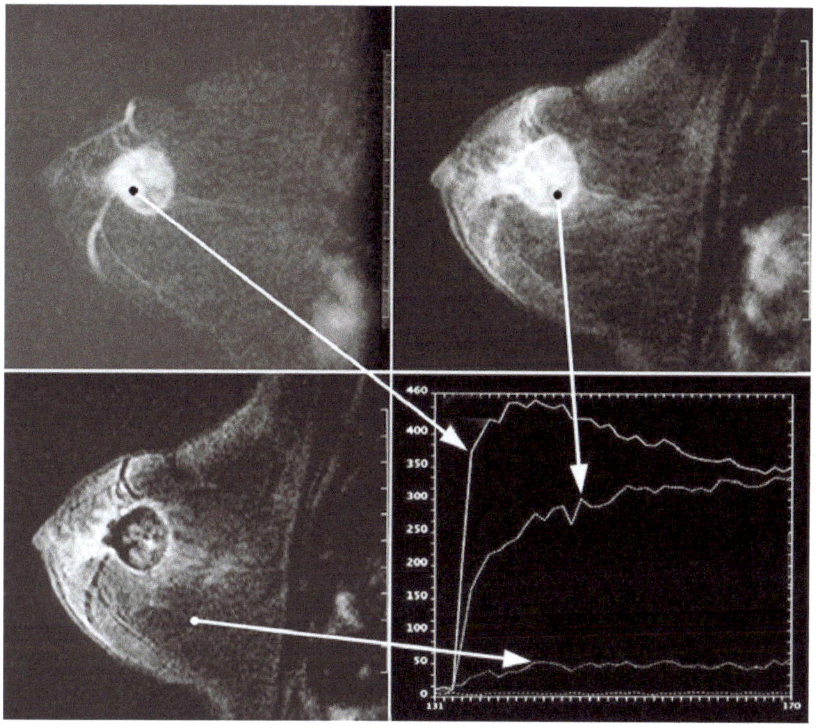

Fig. 10.7a–d. Morphological enhancement patterns and pathophysiological characteristics. (**a**) Early (100 seconds) and (**b**) late (7 minutes) subtraction images of invasive ductal cancer after injection of 0.1 mmol/kg Gd-DTPA contrast medium. (**c**) Subtraction of (**a**) from (**b**) reveals the enhancement that occurs after 100 seconds. Regions of interest and corresponding time signal intensity curves (**d**) show heterogeneity of curve shapes. The pathophysiological correlates of early and late enhancing tissues is discussed in section 10.4.5.

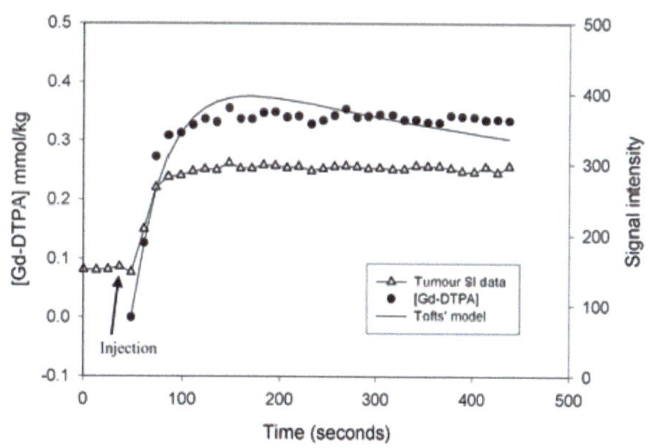

Fig. 10.8. Converting signal intensity into contrast concentration & model fitting. Data obtained from the patient as illustrated in Figures 10.1-10.3, 10.6 and 10.12. Contrast medium injection (11 ml of Gd-DTPA) took place after the third data point. Quantification of time signal intensity data (Δ) from a ROI within the tumour into contrast agent concentration (•) is performed first according to the method described by Parker et al (Parker et al. 1997). The model fitting procedure (continuous line) is done using with the Tofts' model (Tofts et al. 1995). Note that model fitting to contrast agent concentration data is not perfect. Calculated quantified parameters are transfer constant = 0.82 min^{-1}, leakage space 47%, rate constant = 1.74 min^{-1}). Subtraction (100 seconds), transfer constant (colour scale 0-1 min^{-1}) and leakage space (colour scale 0–100%) images are also shown.

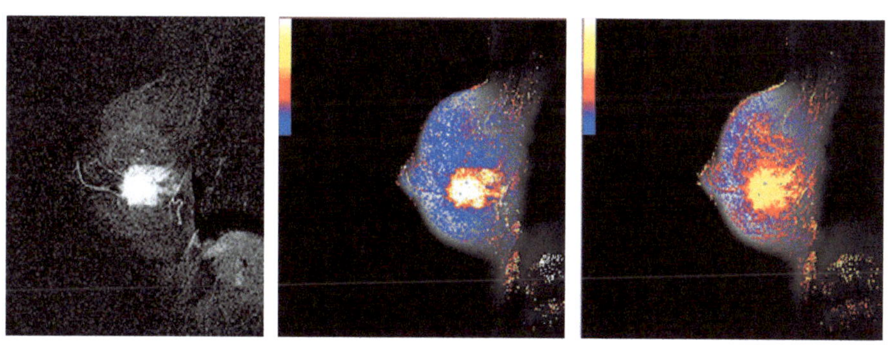

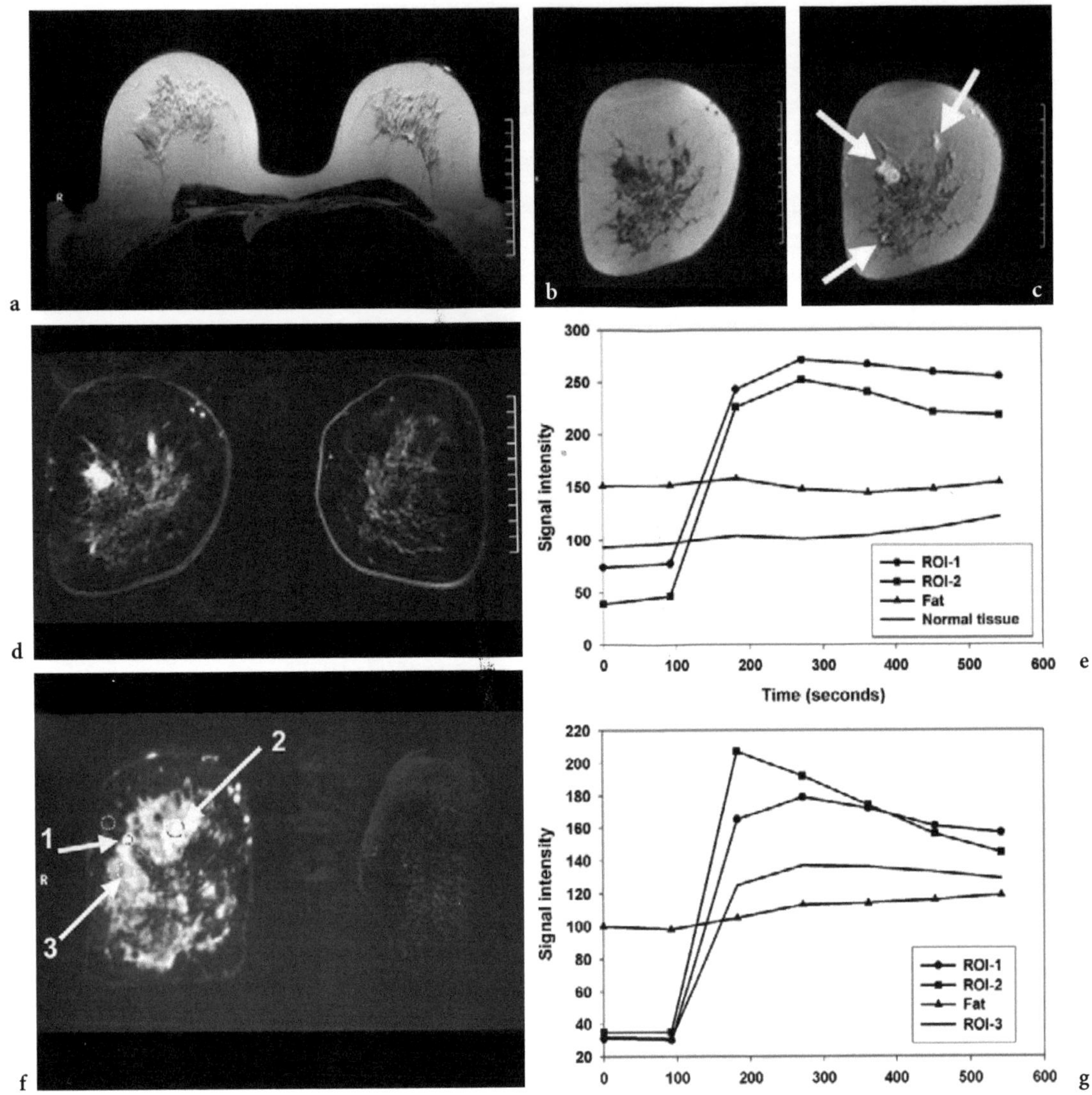

Fig. 10.9a–g. Carcinoma of unknown primary origin. 70 year old woman presenting with metastatic disease in right axillary lymph nodes consistent with breast origin and a normal mammogram (not shown). (**a**) Axial T2-weighted image shows no significant abnormality. (**b** and **c**) T1-weighted coronal images of the right breast before and after 0.2 mmol/kg Gd-DTPA given intravenously shows 3 focal areas of enhancement (*arrows*). (**d**) Subtraction image shows the enhancing lesions well. (**e**) Signal intensity time curves from regions of interest at the 9 o'clock (ROI-1) and 1 o'clock (ROI-2) show fast early enhancement and plateau and well as wash-out in the delayed phase. (**f** and **g**) follow-up examination (3 months later) following unsuccessful chemotherapy shows deterioration of appearances with widespread areas of abnormal enhancement.

factor determining contrast agent kinetics and K^{trans} approximates to tissue blood flow per unit volume (Tofts et al. 1999). The latter situation is commonly found in breast tumours. If tissue perfusion is sufficient and transport out of the vasculature does not deplete intravascular contrast medium concentration (non-flow limited situations – e.g., in areas of breast fibrosis or in the irradiated breast) then transport across the vessel wall is the major factor that determines contrast medium kinetics (K^{trans} then approximates to permeability surface area product). As low molecular weight contrast media do not cross cell membranes, the volume of distribution is effectively the extravascular-extracellular space (EES also called leakage space – v_e). The rate constant or kep describes the rate at which the contrast agent diffuses back into the vasculature from the EES.

The quantitative parameters discussed above are more complicated to derive compared to those derived semi-quantitatively which deters their use at the workbench. The model chosen may not

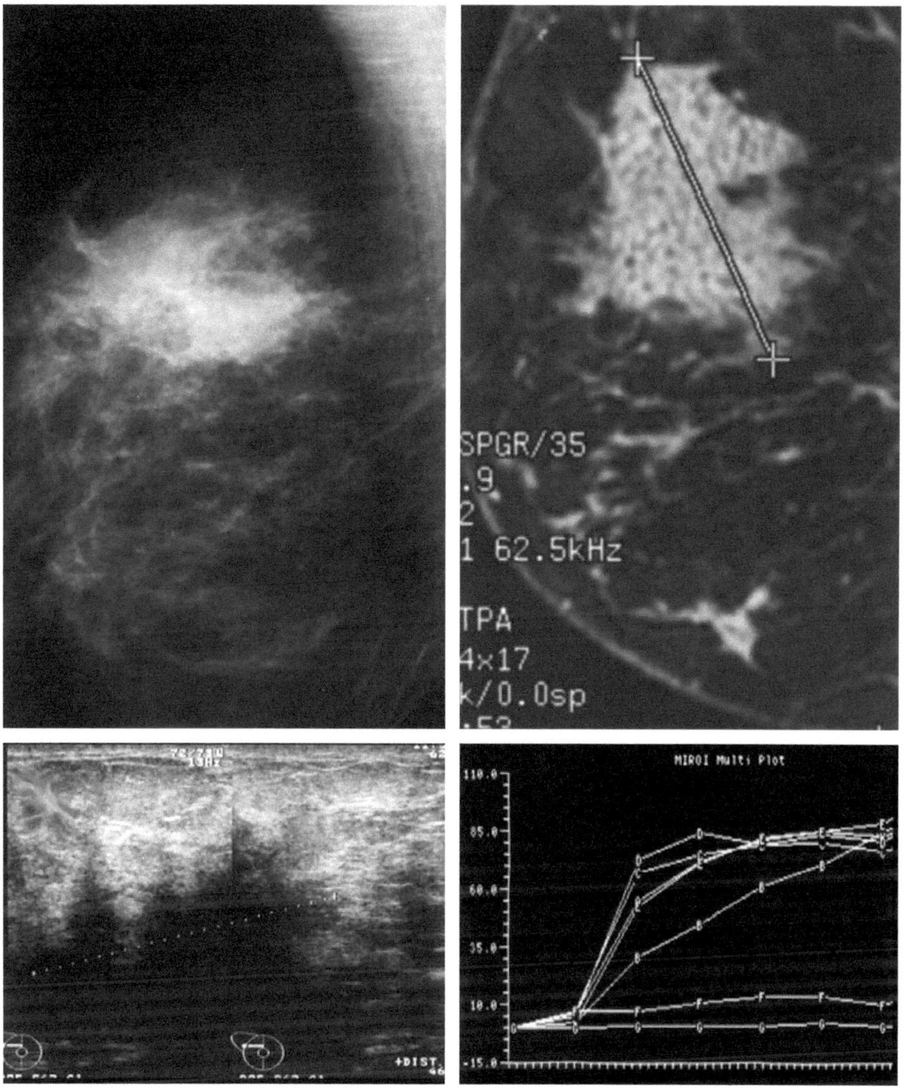

a b c d

Fig. 10.10a–g. Ductal carcinoma in-situ (DCIS). (a) X-ray mammogram of the right breast showing increased density in the upper outer quadrant (microcalcification consistent with DCIS were seen on magnification views – not shown). (**b**) Ultrasound image in the same area shows marked attenuation of the sound beam with a mass whose lesions are not well defined. (c) Post contrast fat suppressed MRI shows regional enhancement and (d) relative signal intensity (%) time curves from several ROIs within the lesion (**b, c, d** and **e**) and from fat (**f**) and background (**g**) shows a variety of patterns of enhancement but no wash-out is seen. Histology showed high grade ductal carcinoma in situ with no evidence of microinvasive disease. We are grateful to Dr. Ruth Warren (Addenbrookes Hospital, Cambridge, UK) for providing this illustration.

exactly fit the data obtained, and each model makes a number of assumptions that may not be valid for every tissue or tumour type (Tofts 1997; Tofts et al. 1999). Nonetheless, if contrast agent concentration can be measured accurately and the type, volume and method of administration are consistent, then it may possible to directly compare pharmacokinetic parameters acquired serially in a given patient and in different patients imaged at the same or different scanning sites (Tofts et al. 1995; den Boer et al. 1997).

10.4.2 Clinical Validation

A number of studies have attempted to correlate the morphological findings on breast MRI with key histological features (Buadu et al. 1996; Knopp et al. 1999; Matsubayashi et al. 2000). Early rim enhancement appears to correlate with high microvessel density particularly in the tumour periphery, increased expression of VEGF, relatively small amounts of fibrosis and small cancer cell nests. Delayed peripheral enhancement appears to correlate with larger amounts of fibrosis and exuberant inflammatory infiltrate around tumours while delayed central enhancement appears to correlate with the presence of internal fibrosis and necrosis (Fig. 10.7).

Vascular density in malignant tissue is higher than normal parenchyma but, as already mentioned, there is an overlap with some benign lesions including fibroadenomas and inflammatory and proliferative processes (Stomper et al. 1997; Weidner 1998). A number of studies have attempted to correlate the kinetic DCE-MRI parameters with immunohistochemical staining in breast cancer. Broad correlations between T_1 kinetic parameters estimates and MVD have been shown by some studies (Buckley et al. 1997; Stomper et al. 1997) while others have found no correlation (Hulka et al. 1997; Su et al. 2003). It appears that factors other than MVD are important in determining the degree of tissue enhancement including VEGF expression (Knopp et al. 1999). Some authors have not found a correlation between VEGF expression and the degree of enhancement in breast cancer (Su et al. 2003).

10.4.3 Clinical Experience

10.4.3.1 Screening

It is estimated that 5%--10% of all breast cancers have a dominantly inherited component. Two high penetrance genes have been identified that predispose to breast cancer, BRCA1 and BRCA2. Germline mutations in these genes are responsible for 2%--3% of all breast cancers and are associated with breast cancer at a younger age. For women carrying either of these two genes, there is a cumulative lifetime risk of developing breast cancer of 40%–85% depending on the type of mutation and population being studied. In addition, there are many familial breast cancers (approximately 70%) which are not attributed to either BRCA1 or BRCA2. Currently, reducing the risk of developing breast cancer in these patients includes prophylactic mastectomy (Hartmann et al. 1999; Rebbeck et al. 2004), ovarian ablation (Rebbeck et al. 1999) and chemoprevention (Fisher et al. 1998b; King et al. 2001) although none of these methods can eliminate the risk entirely. In addition, prophylactic bilateral mastectomy is associated with surgical and physiological morbidity and as a result, screening is now under investigation. Although some studies have demonstrated screening mammography can be efficacious for women as young as 40 years old (Thurfjell and Lindgren 1996; Bjurstam et al. 2003), the value of screening young women at high genetic risk with mammography is incompletely assessed. It has been suggested that mammography has low sensitivity when screening women with BRCA1/2 carriers due to young age and hence increased breast density, a high tumour growth rate and atypical mammographic and histopathological characteristics (Brekelmans et al. 2001). As a result, MRI has been evaluated as a screening test for women at high genetic risk of developing breast cancer.

Many studies of breast MRI in women at high genetic risk have reported while some studies are still ongoing (Kuhl et al. 2000; Tilanus-Linthorst et al. 2000a; Stoutjesdijk et al. 2001a; Warner et al. 2001a; Podo et al. 2002a; Kriege et al. 2004). Details of the sensitivities and specificities of MRI screening in the literature to date are shown in Table 10.2. Breast MRI is not ideal as a screening tool for a number of reasons including the requirement of intravenous of contrast medium, relatively long examination and analysis times and variable specificity (53%–97%). In addition, there are biological reasons that make MRI an imperfect test including the relative inability to detect or properly characterise some lobular and medullary carcinomas and the limited ability to detect DCIS and small invasive cancers (see Sect. 10.4.3.3). Most studies have found high sensitivity for detecting early cancers but lower specificity for MRI screening although MRI remains superior compared to mammography. If MRI screening is to be widely implemented, however, it will involve radiologists of all levels of experience, and so needs to be appropriately robust for this environment (Warren 2001). In addition, it remains to be seen whether early diagnosis of breast cancer in high risk women will lead to a decrease in breast cancer specific mortality. As a result, screening of women at high genetic risk using breast MRI can currently only be recommended in a research setting.

MRI screening may benefit other young women with increased risk of breast cancer from other causes, such as childhood or young adult cancer survivors previously treatment with thoracic irradiation, for example, mantle irradiation in the treatment of Hodgkin's disease (Deniz et al. 2003).

10.4.3.2
Axillary Lymphadenopathy and the Unknown Primary Tumour

Less than 1% of patients with breast cancer present with palpable axillary lymphadenopathy with no clinically or mammographically detectable breast lesion [carcinoma of unknown primary (CUP) syndrome] (FOURQUET et al. 1996). Following the histological confirmation of adenocarcinoma from the axillary disease, treatment usually involves either ipsilateral mastectomy or breast irradiation with axillary node dissection followed by systemic therapy as appropriate (chemotherapy with/without endocrine therapy). In a third of patients who undergo mastectomy, however, no primary tumour is ever identified (ROSEN and KIMMEL 1990; FORTUNATO et al. 1992). Radiotherapy alone to the breast with no surgery has been associated with an increased risk of local recurrence (ELLERBROEK et al. 1990). In light of this, a number of studies have assessed the role of MRI for detecting occult primary breast cancers to aid in the treatment decision process.

OREL et al. (1999) reported on 22 women presenting with malignant axillary lymph nodes and negative clinical and mammographic examinations. MRI identified a primary breast cancer in 19 of the 22 patients (86%) with the lesion sizes measuring between 4 and 30 mm. Two patients with invasive ductal carcinoma (2 mm; 17- and 20-mm) had false negative MRI results. OLSEN et al. (2000) assessed 40 patients with isolated axillary adenocarcinoma from an unknown primary, again with no detectable breast lesion either clinically or mammographically. MRI identified a primary breast lesion in 28 of the 40 women. Of the 12 patients with a negative MRI study, five underwent mastectomy but tumour was found in only one of the specimens. A number of smaller studies have revealed similar results (TILANUS-LINTHORST et al. 1997; SCHORN et al. 1999). Thus, contrast-enhanced MRI can detect occult breast cancer in patients presenting with malignant axillary disease from an unknown primary source where clinically breast examination and mammography are normal (Fig. 10.9). Given the poor performance of mammography in this situation (ELLERBROEK et al. 1990; KNAPPER 1991), MRI of the breast should now be considered the investigation of choice in this clinical situation (Table 10.1). Furthermore, MRI can assist in treatment decisions, particularly with regard to breast conservation.

Table 10.1. Current and potential new indications for breast MRI

Indication	Comments
Problematic mammogram	Mammographic and ultrasound evaluation should be done first and MRI in selected patients with equivocal imaging especially with suspicious clinical findings. May be helpful for mapping suspected extensive DCIS
Staging breast cancer when multiple or bilateral tumours are suspected	MRI improves accuracy of staging, detecting multicentricity and multifocality particularly for large tumours in the radiographically dense breast and for lobular cancers. MRI has a high negative predictive rate. Unsuspected contralateral malignancy can be detected in 5%–10% of cases
High risk screening: breast cancer gene carriers and history of mantle irradiation for lymphoma at young age	MRI may be a valuable adjunct in these high risk patients but MRI is not appropriate for general breast cancer screening
Lobular cancer	MRI shows tumour extent which can be difficult to detect with mammography, is commonly multifocal/multicentric or bilateral (approximately 10%) and is a frequent cause of positive surgical margins
Metastatic cancer in axillary lymph-nodes without detectable lesion on conventional imaging	MRI can locate a primary breast tumour in the majority, enabling breast conservation surgery
Monitoring response to neoadjuvant chemotherapy	Helpful for identifying the extent of residual active disease following completion of chemotherapy for locally advanced breast cancer. Monitoring response to treatment during therapy is experimental
Positive surgical margins following breast conservation surgery	MRI can identify the extent of residual disease
Distinguishing tumour recurrence from surgical scar and other causes of a mass following breast surgery	Recommended 6 months or more after surgery
Suspected implant rupture or suspected cancer in patients with history of liquid silicone injection	

Table 10.2. Comparison of MRI screening for breast cancer in high-risk groups with clinical examination, mammography and ultrasonography

Country	Reference	Number of women reported	Number of BRCA1 or BRCA2 carriers	Number of cancers detected	MRI results		Other modalities tested					
					Sensitivity	Specificity	Clinical examination		Mammography		Ultrasound	
							Sensitivity	Specificity	Sensitivity	Specificity	Sensitivity	Specificity
Canada, Toronto	WARNER et al. 2001b	196	96	6	100%	91%	33%	99.5%	33%	99.5%	60%	93%
Germany, Bonn	KUHL et al. 2000	192	-	15	100%	95%	-	-	33%	93%	33%	80%
Italy, Rome	PODO et al. 2002b	105	-	8	100%	99%	Not Stated	-	Not stated	-	-	-
Netherlands, Nijmegen	STOUDJESDIJK et al. 2001b	179	-	13	100%	93%	-	-	42%	96%	-	-
Netherlands, Rotterdam	TILANUS-LINTHORST et al. 2000b	109	-	3	100%	-	Not stated	-	Not stated	-	-	-
Netherlands	KRIEGE et al. 2004	1909	358	51	80%	90%	18%	98%	33%	95%	-	-

10.4.3.3 Lesion Characterisation Including Lobular Cancer and DCIS Evaluation

Lesion descriptors by morphological and kinetic features have been discussed in Sect. 10.4.1.3. Features that suggest the possibility of malignancy include a mass with irregular or spiculated borders, a mass with peripheral enhancement, internal heterogeneity and ductal enhancement. A rapid initial uptake of contrast agent followed by an early washout or heterogeneous signal intensity curve shapes are also features suggesting malignancy. Features suggesting benign disease include a mass with smooth or lobulated borders, a mass demonstrating no contrast or homogeneous enhancement, a mass with non-enhancing internal septations, and patchy parenchymal enhancement. A slow rate of initial enhancement followed by a continuous increase of thereafter or identical curve shapes regardless of ROI location in a mass are also suggestive of benign lesions. Unfortunately there are many exceptions to these rules. For non-mass lesions, segmental enhancement and linear and ductal enhancement can predict for malignancy (LIBERMAN et al. 2002). Other morphological features for malignant and benign disease are described and interested readers and directed towards the many clinico-pathological studies that describe these in more detail (HARMS et al. 1993; OREL et al. 1994; GILLES et al. 1995; LIBERMAN et al. 2002; QAYYUM et al. 2002; SIEGMANN et al. 2002). Using these features, the sensitivity of MRI for breast cancer detection is high at 94%--100% but specificity still has a wide range from 37%--97% (OREL and SCHNALL 2001; MORRIS 2002).

10.4.3.3.1 Lobular Cancer

As mentioned in Sect. 10.1, imaging with mammography and ultrasound has a low sensitivity for diagnosis of invasive lobular carcinoma (ILC) and often underestimates the extent of disease. MRI for both diagnosis and assessment of disease extent of ILC has demonstrated diagnostic superiority over the conventional imaging modalities. In one series of 17 patients with ILC, MRI detected 30 lesions compared with 21 on conventional imaging, while also identifying multiple lesions in seven cases and contralateral disease in one patient (DIEKMANN et al. 2004). QAYYUM et al. (2002) compared the MRI characteristics of ILC with histopathological finding in 13 patients. They described three distinct patterns for ILC and corresponding histopathological correlates: a solitary mass with irregular margins (n=4) which corresponded with the same morphology on pathological assessment (Fig. 10.4); multiple lesions either connected by enhancing strands (n=6) or separated by non-enhancing intervening tissues (n=2) which correlated with non-contiguous tumour foci with malignant cells streaming in single-file fashion in the breast stroma or small tumour aggregates separated by normal tissue; and enhancing septa only (n=1) corresponding with streaming of tumour cells in the breast stroma.

Yeh et al. (2003) assessed the morphological and kinetic DCE-MRI features in 15 cases of ILC presenting with a palpable mass or an unknown primary. The functional parameter measured was the extraction flow (EF) product, a quantitative measure of contrast medium uptake. They found the range of EF product to be 25--120 with the majority of cases within the 30s (the level for normal breast is 25 or less). In two tumours, however, a substantial part of the tumour demonstrated an EF product within the normal range. A focal mass was found in eight cases (seven irregular and one round; six ill-defined and two spiculated margins) but a mixture of enhancement patterns was seen. Multi-focal disease was demonstrated in four cases and an unsuspected contralateral tumour in one case. They concluded that the combination of morphology and functional characteristics on DCE-MRI is able to detect the majority of cases of ILC despite the variability seen. Caution is still required, however, as some lesions display enhancement features similar to normal breast.

10.4.3.3.2 Ductal Carcinoma In Situ (DCIS)

Ductal carcinoma in situ (DCIS) is usually diagnosed by X-ray mammography due to the presence of microcalcification and currently accounts for approximately 20% of screening-detected cancers. DCIS can be successfully treated with either mastectomy alone or wide local excision followed by adjuvant breast irradiation but relies on complete excision of disease. Often this is only achievable with mastectomy due to the high prevalence of multifocal and multicentric disease (Rosen et al. 1979; Ringberg et al. 1982; Ciatto et al. 1990; Holland et al. 1990). Several investigators have described the morphological and kinetic features of DCIS on MRI although many studies have only had small patient numbers. A large study by Neubauer et al. (2003) retrospectively evaluated DCE-MRI features in 39 consecutive patients with pure DCIS (Neubauer et al. 2003). They defined categories for signal intensity increase (C1, normal; C2, slow, continuous; C3, strong initial and slow further increase; C4, strong initial increase followed by a plateau phenomenon; and C5, strong initial increase followed by a washout phenomenon) and morphological patterns (M0, no pattern observed; M1, linear or linear-branched; M2, segmental dotted or granular; M3, segmental homogeneous; and M4, focal spot-like). C4 and C5 signal intensity-time patterns were considered suspicious of malignancy and 62% of DCIS demonstrated this pattern. In terms of enhancement, segmental enhancement was seen in 82% and segmental dotted or granular patterns (M2) were seen in 73%. When signal increase and morphology were taken together, 70% of non-high grade and 92% of high grade DCIS were correctly identified as being suspicious for malignancy. They conclude that segmental enhancement is a hallmark of DCIS (with hormone effects as the main differential diagnosis) and that the combination of morphology and signal intensity enhancement considerably improves the rate of detection (Fig. 10.10). Gilles et al. (1995) have reported similar findings and also noted that early enhancement on DCE-MRI in patients with DCIS is associated with angiogenesis in the stroma that surrounds ducts involved by DCIS (Gilles et al. 1995). However, it should be remembered that not all ductal enhancement on breast MRI is due to DCIS; Liberman et al. (2003a) reports that ductal enhancement can be caused by DCIS, atypical ductal hyperplasia, LCIS and benign findings such as fibrocystic change, ductal hyperplasia, and fibrosis.

10.4.3.4 Tumour Staging: Tumour Volume Determination, Multicentricity and Multifocality

An accurate assessment of the primary tumour extent is crucial for deciding the best management for breast cancer patients. It is particularly important when considering the feasibility of breast-conserving surgery as inaccurate assessment could lead to incomplete tumour excision will result in the need for a second surgical procedure or a high rate of recurrence following "apparent" curative surgery. Unfortunately, clinical examination, mammography and breast ultrasonography are poor at accurately defining the primary tumour extent and are capable of both overestimating and underestimating the histologically defined tumour size (Pain et al. 1992; Allen et al. 2001; Pritt et al. 2004; Snelling et al. 2004). MRI in this setting has been shown to be superior to both mammography and ultrasonography (Boetes et al. 1995; Davis et al. 1996; Yang et al. 1997; Esserman et al. 1999). When comparing the largest pathological cancer diameter with the tumour size defined by each technique, Davies et al. (1996) found MRI to have the highest correlation coefficient (r=0.98) when compared with ultrasonography (r=0.45) and mammography (r=0.46) (Davis et al. 1996). Esserman et al. (1999) also demonstrated MRI to be superior to mammography with a 98% accuracy for defining anatomic tumour extent compared to only 55% for mammography. In addition, MRI can identify synchronously contralateral occult primary tumours in between 3% and 24% of women presenting with breast cancer (Fischer et al. 1999; Slanetz et al. 2002; Lee et al. 2003; Liberman et al. 2003c).

The presence or absence of tumour multifocality and/or multicentricity is also critical when planning the optimal surgical management of breast cancer. It now well established that MRI can identify additional sites of tumour, not detected using conventional imaging, in between 6% and 34% of women with known breast cancer (Fig. 10.4, 10.9 and 10.11). MRI identifies multifocal disease in 1%--20% of women and multicentric disease in 2%--20% (Harms et al. 1993; Orel et al. 1995; Mumtaz et al. 1997b; Drew et al. 1999; Fischer et al. 1999; Bedrosian et al. 2002; Liberman et al. 2003b). Furthermore, Krämer et al. (1998) demonstrated MRI to have a sensitivity for the diagnosis of multicentricity of 89% compared with 79% for ultrasonography, 66% for mammography and only 47% for clinical examination. Thus, more accurate depiction of tumour size and prediction of the distribution of tumour by MRI may help to achieve definitive surgery in one procedure (Drew et al. 1999; Esserman et al. 1999). This benefit can be expected to be greater in cases where mammography and ultrasound have difficulties, for example in patients with dense breasts or invasive lobular carcinoma. Thus, a strong body of literature data now shows that MRI can play a major role to play in determining patient management (Table 10.1). One must remember, however, that there is the potential to increase the number of mastectomies and so specificity for individual lesions must be high (Tan et al. 1999). If in doubt, biopsy to document additional disease is recommended prior to definitive surgery.

10.4.3.5 Monitoring Response to Treatment and Assessing Residual Disease

It is now well established that wide local excision of breast cancer followed by adjuvant breast irradiation gives the same outcome in terms of survival as mastectomy and, as such, breast conservation is the treatment of choice for many women with breast cancer (Fisher et al. 2002). For some women, however, breast conservation is not possible due to the size, position or multifocal/multicentric disease. Neoadjuvant chemotherapy (NAC), also termed primary systemic therapy, is chemotherapy delivered prior to definitive surgery and aims to both down stage the primary tumour and treat potential micrometastatic disease. It also has the added advantage of allowing an assessment of tumour chemo-responsiveness as the tumour remains in situ during therapy. For inoperable breast cancers (locally advanced or inflammatory) a high tumour regression rate (approximately 70%) has established NAC as the standard treatment option for this patient group (Kaufmann et al. 2003). A number studies have now demonstrated that NAC can also be used in patients with operable breast cancers to downstage the primary tumour and thereby increase the breast-conserving surgery rate while providing an equivalent outcome to adjuvant chemotherapy in terms of disease-free survival and overall survival (Fisher et al. 1998a; Makris et al. 1998; van der Hage et al. 2001). The largest of these randomised studies, the National Surgical Adjuvant Breast and Bowel Project (NSABP) B18 trial demonstrated that pathological response to NAC was a predictor of relapse free survival with a significant increase in 5-year relapse free survival from 77% to 86% in patients achieving a pathological complete response compared to patients achieving a clinical CR but not a pathological CR (Fisher et al. 1998a).

A proportion (2%–30%) of patients receiving NAC will fail to response to treatment and will have a worse outcome (Bonadonna et al. 1998; Fisher et al. 1998a; Makris et al. 1998). The ability to identify non-responders early during treatment would enable the use of alternative therapies which may be more beneficial (Smith et al. 2002). Many studies evaluating NAC in patients with breast cancer have shown that clinical examination, X-ray mammography and ultrasonography are imperfect techniques especially as change in tumour size often only becomes apparent after several cycles of chemotherapy. Assessment of women undergoing NAC has been identified as an area of clinical practice where an immediate impact of breast MRI can be realised (Harms 2001). MRI has two main roles to play in the assessment of patients receiving NAC. Firstly, MRI may help in the early identification of non-responding patients thus allowing treatment intervention that may be more beneficial. Secondly, MRI may provide a more accurate assessment of residual tumour at the end of NAC and thus aid in decisions regarding surgery.

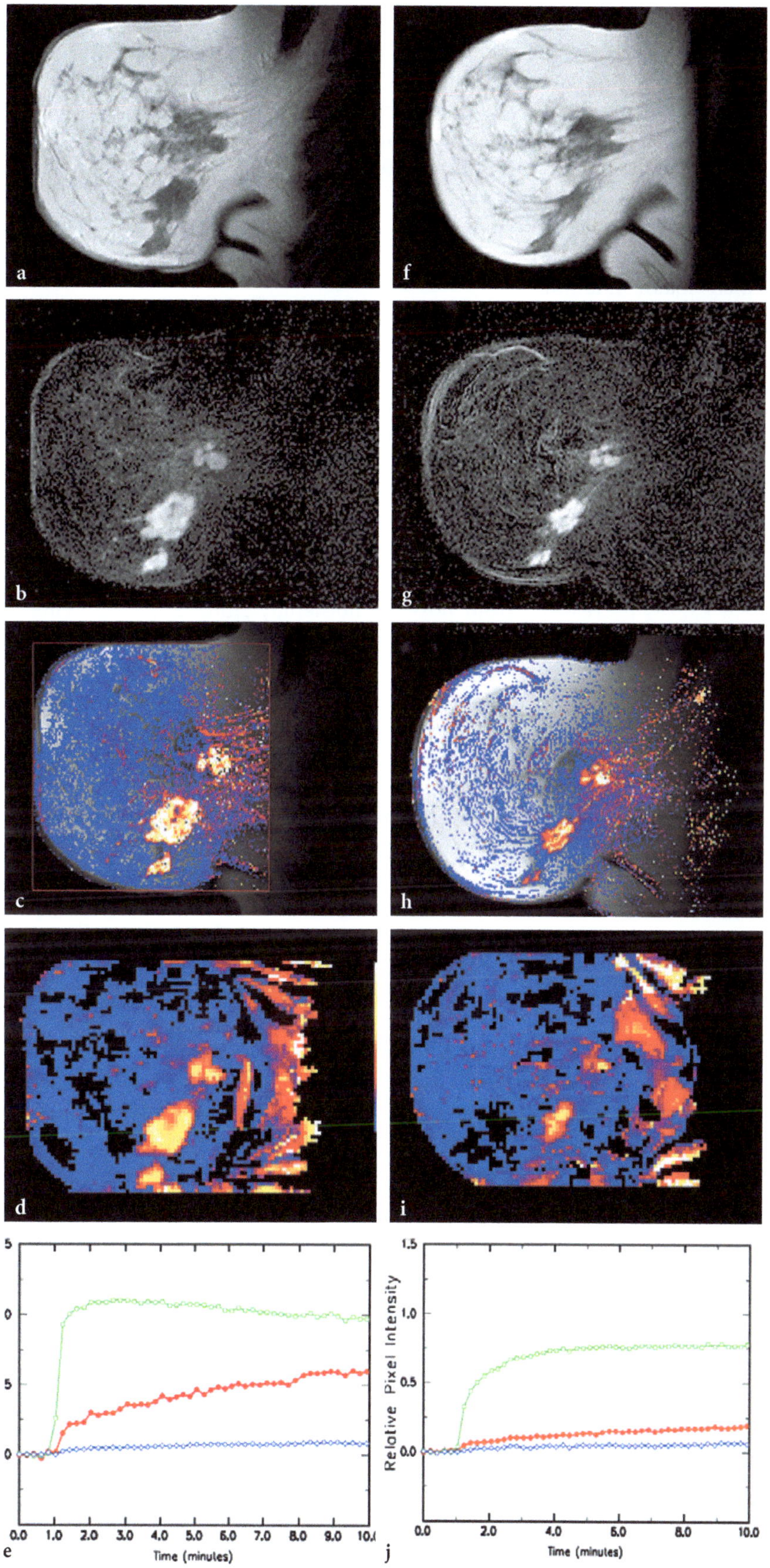

Fig. 10.11. Monitoring chemotherapy response of breast cancer with DCE-MRI. 44-year old post-menopausal woman with a 4 cm, grade 3 invasive ductal cancer of the right breast with palpable ipsilateral axillary nodal enlargement. Rows depict T2-weighted anatomical images, early (100 seconds) subtraction images, transfer constant, relative blood volume (rBV) parametric images and relative signal intensity time curves (ROIs in whole tumour-green, normal tissue-red and fat-blue) at identical slice positions before and after two cycles of FEC (5-Fluorouracil, Epirubicin, Cyclophosphamide) chemotherapy. Transfer constant map (colour range 0-1 min^{-1}) and relative blood volume (colour range 0-500 AU). With treatment, a reduction in the size of individual lesions is seen with a reduction in relative blood volume and transfer constant. This patient had a complete clinical and radiological response to treatment after 6 cycles of chemotherapy with pathology showing microscopic residual disease only.

10.4.3.5.1
Monitoring Tumour Response

Pre-treatment kinetic MRI parameters obtained by T_1 and T_2* weighted techniques as discussed in Sect. 10.3 and 10.4 do not in themselves appear to predict for eventual response to NAC (Ah-See et al. 2003). Recently, it has been suggested that the morphological MRI appearances of breast cancer prior to therapy (circumscribed mass, nodular tissue infiltration, diffuse tissue infiltration, patchy enhancement, and septal spread) can predict likelihood of response but this has not been verified by other workers (Esserman et al. 2001).

MRI evaluates response on the basis of the changes in tumour size and on the amount and speed of enhancement. The value of DCE-MRI as an early predictor of the efficacy of NAC in patients with breast cancer has not been fully assessed. Both Knopp et al. (1994) and Hayes et al. (2002) have shown that successful treatment causes decreases in the rate and magnitude of enhancement and that poor response results in persistent abnormal enhancement (Fig. 10.11 and 10.12).

The ability of T_1 and T_2* DCE-MRI to act as an early predictor of response has been examined in detail by Ah-See et al. (2004) who evaluated 28 patients imaged prior to and following two cycles of NAC. They also performed a reproducibility analysis in order to determine the ability of DCE-MRI to predict pathological non-response on a patient-by-patient basis. Group analysis showed that changes in median K^{trans}, kep, rBV and rBF correlated with both final clinical and pathological response to NAC ($p<0.01$). Receiver operator characteristics (ROC) analysis showed that change in tumour size was ineffective at predicting response and that change in K^{trans} was the best kinetic parameter at predicting non-responsiveness. These data suggest that patients who are destined to fail to respond to NAC can be identified from early changes (after two cycles) in DCE-MRI parameters. A similar result has been reported by Martincich et al. (2004).

The explanation for the changes in DCE-MRI with chemotherapy is likely to be multifactorial, relating to changes in both microvessel density and function due to antiangiogenic effects of the chemotherapy (Kerbel et al. 2002). Makris et al. (1999) reported that fewer tumour microvessels were seen in breast cancer patients treated with chemoendocrine therapy compared with untreated patients. However, they reported no differences in microvessel density counts between responders and non-responders. Additional mechanisms that have been suggested include the hypothesis that successful chemotherapy causes cytotoxic tumour cell death which results in a reduction of tissue VEGF levels and hence apoptosis of immature endothelial cells (Darland and D'Amore 1999) with secondary vascular shutdown.

10.4.3.5.2
Evaluating Residual Disease

Tumour size is used to assess the volume of residual disease after treatment (either by clinical examination, X-ray mammography or ultrasound). Each of these techniques, however, overestimates the extent of residual disease at the end of treatment (Feldman et al. 1986; Helvie et al. 1996; Vinnicombe et al. 1996; Herrada et al. 1997; Fiorentino et al. 2001). Clinical palpation is an imperfect evaluation method because of intervening skin and soft tissue and is unable to distinguish active disease from post-chemotherapy fibrosis. Assessments by X-ray mammography are hampered by radiographic magnification and it may be difficult to distinguish residual tumour from post-chemotherapy fibrosis (Helvie et al. 1996; Vinnicombe et al. 1996). It is also well-recognised that some tumours are simply not visualised in patients with radiographically dense breasts. In some patients, ultrasound appears better in assessing tumour response and Doppler ultrasound of residual disease is being evaluated (Herrada et al. 1997).

Feldman et al. (1986) made the observation that there was a significant difference between the clinical and pathological evaluation of residual disease, such that the improved patient survival achieved in those with complete pathological response ($p=0.002$) would not be appreciated if patients had been judged clinically ($p=0.09$). These results suggest that the usefulness of NAC may not be realised if clinical evaluation alone is undertaken. Most importantly, women with normal examinations may well have residual active disease requiring further treatment. More accurate methods of assessing residual disease are needed as they would enable appropriate selection and increase the precision of subsequent treatments.

Several studies have shown that contrast enhanced MRI is the most accurate technique for evaluating pathological response and volume of residual active disease at the end of treatment (Harms et al. 1993; Gilles et al. 1994; Abraham et al. 1996; Esserman et al. 1999; Drew et al. 2001; Weatherall et al. 2001; Rieber et al. 2002; Delille et al. 2003). Non-

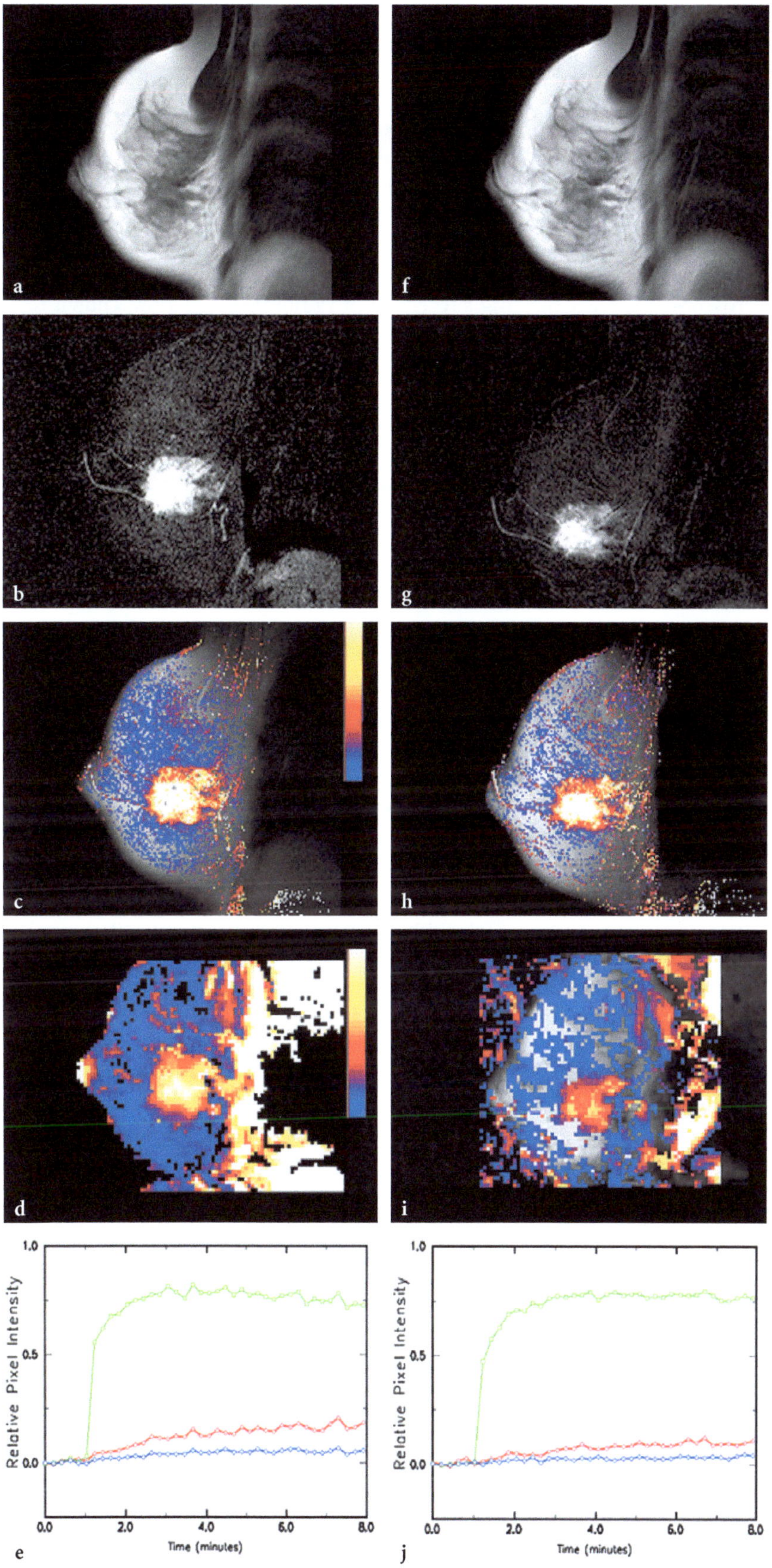

Fig. 10.12a–j. Monitoring chemotherapy non-response of breast cancer with DCE-MRI. The same patient illustrated in Figures 10.1-10.3, 10.6 and 10.8. Rows depict T2-weighted anatomical images, early (100 seconds) subtraction images, transfer constant, relative blood volume (rBV) parametric images and relative signal intensity time curves (ROIs in whole tumour-green, normal tissue-red and fat-blue) at identical slice positions before (**a–e**) and after (**f–j**) two cycles of FEC (5-Fluorouracil, Epirubicin, Cyclophosphamide) chemotherapy. Transfer constant map (colour range 0-1 min^{-1}) and relative blood volume (colour range 0-500 AU). With treatment, no discernable changes are seen in tumour size, transfer constant or in the relative signal intensity time curves. There is a tendency for relative blood volume (rBV) to decrease. This patient had no clinical and radiological response to treatment after 6 cycles of chemotherapy and macroscopic, invasive grade 3, ductal cancer was noted at pathological evaluation.

dynamic fat suppressed breast MRI assesses the presence of residual disease better than mammography when the presence of residual abnormal enhancement on post-contrast images is used (Abraham et al. 1996). Interestingly, many have observed that the fibrosis observed after chemotherapy is hyalinized and does not arise from granulation tissue (seen post-operatively) (Bonadonna et al. 1990) and this enables residual active disease to be depicted as an area of persistent abnormal enhancement (Pierce et al. 1991; Davis et al. 1996). Microscopic disease can of course not be detected, but MRI can indicate the extent of remaining tumour tissue on which to plan breast conserving surgery.

10.4.3.6 Assessing Breast Induration Following Radiation Therapy

Fibrosis is assumed to be the usual explanation of palpable induration (firmness or hardness) in the breast developing several years post-radiotherapy in the absence of features suggesting tumour recurrence. Fibrosis is a well-recognised late consequence of high dose radiotherapy elsewhere in the body, where it contributes to dose-limiting complications and reduced compliance of organs with important elastic properties. Clinico-pathological studies of induration in the early years following breast radiotherapy suggest that fat necrosis is sometimes a contributory factor (Clarke et al. 1983; Rostom and el-Sayed 1987; Boyages et al. 1988). However, in a recent report, Padhani et al. (2002) showed that radiotherapy-induced induration in the breast many years after treatment is seldom caused by fat necrosis but that there was a close correspondence between breast oedema demonstrated on MRI and the severity of clinical induration (Padhani et al. 2002). They suggested that the parenchymal oedema might be a manifestation of a persistent vascular leak related to radiation induced vascular injury. In another publication, the same authors showed that increased parenchymal oedema contributing to palpable induration was associated with an increased extravascular-extracellular leakage space demonstrated on DCE-MRI (Padhani et al. 2003). They did not demonstrate an increase in transfer constant but did observe increased enhancement of irradiated breasts compared to contralateral non-irradiated breast tissue. They hypothesised that increased enhancement was due to increased number of microvessels with normal permeability and that impaired drainage may be a further contributing factor for hardness of the breasts following radiotherapy.

10.4.3.7 Detecting Relapse

The surgical management of primary breast cancer depends on a number of factors including tumour size, position, multicentricity, patient choice and contraindications to radiotherapy. Breast-conserving surgery with radiotherapy is now known to be equivalent to mastectomy in terms of survival and is generally the treatment of choice for early stage breast cancer (van Dongen et al. 2000; Fisher et al. 2002; Veronesi et al. 2002). Breast-conservation is, however, associated with a higher rate of local recurrence compared to mastectomy (van Dongen et al. 2000; Veronesi et al. 2002). Early detection of local recurrence is important as many patients can be successfully treated by salvage mastectomy (Kurtz et al. 1988; Osborne et al. 1992).

Currently, the recommended follow-up for local recurrence is involves routine clinical examination and mammography at least every 2 years. Ultrasonography and core biopsy are used to confirm the diagnosis of recurrence. In the first 6 months after surgery, mammograms show a general increased density, architectural distortion, skin thickening, asymmetric densities, dystrophic calcification and features of fat necrosis (Sickles and Herzog 1981; Paulus 1984; Krishnamurthy et al. 1999). Mammographic changes are most helpful in suggesting recurrence 6 months to 1 year later after surgery. It is well recognised that features of scar tissue can mimic those of recurrent cancer on mammography (Stomper et al. 1987) and, therefore, any increase in the extent of the post-surgical abnormality or the development of a mass within the area of distortion must be regarded as suspicious. The overall positive predictive value of cancer detection in the treated breast using mammography is thought to be around 72% (Orel et al. 1992) with a false negative rate of 5%–15% (Homer 1991).

As a result, the role of MRI in detecting tumour recurrence in the surgically treated breast has been assessed. Some studies previously reported that non-enhanced T_2 weighted sequences alone could be used to differentiate tumour from scar tissue as recurrent tumour demonstrated a higher signal intensity than dysplastic tissue and fibrosis (Revel et al. 1986; Lewis-Jones et al. 1991). However, other groups have failed to show this (Dao et al. 1993; Gilles et al. 1993; Kerslake et al. 1994; Orel et al.

1994) and therefore non-enhanced MRI scans alone must be considered unreliable.

In contrast, a number of studies have now shown the benefit of DCE-MRI for detecting recurrent tumour in the previously treated breast relying on the different enhancement pattern of malignant disease compared with benign post-treatment changes such as scarring and fat necrosis. On T_1 weighted sequences following Gd-DTPA contrast medium rapid, early (within 3 min) and avid enhancement is seen in recurrent tumour compared with slower enhancement of lower magnitude in benign disease (Dao et al. 1993; Gilles et al. 1993; Kerslake et al. 1994; Whitehouse and Moore 1994; Murray et al. 1996; Mussurakis et al. 1995; Viehweg et al. 1998). Gilles et al. (1993) also noted nodular enhancement to be associated with invasive carcinoma while linear enhancement was seen with ductal carcinoma in situ. Mussurakis et al. (1995) reported significant differences between benign and malignant lesions for enhancement index, maximum uptake, amplitude of uptake, wash-in and wash-out rates.

The timing of DCE-MRI following surgery is an important factor in the accuracy of the technique, because benign proliferative disease and recent fat necrosis may mimic malignant recurrence within 6 months of surgery and 9 months of breast irradiation (Heywang et al. 1990; Whitehouse and Moore 1994). Indeed, in some patients post-operative scarring may continue to enhance up to 18 months post-treatment although beyond that time no significant enhancement is seen (Heywang-Kobrunner et al. 1993). The absence of focal enhancement, even in the first year post-treatment, is thought to be able to exclude recurrent malignancy with a negative predictive value of 100% (Müller et al. 1998).

When compared with other techniques for detecting local recurrence, DCE-MRI has a consistently high sensitivity (91%--100%) with an associated high specificity (88%--93%) (Mumtaz et al. 1997a; Drew et al. 1998; Krämer et al. 1998; Belli et al. 2002). This compares with sensitivities for other techniques as follows: clinical examination alone, 51%–89%; mammography alone 50%–67%; ultrasound 85%; and cytology 79% (Mumtaz et al. 1997a; Drew et al. 1998; Krämer et al. 1998). Of note, Drew et al. (1998) showed clinical examination combined with mammography to have 100% sensitivity but a poor specificity at 67% (Drew et al. 1998). MRI may reveal other lesions in the post-surgical breast, which can confound evaluation by clinical examination, mammography or ultrasound. These include seromas, hematomas, oil cysts, fat necrosis and areas of residual or recurrent DCIS. Many of these have characteristic appearances on MRI and hence definitive diagnoses can be made.

10.5 Conclusions

DCE-MRI has a well established role in the evaluation of patients with diseases when cancer is strongly suspected or has been diagnosed. MRI is used as a supplementary tool to complement conventional methods of breast evaluation because it has excellent problem solving capabilities. Many indications are recognised as discussed in this chapter. Breast MRI is also the method of choice for the evaluation of the augmented breast both for the implant itself and also for the breast tissue around implants. There is encouraging ongoing research evaluating its role for the assessment of patients at high risk of breast cancer, for primary staging of cancers in the radiographically dense breasts and for the assessment response to chemotherapy. However, it should be noted that there is currently insufficient evidence to permit conclusions on the effects of breast MRI on health outcomes in patients with low-suspicion findings on conventional testing.

Acknowledgements

We are grateful to Dr. Jane Taylor and Mr. James Stirling for their assistance in the preparation of the illustrative material for this review. Parametric calculations and images were produced by Magnetic Resonance Imaging Software (MRIW) developed at the Institute of Cancer Research, Royal Marsden Hospital, London.

The support of the Breast Cancer Research Trust, Cancer Research UK and the Childwick Trust who support the research work at the Paul Strickland Scanner Centre, Mount Vernon Hospital respectively is gratefully acknowledged.

References

Abraham DC, Jones RC, Jones SE, et al. (1996) Evaluation of neoadjuvant chemotherapeutic response of locally advanced breast cancer by magnetic resonance imaging. Cancer 78: 91–100

Adams J, Carder PJ, Downey S, et al. (2000) Vascular endothelial growth factor (VEGF) in breast cancer: comparison of plasma, serum, and tissue VEGF and microvessel density and effects of tamoxifen. Cancer Res 60: 2898–2905

Ah-See MW, Makris A, Taylor NJ, et al. (2004) Does vascular imaging with MRI predict response to neoadjuvant chemotherapy in primary breast cancer. Meeting Proceedings of American Society of Clinical Oncology Vol. 23: 22 , abstract 582

Ah-See MW, Makris A, Taylor NJ, et al. (2003) Multifunctional magnetic resonance imaging predicts for clinicopathological response to neoadjuvant chemotherapy in primary breast cancer. In: 26th Annual San Antonio Breast Cancer Symposium, San Antonio, p 252

Allen SA, Cunliffe WJ, Gray J, et al. (2001) Pre-operative estimation of primary breast cancer size: a comparison of clinical assessment, mammography and ultrasound. Breast 10: 299–305

Barbier EL, Lamalle L, Decorps M (2001) Methodology of brain perfusion imaging. J Magn Reson Imaging 13: 496–520

Bedrosian I, Schlencker J, Spitz FR, et al. (2002) Magnetic resonance imaging-guided biopsy of mammographically and clinically occult breast lesions. Ann Surg Oncol 9: 457–461

Belli P, Costantini M, Romani M, et al. (2002) Magnetic resonance imaging in breast cancer recurrence. Breast Cancer Res Treat 73: 223–235

Bernsen HJ, Rijken PF, Peters JP, et al. (1998) Delayed vascular changes after antiangiogenic therapy with antivascular endothelial growth factor antibodies in human glioma xenografts in nude mice. Neurosurgery 43: 570-575; discussion 575–576

Bjurstam N, Bjorneld L, Warwick J, et al. (2003) The Gothenburg Breast Screening Trial. Cancer 97: 2387–2396

Boetes C, Barentsz JO, Mus RD, et al. (1994) MR characterization of suspicious breast lesions with a gadolinium-enhanced TurboFLASH subtraction technique. Radiology 193: 777–781

Boetes C, Mus RD, Holland R, et al. (1995) Breast tumors: comparative accuracy of MR imaging relative to mammography and US for demonstrating extent. Radiology 197: 743–747

Bonadonna G, Valagussa P, Brambilla C, et al. (1998) Primary chemotherapy in operable breast cancer: eight-year experience at the Milan Cancer Institute. J Clin Oncol 16: 93–100

Bonadonna G, Veronesi U, Brambilla C, et al. (1990) Primary chemotherapy to avoid mastectomy in tumors with diameters of three centimeters or more. J Natl Cancer Inst 82: 1539–1545

Bose S, Lesser ML, Norton L, et al. (1996) Immunophenotype of intraductal carcinoma. Arch Pathol Lab Med 120: 81–85

Boyages J, Bilous M, Barraclough B, et al. (1988) Fat necrosis of the breast following lumpectomy and radiation therapy for early breast cancer. Radiother Oncol 13: 69–74

Brekelmans CT, Seynaeve C, Bartels CC, et al. (2001) Effectiveness of breast cancer surveillance in BRCA1/2 gene mutation carriers and women with high familial risk. J Clin Oncol 19: 924–930

Brown J, Coulthard A, Dixon AK, et al. (2000) Protocol for a national multi-centre study of magnetic resonance imaging screening in women at genetic risk of breast cancer. Breast 9: 78–82

Bruening R, Berchtenbreiter C, Holzknecht N, et al. (2000) Effects of three different doses of a bolus injection of gadodiamide: assessment of regional cerebral blood volume maps in a blinded reader study. AJNR Am J Neuroradiol 21: 1603–1610

Buadu LD, Murakami J, Murayama S, et al. (1996) Breast lesions: correlation of contrast medium enhancement patterns on MR images with histopathologic findings and tumor angiogenesis. Radiology 200: 639–649

Buckley DL, Drew PJ, Mussurakis S, et al. (1997) Microvessel density of invasive breast cancer assessed by dynamic Gd-DTPA enhanced MRI. J Magn Reson Imaging 7: 461–464

Burcombe RJ, Makris A, Davies C, et al. (2002) Vascular endothelial growth factor (VEGF) fails to predict response to neoadjuvant chemotherapy for primary breast cancer. Breast Cancer Res Treat 76: S122 abstract 475

Ciatto S, Grazzini G, Iossa A, et al. (1990) In situ ductal carcinoma of the breast--analysis of clinical presentation and outcome in 156 consecutive cases. Eur J Surg Oncol 16: 220–224

Clarke D, Curtis JL, Martinez A, et al. (1983) Fat necrosis of the breast simulating recurrent carcinoma after primary radiotherapy in the management of early stage breast carcinoma. Cancer 52: 442–445

Crystal P, Strano SD, Shcharynski S, et al. (2003) Using sonography to screen women with mammographically dense breasts. AJR Am J Roentgenol 181: 177–182

Dao TH, Rahmouni A, Campana F, et al. (1993) Tumor recurrence versus fibrosis in the irradiated breast: differentiation with dynamic gadolinium-enhanced MR imaging. Radiology 187: 751–755

D'Arcy J, Collins D, Padhani A, et al. (2004) Magnetic Resonance Imaging Workbench (MRIW): Dynamic contrast enhanced MRI data analysis and visualisation. Radiographics - submitted

Darland DC, D'Amore PA (1999) Blood vessel maturation: vascular development comes of age. J Clin Invest 103: 157–158

Davis PL, Staiger MJ, Harris KB, et al. (1996) Breast cancer measurements with magnetic resonance imaging, ultrasonography, and mammography. Breast Cancer Res Treat 37: 1–9

Delille JP, Slanetz PJ, Yeh ED, et al. (2003) Invasive ductal breast carcinoma response to neoadjuvant chemotherapy: noninvasive monitoring with functional MR imaging pilot study. Radiology 228: 63–69

den Boer JA, Hoenderop RK, Smink J, et al. (1997) Pharmacokinetic analysis of Gd-DTPA enhancement in dynamic three-dimensional MRI of breast lesions. J Magn Reson Imaging 7: 702-715

Deniz K, O'Mahony S, Ross G, et al. (2003) Breast cancer in women after treatment for Hodgkin's disease. Lancet Oncol 4: 207–214

Dennie J, Mandeville JB, Boxerman JL, et al. (1998) NMR imaging of changes in vascular morphology due to tumor angiogenesis. Magn Reson Med 40: 793–799

Denton ER, Sonoda LI, Rueckert D, et al. (1999) Comparison and evaluation of rigid, affine, and nonrigid registration of breast MR images. J Comput Assist Tomogr 23: 800–805

Diekmann F, Diekmann S, Beljavskaja M, et al. (2004) [Preoperative MRT of the breast in invasive lobular carcinoma in comparison with invasive ductal carcinoma].

Rofo Fortschr Geb Rontgenstr Neuen Bildgeb Verfahr 176: 544–549

Drew PJ, Chatterjee S, Turnbull LW, et al. (1999) Dynamic contrast enhanced magnetic resonance imaging of the breast is superior to triple assessment for the pre-operative detection of multifocal breast cancer. Ann Surg Oncol 6: 599–603

Drew PJ, Kerin MJ, Mahapatra T, et al. (2001) Evaluation of response to neoadjuvant chemoradiotherapy for locally advanced breast cancer with dynamic contrast-enhanced MRI of the breast. Eur J Surg Oncol 27: 617–620

Drew PJ, Kerin MJ, Turnbull LW, et al. (1998) Routine screening for local recurrence following breast-conserving therapy for cancer with dynamic contrast-enhanced magnetic resonance imaging of the breast. Ann Surg Oncol 5: 265–270

Ellerbroek N, Holmes F, Singletary E, et al. (1990) Treatment of patients with isolated axillary nodal metastases from an occult primary carcinoma consistent with breast origin. Cancer 66: 1461–1467

Esserman L, Hylton N, Yassa L, et al. (1999) Utility of magnetic resonance imaging in the management of breast cancer: evidence for improved preoperative staging. J Clin Oncol 17: 110–119

Esserman L, Kaplan E, Partridge S, et al. (2001) MRI phenotype is associated with response to doxorubicin and cyclophosphamide neoadjuvant chemotherapy in stage III breast cancer. Ann Surg Oncol 8: 549–559

Evelhoch JL (1999) Key factors in the acquisition of contrast kinetic data for oncology. J Magn Reson Imaging 10: 254–259

Feldman LD, Hortobagyi GN, Buzdar AU, et al. (1986) Pathological assessment of response to induction chemotherapy in breast cancer. Cancer Res 46: 2578–2581

Fiorentino C, Berruti A, Bottini A, et al. (2001) Accuracy of mammography and echography versus clinical palpation in the assessment of response to primary chemotherapy in breast cancer patients with operable disease. Breast Cancer Res Treat 69: 143–151

Fischer U, Kopka L, Grabbe E (1999) Breast carcinoma: effect of preoperative contrast-enhanced MR imaging on the therapeutic approach. Radiology 213: 881–888

Fisher B, Anderson S, Bryant J, et al. (2002) Twenty-year follow-up of a randomized trial comparing total mastectomy, lumpectomy, and lumpectomy plus irradiation for the treatment of invasive breast cancer. N Engl J Med 347: 1233–1241

Fisher B, Bryant J, Wolmark N, et al. (1998a) Effect of preoperative chemotherapy on the outcome of women with operable breast cancer. J Clin Oncol 16: 2672-2685

Fisher B, Costantino JP, Wickerham DL, et al. (1998b) Tamoxifen for prevention of breast cancer: report of the National Surgical Adjuvant Breast and Bowel Project P-1 Study. J Natl Cancer Inst 90: 1371-1388

Fitzgibbons PL, Page DL, Weaver D, et al. (2000) Prognostic factors in breast cancer. College of American Pathologists Consensus Statement 1999. Arch Pathol Lab Med 124: 966–978

Flickinger FW, Allison JD, Sherry RM, et al. (1993) Differentiation of benign from malignant breast masses by time-intensity evaluation of contrast enhanced MRI. Magn Reson Imaging 11: 617–620

Fobben ES, Rubin CZ, Kalisher L, et al. (1995) Breast MR imaging with commercially available techniques: radiologic-pathologic correlation. Radiology 196: 143–152

Folkman J (1996) New perspectives in clinical oncology from angiogenesis research. Eur J Cancer 32A: 2534–2539

Fortunato L, Sorrento JJ, Golub RA, et al. (1992) Occult breast cancer. A case report and review of the literature. N Y State J Med 92: 555–557

Fourquet A, De La Rochefordiere A, Campana F (1996) Occult primary cancer with axillary metastases. In: Harris JR LM, Morrow M, Hellman S (Eds) (ed) Diseases of the breast. Lippincott-Raven, Philadelphia, Pa, pp 892–896

Fox SB (1997) Tumour angiogenesis and prognosis. Histopathology 30: 294–301

Fox SB, Leek RD, Weekes MP, et al. (1995) Quantitation and prognostic value of breast cancer angiogenesis: comparison of microvessel density, Chalkley count, and computer image analysis. J Pathol 177: 275–283

Gasparini G, Toi M, Gion M, et al. (1997) Prognostic significance of vascular endothelial growth factor protein in node-negative breast carcinoma. J Natl Cancer Inst 89: 139–147

Gilles R, Guinebretiere JM, Shapeero LG, et al. (1993) Assessment of breast cancer recurrence with contrast-enhanced subtraction MR imaging: preliminary results in 26 patients. Radiology 188: 473–478

Gilles R, Guinebretiere JM, Toussaint C, et al. (1994) Locally advanced breast cancer: contrast-enhanced subtraction MR imaging of response to preoperative chemotherapy. Radiology 191: 633–638

Gilles R, Zafrani B, Guinebretiere JM, et al. (1995) Ductal carcinoma in situ: MR imaging-histopathologic correlation. Radiology 196: 415–419

Gribbestad IS, Nilsen G, Fjosne HE, et al. (1994) Comparative signal intensity measurements in dynamic gadolinium-enhanced MR mammography. J Magn Reson Imaging 4: 477–480

Harms SE (2001) Integration of breast MRI in clinical trials. J Magn Reson Imaging 13: 830–836

Harms SE, Flamig DP, Hesley KL, et al. (1993) MR imaging of the breast with rotating delivery of excitation off resonance: clinical experience with pathologic correlation. Radiology 187: 493–501

Hartmann LC, Schaid DJ, Woods JE, et al. (1999) Efficacy of bilateral prophylactic mastectomy in women with a family history of breast cancer. N Engl J Med 340: 77–84

Hasan J, Byers R, Jayson GC (2002) Intra-tumoural microvessel density in human solid tumours. Br J Cancer 86: 1566-1577

Hayes C, Padhani AR, Leach MO (2002) Assessing changes in tumour vascular function using dynamic contrast-enhanced magnetic resonance imaging. NMR Biomed 15: 154-163

Helvie MA, Joynt LK, Cody RL, et al. (1996) Locally advanced breast carcinoma: accuracy of mammography versus clinical examination in the prediction of residual disease after chemotherapy. Radiology 198: 327-332

Herrada J, Iyer RB, Atkinson EN, et al. (1997) Relative value of physical examination, mammography, and breast sonography in evaluating the size of the primary tumor and regional lymph node metastases in women receiving neoadjuvant chemotherapy for locally advanced breast carcinoma. Clin Cancer Res 3: 1565–1569

Heywang SH, Hilbertz T, Beck R, et al. (1990) Gd-DTPA

enhanced MR imaging of the breast in patients with postoperative scarring and silicon implants. J Comput Assist Tomogr 14: 348–356
Heywang SH, Wolf A, Pruss E, et al. (1989) MR imaging of the breast with Gd-DTPA: use and limitations. Radiology 171: 95-103
Heywang-Kobrunner SH, Haustein J, Pohl C, et al. (1994) Contrast-enhanced MR imaging of the breast: comparison of two different doses of gadopentetate dimeglumine. Radiology 191: 639–646
Heywang-Kobrunner SH, Schlegel A, Beck R, et al. (1993) Contrast-enhanced MRI of the breast after limited surgery and radiation therapy. J Comput Assist Tomogr 17: 891–900
Holland R, Hendriks JH, Vebeek AL, et al. (1990) Extent, distribution, and mammographic/histological correlations of breast ductal carcinoma in situ. Lancet 335: 519–522
Homer MJ (1991) Mammographic interpretation: a practical approach. New York, NY: McGraw-Hill 4–5
Hulka CA, Edmister WB, Smith BL, et al. (1997) Dynamic echo-planar imaging of the breast: experience in diagnosing breast carcinoma and correlation with tumor angiogenesis. Radiology 205: 837–842
Ikeda DM, Hylton NM, Kinkel K, et al. (2001) Development, standardization, and testing of a lexicon for reporting contrast-enhanced breast magnetic resonance imaging studies. J Magn Reson Imaging 13: 889–895
Jain RK (2003) Molecular regulation of vessel maturation. Nat Med 9: 685-693
Jitsuiki Y, Hasebe T, Tsuda H, et al. (1999) Optimizing microvessel counts according to tumor zone in invasive ductal carcinoma of the breast. Mod Pathol 12: 492–498
Kaiser WA, Zeitler E (1989) MR imaging of the breast: fast imaging sequences with and without Gd-DTPA. Preliminary observations. Radiology 170: 681–686
Kaufmann M, von Minckwitz G, Smith R, et al. (2003) International expert panel on the use of primary (preoperative) systemic treatment of operable breast cancer: review and recommendations. J Clin Oncol 21: 2600–2608
Kerbel RS, Klement G, Pritchard KI, et al. (2002) Continuous low-dose anti-angiogenic/ metronomic chemotherapy: from the research laboratory into the oncology clinic. Ann Oncol 13: 12–15
Kerslake RW, Fox JN, Carleton PJ, et al. (1994) Dynamic contrast-enhanced and fat suppressed magnetic resonance imaging in suspected recurrent carcinoma of the breast: preliminary experience. Br J Radiol 67: 1158–1168
King MC, Wieand S, Hale K, et al. (2001) Tamoxifen and breast cancer incidence among women with inherited mutations in BRCA1 and BRCA2: National Surgical Adjuvant Breast and Bowel Project (NSABP-P1) Breast Cancer Prevention Trial. Jama 286: 2251-2256
Knapper WH (1991) Management of occult breast cancer presenting as an axillary metastasis. Semin Surg Oncol 7: 311–313
Knopp MV, Bourne MW, Sardanelli F, et al. (2003) Gadobenate dimeglumine-enhanced MRI of the breast: analysis of dose response and comparison with gadopentetate dimeglumine. AJR Am J Roentgenol 181: 663–676
Knopp MV, Brix G, Junkermann HJ, et al. (1994) MR mammography with pharmacokinetic mapping for monitoring of breast cancer treatment during neoadjuvant therapy. Magn Reson Imaging Clin N Am 2: 633–658
Knopp MV, Weiss E, Sinn HP, et al. (1999) Pathophysiologic basis of contrast enhancement in breast tumors. J Magn Reson Imaging 10: 260–266
Krämer S, Schulz-Wendtland R, Hagedorn K, et al. (1998) Magnetic resonance imaging and its role in the diagnosis of multicentric breast cancer. Anticancer Res 18: 2163–2164
Kriege M, Brekelmans CT, Boetes C, et al. (2004) Efficacy of MRI and mammography for breast-cancer screening in women with a familial or genetic predisposition. N Engl J Med 351: 427–437
Krishnamurthy R, Whitman GJ, Stelling CB, et al. (1999) Mammographic findings after breast conservation therapy. Radiographics 19 Spec No: S53-62; quiz S262–263
Kuhl CK, Bieling H, Gieseke J, et al. (1997a) Breast neoplasms: T2* susceptibility-contrast, first-pass perfusion MR imaging. Radiology 202: 87–95
Kuhl CK, Bieling HB, Gieseke J, et al. (1997b) Healthy premenopausal breast parenchyma in dynamic contrast-enhanced MR imaging of the breast: normal contrast medium enhancement and cyclical-phase dependency. Radiology 203: 137–144
Kuhl CK, Mielcareck P, Klaschik S, et al. (1999) Dynamic breast MR imaging: are signal intensity time course data useful for differential diagnosis of enhancing lesions? Radiology 211: 101–110
Kuhl CK, Schmutzler RK, Leutner CC, et al. (2000) Breast MR imaging screening in 192 women proved or suspected to be carriers of a breast cancer susceptibility gene: preliminary results. Radiology 215: 267–279
Kurtz JM, Amalric R, Brandone H, et al. (1988) Results of salvage surgery for mammary recurrence following breast-conserving therapy. Ann Surg 207: 347–351
Kvistad KA, Lundgren S, Fjosne HE, et al. (1999) Differentiating benign and malignant breast lesions with T2*-weighted first pass perfusion imaging. Acta Radiol 40: 45–51
Lantzsch T, Hefler L, Krause U, et al. (2002) The correlation between immunohistochemically-detected markers of angiogenesis and serum vascular endothelial growth factor in patients with breast cancer. Anticancer Res 22: 1925–1928
Lee SG, Orel SG, Woo IJ, et al. (2003) MR imaging screening of the contralateral breast in patients with newly diagnosed breast cancer: preliminary results. Radiology 226: 773–778
Lewis-Jones HG, Whitehouse GH, Leinster SJ (1991) The role of magnetic resonance imaging in the assessment of local recurrent breast carcinoma. Clin Radiol 43: 197–204
Liberman L, Morris EA, Dershaw DD, et al. (2003a) Ductal enhancement on MR imaging of the breast. AJR Am J Roentgenol 181: 519-525
Liberman L, Morris EA, Dershaw DD, et al. (2003b) MR imaging of the ipsilateral breast in women with percutaneously proven breast cancer. AJR Am J Roentgenol 180: 901–910
Liberman L, Morris EA, Kim CM, et al. (2003c) MR imaging findings in the contralateral breast of women with recently diagnosed breast cancer. AJR Am J Roentgenol 180: 333–341
Liberman L, Morris EA, Lee MJ, et al. (2002) Breast lesions detected on MR imaging: features and positive predictive value. AJR Am J Roentgenol 179: 171–178
Linderholm B, Lindh B, Tavelin B, et al. (2000) p53 and

vascular-endothelial-growth-factor (VEGF) expression predicts outcome in 833 patients with primary breast carcinoma. Int J Cancer 89: 51-62

Lissoni P, Fugamalli E, Malugani F, et al. (2000) Chemotherapy and angiogenesis in advanced cancer: vascular endothelial growth factor (VEGF) decline as predictor of disease control during taxol therapy in metastatic breast cancer. Int J Biol Markers 15: 308-311

Liu PF, Debatin JF, Caduff RF, et al. (1998) Improved diagnostic accuracy in dynamic contrast enhanced MRI of the breast by combined quantitative and qualitative analysis. Br J Radiol 71: 501-509

Liu YJ, Chung HW, Huang IJ, et al. (2002) A reinvestigation of maximal signal drop in dynamic susceptibility contrast magnetic resonance imaging. J Neuroimaging 12: 330-338

Makris A, Powles TJ, Ashley SE, et al. (1998) A reduction in the requirements for mastectomy in a randomized trial of neoadjuvant chemoendocrine therapy in primary breast cancer. Ann Oncol 9: 1179-1184

Makris A, Powles TJ, Kakolyris S, et al. (1999) Reduction in angiogenesis after neoadjuvant chemoendocrine therapy in patients with operable breast carcinoma. Cancer 85: 1996-2000

Mandelson MT, Oestreicher N, Porter PL, et al. (2000) Breast density as a predictor of mammographic detection: comparison of interval- and screen-detected cancers. J Natl Cancer Inst 92: 1081-1087

Martincich L, Montemurro F, De Rosa G, et al. (2004) Monitoring response to primary chemotherapy in breast cancer using dynamic contrast-enhanced magnetic resonance imaging. Breast Cancer Res Treat 83: 67-76

Matsubayashi R, Matsuo Y, Edakuni G, et al. (2000) Breast masses with peripheral rim enhancement on dynamic contrast-enhanced MR images: correlation of MR findings with histologic features and expression of growth factors. Radiology 217: 841-848

Morris EA (2002) Breast cancer imaging with MRI. Radiol Clin North Am 40: 443-466

Mumtaz H, Davidson T, Hall-Craggs MA, et al. (1997a) Comparison of magnetic resonance imaging and conventional triple assessment in locally recurrent breast cancer. Br J Surg 84: 1147-1151

Mumtaz H, Hall-Craggs MA, Davidson T, et al. (1997b) Staging of symptomatic primary breast cancer with MR imaging. AJR Am J Roentgenol 169: 417-424

Murray AD, Redpath TW, Needham G, et al. (1996) Dynamic magnetic resonance mammography of both breasts following local excision and radiotherapy for breast carcinoma. Br J Radiol 69: 594-600

Mussurakis S, Buckley DL, Bowsley SJ, et al. (1995) Dynamic contrast-enhanced magnetic resonance imaging of the breast combined with pharmacokinetic analysis of gadolinium-DTPA uptake in the diagnosis of local recurrence of early stage breast carcinoma. Invest Radiol 30: 650-662

Muüller RD, Barkhausen J, Sauerwein W, et al. (1998) Assessment of local recurrence after breast-conserving therapy with MRI. J Comput Assist Tomogr 22: 408-412

Neubauer H, Li M, Kuehne-Heid R, et al. (2003) High grade and non-high grade ductal carcinoma in situ on dynamic MR mammography: characteristic findings for signal increase and morphological pattern of enhancement. Br J Radiol 76: 3-12

Nieuwenhuis B, Van Assen-Bolt AJ, Van Waarde-Verhagen MA, et al. (2002) BRCA1 and BRCA2 heterozygosity and repair of X-ray-induced DNA damage. Int J Radiat Biol 78: 285-295

Olson JA, Jr., Morris EA, Van Zee KJ, et al. (2000) Magnetic resonance imaging facilitates breast conservation for occult breast cancer. Ann Surg Oncol 7: 411-415

Orel SG, Mendonca MH, Reynolds C, et al. (1997) MR imaging of ductal carcinoma in situ. Radiology 202: 413-420

Orel SG, Schnall MD (2001) MR imaging of the breast for the detection, diagnosis, and staging of breast cancer. Radiology 220: 13-30

Orel SG, Schnall MD, LiVolsi VA, et al. (1994) Suspicious breast lesions: MR imaging with radiologic-pathologic correlation. Radiology 190: 485-493

Orel SG, Schnall MD, Powell CM, et al. (1995) Staging of suspected breast cancer: effect of MR imaging and MR-guided biopsy. Radiology 196: 115-122

Orel SG, Troupin RH, Patterson EA, et al. (1992) Breast cancer recurrence after lumpectomy and irradiation: role of mammography in detection. Radiology 183: 201-206

Orel SG, Weinstein SP, Schnall MD, et al. (1999) Breast MR imaging in patients with axillary node metastases and unknown primary malignancy. Radiology 212: 543-549

Osborne MP, Borgen PI, Wong GY, et al. (1992) Salvage mastectomy for local and regional recurrence after breast-conserving operation and radiation therapy. Surg Gynecol Obstet 174: 189-194

Ostergaard L, Sorensen AG, Kwong KK, et al. (1996) High resolution measurement of cerebral blood flow using intravascular tracer bolus passages. Part II: Experimental comparison and preliminary results. Magn Reson Med 36: 726-736

Padhani A, Yarnold J, Regan J, et al. (2002) Magnetic resonance imaging of induration in the irradiated breast. Radiother Oncol 64: 157

Padhani AR, Dzik-Jurasz A (2004) Perfusion MR imaging of extracranial tumor angiogenesis. Top Magn Reson Imaging 15: 41-57

Padhani AR, Yarnold J, Regan J, et al. (2003) Dynamic MRI of breast hardness following radiation treatment. J Magn Reson Imaging 17: 427-434

Pain JA, Ebbs SR, Hern RP, et al. (1992) Assessment of breast cancer size: a comparison of methods. Eur J Surg Oncol 18: 44-48

Parker GJ, Baustert I, Tanner SF, et al. (2000) Improving image quality and T(1) measurements using saturation recovery turboFLASH with an approximate K-space normalisation filter. Magn Reson Imaging 18: 157-167

Parker GJ, Suckling J, Tanner SF, et al. (1997) Probing tumor microvascularity by measurement, analysis and display of contrast agent uptake kinetics. J Magn Reson Imaging 7: 564-574

Paulus DD (1984) Conservative treatment of breast cancer: mammography in patient selection and follow-up. AJR Am J Roentgenol 143: 483-487

Pierce WB, Harms SE, Flamig DP, et al. (1991) Three-dimensional gadolinium-enhanced MR imaging of the breast: pulse sequence with fat suppression and magnetization transfer contrast. Work in progress. Radiology 181: 757-763

Podo F, Sardanelli F, Canese R, et al. (2002a) The Italian multi-centre project on evaluation of MRI and other imaging modalities in early detection of breast cancer in subjects at high genetic risk. In: J Exp Clin Cancer Res, vol 21, pp 115–124

Podo F, Sardanelli F, Canese R, et al. (2002b) The Italian multi-centre project on evaluation of MRI and other imaging modalities in early detection of breast cancer in subjects at high genetic risk. J Exp Clin Cancer Res 21: 115–124

Pritt B, Ashikaga T, Oppenheimer RG, et al. (2004) Influence of breast cancer histology on the relationship between ultrasound and pathology tumor size measurements. Mod Pathol

Qayyum A, Birdwell RL, Daniel BL, et al. (2002) MR imaging features of infiltrating lobular carcinoma of the breast: histopathologic correlation. AJR Am J Roentgenol 178: 1227–1232

Rak JW, St Croix BD, Kerbel RS (1995) Consequences of angiogenesis for tumor progression, metastasis and cancer therapy. Anticancer Drugs 6: 3-18

Rebbeck TR, Friebel T, Lynch HT, et al. (2004) Bilateral prophylactic mastectomy reduces breast cancer risk in BRCA1 and BRCA2 mutation carriers: the PROSE Study Group. J Clin Oncol 22: 1055–1062

Rebbeck TR, Levin AM, Eisen A, et al. (1999) Breast cancer risk after bilateral prophylactic oophorectomy in BRCA1 mutation carriers. J Natl Cancer Inst 91: 1475–1479

Revel D, Brasch RC, Paajanen H, et al. (1986) Gd-DTPA contrast enhancement and tissue differentiation in MR imaging of experimental breast carcinoma. Radiology 158: 319–323

Rieber A, Brambs HJ, Gabelmann A, et al. (2002) Breast MRI for monitoring response of primary breast cancer to neo-adjuvant chemotherapy. Eur Radiol 12: 1711–1719

Ringberg A, Palmer B, Linell F (1982) The contralateral breast at reconstructive surgery after breast cancer operation--a histopathological study. Breast Cancer Res Treat 2: 151–161

Roberts TP (1997) Physiologic measurements by contrast-enhanced MR imaging: expectations and limitations. J Magn Reson Imaging 7: 82–90

Rosen BR, Belliveau JW, Buchbinder BR, et al. (1991) Contrast agents and cerebral hemodynamics. Magn Reson Med 19: 285–292

Rosen PP, Kimmel M (1990) Occult breast carcinoma presenting with axillary lymph node metastases: a follow-up study of 48 patients. Hum Pathol 21: 518-523

Rosen PP, Senie R, Schottenfeld D, et al. (1979) Noninvasive breast carcinoma: frequency of unsuspected invasion and implications for treatment. Ann Surg 189: 377–382

Rostom AY, el-Sayed ME (1987) Fat necrosis of the breast: an unusual complication of lumpectomy and radiotherapy in breast cancer. Clin Radiol 38: 31

Schnall MD, Ikeda DM (1999) Lesion Diagnosis Working Group report. J Magn Reson Imaging 10: 982–990

Schorn C, Fischer U, Luftner-Nagel S, et al. (1999) MRI of the breast in patients with metastatic disease of unknown primary. Eur Radiol 9: 470–473

Senger DR, Galli SJ, Dvorak AM, et al. (1983) Tumor cells secrete a vascular permeability factor that promotes accumulation of ascites fluid. Science 219: 983–985

Sickles EA, Herzog KA (1981) Mammography of the postsurgical breast. AJR Am J Roentgenol 136: 585–588

Siegmann KC, Muller-Schimpfle M, Schick F, et al. (2002) MR imaging-detected breast lesions: histopathologic correlation of lesion characteristics and signal intensity data. AJR Am J Roentgenol 178: 1403–1409

Slanetz PJ, Edmister WB, Yeh ED, et al. (2002) Occult contralateral breast carcinoma incidentally detected by breast magnetic resonance imaging. Breast J 8: 145–148

Smith IC, Heys SD, Hutcheon AW, et al. (2002) Neoadjuvant chemotherapy in breast cancer: significantly enhanced response with docetaxel. J Clin Oncol 20: 1456–1466

Snelling JD, Abdullah N, Brown G, et al. (2004) Measurement of tumour size in case selection for breast cancer therapy by clinical assessment and ultrasound. Eur J Surg Oncol 30: 5–9

Sorensen AG, Tievsky AL, Ostergaard L, et al. (1997) Contrast agents in functional MR imaging. J Magn Reson Imaging 7: 47–55

Stomper PC, Herman S, Klippenstein DL, et al. (1995) Suspect breast lesions: findings at dynamic gadolinium-enhanced MR imaging correlated with mammographic and pathologic features. Radiology 197: 387–395

Stomper PC, Recht A, Berenberg AL, et al. (1987) Mammographic detection of recurrent cancer in the irradiated breast. AJR Am J Roentgenol 148: 39–43

Stomper PC, Winston JS, Herman S, et al. (1997) Angiogenesis and dynamic MR imaging gadolinium enhancement of malignant and benign breast lesions. Breast Cancer Res Treat 45: 39–46

Stoutjesdijk MJ, Boetes C, Jager GJ, et al. (2001a) Magnetic resonance imaging and mammography in women with a hereditary risk of breast cancer. In: J Natl Cancer Inst, vol 93, pp 1095–1102.

Stoutjesdijk MJ, Boetes C, Jager GJ, et al. (2001b) Magnetic resonance imaging and mammography in women with a hereditary risk of breast cancer. J Natl Cancer Inst 93: 1095–1102.

Su MY, Cheung YC, Fruehauf JP, et al. (2003) Correlation of dynamic contrast enhancement MRI parameters with microvessel density and VEGF for assessment of angiogenesis in breast cancer. J Magn Reson Imaging 18: 467–477

Takehara M, Tamura M, Kameda H, et al. (2004) Examination of breast conserving therapy in lobular carcinoma. Breast Cancer 11: 69–72

Tan JE, Orel SG, Schnall MD, et al. (1999) Role of magnetic resonance imaging and magnetic resonance imaging--guided surgery in the evaluation of patients with early-stage breast cancer for breast conservation treatment. Am J Clin Oncol 22: 414–418

Teh W, Wilson AR (1998) The role of ultrasound in breast cancer screening. A consensus statement by the European Group for Breast Cancer Screening. Eur J Cancer 34: 449–450

Thurfjell EL, Lindgren JA (1996) Breast cancer survival rates with mammographic screening: similar favorable survival rates for women younger and those older than 50 years. Radiology 201: 421–426

Tilanus-Linthorst MM, Obdeijn AI, Bontenbal M, et al. (1997) MRI in patients with axillary metastases of occult breast carcinoma. Breast Cancer Res Treat 44: 179–182

Tilanus-Linthorst MM, Obdeijn IM, Bartels KC, et al. (2000a) First experiences in screening women at high risk for breast cancer with MR imaging. In: Breast Cancer Res Treat, vol 63, pp 53–60.

Tilanus-Linthorst MM, Obdeijn IM, Bartels KC, et al. (2000b)

First experiences in screening women at high risk for breast cancer with MR imaging. Breast Cancer Res Treat 63: 53–60.
Tofts PS (1997) Modeling tracer kinetics in dynamic Gd-DTPA MR imaging. J Magn Reson Imaging 7: 91–101
Tofts PS, Berkowitz B, Schnall MD (1995) Quantitative analysis of dynamic Gd-DTPA enhancement in breast tumors using a permeability model. Magn Reson Med 33: 564–568
Tofts PS, Brix G, Buckley DL, et al. (1999) Estimating kinetic parameters from dynamic contrast-enhanced T(1)-weighted MRI of a diffusable tracer: standardized quantities and symbols. J Magn Reson Imaging 10: 223–232
Toi M, Inada K, Suzuki H, et al. (1995) Tumor angiogenesis in breast cancer: its importance as a prognostic indicator and the association with vascular endothelial growth factor expression. Breast Cancer Res Treat 36: 193–204
Toi M, Matsumoto T, Bando H (2001) Vascular endothelial growth factor: its prognostic, predictive, and therapeutic implications. Lancet Oncol 2: 667–673
van der Hage JA, van de Velde CJ, Julien JP, et al. (2001) Preoperative chemotherapy in primary operable breast cancer: results from the European Organization for Research and Treatment of Cancer trial 10902. J Clin Oncol 19: 4224–4237
van Dongen JA, Voogd AC, Fentiman IS, et al. (2000) Long-term results of a randomized trial comparing breast-conserving therapy with mastectomy: European Organization for Research and Treatment of Cancer 10801 trial. J Natl Cancer Inst 92: 1143–1150
Vermeulen PB, Gasparini G, Fox SB, et al. (2002) Second international consensus on the methodology and criteria of evaluation of angiogenesis quantification in solid human tumours. Eur J Cancer 38: 1564–1579
Vermeulen PB, Gasparini G, Fox SB, et al. (1996) Quantification of angiogenesis in solid human tumours: an international consensus on the methodology and criteria of evaluation. Eur J Cancer 32A: 2474–2484
Veronesi U, Cascinelli N, Mariani L, et al. (2002) Twenty-year follow-up of a randomized study comparing breast-conserving surgery with radical mastectomy for early breast cancer. N Engl J Med 347: 1227–1232
Viehweg P, Heinig A, Lampe D, et al. (1998) Retrospective analysis for evaluation of the value of contrast-enhanced MRI in patients treated with breast conservative therapy. Magma 7: 141–152
Vinnicombe SJ, MacVicar AD, Guy RL, et al. (1996) Primary breast cancer: mammographic changes after neoadjuvant chemotherapy, with pathologic correlation. Radiology 198: 333–340
Warner E, Plewes DB, Shumak RS, et al. (2001a) Comparison of breast magnetic resonance imaging, mammography, and ultrasound for surveillance of women at high risk for hereditary breast cancer. In: J Clin Oncol, vol 19, pp 3524–3531.
Warner E, Plewes DB, Shumak RS, et al. (2001b) Comparison of breast magnetic resonance imaging, mammography, and ultrasound for surveillance of women at high risk for hereditary breast cancer. J Clin Oncol 19: 3524–3531.
Warren R (2001) Is breast MRI mature enough to be recommended for general use? Lancet 358: 1745–1746.
Weatherall PT, Evans GF, Metzger GJ, et al. (2001) MRI vs. histologic measurement of breast cancer following chemotherapy: comparison with x-ray mammography and palpation. J Magn Reson Imaging 13: 868–875
Weidner N (1998) Tumoural vascularity as a prognostic factor in cancer patients: the evidence continues to grow. J Pathol 184: 119–122
Weind KL, Maier CF, Rutt BK, et al. (1998) Invasive carcinomas and fibroadenomas of the breast: comparison of microvessel distributions--implications for imaging modalities. Radiology 208: 477–483
Whitehouse GH, Moore NR (1994) MR imaging of the breast after surgery for breast cancer. Magn Reson Imaging Clin N Am 2: 591–603
Yang WT, Lam WW, Cheung H, et al. (1997) Sonographic, magnetic resonance imaging, and mammographic assessments of preoperative size of breast cancer. J Ultrasound Med 16: 791–797
Yeh ED, Slanetz PJ, Edmister WB, et al. (2003) Invasive lobular carcinoma: spectrum of enhancement and morphology on magnetic resonance imaging. Breast J 9: 13–18

11 Dynamic Contrast-Enhanced MR Imaging for Predicting Tumor Control in Patients with Cervical Cancer

Joseph F. Montebello, Nina A. Mayr, William T. C. Yuh, D. Scott McMeekin, Dee. H. Wu, and Michael W. Knopp

CONTENTS

11.1 Introduction

Carcinoma of the cervix is a disease which is having a significant impact on healthcare worldwide, being a leading cause of death among women in some countries. Factors reported to influence outcome in cervical cancer include stage, extent of disease, histological type, lymphatic spread, and vascular invasion (Zaino et al. 1992). Prognosis, however, is still difficult to predict due to the heterogeneity of this tumor. Cytotoxic therapy – largely radiation therapy with concurrent cisplatin-based chemotherapy – remains the treatment of choice for all except for those with early operable stages (Rose et al. 1999).

Factors specific to tumor and tumor micro-environment that influence the success of cytotoxic therapy have been studied for many decades. Hypoxia and poor tumor blood supply have long been implicated as detrimental to tumor control by radiation therapy (Dische et al. 1983; Evans and Bergso 1965; Gatenby et al. 1988; Kallinowski et al. 1990; Kolstad 1968; Lartigau et al. 1997; Tannock 1998; Thomlinson and Gray 1955; Urtasun et al. 1986). This correlation has been particularly true for cervical cancer (Bush et al. 1978; Dische et al. 1983; Evans and Bergso 1965; Höckel et al. 1996a; Kolstad 1968). The availability of these parameters in the clinical setting have, until recently, remained very elusive because of the lack of reliable and widely applicable methods to assess tumor oxygenation and perfusion clinically in patients undergoing therapy for cervical cancer. With the advancement of imaging techniques, particularly perfusion (microcirculation) imaging techniques, the assessment of tumor blood supply and oxygenation has become a reality and is ever more available in the current clinical setting. Recent studies have suggested that these microcirculation imaging techniques can provide information relating to tumor blood supply and may indirectly reflect the status of tumor hypoxia (Mayr et al. 1996a,b; Hawighorst et al. 1998; Gong et al. 1999; Cooper et al. 2000).

11.2 Tumor Blood Supply and Hypoxia

Hypoxia results when there is a discrepancy between oxygen supply and oxygen consumption. There are

J. F. Montebello, MD; W. T. C. Yuh, MD, MSEE
Radiation Oncology Center, Oklahoma University Health Sciences Center, Oklahoma City, OK 73190, USA
D. S. McMeekin, MD
Department of Obstetrics and Gynecology, Oklahoma University Health Sciences Center, Oklahoma City, OK 73190, USA
D. H. Wu, PhD
Department of Radiology, Oklahoma University Health Sciences Center, Oklahoma City, OK 73190, USA
N. A. Mayr, MD
Professor, Director Radiation Oncology Center, Department of Radiology, Oklahoma University Health Sciences Center, University Hospital, 1200 N. Everett Drive, Rm.BNP 603, Oklahoma City, OK 73190, USA
M. W. Knopp, MD, PhD
Division of Imaging Research, Department of Radiology, Ohio State University, 657 Means Hall, 1654 Upham Dr., Columbus, OH 43210-1250, USA

two types of hypoxia: chronic diffusion-limited hypoxia, which results when tumor cells are too far from a functioning blood vessel (Tannock 1972), and acute/intermittent or perfusion-limited hypoxia, which occurs when the abnormal periodic flow and arteriovenous shunting within tumor vessels renders some cells under-perfused for a period of time (Chaplin et al. 1987). Acute hypoxia is based on arteriovenous shunting of tumor vessels in which, even in the presence of vessel perfusion, is inadequate. The potency of tumor vessels is a dynamic process that is continuously changing depending on the metabolic demands of the tumor (Chaplin et al. 1987). Acute hypoxia may be morphologically undetectable. It is more likely to be detected by functional imaging than by histological assessment.

Within a tumor, chronic and acute hypoxia is distributed inhomogeneously on a microscopic level and varies over time. Studies involving animal models and clinical investigations have suggested an association between hypoxia and rapid proliferation (Nordsmark et al. 1996). Hypoxia is the most important stimulus for the expression of angiogenic cytokines, in particular VEGF, and thereby triggering neoangiogenesis in tumors (Höckel 1996b). In histological specimens, angiogenesis has been measured as vascular density (Weidner et al. 1991; Vermeulen et al. 1996; Sundfør et al. 1998). The functional aspect of these vessels, such as permeability, remained elusive until recently (Mayr et al. 1999, 2000; Knopp et al. 1995).

Poor oxygenation has been associated with a decrease in local control in patients treated with radiotherapy. The oxygen enhancement ratio of radiation for mammalian cells in culture ranges from two to three for sparsely ionizing radiations such as X-rays (Palcic and Skarsgard 1984). In 1955 Thomlinson and Gray published an often cited paper in which it was postulated, on the basis of histological sections of lung cancers and the relationship of tumor cells to blood capillaries, that between necrotic tumor and healthy tumor, there are tumor cells that would be protected from radiation therapy by the low oxygen tension in this region. These tumor cells could subsequently lead to radiation failures. In subsequent years much effort was expended to overcome hypoxia-induced radioresistance. These efforts have included blood transfusions (Bush et al. 1978; Fyles et al. 2000), hyperbaric oxygen (Brady et al. 1981), oxygen mimetic sensitizers (Overgaard and Horsman 1996; Grigsby et al. 1999), and high LET radiation (Maor et al. 1988; Maruyama et al. 1991).

Recent studies suggest that, in addition to hypoxia-imparted radioresistance, hypoxia has been implicated in the malignant progression of neoplasm resulting in increased invasiveness and metastatic potential (Höckel and Vaupel 2001). These observations may be viewed to be simply the result of more aggressive neoplasms outgrowing their blood supply. However, hypoxia itself may induce cellular changes as well as exert clonal selection pressure which results in a more clinically aggressive phenotype (Semenza et al. 2000 a). Hypoxia-inducible factor-1 (HIF-1) is known to control gene expression in response to changes in oxygen tension. HIF-1á, a subunit of HIF-1, is postulated to stimulate angiogenesis by the trans-activation of the vascular endothelial growth factor (VEGF) gene (Shweiki et al. 1992). An explanation for the increased metastatic potential of hypoxic tumors is the association of hypoxia and angiogenesis (Weidner et al. 1991). Under hypoxic conditions the level of P53 increases in cells that induce apoptosis. However, tumor cells expressing mutant P53 will demonstrate reduced hypoxia-mediated apoptosis. In vitro studies have demonstrated that under hypoxic conditions these mutant cells will overtake cells expressing wild-type P53 (Graeber et al. 1996). Theoretically, in the clinical setting, hypoxia would be able to promote tumor progression by exerting a clonal selective pressure for a more aggressive phenotype.

There have been many efforts to develop predictive assays based on tumor hypoxia (Höckel et al. 2001). However, methods for accomplishing this have been difficult to implement in daily clinical practice. These methods may be invasive (oxygen-electrode placement, in vitro tumor labeling), fraught with time delays to obtain results from tissue culture-based assays, or associated with difficulties in defining anatomic tissue (PET) – making them unsuitable to be widely applied in large patient populations. Dynamic contrast-enhanced (DCE) MRI provides a new tool to study these parameters non-invasively and employ them in clinical patient populations.

11.3 Dynamic Contrast-Enhanced (DCE) MR Imaging

The terms "MR perfusion imaging," and "dynamic MR imaging," which are frequently used interchangeaEarly clinical data suggests that DCE MRI can provide functional assessment of tumor microcirculation and angiogenesis that may be used to monitor therapy and/or predict therapy outcome in cervical cancer (Figs. 11.1, 11.2). DCE tumor imaging requires

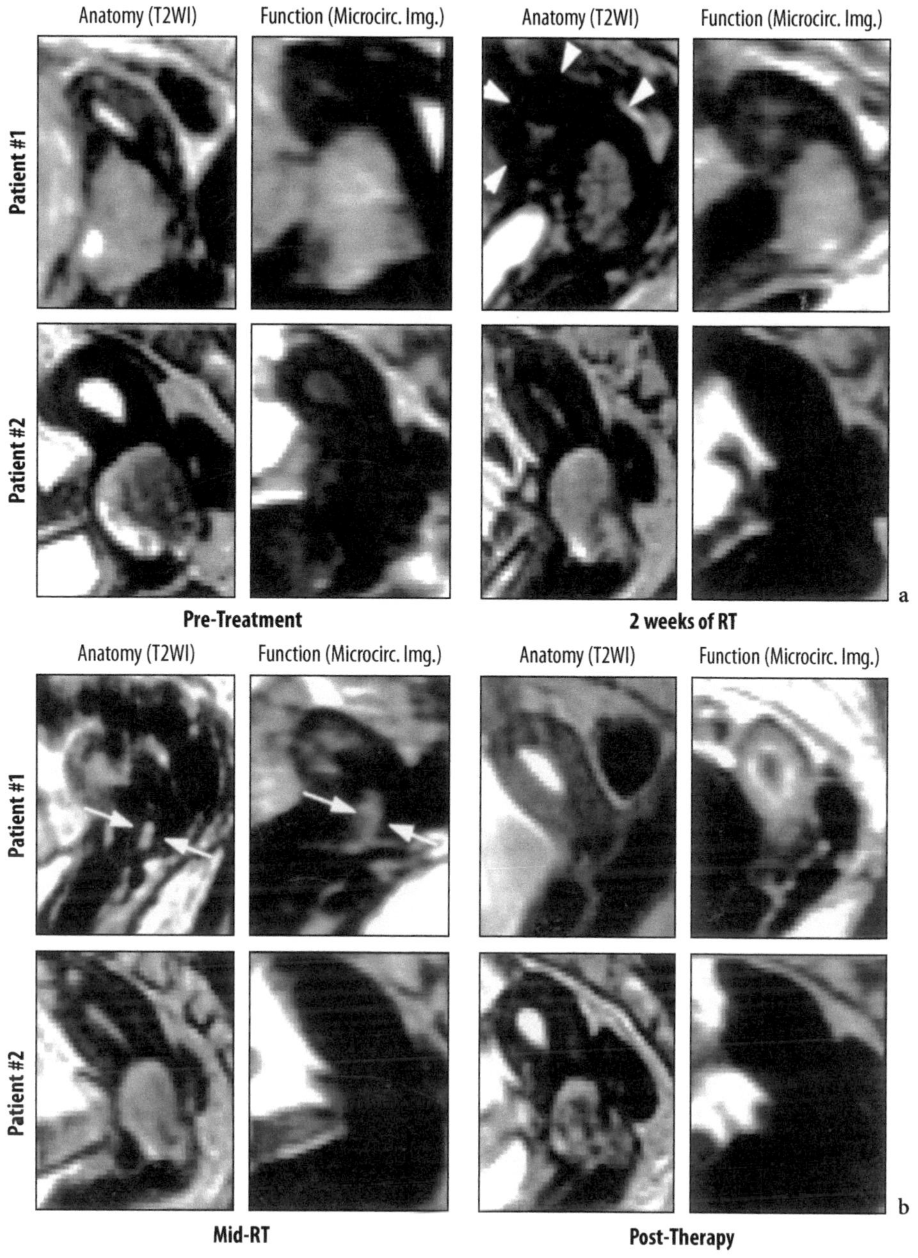

Fig. 11.1a,b. High vs. low dynamic enhancement pattern and therapy outcome. Anatomical (T2-weighted images) and microcirculation imaging (dynamic contrast-enhanced imaging) of two patients with similar-size cervical cancers. **a** Imaging obtained pre-radiation therapy and at 2 weeks of radiation therapy. Patient #1 (*top*) and patient #2 (*bottom*) have similar-size tumors on the pre-therapy T2-weighted images. In patient #1 (*top*), the dynamic contrast image shows intense enhancement. However, in patient #2 (*bottom*), there is poor enhancement in the tumor region. In the early-radiation therapy studies (at 2 weeks of RT), tumor size decreases in both patients. Enhancement is again intense in patent #1 (*top*) and poor in patient #2 (*bottom*). **b** Imaging in the same patients obtained later in radiation therapy (RT) (mid-RT, 4–5 weeks) and post-therapy (1–2 months follow-up). In patient #1 (*top*), the tumor has almost completely resolved on the T2-weighted images (*arrows*) at mid-RT and the dynamic contrast image shows a small amount of residual intense enhancement (*arrows*). In patient #2, tumor size has not further decreased, and poor dynamic enhancement within the tumor region persists. In the post-therapy study, the tumor in patient #1 is no longer visible on the T2-weighted images, consistent with a complete clinical response. The lack of tumor enhancement in the post-therapy study did not adversely influence outcome. This patient is alive and well without evidence of recurrence 7 years after radiation therapy. These findings support the concept that blood and oxygen supply *before and during* radiation therapy is crucial for radiation therapy outcome. The tumor in patient #2 is still present in the post-therapy T2-weighted study, although no tumor was present on clinical palpation, and thus the patient was classified as a complete clinical response by pelvic exam findings. The dynamic contrast image again shows poor dynamic enhancement. Tumor persistence was demonstrated by biopsy 4 months after completion of RT and subsequent pelvic exenteration histologically confirmed persistent tumor in the cervix consistent with the MRI findings. [Adapted and reprinted with permission from Mayr et al. (1996b)]

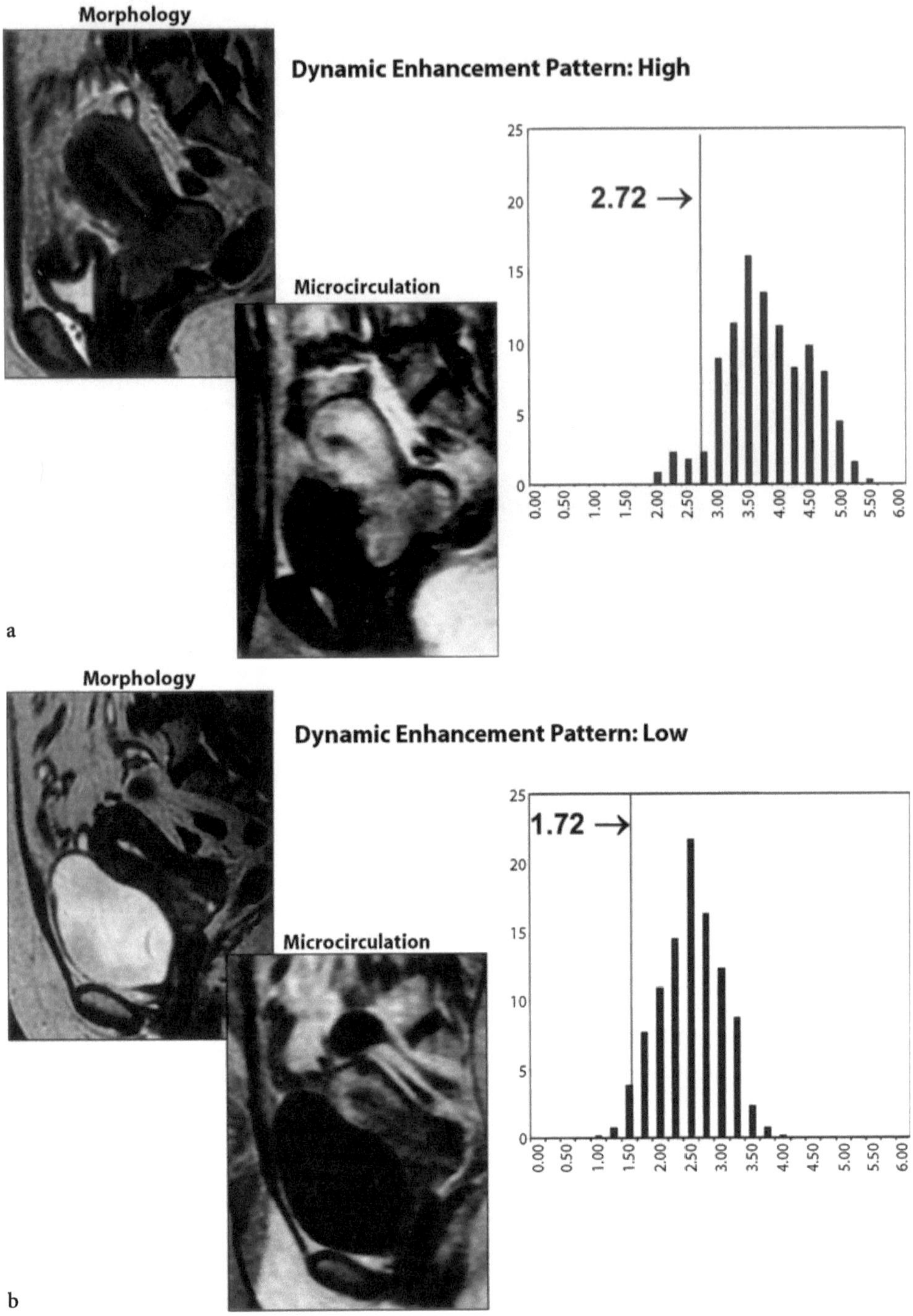

Fig. 11.2a,b. Pixel distribution of dynamic enhancement pattern and therapy outcome. **a** High dynamic enhancement and high 10th percentile RSI. MR image and pixel distribution in a 44-year-old woman with a stage IIB squamous cell carcinoma of the cervix. Parasagittal T2-weighted MR image (*left*) shows a well-delineated tumor replacing the cervix. Dynamic contrast-enhanced image (*middle*) suggests high enhancement within the tumor. Pixel histogram (*right*) demonstrates a wide range of pixel RSI values (range, 3.43), confirms high enhancement (mean RSI, 3.68; median RSI, 3.66) and a high 10th percentile RSI of 2.84. The patient's tumor remains controlled and she is alive and well 6 years after radiation therapy completion. **b** Low dynamic enhancement and low 10th percentile RSI. MR image and pixel distribution in a 45-year-old woman with a stage IIB squamous cell carcinoma of the cervix. Parasagittal T2-weighted MR image (*left*) shows a well-delineated tumor replacing the cervix. Dynamic contrast-enhanced image (*right*) suggests poor enhancement within the tumor region. Pixel histogram (*right*) shows a wide range of RSI values (RSI range, 2.87) overlapping that of the patient in (**a**). The histogram confirms low enhancement (mean RSI, 2.38; median RSI, 2.41) and shows a low 10th percentile RSI of 1.72. The patient had tumor recurrence 3 months after completion of therapy and subsequently died. [Adapted and reprinted with permission from Mayr et al. (2000)]

the injection of gadolinium (Gd) chelated contrast agents. These contrast agents are widely employed routinely and are safely used in clinical practice. Contrary to more invasive techniques to study tissue microcirculation, DCE imaging techniques can be easily repeated in the same patient allowing for monitoring of tumor response during therapy and assessment of specific changes in the tumor/tumor microenvironment in response to cytotoxic therapy. The combination of ease of implementation, availability and repeatability ultimately makes this technique feasible in community oncology centers where it can become available to large patient populations.

11.3.1 Techniques

DCE MRI techniques describe the imaging acquisition and processing methods used to evaluate the passage of a contrast agent from successively acquired MR images after controlled injection. Signal enhancement can be demonstrated on T1-weighted imaging 2D or 3D gradient echo sequences such as fast low angle shot (FLASH), snapshot-FLASH, and/or spoiled gradient recalled at steady state (SPGR) techniques (Hoehn-Berlage et al. 1992; Nagele et al. 1993; Evelhoch 1999). 3D volumetric methods can provide better slice coverage (typically thinner slices and a greater number of slices) with higher signal-to-noise ratio at the expense of temporal efficiency. Alternatively, spin echo techniques such as T1-fast spin echo (T1-FSE) can be used to evaluate dynamic contrast-related signal changes (Mayr et al. 1999). These spin echo techniques have reduced sensitivity to local field variations and may be more desirable in the pelvic region where large susceptibility differences may exist; however, gradient echo techniques may be more amenable to volumetric imaging.

The delivery methods of contrast agents, such as gadopentetate dimeglumine (Gd-dimeglumine), can be classified into two major categories. One group of methods uses a rapid bolus injection (typically less than 10 s), while the other class of methods employs a prolonged injection (in the order of a half a minute). A rapid bolus can aid uniform blood and contrast agent mixing and additionally can reduce the impact of transit delays from the injection. The second type of contrast bolus injection, which uses a prolonged injection time, enables elongated sampling during the uptake phase of the contrast agent that can improve estimation and reduce sensitivity to variations in the input function. While the exact mechanism of the degree of enhancement is yet to be fully determined, resultant signal changes are thought to reflect blood flow, extraction factors, and/or equilibrium distribution of contrast agent between tissue and blood. The optimal type of bolus will depend on the parameters that the investigators are trying to extract, the type of cancer, and its micro-environment (pathophysiology).

The tracer models and their associated parametric models can be further classified into three types, depending upon whether observation of the effect is sought in the early (uptake) phase (Type I), in the late phase (Type II), or during the entire curve of the contrast passage (Type III) (Tofts et al. 1999; Yuh 1999). ***Type I*** imaging, commonly known as the "first-pass method," relies on measurements that attempt to evaluate the early part of the dynamic contrast study. It is generally accepted that the signal intensity changes seen in the early phase of contrast injection represent the concentration of contrast agent in both the intravascular and extravascular/interstitial space, and is strongly influenced by the arterial phase of the circulation.

Type II imaging ("equilibrium method") seeks to maximize the evaluation of the later phase where extraction factors and compartmental equilibrium conditions play a greater role. In the equilibrium method (Tofts et al. 1999; Knopp et al. 2001), the regenerated signal–time curve will rely more heavily on the local microvessel density, regional blood flow, microvessel permeability of the contrast agent, and size and physiochemical nature of the extracellular space accessible for Gd-based contrast agents.

Type III perfusion studies can require in the order of 10–30 min to satisfactorily evaluate the effects; this requirement may make type III studies more appropriate for animal studies or highly motivated human subjects who can better tolerate the longer imaging conditions. In addition, if recirculation of the contrast agent is not carefully accounted for, it can confound accurate assessment in the equilibrium phase. Nevertheless, some parameters (such as peak signal of the DCE time intensity curve) will exhibit features that reflect to first-order the concentration of contrast agent in both the intravascular and extravascular interstitial space and can be easily estimated.

The basis for interpretation of the equilibrium methods (Type II) rests on the pharmacokinetic open two-compartment model described by Brix (Brix et al. 1991). In this model, highly permeable microvessels allow Gd-based agents to enter the interstitial space. Two parameters are calculated from the time-signal intensity curve: peak enhancement or amplitude (A), which reflects the extracellular volume, and the exchange rate constant (k_{21}) which designates the rate of exchange of contrast between the plasma compartment and interstitial compartment. The later phase is thought to hold critical information pertaining to tumor vascularity and thus potentially has a significant impact on tumor oxygenation (Mayr et al. 1999; Brix et al. 1991; Tofts and Kermode 1991).

11.3.2 Imaging Protocols

In the early clinical studies conducted at the University of Iowa, the "first-pass method" (Type I model) was employed. Pre-contrast MR examination used 5-mm sagittal T2-weighted images and 10-mm axial T1-weighted images for optimal localization and delineation of the tumor. Using the findings of the pre-contrast study, a sagittal slice with 1-cm thickness through the epicenter of the tumor mass was delineated for the dynamic studies. DCE imaging was obtained using T1-weighted fast spin-echo sequences at 3-s intervals over a total of 120 s. Precisely 30 s after the initiation of image acquisition, a bolus of 0.1 mmol/kg of gadopentetate dimeglumine was injected at a rate of 9 ml/s using an MR-compatible power injector. More recently, volumetric gradient echo acquisitions were used to cover the entire tumor region and allow higher spatial coverage at the expense of temporal resolution.

For the quantitative analysis, a region of interest (ROI) was determined from the pre-contrast tumor delineation image. From the time/signal-intensity curve, two parameters, relative signal intensity (RSI) and slope (M), were generated to quantify the degree of dynamic tumor enhancement. The RSI is the mean intensity change at the plateau phase of the signal intensity averaged over the tumor ROI. The RSI has been used as a parameter for tissue microvascularity, and has shown correlation with radiation therapy outcome in cervical cancer and appears to be the most useful predictive parameter.

The slope M consists of the incremental rate of change in the time-intensity curve from the baseline level to the peak of contrast enhancement after the arrival of the contrast bolus. The slope has also been analyzed for cervical cancer (Mayr et al. 1996b; Cooper et al. 2000); however, is thought to be more dependent on local hemodynamics and other physiologic factors that would result in lesser reliability and reproducibility (Mayr et al. 1996b, 1999). The slope also shows a less reliable correlation with in vivo parameters of tumor oxygenation (Cooper et al. 2000).

In the studies conducted at the German Cancer Research Center/University of Heidelberg, the "equilibrium method" (Type II, two-compartment model) was employed. Following tumor delineation with pre-contrast imaging, the MR examination, and the dynamic MRI protocol includes a sequence with a nearly linear relation between the contrast concentration and signal intensity (Knopp et al. 2001; Hawighorst et al. 1997). In the study by Hawighorst et al. (1997) an ultra fast T1-weighted saturation-recovery turbo fast low-angle shot (SRTF) with temporal resolution time of 1.4 s per section or a turbo FLASH sequence with a resolution of 1.3 s is used (Hawighorst et al. 1996). Gadolinium contrast is administered at a dose of 0.1 mmol Gd-DTPA/kg a slower rate than for the first-pass method, typically over a period of 60 s using a controlled short constant rate infusion with a variable-speed infusion pump. At least three image acquisitions are performed during the contrast infusion, and at least five after completion of the infusion to consistently assess contrast wash-out (Knopp et al. 2001).

For the pharmacokinetic analysis, the maximal amplitude of enhancement (A) and of the tissue exchange rate constant (k21) is computed using the two-compartment pharmacokinetic model (Knopp et al. 1995, 2001; Hoffmann et al. 1995; Brix et al. 1991; Tofts et al. 1999; Hawighorst et al. 1997). The lag-time between injection of contrast material and the individual arrival at the tumor is estimated from the signal-time course measured in the common iliac arteries to generate an arterial input function. Quantitative data analysis is performed at a high in-plane resolution (0.9–1.4×0.9–1.4 mm) of a dynamic time course by a nonlinear least-squares fitting through an automated processing of the measured signal intensity-time curves in each pixel (Hawighorst et al. 1997; Knopp et al. 2001). Calculated MR images are color-coded to allow identification and graphic display of intra-tumoral variation of the microcirculation parameters (amplitude A and exchange rate constant k_{21}). These parameter images can be overlaid on high-resolution 2- or 3-D anatomic images resulting in parameterized color-coded functional tumor maps. This tumor mapping allows comparisons with histologic parameters in whole section mount specimens for histo-morphometric correlation (Hawighorst et al. 1997).

11.4 Clinical Application of DCE MRI in Cervical Cancer

A great variety of microcirculation imaging and imaging analysis methods (types I–III; first pass, equilibrium methods), as discussed in the previous sections (Sects. 11.3.1 and 11.3.2), of patient outcome assessment, survival, tumor control, tumor response), and of study design (prospective, retrospective) have been employed to assess the clinical role of microcirculation imaging in cervical carcinoma (Gong et al. 1999; Hawighorst et al. 1997; Mayr et al. 1996b, 1998, 1999; Postema et al. 1999; Yamashita et al. 2000).

11.4.1 Tumor Microcirculation Studies

In 1996 Mayr and colleagues reported on a prospective study investigating temporal changes in the tumor perfusion pattern before, during, and after radiotherapy using fast MRI techniques as a predictor of outcome of patients with advanced cervical cancer (Mayr et al. 1996b). This study included 17 patients with stages IB2-IVA cervical carcinoma, who were treated with primary radiation therapy and imaged before, during, and after the course of radiation therapy. Four MR studies were obtained serially: pre-therapy, early therapy (after a radiation dose of 20–22 Gy in 2–2.5 weeks), mid-therapy (after 40–45 Gy, 4–5 weeks) and post-therapy (4–6 weeks after completion of the entire course of radiation therapy). Following standard pre-contrast imaging, the dynamic contrast study using the first-pass method was employed using a 1-cm parasagittal slice (based on the precontrast MR study) for maximal sampling of the tumor mass. This provides a sampling of approximately 30% of the entire tumor (Mayr et al. 1998).

The incidence of pelvic tumor recurrence correlated with the mean enhancement (RSI) over the tumor region in the pre-therapy dynamic MRI ($p=0.05$), and particularly in the MR performed early in the course of radiation therapy (at 20–25 Gy at 2 weeks of radiation; $p=0.002$). Using an RSI of 2.8 as a critical threshold level, tumors with RSI ≥ 2.8 in the early-therapy study had a pelvic recurrence rate of 0% versus 78% for those patients with an RSI value <2.8 ($p=0.002$). Tumor recurrence was less common in patients with an RSI that increased during the early therapy study when compared to the pretreatment study (25% vs. 80%; $p=0.101$). An increase in enhancement particularly to or beyond an RSI value of 2.8 was a favorable prognostic factor for local control.

Dynamic enhancement pattern later in the course of radiation therapy or after treatment completion was less useful. In the mid-therapy study (after 40–45 Gy in 4–5 weeks) the correlation between degree of enhancement and local recurrence remained valid, but less significant than that of the early therapy study ($p=0.01$). In the post-therapy follow-up perfusion studies (4–6 weeks after completion of all therapy), there was no significant association between the degree of enhancement and local failure.

The slope of the time-intensity curve obtained in the pre-therapy study also correlated with local failure. Slope values ranged from 1.61–7.81/s (mean = 3.44/s). A steeper slope ($\geq$3/s) in the pre-therapy, early-therapy, and mid-therapy studies was associated with better treatment outcome (lower incidence of tumor recurrence). Other investigators have also found a correlation with slope and tumor control. However, there is concern that the slope is influenced by variable physiologic factors that may cause it to be less reliable than the plateau phase data.

Because traditionally response assessment in cervical cancer is assessed by clinical palpation of the tumor (pelvic examination), the results of clinical examinations performed in the early-therapy phase (2–2.5 weeks after the initiation of therapy) were compared with the results of the pre-therapy and early therapy perfusion imaging (Mayr et al. 1996b). Indicators of tumor response by dynamic enhancement imaging (RSI $\geq$2.8) were more predictive of local control than clinical palpation findings.

Others have found variable correlation of microcirculation parameters with outcome (Table 11.1).

Gong et al. (1999) reported on the use of contrast dynamic MRI of cervical carcinoma during radiation therapy for predicting tumor regression rate. This study involved seven patients with carcinoma of the cervix. All patients underwent external beam irradiation to the pelvis ranging from 40–55 Gy (mean 47 Gy). A baseline MRI was obtained at the beginning of radiation therapy. DCE MRI scans were obtained before the start of radiation therapy and after the first 2 weeks of radiotherapy. Using a region of interest and signal-to-noise ratio method, tumor enhancement was obtained on the dynamic images. Serial tumor volumes were correlated with tumor regression rate based on weekly MRIs using the Cavalieri method. Dynamic enhancement parameters prior to and after the course of radiation therapy did not show any correlation. However, in the MRI in the early phase of radiation therapy (at 2 weeks), there was a statistically significant correlation between the change in mean enhancement (p=0.006) and peak enhancement (p=0.004) and tumor regression rate, similar to the study by Mayr et al. (1996b). There was no correlation between initial tumor volume and tumor enhancement either pre-therapy or after 2 weeks of radiation therapy. There was no correlation with the change in enhancement between these time points (p>0.05). Tumor enhancement did not correlate with either the relative or the absolute change in tumor volume after the first 2 weeks of radiotherapy (p>0.05) indicating that tumor volume criteria alone may not be as sensitive to predict outcome as microcirculation characteristics. Tumor volumes decreased exponentially with time with regression rates ranging from 2.0% to 15.2% per day (Knopp et al. 2001).

Yamashita et al. (2000) investigated pre-therapy microcirculation parameters in 36 patients undergoing radiation therapy for cervical cancer. Imaging was performed according to the equilibrium method using a turbo spin echo technique with five acquisitions and a contrast injection rate of 1 cc/s. Imaging was started at the end of the injection. The dynamic enhancement imaging was evaluated qualitatively with respect to overall intensity and homogeneity of enhancement and quantitatively by computing the capillary permeability from the signal-intensity curve. Patients were followed for outcome, although the follow-up time was relatively short.

Patients with homogenously high enhancement (i.e., intense enhancement over more than 70% of the imaged tumor area) had less tumor recurrence after radiation therapy than those with poor or irregular ring-like enhancement (p<0.1). Patients with tumors showing decreased permeability tended to have a higher risk of recurrence. These early results again indicate that tumors with high dynamic enhancement – although studied with different technique and parameters – tend to have better response to and outcome after radiation therapy than those with low enhancement. These findings are in keeping with earlier observations (Mayr et al. 1996b, 1998) and further support the claim that dynamic MRI can assess parameters related to tumor blood supply, perfusion and oxygenation clinically in cervical cancer patients undergoing cytotoxic therapy.

11.4.2 Combination of Microcirculation with Other Parameters

The dynamic enhancement parameters have been combined with other well-established prognostic factors to evaluate them for any additive effect in predicting treatment outcome. Among these, tumor size has long been recognized as a highly significant factor predicting outcome. Imaging-based 3-D assessment of tumor volume has greatly increased the sensitivity of tumor size in predicting radiation therapy outcome (Mayr et al. 1998). The addition of dynamic enhancement parameters to tumor volumetry has been shown to complement the morphologic/anatomic parameter. Both tumor volumetry and dynamic enhancement pattern analysis can easily be obtained in a single MR examination. Outcome correlation data investigating the combination of both analyses have shown that the addition of a dynamic enhancement pattern substantially improved the prediction of tumor recurrence in patients with intermediate-sized tumors (40–99 cm^3) and maybe also useful in patients with large-size tumors. This combined analysis improved the capacity of differentiating between high-risk patients with an 80% recurrence rate (p=0.010) and low-risk patients with a 0% recurrence rate within the intermediate-size tumor groups (40–99 cm^3) from an overall recurrence rate of 33%. In patients with large-size tumors the combined analyses enable classification of patients into high risk (75% recurrence) and low-risk (0% recurrence) groups. The group of patients with the highest risk of local recurrence could thus be identified as those with intermediate- and large-sized tumor with a low dynamic enhancement pattern.

A recent study by Loncaster et al. et al. (2002) in 50 patients with cervical cancer also demonstrated that the combination of microcirculation parameters with tumor volume analysis correlated with outcome. Patients, who were treated with standard external beam irradiation and brachytherapy, underwent DCE MRI. MR was performed to obtain tumor stage, volume, and nodal status and DCE MR. Analysis of the data from the whole tumor volume was performed using a pharmacokinetic model describing the enhancement in terms of two parameters, amplitude and rate (Hoffmann et al. 1995). Patients with poorly enhancing large tumors had a significantly worse disease-specific survival than those with small well-enhanced tumors. Disease-specific survival was significantly improved (p=0.024) in patients with values greater than the median of amplitude.

Improved metastasis-free survival and local control were associated with a high value for amplitude, but were not statistically significant. By combining amplitude and tumor volume, large differences in treatment outcome probabilities were noted. Patients with large poorly enhancing tumors had a survival rate of 55% in comparison to a survival rate of 92% for patients with small well-enhancing tumors (p=0.0054).

11.4.3 Pixel Analysis of MR Microcirculation Imaging

In several studies on investigating microcirculation parameters in cervical cancer (Gong et al. 1999; Mayr et al. 1996b; Cooper et al. 2000; Yamashita et al. 2000), the parameter of signal intensity was averaged over the imaged tumor region resulting in a mean dynamic enhancement pattern, not accounting for regional variation in tumor microcirculation. However, the heterogeneity of tumor blood supply and oxygenation is a well-described phenomenon in cancer. Using the mean enhancement value does not account for the variable degrees of dynamic enhancement throughout the tumor. It is likely that hypoxic cells in poorly perfused tumor regions are responsible for radiation therapy failure and these regions may present as low dynamic enhancement regions. The ability to assess the variability of enhancement patterns thus has potential value in predicting radiation therapy response and treatment outcome. Pixel-by-pixel analysis of the enhancement pattern provides an opportunity to assess tumor heterogeneity. The correlation of microcirculatory imaging and outcome using pixel analysis was studied by Mayr et al. (2000). The parameters characterizing the pixel-histogram distribution of the dynamic enhancement pattern were correlated with tumor control in patients with cervical cancer treated with radiation therapy. Nineteen patients with advanced cervical cancer (stages IB2-IVA/recurrent) underwent dynamic MRI at 2 weeks of fractionated radiation therapy.

For the pixel-by-pixel analysis the tumor region was drawn on the contrast-enhanced images with reference to the tumor location defined by the precontrast T2-weighted images. For each pixel in the tumor volume, a first pass time-signal intensity curve was generated. The ratio of signal intensity in the plateau phase to signal intensity in the baseline defined the relative signal intensity (RSI) of the dynamic contrast enhancement of each pixel in the tumor region (Fig. 11.2). An RSI pixel histogram was formed by the tabulation of the RSI values. The pixel histogram was analyzed with respect to number of pixels, mean RSI value, median RSI value, and RSI percentiles in increments of 10% to quantitate the degree and proportion of lowest dynamic enhancement within the tumor region. The RSI value, below which a given percentage of pixels in the histogram fell, defined the RSI percentile. Each parameter was examined by correlation with tumor recurrence versus control based on cancer follow-up data.

Pixel distribution histograms displayed a wide variation in the dynamic contrast enhancement patterns within the tumor suggesting extensive tumor heterogeneity in cervical cancer (Fig. 11.2). Within individual tumors, pixel values ranged from an RSI of 1.63–4.75 (mean 2.83). RSI pixel values widely overlapped between different patient outcome groups, those with subsequent tumor control and recurrence. However, statistical analysis of the individual imaging parameters characterizing the RSI pixel histogram showed significant differences between patient groups with tumor recurrence and those with tumor control. The most significant imaging predictor of tumor recurrence was the 10th percentile RSI value (p=0.00001), i.e., the RSI value below which fall the lowest 10% of the tumor pixels. The tumor recurrence rate was 89% among patients with a 10th percentile RSI of $<$2.5 versus 0% for those with a 10th percentile RSI of $\geq$2.5 (p=0.0004). All tumor recurrences could be predicted if a 10th percentile RSI of 2.0 was used as a threshold value. The tumor recurrence was 100% for patients with a 10th percentile RSI of less than 2.0 versus 18% for those patients with a 10th percentile RSI of greater than 2.0 ($p = 0.0022$). Mean ($p = 0.0001$) and median ($p = 0.0001$) RSI were the next best imaging parameters to predict tumor control. The pixel number (N) reflecting tumor was of borderline significance as an individual parameter ($p = 0.05$). Larger pixel numbers were associated with an increase recurrence rate (58% vs. 0% in $N > 350$ vs. $N < 350$; $p = 0.068$).

The 10th percentile RSI (Fig. 11.2) and the pixel number were the two best independent imaging parameters. Their combination mutually enhanced predictive power. Using both parameters, all patients could be separated by the discrimination line into those with subsequent tumor recurrence vs. those with long-term tumor control.

11.5 Correlation of Dynamic Enhancement with Histological Parameters of Angiogenesis

The studies previously described largely focused on the correlation of imaging parameters with patient outcome measures, e.g., tumor regression, tumor control, or survival. Because these tumors are not resected, there is no correlation with histopathological parameters. The correlation of imaging with histologic parameters has been investigated at the German Cancer Research Center in Heidelberg, Germany. Hawighorst et al. (1998) have examined the association of dynamic MRI parameters using pharmacokinetic analysis with histomorphological markers of tumor angiogenesis including microvessel density (MVD) and vascular endothelial growth factor (VEGF), and have developed a histomorphological and dynamic MRI approach to correlate those data with therapy outcome in patients with cervical The study, undertaken by 37 patients with stages IB-IVA cervical cancer, consisted of dynamic MRI prior to supra-radical surgery including radical hysterectomy (n=27) and pelvic exenteration (n=10) with pelvic lymph node dissection. Quantitative DCE MRI examination was performed prospectively prior to surgery using the equilibrium method described above. Histopathological and MRI correlations were performed by matching whole-mounted tumor specimens with the corresponding MRI-derived maps of MRI sections in identical medioaxial or mediosagittal plane to achieve maximal co-localization of the imaging and the histology. VEGF expression and MVD were correlated with the MRI parameters amplitude (A) and exchange rate constant (k_{21}) (Fig. 11.3), as well as with patient survival.

There was a statistically significant association between MVD and the MRI parameters A ($p<0.001$) and k_{21} ($p<0.05$). There was no association between VEGF expression, amplitude A and k_{21}. Paradoxically, areas of high VEGF expression were commonly associated with a low MVD, and those areas with a low VEGF expression were associated with a high MVD. This inverse association was statistically significant ($p<0.02$). Neither MVD nor VEGF expression showed predictive power (p=0.3 and 0.4, respectively) for survival. Only the functional imaging derived pharmacokinetic parameter k_{21} was significant in predicting outcome. A high median k_{21} predicted poor survival ($p<0.05$).

This experience demonstrated that imaging histologic correlations may not show predictable findings. This is likely related to the difference in the parameters measured: the "static" morphologic vs. the dynamic functional imaging parameters. It may also be related to sampling errors despite meticulous histologic processing.

It was concluded that the inverse correlation between angiogenesis and DCE MRI parameters does not negate an association between them, but may represent the upregulation of overall angiogenic activity of tumor cells according to its metabolic demands (Neeman et al. 1997; Shweiki et al. 1995; Stein et al. 1995). Tumor tissue with insufficient nutrient supply (e.g. poor perfusion, hypoxia) may elicit compensatory angiogenesis to match supply to demand and that VEGF plays a key role as a mediator of this feedback response (Shweiki et al. 1995), explaining the inverse relationship between MVD and VEGF expression. In addition, it is possible that a "static" view of VEGF expression on a histological specimen alone

a

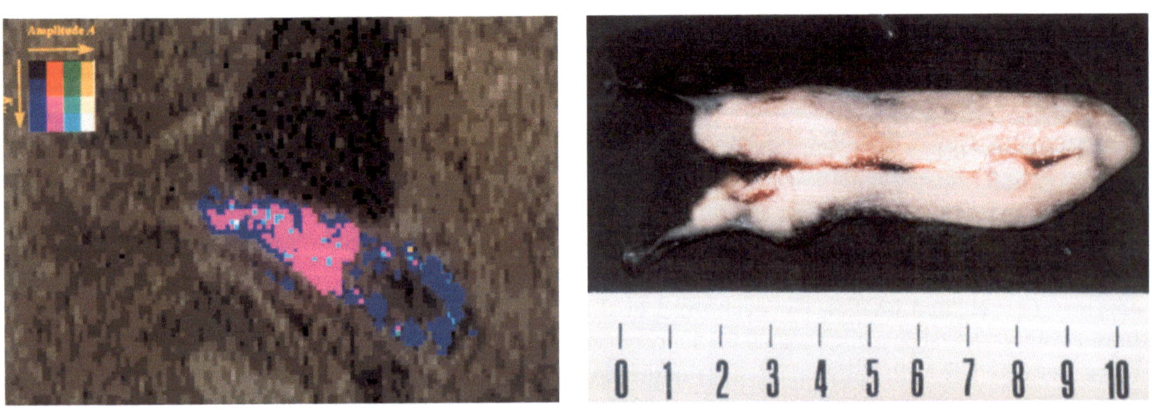

b

Fig. 11.3a,b. Comparison of histo-morphometric parameters of angiogenesis and microcirculation imaging. Stage IIIB cervical cancer. Pharmacokinetic MR imaging (**a**) of a sagittal section through the tumor region demonstrates a homogenously colored map encoding areas with fast (k_{21}=23.1 min^{-1}) but low amplitude (A=0.7). This section corresponds to the whole mount specimen of the uterus/tumor shown in (**b**). Histomorphometric analysis showed homogenous distribution of microvessel density and VGEF expression (not shown). [Reprinted with permission from Mayr et al. (1999)]

may not represent the "dynamic" continuously modulated microvascular function that can be assessed by in vivo imaging. Functional microvessel properties may present at a particular time and location, but the VEGF expression may have already been down regulated. Microcirculation imaging thus provides insight into a highly dynamic process of micro-environment tumor.

Yamashita et al. (2000) also correlated pre-therapy microcirculation parameters in 23 patients imaged prior to surgery and correlated dynamic imaging findings spatially with morphologic factors of tumor angiogenesis. The authors used an imaging protocol according to the equilibrium method. The dynamic enhancement imaging was evaluated qualitatively with respect to overall intensity and homogeneity of enhancement and quantitatively by computing the capillary permeability data from the signal-intensity curve. Tissue planes from the surgical specimens that were consistent with the imaging planes were studied for histopathologic findings including cell and microvessel density.

Areas of homogenously or peripheral ring-like high enhancement were composed predominately of cancer cells surrounded by connective tissue with numerous capillaries. Conversely, tumors or tumor regions with predominately poor enhancement consisted largely of fibrous tissue with scattered tumor cells and only scant numbers of capillaries.

These findings again overall suggest a possible correlation of histologic parameters related to tumor blood supply and dynamic MRI findings. However, further study in this area is needed to confirm these early results. Particularly the correlation of DCE MR parameters with gene expression of angiogenesis modulators is an area of ongoing investigation (Knopp et al. 2001; Costouros et al. 2002).

11.6 Correction of Dynamic Enhancement with In Vivo Measurement of Oxygenation

At present the measurement of oxygen tension in tissue by needle oximetry (pO_2-histograph) is regarded as the "gold standard" method for measuring tumor oxygenation and has been extensively studied in cervical cancer (Dunst et al. 1999; Höckel et al. 1996a; Lyng et al. 1997; Fyles et al. 2002; Knockle et al. 1999; Rofstad et al. 2000). Poor oxygenation has been shown to correlate with poor outcome (Dunst et al. 1999; Fyles et al. 2002; Höckel et al. 1996b; Knockle et al. 1999; Rofstad et al. 2000). The correlation of in vivo oximetry measurement with the imaging-derived dynamic enhancement parameters is of great interest because it may validate the non-invasive imaging observation with pathophysiological data. While needle oximetry provides a "true" in vivo measurement of tumor oxygenation, it is hampered by the limitation in tumor sampling to small microscopic needle tracts. The non-invasive imaging-based method can assess the *entire* tumor region; however, the information gained is not a *direct* measurement of tumor oxygenation, but only an indirect parameter relating to tissue microcirculation and thus to oxygenation. Both methods – in-vivo needle oximetry and microcirculation imaging – may thus complement each other.

Cooper et al. (2000) investigated the relationship between tumor oxygenation and DCE MRI parameters. This study included 30 patients with locally advanced cervix cancer, who underwent tumor oxygenation measurements with the Eppendorf pO_2 histograph system and DCE MRI prior to radiation therapy. The imaging protocol was consistent with the fist-pass method. From the time-signal intensity curve, the maximum enhancement over baseline and the steepest incremental rate of enhancement were computed.

Well-oxygenated tumors showed a higher maximum enhancement than poorly-oxygenated tumors. A significant correlation was found between maximum enhancement and the in vivo oximetry parameters, median pO_2 ($p<0.001$) and HP5 (the percentage of pO_2 reading of <5 mm Hg) ($p=0.037$). There was no correlation between the rate of enhancement with the oximetry parameters.

There was a significant relationship between pre-treatment tumor size and maximum enhancement ($p<0.001$). As expected, large tumors had a lower maximum enhancement than small tumors; $p=0.04$). Again, rate of enhancement showed no correlation. There was an association between maximum tumor diameter and median pO_2 ($p=0.04$) but not with HP5.

Data correlating imaging-derived parameters with oximetry measurements is difficult to establish and difficult to interpret. Two vastly different methods are compared. The tissue sampling in in vivo needle oximetry is limited to a small number of needle tracts. These can only sample a small fraction of tumors, which are known to be vastly heterogeneous (Fig. 11.2) (Mayr et al. 2000). Imaging samples a much larger proportion of the tumor. Considering these limitations, the findings of the study by Cooper et al. (2000) are encouraging as they suggest that a correlation exists between microcirculation imaging and tumor oxygenation in a clinical patient population of cervical cancer patients. A recent update of the series and outcome correlation has suggested that, although oximetry data correlated with pharmacokinetic parameters, only the pharmacokinetic parameters, not the oximetry data were of prognostic value for treatment outcome (Loncaster et al. 2002).

11.7 Discussion

New dynamic MRI methods provide a functional assessment of tumor microcirculation and may provide information on tumor oxygenation, which is known to greatly influence the success of cytotoxic therapy.

Tumor oxygenation is dependent on blood flow, which will impact radiation response. Poor blood supply to the tumor would indicate decreased oxygenation and reduce the response to radiation therapy. DCE MRI allows estimation of tissue microcirculation, blood volume, and permeability of vessels. Imaging parameters derived from the DCE MR correlate with tumor blood flow and microvessel density. This allows the correlation of MR parameters and tumor oxygenation.

Evidence for such a correlation is now emerging from studies correlating MRI-based microcirculation parameters with in vivo tumor oxygenation measurement (Cooper et al. 2000) and with histologic markers of tumor angiogenesis (Hawighorst et al. 1997; Cooper et al. 2000). The study by Cooper et al. (2000) showed a correlation between tumor oxygenation and degree of dynamic enhancement. Hawighorst et al. (1997) showed a significant association between microvessel density and the MR parameters of amplitude (p=0.001) and exchange rate constant (p<0.05). At the same time, outcome studies have shown that these MR-based parameters indeed predict therapy outcome (Mayr et al. 1996b, 2000; Gong et al. 1999; Yamashita et al. 2000). These observations begin to validate the long-advocated radiobiological concept of inter-relation between poor perfusion, hypoxia tumor bulk, and therapy failure in cervical cancer (Perez et al. 1992; Thoms et al. 1992). With the advanced imaging techniques we now have the opportunity to use this information in the clinical setting to adjust therapy.

The expected inverse relationship between tumor bulk (size) and tumor oxygenation/blood supply is also beginning to emerge from quantitative and dynamic enhanced MRI studies. The results of these studies appear to reflect what was postulated by radiobiologists for many decades: the association between tumor bulk and development of hypoxia as the tumor outgrows its vascular supply. Thus large tumors would be expected to be not as well oxygenated as small tumors (Cooper et al. 2000, Fyles et al. 2002, Mayr et al. 1998). In the study by Cooper et al. (2000) large tumor size clearly correlated with low dynamic enhancement *and* hypoxia. Mayr et al. (1998) also found that in very large-sized tumors (>100 cm^3), dynamic contrast enhancement was generally low and in very small tumors (<40 cm^3) generally high. These findings correlated with unfavorable vs. favorable radiation therapy outcome.

However, at the same time, in intermediate-size tumors (40–99 cm^3), which constitute the majority of tumors, microcirculation imaging provides additional predictive information that can classify patients into high-risk vs. low-risk groups for therapy failure. Mayr et al. (1998) found that microcirculation was highly variable in these tumors and predicted outcome independent of size. This methodology is particularly powerful if tumor heterogeneity is assessed using pixel-by-pixel analysis of the dynamic enhancement pattern to assess the lowest-enhancement regions within the tumor, which likely contain radio-resistant hypoxic cells. Pixel histogram analysis of tumor microcirculation for tumors has shown that the distribution of excess low microcirculation regions within the tumor correlated with local recurrence (Fig. 11.2) (Mayr et al. 2000).

One of the most interesting aspects is to investigate the changes occurring in tumor microcirculation during cytotoxic or anti-angiogenic therapy. Serial studies of tumor microcirculation in cervical cancer have shown that an increase in dynamic enhancement during radiation therapy predicts favorable therapy outcome (Mayr et al. 1996b; Gong et al. 1999). Mayr et al. (1996b) showed improved tumor control in patients with an increase in dynamic enhancement early in the course of radiation therapy (20–22 Gy). Similarly Gong et al. (1999) showed that an increase in dynamic enhancement during the first 2 weeks of radiation therapy significantly correlated with tumor regression rate in cervical carcinoma. The results of these serial imaging–outcome studies are paralleled by serial in vivo tumor oxygenation measurements during radiation therapy (Dunst et al. 1999). Dunst et al. (1999) showed that changes in oxygenation of cervical cancers do occur during radiation therapy. Among patients with initially hypoxic tumors and an increase in median pO_2 above 10 mmHg early in the course of radiation therapy (at 2 weeks, 19.8 Gy), all had complete remission. These studies support the radiobiological concept that an increase in signal enhancement in the early phase of radiation therapy may indicate improvement of tumor blood supply with resulting re-oxygenation of previously hypoxic tumor cells.

The elucidation of the nature of hypoxia in cervical cancer is ongoing. Early prospective outcome studies suggest the clinical validity the correlations of poor perfusion microcirculation and poor outcome – providing a non-invasive predictive assay for patients with cervical cancer that can predict therapy response early enough in the course of treatment to implement changes in the therapy regimen that can ultimately improve outcome for women with cervical cancer. Hypoxic tumors might need more aggressive treatment. Further studies are required to refine these parameters, investigate the pathophysiology related to these parameters, and to refine their application, so they can be most efficiently implemented in the therapy of women with cervical cancer.

11.8 Conclusion

DCE MRI and its quantitative analysis appears to have value in providing essential information that reflects the underlying pathophysiology in cervical cancer (Cooper et al. 2000; Mayr et al. 1999; Knopp et al. 1995), and therefore improves patient management and treatment outcome. The critical question remains: how will this methodology realistically impact medical care of this specific patient population suffering with advanced cervical cancer? The answer to this question depends upon at least two factors: (1) the availability of DCE MRI to the general patient population, and (2) the realistic approach to deriving meaningful information from the rather massive DCE MR data sets in a busy clinical setting. The DCE MR techniques have become more readily available in the industrial countries. Unfortunately, the prevalence of cervical cancer remains highest in Third World countries, where MRI is not available to most patients. This factor (#1) will be a major limitation of MR to the overall cervical cancer patient population worldwide.

In the situation where DCE MRI is available, a simple approach becomes the key factor (#2) to attain a realistic impact on patient management and treatment outcome. For now, our simple but practical approach (one-compartmental analysis) appears to be effective in the management of advanced cervical cancer although the optimal approach or approaches remain to be investigated and validated in future studies. Our approach consists of a simple estimation of the amplitude (SI) during the plateau phase of the dynamic contrast curve, which is readily available on a standard MR scanner. Such an approach does not require the busy clinician or scientist to be present during the MR study and extensive high level computation and analysis. For now, this approach can be realistically implemented and be potentially beneficial for the treatment outcome.

References

Brady LW, Plenk HP, Hanley JA et al. (1981) Hyperbaric oxygen therapy for carcinoma of the cervix – Stages IIB IIIA IIIB and IVA results of a randomized study by the Radiation Therapy Oncology Group. Int J Radiat Oncol Biol Phys 7: 991–998

Brix G, Semmler W, Port R et al. (1991) Pharmacokinetic parameters in CNS Gd-DTPA enhanced MR imaging. J Comput Assist Tomogr 15:621–727

Bush R, Jenkin R, Allt W (1978) Definitive evidence for hypoxic cells influencing cure in cancer therapy. Br J Cancer 37[Suppl 3]:302–306

Chaplin D, Olive P, Durand R (1987) Intermittent blood flow in a murine tumor: radiobiological effects. Cancer Res 47:597–601

Cooper RA, Carrington BM, Loncaster JA et al. (2000) Tumor oxygenation levels correlate with dynamic contrast-enhanced magnetic resonance imaging parameters in carcinoma of the cervix. Radiother Oncol 57:53–59

Costouros NG, Lorang D, Zhang Y, Miller MS, Diehn FE, Hewitt SM, Knopp MV, Li KC, Choyke PL, Alexander HR Libutti SK (2002) Microarray gene expression analysis of murine tumor heterogeneity defined by dynamic contrast-enhanced MRI. Mol Imaging 1:301–308

Dische S, Anderson P, Sealy R et al. (1983) Carcinoma of the cervix-anemia, radiotherapy and hyperbaric oxygen. Br J Radiol 56:251–255

Dunst J, Hansgen G, Laytenschlager C et al. (1999) Oxygenation of cervical cancers during radiotherapy and radiotherapy + cis-retinoic acid/interferon. Int J Radiat Oncol Biol Phys 43:367–373

Evans J, Bergso P (1965) The influence of animia and results of radiotherapy in carcinoma of the cervix. Radiology 48:709–717

Evelhoch J (1999) Key factors in the acquisition of contrast kinetic data for oncology. J Magn Reson Imaging 10:254–259

Fyles WF, Milosevic M, Pintilie M et al. (2000) Anemia, hypoxia

and transfusion in patients with cervix cancer: a review. Radiother Oncol 57:13–19
Fyles A, Milosevic M, Hedley D et al. (2002) Tumor hypoxia has independent predictor impact only in patients with node-negative cervix cancer. J Clin Oncol 20:680–687
Gatenby RA, Kessler HB, Rosenblum JS et al. (1988) Oxygen distribution in squamous cell carcinoma metastasis and its relationship to outcome of radiation therapy. Int J Radiat Oncol Biol Phys 14:831–838
Giaccia AJ (1996) Hypoxia stress proteins: survival of the fittest. Sem Radiat Oncol 6:46–58
Gong QY, Brunt JNH, Romaniuk CS et al. (1999) Contrast enhanced dynamic MRI of cervical carcinoma during radiotherapy: early prediction of tumour regression rate. Br J Radiol 72:1177–1184
Graeber TG, Osmanian C, Jacks T et al. (1996) Hypoxia-mediated selection of cells with diminished apoptotic potential in solid tumors. Nature 379: 88–91
Grigsby PW, Winter K, Wasserman TH et al. (1999) Irradiation with or without misonidazole for patients with stages IIIB and IVA carcinoma of the cervix: Final results of RTOG 80-05. Int J Radiat Oncol Phys 44:513–519
Hawighorst H, Knapstein P, Weikel W et al. (1996) Cervical carcinoma: comparison of standard and pharmacokinetic MR imaging. Radiology 201:531–539
Hawighorst H, Knapstein P, Weikel W et al. (1997) Angiogenesis of uterine cervical carcinoma: characterization by pharmacokinetic magnetic resonance parameters and histological microvessel density with correlation to lymphatic involvement. Cancer Research. 57:4777–4786
Hawighorst H, Block M, Knopp M et al. (1998) Magnetically labeled water perfusion imaging of the uterine arteries and of normal and malignant cervical tissue: initial experiences. Magn Reson Imaging 16:225–234
Hoehn-Berlage M et al.(1992) T1 snapshot FLASH measurement of rat brain glioma: kinetics of the tumor-enhancing contrast agent manganese (III) tetraphenylporphine sulfonate. Magn Reson Med 27:201–213
Höckel M, Schlenger K, Billur A et al. (1996a) Association between tumor hypoxia and malignant progression in advanced cancer of the uterine cervix. Cancer Res 56:4509–4515
Höckel M, Schlenger K, Mitze M et al. (1996b) Hypoxia and radiation response in human tumors. Sem Radiat Oncol 6:3–9
Höckel M, Vaupel P (2001) Tumor hypoxia: definitions and current clinical, biologic, and molecular aspects. J Natl Cancer Inst 93:266–276
Hoffmann U, Brix G, Knopp MV et al. (1995) Pharmacokinetic mapping of the breast: a new method for dynamic mammography. Magn Res Med 33:506–514
Kallinowski F, Zander R, Hoeckel M et al. (1990) Tumor tissue oxygenation as evaluated by computerized-pO2-histography. Int J Radiat Oncol Biol Phys 19:953–961
Knockle TH, Weitmann HD, Feldmann HJ et al. (1999) Intratumoral pO2-measurements as predictive assay in the treatment of carcinoma of the uterine cervix. Radiother Oncol 53:99–104
Knopp MV, Hoffmann U, Brix G et al. (1995) Schnelle Kontrastmitteldynamik zur Charakterisierung von Tumoren. Radiologe 35:964–972
Knopp MV, Giesel FL, Marcos H et al. (2001) Dynamic contrast-enhanced magnetic resonance imaging in oncology. Top Magn Reson Imaging 12:301–308
Kolstad P (1968) Intercapillary distance, oxygen tension and local recurrence in cervix cancer. Scand J Clin Lab Invest 106[suppl]:145–157
Lartigau E, Randrianarivelo H, Avril MF et al. (1997) Intratumoral oxygen tension in metastatic melanoma. Melanoma Res 7:400–406
Loncaster JA, Carrington BM, Sykes JR et al. (2002) Prediction of radiotherapy outcome using dynamic contrst enhanced MRI of carcinoma of the cervix. Int J Radiat Oncol Biol Phys 54:759–767
Lyng H, Sundfør K, Rofstad EK (1997) Oxygen tension in human tumours measured with polarographic needle electrodes and its relationship to vascular density, necrosis and hypoxia. Radiother Oncol 44:163–169
Maor MH, Gillespie BW, Peters LJ et al. (1988) Neutron therapy in cervical cancer: results of a phase III RTOG study. Int J Radiat Oncol Phys 14:885–891
Maruyama Y, van Nagell JR, Yoneda J et al. (1991) A review of californium-252 neutron brachytherapy for cervical cancer. Cancer 68:1189-1197
Mayr NA, Magnotta VA, Ehrhardt JC et al. (1996a) Usefulness of tumor volumetry by magnetic resonance imaging in assessing response to radiation therapy in carcinoma of the uterine cervix. Int J Radiat Oncol Biol Phys 35:915–924
Mayr NA, Yuh WTC, Magnotta VA et al. (1996b) Tumor perfusion studies using fast magnetic resonance imaging technique in advanced cervical cancer: a new noninvasive predictive assay. Int J Radiat Oncol Biol Phys 36:623–633
Mayr NA, Yuh WTC, Zheng J et al. (1998) Prediction of tumor control in patients with cervical cancer: analysis of combined volume and dynamic enhancement pattern by MR imaging. Am J Roentgenol 170:177–182
Mayr NA, Hawighorst H, Yuh WTC et al. (1999) MR microcirculation assessment in cervical cancer: correlations with histomorphological tumor markers and clinical outcome. J Magn Reson Imaging 10:267–276
Mayr NA, Yuh WTC, Arnholt JC et al. (2000) Pixel analysis of MR perfusion imaging in predicting radiation therapy outcome in cervical cancer. J Magn Reson Imaging 12:1027–1033
Nagele T, Petersen D, Klose U et al. (1993) Dynamic contrast enhancement of intracranial tumors with snapshot-FLASH MR imaging. AJNR 14:89–98
Neeman M, Abramovitch R, Schiffenbauer Y et al. (1997) Regulation of angiogenesis by hypoxic stress: from solid tumors to the ovarian follicle. Int J Exp Pathol 78:57–70
Nordsmark M, Hoyer M, Keller J et al. (1996) The relationship between tumor oxygenation and cell proliferation in human soft tissue sarcomas. Int J Radiat Oncol Biol Phys 35:701–708
Overgaard J, Horsman MR (1996) Modification of hypoxia-induced radioresistance in tumors by the use of oxygen and sensitizers. Semin Radiat Oncol 6:10–21
Palcic B, Skarsgard LD (1984) Reduced oxygen enhancement ratio at low doses of ionizing radiation. Radiat Res 100:328–339
Perez CA, Grisby PW, Nene SM et al. (1992) Effect of tumor size on the prognosis of carcinoma of the uterine cervix treated with irradiation alone. Cancer 69:2796–2806
Postema S, Pattynama P, van Rijswijk C et al. (1999) Cervical carcinoma: can dynamic contrast-enhanced MR imaging help predict tumor aggressiveness? Radiology 210:217–220

Rofstad EK, Sundfør K, Lyng H et al. (2000) Hypoxic-induced treatment failure in advanced squamous cell carcinoma of the uterine cervix is primarily due to hypoxia-induced radiation resistance rather than hypoxia-induced metastasis. Br J Cancer 83:354–359

Rose PG, Bundy BN, Watkins EB et al. (1999) Concurrent cisplatin-based radiotherapy and chemotherapy for locally advanced cervical cancer. N Engl J Med 340:1145–1153

Semenza GL (2000a) Hypoxia, clonal selection, and the role of HIF-1 in tumor progression. Crit Rev Biochem Mol Biol 35:71–103

Semenza Gl (2000 b) HIF-1 mediator of physiological and pathophysiological responses to hypoxia. J Appl Physiol 88:1474–1480

Shweiki D, Itin A, Soffer D et al. (1992) Vascular endothelial growth factor induced by hypoxia may mediate hypoxia-initiated angiogenesis. Nature 359:843–845

Shweiki D, Neeman M, Itin A et al. (1995) Induction of vascular endothelial growth factor expression by hypoxia and by glucose deficiency in multiple spheroids: implications for tumor angiogenesis. Proc Natl Acad Sci USA 92:768–772

Stein I, Neeman M, Shweiki D et al. (1995) Stabilization of vascular endothelial growth factor mRNA by hypoxia and hypoglycemia and coregulation with other ischemia-induced genes. Mol Cell Biol 15:5363–5368

Sundfør K, Lyng H, Rofstad EK (1998) Tumor hypoxia and vascular density as predictors of metastasis in squamous cell carcinoma of the uterine cervix. Br J Cancer 78:822–827

Tannock I (1972) Oxygen diffusion and the distribution of cellular radiosensitivity in tumors. Br J Radiol 52:650–656

Tannock IF (1998) Conventional cancer therapy: promise broken or promise delayed? Lancet 351[Suppl 2]:SII9-16 (reveiw)

Thomlinson RH, Gray LH (1955) The histological structure of some human lung cancers and the possible implications for radiotherapy. Br J Cancer 9:539–549

Thoms WW, Eifel PJ, Smith TL et al. (1992) Bulky endocervical carcinoma: a 23-year experience. Int J Radiat Oncol Biol Phys 23:491–499

Tofts PS, Kermode AG (1991) Measurement of the blood–brain barrier permeability and leakage space using dynamic MR imaging. 1. Fundamental concepts. Magn Reson Med 17:357–367

Tofts PS, Brix G, Buckley DL et al. (1999) Estimating kinetic parameters from dynamic contrast-enhanced T(1)-weighted MRI of a diffusible tracer: standardized quantities and symbols. J Magn Reson Imaging 10:223–232

Urtasun R, Chapman J, Raleigh J et al. (1986) Binding of H-3-misonidazole to solid human tumors as a measure of tumor hypoxia. Int J Radiat Oncol Biol Phys 12:1263–1267

Vermeulen PB, Gasparini G, Toi M et al. (1996) Quantification of angiogenesis in solid human tumors: an international consensus on the methodology and criteria of evaluation. Eur J Cancer 32A:2474–2484

Weidner N, Semple JP, Welch WR (1991) Tumor angiogenesis and metastasis – correlation in invasive breast carcinoma. N Engl J Med 324:1–8

Yamashita Y, Baba T, Baba Y et al. (2000) Dynamic contrast-enhanced MR imaging of uterine cervical cancer: pharmacokinetic analysis with histopathologic correlation and its importance in predicting the outcome of radiation therapy. Radiology 216: 803–809

Yuh WT (1999) An exciting and challenging role for the advanced contrast MR imaging. J Magn Reson Imaging 10:221–222

Zaino RJ, Ward S, Delgado G et al. (1992) Histopathologic predictors of the behavior of surgically treated stage IB squamous cell carcinoma of the cervix. Cancer 69:1750–1758

12 Dynamic Contrast-Enhanced MRI of Prostate Cancer

ANWAR R. PADHANI

CONTENTS

12.1 Introduction and Role of Imaging

Prostate cancer is the most commonly diagnosed cancer in men in the United States of America with an estimated 198,000 new cases in 2001 (GREENLEE et al. 2001). In the United Kingdom and the European Union, prostate cancer is the second most common cancer in men, with 20,000 new cases in the UK in 1997 (CRC 2001) and an estimated 134,000 new cases in the EU in 1996 (EUCAN). Substantial increases in incidence have been reported in recent years around the world, some of which can be attributed to frequent use of transurethral resection of the prostate (TURP) to treat symptoms of obstructive benign prostatic hyperplasia (BPH) and serum prostate-specific antigen (PSA) testing to screen for prostate cancer. While there is some debate on whether there is a real increase in incidence of prostate cancer, what is clear is that the population at risk (old men) continues to increase in size with lengthening of life expectancy. These factors make prostate cancer a large and growing healthcare problem for men.

There are three major areas where imaging techniques may lead to improvements in the management of patients with suspected or proven prostate cancer (THORNBURY et al. 2001). These are detection and localisation of early prostate cancer, identification of men for whom treatment is likely to be curative and early detection of the site of recurrent disease: (1) Many men with a raised PSA level detected at screening do not have an underlying prostate cancer. Approximately 70%-80% of men do not have a prostate cancer diagnosed at the time of the raised PSA test but on long-term follow-up, 38% of men will develop prostate cancer if the presenting PSA lies between 4.1–10 ng/ml (SMITH et al. 1996). (2) Once the diagnosis of cancer is made, it can be difficult to determine which patients will benefit from treatment. The most appropriate treatment for localised prostate cancer remains controversial. Prostate cancer can be an indolent malignancy and yet contributes substantially to cancer mortality. Whilst a number of prognostic factors are recognised, new indices of biological activity are needed to help distinguish between clinically indolent and potentially life-threatening carcinomas (BOSTWICK et al. 2000). (3) Clinical evaluation of suspected recurrent cancer after radical local treatment can be challenging. An elevated PSA level may be the only evidence of treatment failure after local treatment. Determining the site of recurrence is important because men with an isolated local recurrence can benefit from further treatments such as radiotherapy to a prostatectomy resection bed.

The identification of prostate cancer with a view to a targeted biopsy is the major current role for transrectal ultrasound (TRUS) (CLEMENTS 2001). Using TRUS, prostate cancer can be visualised as a hypoechoic lesion in the peripheral gland; however, lesions can appear hyperechoic or isoechoic. It should be noted that hypoechoic lesions in the peripheral gland are not necessarily cancers (41% are cancers overall, 52% if the PSA is raised and 71% if the PSA is raised with a palpable abnormality; LEE et al. 1989). With the increasing use of PSA screening there has been downward stage migration of diagnosed prostate cancer. As a result, it has become increasingly difficult

A. R. PADHANI MRCP, FRCR
Consultant Radiologist and Lead in MRI, Paul Strickland Scanner Centre, Mount Vernon Hospital, Rickmansworth Road, Northwood, Middlesex, HA6 2RN, UK

for TRUS to identify small volume, low-grade tumours (SANDERS and EL-GALLEY 1997). Furthermore, small central gland cancers cannot be distinguished from nodules of BPH. Thus, a normal prostate ultrasound in a man with a raised PSA cannot exclude the presence of malignancy. Colour Doppler TRUS has not been shown to be superior to grey-scale ultrasound for the overall detection of prostate cancer although it may have a role in identifying areas for targeted biopsy (PATEL and RICKARDS 1994; ALEXANDER 1995; NEWMAN et al. 1995; BREE 1997). Power Doppler is considered more sensitive for detecting flow in smaller vessels but early results have been inconclusive with regard to its ability to improve the detection of early prostate cancer (CHO et al. 1998; OKIHARA et al. 2002). Ultrasound contrast agents which have the ability to improve the visibility of small blood vessels are currently being investigated for their potential role in the detection of prostate cancer (FRAUSCHER et al. 2001, 2002). Other indications for TRUS include tumour staging, to guide the placement of brachytherapy seeds and for monitoring ablative treatments such as cryotherapy and high-frequency ultrasound (BEERLAGE et al. 2000). Limitations of TRUS include high operator dependence and poor overall staging accuracy (50%-80%). Seminal vesicle invasion is inadequately assessed with TRUS and MRI is the imaging technique of choice for making this determination.

MRI provides an effective means of depicting the internal structure of the prostate gland (SOMMER et al. 1986; HRICAK et al. 1987; PHILLIPS et al. 1987) and is able to demonstrate the relationship of the gland to surrounding structures. On T_2-weighted MRI, the normal prostate gland demonstrates two distinct regions; the central gland and peripheral prostate. The central gland, which is of intermediate signal intensity and often heterogeneous in texture, is histologically constituted by the anterior prostate and inner prostate. The anterior prostate is non-glandular fibromuscular stroma thickened anteriorly which thins as it surrounds the prostate posteriorly and laterally, forming the „capsule". The inner prostate is a combination of the smooth muscle of the internal sphincter, a thin lining of periurethral glandular tissue, the verumontanum and the transition zone. BPH develops exclusively from the inner prostate, 95% originating from the transition zone and the remainder from the periurethral glandular tissue. The peripheral prostate is composed entirely of glandular tissue and comprises the central and peripheral zones and is of high signal intensity on T_2-weighted MRI. This high signal intensity region of peripheral prostate gland is often called „peripheral zone" on MRI. The importance of zonal anatomy is that it correlates well with sites of origin of disease. In total, 80% of prostatic carcinomas develop in the peripheral prostate (70% in the peripheral zone, 10% in the central zone). The remaining 20% of prostatic carcinomas originate in the transition zone of the inner gland.

MRI shows prostate cancer as a low signal abnormality on T_2-weighted images. MRI is a valuable technique for staging patients with prostate cancer and is helpful for selecting patients with surgically resectable disease (D'AMICO et al. 1995; JAGER et al. 1996). MRI assessment of prostate cancer has a number of important limitations including a restricted ability to demonstrate microscopic and early macroscopic capsular penetration. Furthermore, it is not possible using conventional imaging criteria to reliably distinguish tumours from other causes of reduced signal in the peripheral gland such as scars, haemorrhage, areas of prostatitis and treatment effects (LOVETT et al. 1992). Central gland tumours are not well delineated on T_2-weighted images particularly in the presence of BPH (SCHIEBLER et al. 1989). In addition, tumour volume is often under estimated when compared with pathological specimens (BEZZI et al. 1988; KAHN et al. 1989; MCSHERRY et al. 1991; QUINT et al. 1991; SCHNALL et al. 1991; SOMMER et al. 1993; BRAWER et al. 1994; LENCIONI et al. 1997). Other limitations of conventional MRI include the lack of information on tumour grade or vascularity, both of which are known to be useful predictors of patient prognosis (BRAWER et al. 1994; BOSTWICK and ICZKOWSKI 1998).

Before discussing dynamic contrast enhanced MRI (DCE-MRI) it is important to mention ^{1}H chemical shift spectroscopic imaging (MRSI) which is becoming an important tool for the evaluation of patients with prostate cancer (SWANSON et al. 2001). With MRSI, tumours show elevated levels of choline (significantly increased in cancerous regions compared with normal prostatic tissue), which are thought to arise from cellular proliferation, and decreased citrate resonances that occur due to the displacement of normal prostatic tissues and decreased production by cancerous tissues. The ratio of these metabolites (choline/citrate) has demonstrated high specificity in discriminating cancer from normal peripheral zone (KURHANEWICZ et al. 1996). MRSI has been shown to significantly improve the detection and localisation of prostate cancer before and after androgen deprivation therapy (KURHANEWICZ et al. 1996; WEFER et al. 2000; MUELLER-LISSE et al. 2001). Furthermore, MRSI data have been shown to correlate with the histological grade of prostate cancer (Gleason score) (Vigneron et al. 1998). The choline/normal-choline and (choline+creatine)/citrate ratios have been demonstrated to increase with increasing tumour grade and the citrate/normal-citrate ratio to decrease; the values of the choline/normal choline ratio between low grade (5+6) and high grade (7+8) tumours have been found to be significantly different ($p<0.0001$, $n=26$). There are also early indications that MRSI may aid in the staging of prostate cancer for less experienced radiologists by helping to predict the presence of extracapsular extension of disease (Kurhanewicz et al. 1996; Yu et al. 1999).

DCE-MRI has successfully made the transition from methodological development to pre-clinical and clinical validation and is now rapidly becoming a mainstream clinical tool. DCE-MRI is usually performed after the bolus administration of intravenous contrast medium to access tumour vascular characteristics non-invasively. The technical details concerning contrast agent kinetics, data acquisition, mathematical modelling of kinetic data and general pathophysiological correlates of DCE-MRI are discussed elsewhere in this book (Chaps. 5 and 6). The success of DCE-MRI techniques depends on their ability to demonstrate quantitative differences of contrast medium behaviour in a variety of tissues. Evidence is mounting that kinetic parameters derived from DCE-MRI correlate with immunohistochemical markers of tumour angiogenesis and with pathologic tumour grade (Padhani 2002). In this chapter, an appraisal of recognised and potential clinical applications of DCE-MRI using T_2*-weighted and T_1-weighted techniques for the evaluation of prostate cancer will be made.

12.2 Prostate Cancer Angiogenesis

Tumour hypoxia is thought to be the likely explanation for the induction of angiogenesis in prostate cancer (Izawa and Dinney 2001). Hypoxia induces vascular endothelial growth factor (VEGF) transcription via hypoxia-inducible factor-1 (Zhong et al. 1999). VEGF is a recognised stimulus of neoangiogenesis in tumours, and is also a potent tissue permeability factor (Dvorak et al. 1995). Androgens seem to regulate VEGF expression in prostate cancer cells and prostatic fibroblasts (Joseph et al. 1997; Levine et al. 1998). It has been shown that VEGF is produced in abundance by the secretory epithelium of normal, hyperplastic, and tumorous prostate glands (Jackson et al. 1997; Ferrer et al. 1998). The physiological role(s) of VEGF in the prostate is poorly understood and target cells may include cells other than the vascular endothelium. With respect to the vasculature, it is clear that VEGF is required for vascular homeostasis in the prostate gland and maintains the high fraction of immature vessels in prostate cancers. Immature vessels (those without investing pericytes/smooth muscle cells) (Eberhard et al. 2000) are highly dependent on exogenous survival factors including VEGF (Benjamin et al. 1999). In the prostate, VEGF production requires continual stimulation by androgens (Haggstrom et al. 1999); VEGF expression in androgen-dependent cell lines is down regulated upon androgen withdrawal, and prostate tumours from these cell lines undergo vascular regression prior to tumour cell death (Jain et al. 1998).

Both microvessel density (MVD) and the expression of angiogenic factors have been evaluated as prognostic factors in patients with prostate cancer. MVD is a potential prognostic factor that has been correlated with clinical and pathological stage, metastasis and histological grade in prostate cancer (Fregene et al. 1993; Weidner et al. 1993; Brawer et al. 1994; Silberman et al. 1997; Borre et al. 1998; Strohmeyer et al. 2000). MVD has also been correlated with disease-specific survival and progression after treatment. MVD has not however been shown to correlate consistently with outcome after radical prostatectomy (Hall et al. 1994; Bettencourt et al. 1998; Gettman et al. 1999; Moul 1999; Rubin et al. 1999). Both serum VEGF and beta fibroblastic growth actor (bFGF) have also been evaluated as prognostic factors in prostate cancer. Neither is prognostic which may be a reflection of coexistent benign prostatic hyperplasia (Meyer et al. 1995; Weingartner et al. 1998; Duque et al. 1999; Walsh et al. 1999).

12.3 T_2*-Weighted DCE-MRI of the Prostate

When a bolus of paramagnetic, low molecular weight contrast agent passes through a capillary bed, it is often assumed that it is initially confined within the vascular space. Concentrated intravascular contrast media produces magnetic field (Bo) inhomogeneities that reduce the signal intensity of surrounding tissues when using T_2*-weighted imaging. Perfusion-weighted images can be obtained with „bolus-tracking techniques" that monitor the passage of contrast material through a capillary bed (Sorensen et al. 1997; Barbier et al. 2001). MRI systems capable of rapid image acquisition are required to adequately characterise these effects. High specification, echo-planar capable systems allow rapid, multi-slice data acquisition, although such studies are also possible on conventional MRI systems using standard gradient-echo sequences but are limited to a fewer slices. When examining the prostate, even small amounts of rectal air and rectal movement can cause marked susceptibility effects thus spoiling dynamic T_2*-weighted DCE-MRI examinations.

Tracer kinetic principles can be used to provide estimates of relative blood volume (rBV), relative blood flow (rBF) and mean transit time (MTT) derived from the first-pass of contrast agent through the microcirculation (Rosen et al. 1991; Sorensen et al. 1997; Barbier et al. 2001). Quantification techniques only provide accurate measurements of perfusion parameters in the brain because the intact blood brain barrier retains the contrast medium within the vasculature. Errors occur in visceral tissues such as the prostate because of marked capillary leakage of contrast media in normal and pathologic tissues[1]. The loss of compartmentalisation of the bolus injection and the T_1 enhancing effects of contrast agent in the extravascular-extracellular compartment of tissues counters T_2* signal-lowering effects, resulting in falsely lowered blood volume computations. Solutions for obtaining more reliable perfusion data under these circumstances are currently being investigated (Barbier et al. 2001).

12.3.1 Clinical Experience

A limited number of studies have reported in abstract form on the feasibility of using susceptibility weighted DCE-MRI to examine the prostate gland (Noseworthy et al. 1999; Gibbs et al. 2001). Their observations show that it is possible to demonstrate first pass, signal lowering effects using echo-planar T_2*-weighted sequences and that limited tissue characterisation is possible. That is, significant differences have been noted between peripheral gland and tumour with respect to signal intensity change (Gibbs et al. 2001). However, no systematic differences between central gland and tumour enhancement values have been observed. We too have performed a limited number of T_2* weighted DCE-MRI studies and have found that prostate cancer has relatively low blood volume levels compared to pericapsular and neurovascular bundle vessels and other tumours such as breast and rectal cancers. Occasionally, strong susceptibility effects can be recorded from prostatic tumours (Fig. 12.1). No studies in prostate cancer have as yet correlated T_2* derived kinetic parameters with tumour stage, Gleason score, serum PSA levels or tumour MVD.

12.4 T_1-Weighted DCE-MRI of the Prostate

After intravenous contrast medium administration, T_1-weighted images can demonstrate prostatic zonal anatomy but in general, unenhanced T_2-weighted spin-echo images are better in this regard. On MRI, after the administration of intravenous contrast medium, the normal central gland enhances more than the peripheral prostate; both enhancing homogeneously. In the presence of benign prostatic hyperplasia (BPH), enhancement of the central gland becomes heterogeneous (Mirowitz et al. 1993; Brown et al. 1995). Prostate cancer also enhances following contrast medium administration (Brown et al. 1995; Jager et al. 1997). The role of contrast enhancement for evaluating patients with prostate cancer has not been completely defined. Early studies suggested no additional role of contrast enhancement compared to conventional T_2-weighted imaging (Mirowitz et al. 1993; Quinn et al. 1994). However, Brown et al. (1995) showed improved depiction of the tumour when MR images are obtained early after contrast enhancement (Fig. 12.2), and it has been reported that contrast enhancement can improve the detection of minimal seminal vesicle invasion by tumour (Huch Boni et al. 1995a). The specific value of DCE-MRI techniques is discussed below.

When a bolus of paramagnetic, low molecular weight contrast agent passes through a capillary bed, it is transiently confined within the vascular space (see comments in previous sections however). In most tissues except the brain, testes and retina, the contrast agent subsequently rapidly passes into the extravascular-extracellular space (EES also called leakage space – v_e) at a rate determined by the per-

[1] As discussed in Chap. 4 the assumptions necessary for reliable bolus tracking imaging are often not justified in tissues outside the brain. Indeed, near complete extraction of small molecular weight contrast agents may be observed on first passage through some tissues.

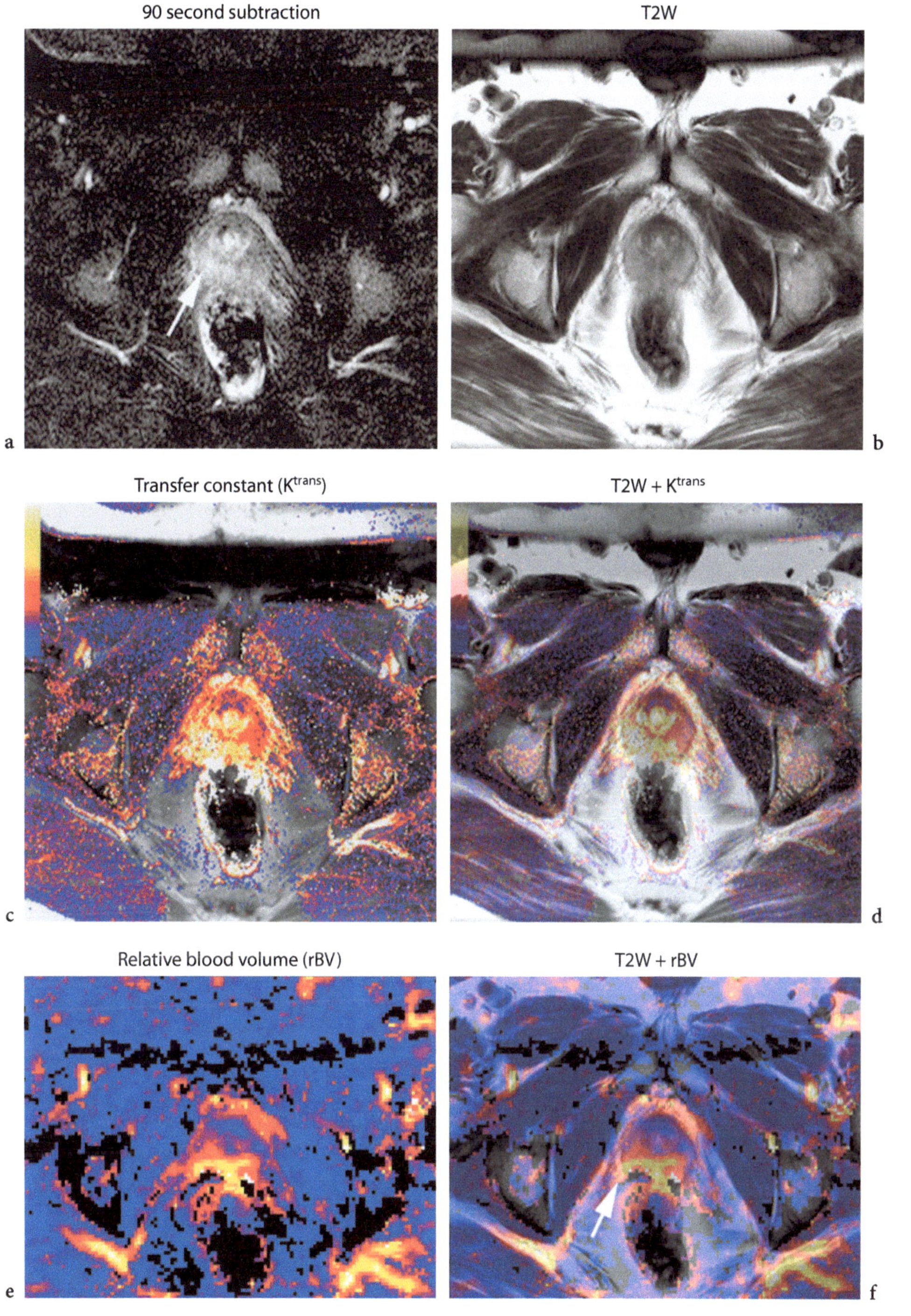

Fig. 12.1a–f. Blood volume and transfer constant imaging of prostate cancer. A 58-year-old man with prostate cancer (Gleason grade 3+4 and PSA 6.9 ng/ml). The prostatectomy specimen showed that the tumour was on the right side posteriorly and measured 17 mm. **a** T_2-weighted turbo spin-echo image showing low signal intensity in the peripheral gland bilaterally (right to left) compatible with tumour infiltration. The central gland is morphologically normal. **b** A 90-s subtraction image after injection of contrast medium at the same level showing enhancement of the tumour (*arrow*) and the central gland. **c** Transfer constant (K^{trans}) map (maximum transfer constant depicted 1 min^{-1}). **d** Transfer constant map superimposed on the coregistered T_2-weighted image shows the anatomic distribution of transfer constant. High transfer constant levels are seen in the tumour and in the central gland. **e** Relative blood volume (rBV) map at the same anatomical position. **f** Relative blood volume (rBV) map superimposed on the coregistered T_2-weighted image shows high rBV localized to the tumour. High blood volume is also seen related to capsular and neurovascular bundle vessels (*arrow*)

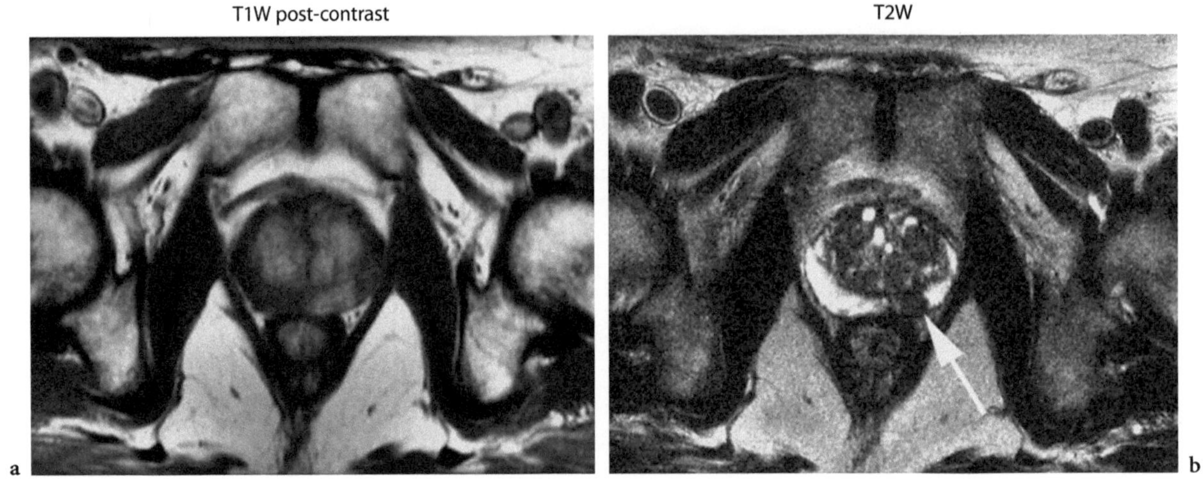

Fig. 12.2a,b. Enhancing prostate cancer. **a** T_2-weighted turbo spin-echo image showing a low signal intensity mass in the peripheral gland (*arrow*). The central gland is morphologically normal. **b** T_1-weighted post-contrast medium enhanced MR image showing clear, well defined prostate tumour enhancement and heterogeneous enhancement of benign prostatic hyperplasia in the central gland

meability of the microvessels, their surface area and by blood flow (Crone 1963; Taylor and Reddick 2000). In tumours, a variable proportion (possibly close to 100%) of the contrast media can leak into the EES during the first pass (Daldrup et al. 1998). The transfer constant (K^{trans}) describes the transendothelial transport of the contrast medium by diffusion. Over a period typically lasting several minutes to hours, the contrast agent diffuses back into the vasculature from where it is excreted (usually by the kidneys although some extracellular fluid contrast media have significant hepatic excretion). T_1-weighted sequences are used to detect the presence of contrast medium in the EES and so can be employed to estimate transfer constant (K^{trans}) and leakage space (v_e).

12.4.1
Data Acquisition

To monitor the tissue enhancing effects of contrast agents on T_1-weighted prostate DCE-MRI, confounding T_2 and T_2* signal lowering effects must be controlled. T_1-weighted gradient-echo, saturation recovery/inversion recovery snapshot sequences (e.g., turboFLASH) have been used in prostate imaging (Jager et al. 1997; Parker et al. 1997, 2000; Namimoto et al. 1998; Liney et al. 1999; Tanaka et al. 1999; Turnbull et al. 1999; Huisman et al. 2001; Ogura et al. 2001). The choice of sequence and parameters used is dependent on intrinsic advantages and disadvantages of the sequences taking into account T_1 sensitivity, anatomical coverage, acquisition times, susceptibility to artefacts arising from sources of magnetic field inhomogeneities (e.g., from rectal gas, hip prostheses, etc.) and the need for quantification (Parker and Tofts 1999). It is recognised that high-resolution and short imaging-time are competing requirements, and are restricted by the capabilities of current equipment and software. Higher temporal resolution imaging necessitates reduced spatial resolution, decreased anatomic coverage or a combination thereof. Higher temporal resolution techniques are essential for T_2*-weighted techniques and may improve specificity of T_1-weighted DCE-MRI because of better characterisation of time signal intensity/contrast agent concentration curves; one study has suggested that characterisation of prostate lesions is optimal using image acquisition times of 2 s (Engelbrecht et al. 2003).

In order to model tissue contrast agent behaviour the contrast agent concentration at each time point during the imaging procedure needs to be known. Some workers assume that the change in signal intensity or relative signal intensity is directly proportional to tissue contrast agent concentration (Brix et al. 1991; Buckley et al. 1994; Hoffmann et al. 1995). However, when contrast agent concentrations become large (e.g., within vessels) this may become a poor approximation, because signal intensity varies nonlinearly with contrast agent concentration. If the rapidly changing T_1 relaxation time can be accurately estimated over a large range of T_1 values (Brookes et al. 1996; Parker et al. 1997, 2000; Huisman et al. 2001; Engelbrecht et al. 2003), then tissue concentration of the contrast agent and its time course can be calculated (Donahue et al. 1994).

Multislice spoiled gradient-echo or saturation recovery turboFLASH sequences are often used to examine the prostate gland (Parker et al. 1997, 2000). A single proton density weighted measurement is usually acquired as a reference before the series of T_1-weighted gradient-echo images (Fig. 12.3). Images are typically obtained sequentially every few seconds for 6-7 min. Contrast medium is injected intravenously as a bolus through a peripherally placed cannula after 3-4 baseline data points (dose 0.1-mmol/kg body weight, injected within 10 s followed by a 20-ml flush of normal saline) using a power injector. Common causes for failed examinations include poor contrast medium bolus particularly if manual injection techniques are used (Fig. 12.4), the presence of hip prostheses causing susceptibility artefacts and technical failures (poor signal-to-noise ratio of images or unexpected machine gain changes during the data acquisition). Patient or internal organ movements (rectal movement or bladder filling affecting prostatic position) are also not uncommon during the DCE-MRI examinations and can lead to modelling failures (Fig. 12.5) (Padhani et al. 1999; Engelbrecht et al. 2003). Padhani et al. (1999) have reported that 16% of 55 patients had anterior prostatic displacements of >5 mm due to rectal motion during DCE-MRI examinations lasting 7 min and have noted an inverse correlation between rectal distension of the frequency of rectal movements.

12.4.2 Quantification

Signal enhancement on T_1-weighted DCE-MRI images can be assessed in two ways: by the analysis of signal intensity changes (semi-quantitative) and/or by quantifying contrast agent concentration change or R_1 ($R_1 = 1/T_1$) using pharmacokinetic modelling techniques. Semi-quantitative kinetic parameters describe tissue enhancement using of a number of descriptors derived from signal intensity-time curves. These parameters include onset time (time from injection or first appearance in a pelvic artery to the first increase in tissue signal enhancement; Parker et al. 1998; Engelbrecht et al. 2003), initial and mean gradient of the upsweep of enhancement curves, maximum signal intensity, and washout gradient. Clinical practice has shown that the rate of enhancement is also important for improving the specificity of DCE-MRI and parameters that include a timing element are widely used in non-prostatic examinations, e.g. maximum intensity time ratio (MITR) (Flickinger et al. 1993) and

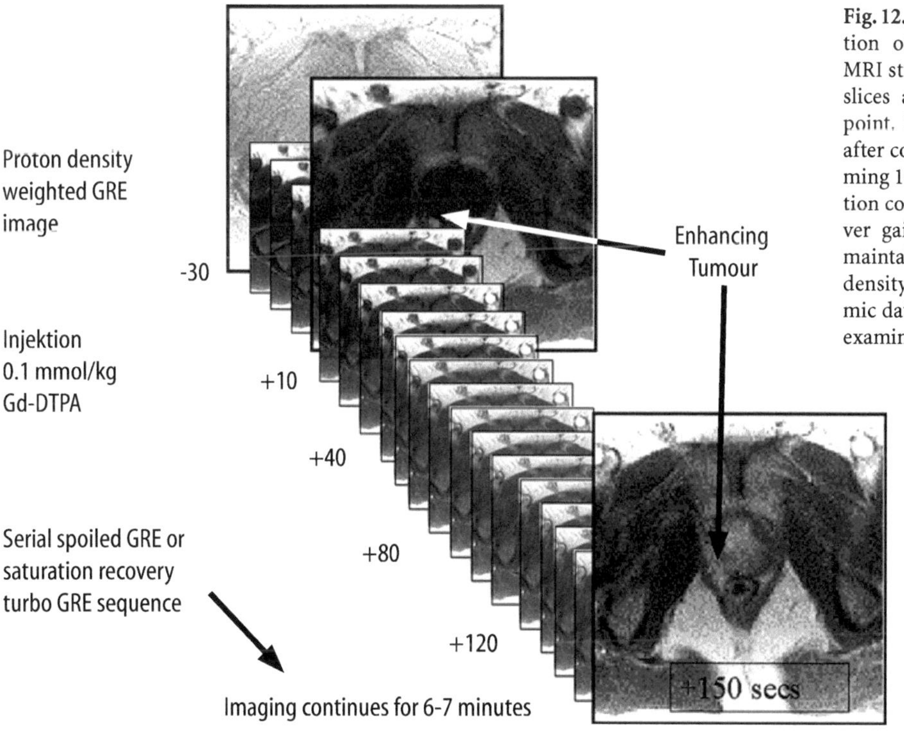

Fig. 12.3. Schematic representation of a T_1-weighted dynamic MRI study. Between three and five slices are acquired at each data point. Numbers indicate seconds after contrast enhancement (assuming 10 s per data point). Acquisition continues for 6–7 min. Receiver gain and scaling factors are maintained between the proton density acquisition and the dynamic data acquisition phases of the examinations

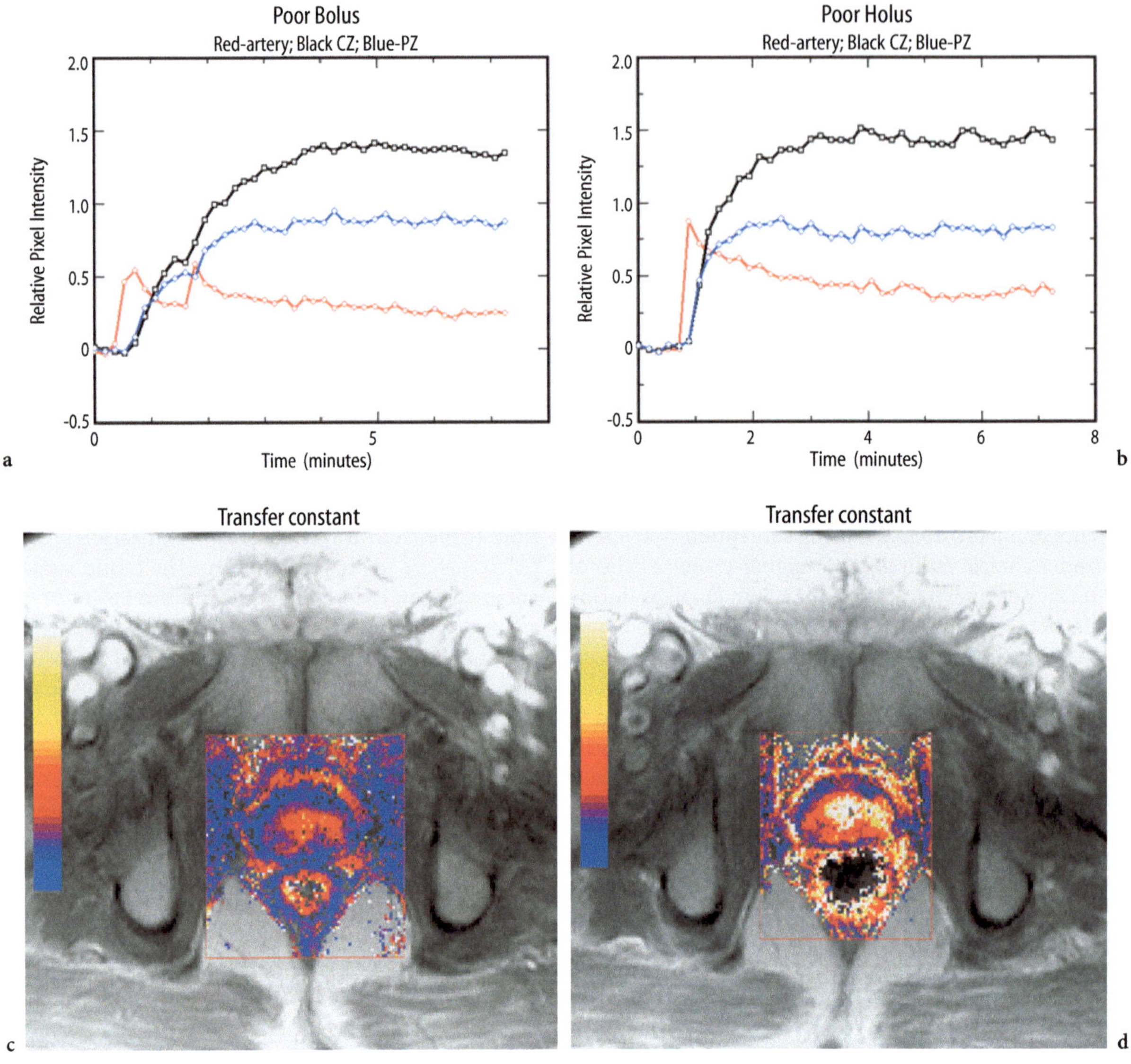

Fig. 12.4a–d. Poor manual injection technique. Relative signal intensity time curves from a patient with prostate cancer obtained 14 days apart. Note the difference in curve shapes of the arterial curve and prostatic tissues between the two examinations. A double peak on the arterial curve is seen in (**a**) due to poor manual injection technique. Poor injection technique can markedly alter calculations of kinetic parameters as seen on the corresponding transfer constant maps (**c**,**d**) (colour scale 0–1 min^{-1}). Such errors can be minimised by the use of a power injector or by the utilisation of a modelling technique that explicitly accounts for any given arterial input function (**b**)

maximum focal enhancement at 1 min (Kaiser and Zeitler 1989; Gribbestad et al. 1994) in breast examinations. Some studies have correlated the shape of time signal intensity curves with prostatic tissue characteristics and response to treatment (Padhani et al. 2000, 2001). Semi-quantitative parameters have the advantage of being relatively straightforward to calculate but have a number of limitations. These limitations include the fact that they do not accurately reflect contrast agent concentration in the tissue of interest and can be influenced by scanner settings (including TR, TE, flip angle, gain and scaling factors). Quantitative parameters are more complicated to derive compared to those derived semi-quantitatively which deters their use at the workbench. The model chosen may not adequately reflect the acquired data: each model makes a number of assumptions that may not be valid for every tissue or tumour type and software for data analysis is not widely available

a

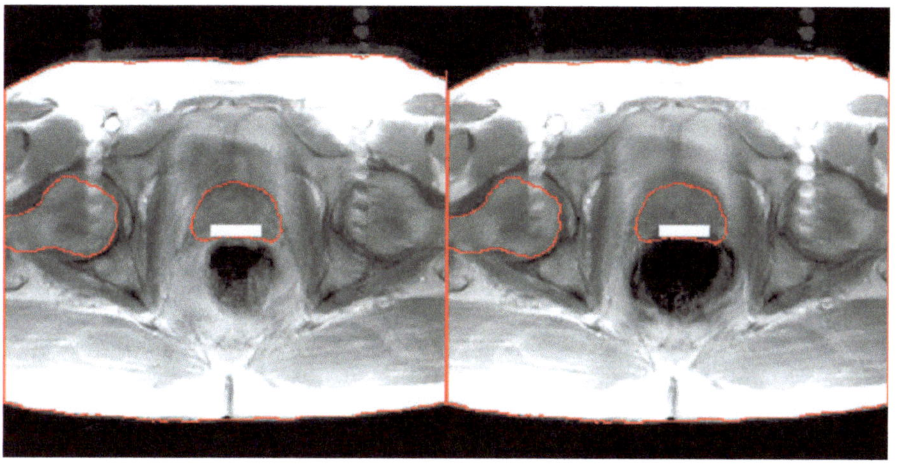

b

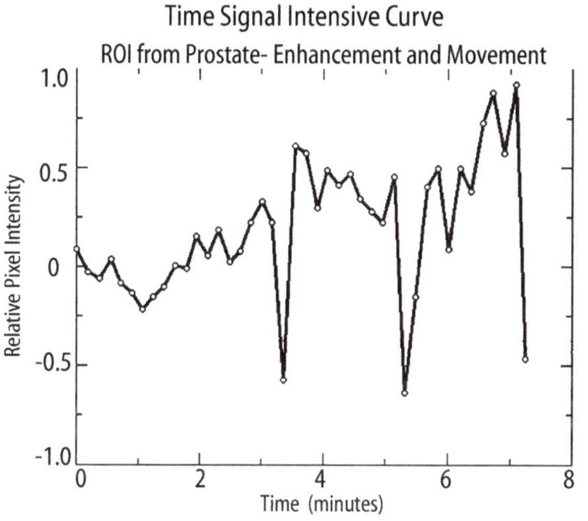

Fig. 12.5a,b. Marked prostatic movements during DCE-MRI study. **a** Sample images obtained during a T_1-weighted DCE-MRI study showing marked rectal movements and corresponding anterior displacements of the prostate gland. The prostate gland and bony landmarks are outlined in and maintained in position. The rectangular boxes in (**a**) represent the region of interest (ROI) from which signal intensity time curves are generated in (**b**). A number of minor and large prostatic movements are seen illustrated as drops in signal enhancement as the air in the rectum encroaches into the fixed ROI. The general signal intensity increases because contrast medium has been administered for the DCE-MRI study

(Tofts 1997; Tofts et al. 1999). Further discussion of modelling techniques can be found elsewhere in this book (Chap. 6).

12.4.3 Clinical Validation

It is possible to show characteristic differences in the enhancement patterns of peripheral prostate gland compared to the central gland and/or tumours using both semi-quantitative and quantitative kinetic parameters derived from T_1-weighted DCE-MRI (Jager et al. 1997; Liney et al. 1999; Turnbull et al. 1999; Padhani et al. 2000; Engelbrecht et al. 2003) (Table 12.1) (Figs. 12.6, 12.7). These differences are likely to be related to underlying variations in tissue perfusion, MVD and tissue VEGF expression. Immunohistochemical studies have found that MVD in prostate cancer and BPH is higher than in the peripheral zone (Bigler et al. 1993; Offersen et al. 1998). These studies also show that there is an overlap in MVD counts between tumours and BPH (Deering et al. 1995). Many clinical studies have correlated tissue MRI enhancement with immunohistochemical MVD measurements [see Padhani (2002) for a review], but no studies have been performed in prostate cancer. There are also no clinical studies correlating tissue MRI enhancement with immunohistochemical VEGF staining in prostate cancer.

Jager et al. have noted that poorly differentiated prostate cancer showed earlier onset and faster rate of enhancement compared to other histological grades in five patients but made no formal correlation with histological grade or tumour stage (Jager et al. 1997). Pfleiderer et al. in an Inter-

Table 12.1. Prostatic tissue characterisation using enhancement parameters. [Table reproduced from PADHANI et al. (2000), with kind permission]

Enhancement parameters	Tissue regions of interest			
	Peripheral	Central gland gland[a]	Whole tumour outline	Tumour: - fastest enhancing area
Time signal intensity parameters				
Number of observations	33	30	39	45
Onset time (min)	1.02[b] (0.93–1.11)	0.92 (0.86–0.97)	0.94 (0.88–1.01)	0.93 (0.87–1.00)
Mean gradient	66 (43–89)	260 (164–357)	164 (118–209)	332 (231–433)
Maximum enhancement (% from baseline)	88 (76–99)	145 (120–170)	125 (111–139)	142 (126–157)
Washout patterns (benign: suspicious: malignant)	23 : 9 : 1	5 : 8 : 17	2 : 18 : 19	3 : 10 : 32
Modelling Parameters				
Number of observations	32	29	38	43
Transfer constant (K^{trans}) (min^{-1})	0.22 (0.15–0.29)	1.08 (0.68–1.48)	0.79 (0.62–0.96)	1.10 (0.78–1.41)
Tissue leakage space (v_e) (%)	26 (22–31)	51 (45–56)	45 (42–48)	49 (46–53)
Maximum gadolinium concentration (mmol/kg)	0.20 (0.17–0.24)	0.38 (0.34–0.42)	0.33 (0.31–0.35)	0.38 (0.36–0.40)

Mean values and 95% confidence intervals in parentheses except for washout patterns where the number of patients in each category is indicated. Washout patterns are scored as „benign“ when a slow monotonic increase in signal intensity was seen through the observation period; as „suspicious“ if the peak signal intensity was achieved within the first 2 min and was sustained or if there was late decrease in signal intensity (washout); and as „malignant“ when an early peak of enhancement was seen followed immediately by a decrease in signal intensity.
[a] Kruskal-Wallis test p = 0.0001 for all parameters except for onset time (b) where p = not significant

net-only report (http://medweb.uni-muenster.de/institute/ikr/mrs/pfleide/poster2.htm) have shown strong correlation between tumour enhancement patterns and histopathological tumour grading but this report has never appeared in the peer-reviewed literature. However, both PADHANI et al. (2000) and ENGELBRECHT et al. (2003) have found poor correlations between enhancement parameters and histological grade measured by the Gleason score. A correlation may be expected because Gleason score has been shown to correlate with microvessel density measurements (BOSTWICK et al. 1996). PADHANI et al. (2000) commented that the lack of correlation might be explained by histological sampling errors inherent in TRUS needle biopsy techniques. However, ENGELBRECHT et al. (2003) using whole mount prostatectomy sectioning also demonstrated poor correlations. Both groups had relatively few patients with well or poorly differentiated cancers and this may also have contributed to the lack of correlation. It should be noted that MRSI may be able to grade prostate cancers non-invasively based on early clinical data (vide supra) (VIGNERON et al. 1998).

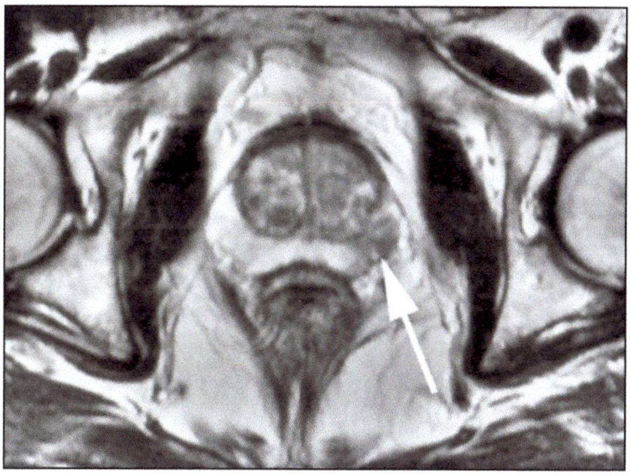

a

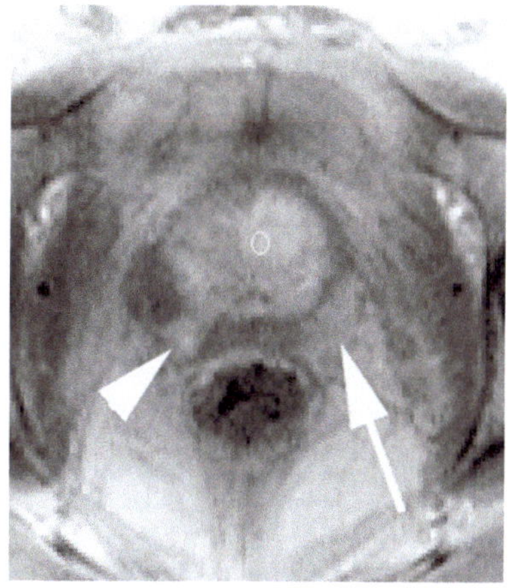

b

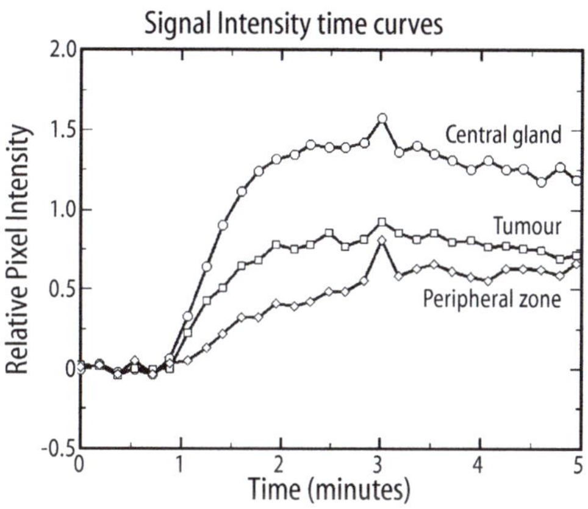

c

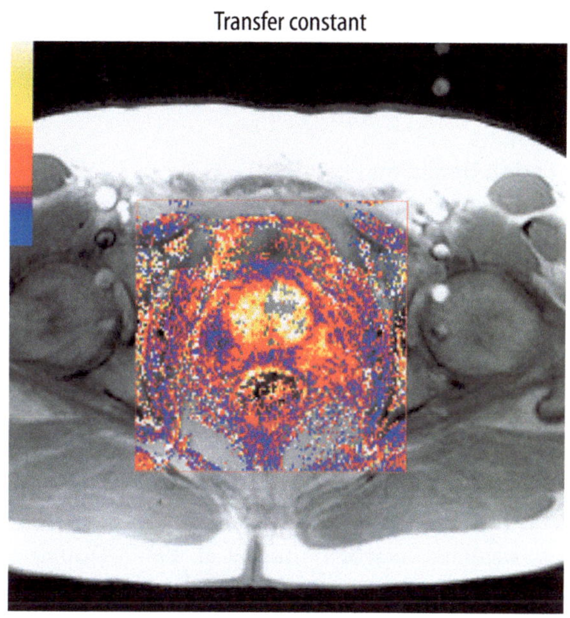

d

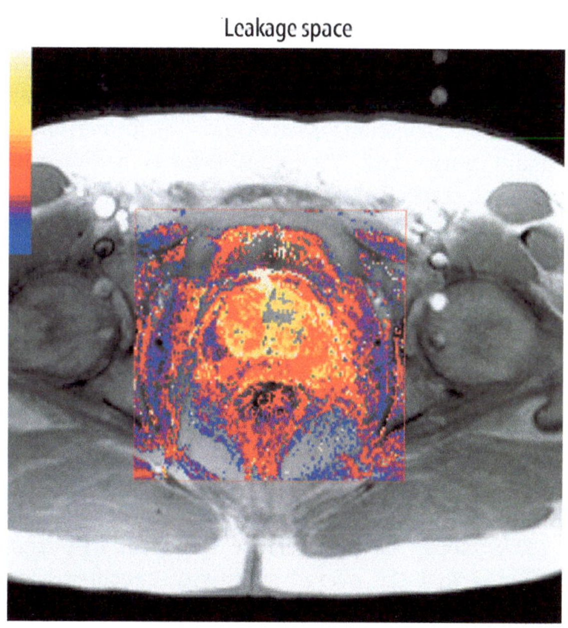

e

Fig. 12.6a-e. Typical dynamic contrast enhanced MRI study. A 62-year-old man with prostate cancer biopsied 31 days before MR imaging (Gleason grade 2+2 and PSA 11.5 ng/ml). **a** T_2-weighted turbo spin-echo image showing a low signal intensity mass in the left peripheral zone (*arrow*) compatible with tumour. The peripheral zone shows homogeneous intermediate to high signal. The central gland is morphologically normal. **b** T_1-weighted gradient-echo FLASH image at the same slice position as (**a**) obtained 30 s after injection of contrast medium shows enhancement of the tumour (*arrow*) and the central gland. The peripheral zone shows minimal enhancement. An area of post-biopsy haemorrhage in the right peripheral zone (*arrowhead*) was seen on precontrast images. **c** Time relative signal intensity curves for the regions of interest placed in the peripheral zone (*diamonds*), tumour (*squares*) and the central gland (*circles*). The peripheral zone shows a slow rising curve compared to the tumour or central gland. The tumour curve shows a faster rise and a higher maximum enhancement compared to the peripheral zone. The central gland shows the steepest rise and highest peak in enhancement; some washout is seen in the tumour and the central gland. **d** Transfer constant map (maximum transfer constant depicted=1 min^{-1}) and **e** Tissue leakage space map (maximum leakage space depicted=100%). High levels of transfer constant and leakage space are seen in the tumour (0.66/min and 45%) and central gland (1.14/min and 51%) compared to the peripheral zone (0.11/min and 17%). Note that some pixels do not display a colour because there was a poor fit of the multi-compartment model to the data observed. [Images reproduced from Padhani et al. (2000) with kind permission]

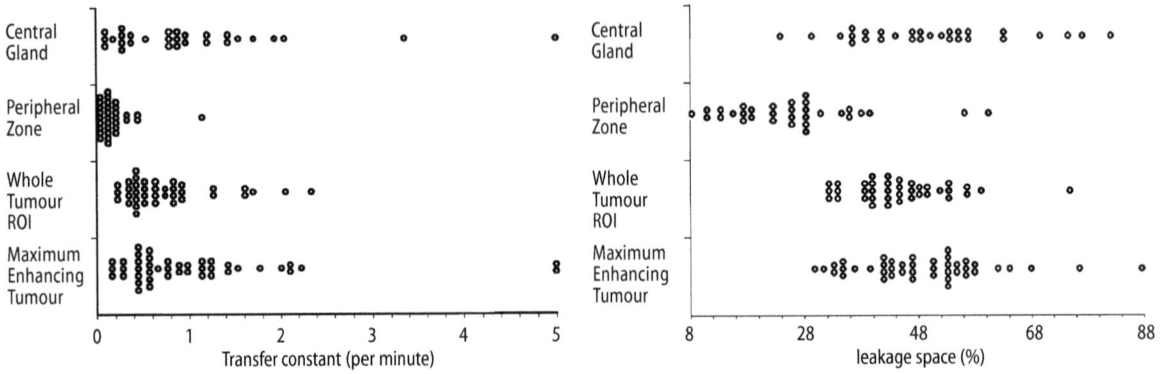

Fig. 12.7. Transfer constant (*left*) and leakage space (*right*) in different prostatic tissues. [Images reproduced from PADHANI et al. (2000) with kind permission]

12.4.4 Clinical Experience

12.4.4.1 Lesion Detection

A number of studies have compared DCE-MRI with spin-echo T_2-weighted images in patients with known prostate cancer and have found that there is a modest advantage in the detection of tumours (Table 12.2) (JAGER et al. 1997; NAMIMOTO et al. 1998; TANAKA et al. 1999; OGURA et al. 2001). For example, JAGER et al. noted that the average sensitivity, specificity and accuracy for detection of tumours for two readers with TurboFLASH images were 74%, 81% and 78 % compared to 58%, 81% and 72% for fast SE T_2-weighted images (JAGER et al. 1997). OGURA et al. (2001) made the specific point that DCE-MRI was more accurate in detecting cancers in the peripheral gland where the overall accuracy rate was 80% compared with transitional zone where tumour detection accuracy was only 63%. The sensitivity and specificity of tumour detection was 81% and 79% for peripheral gland cancers and 37% and 97% for transition zone cancers, respectively (OGURA et al. 2001).

The role of DCE-MRI in detecting prostate cancer in men with a raised PSA level without an abnormality on digital rectal examination and TRUS and with negative systematic biopsy has not been formally tested. However, anecdotal experience suggests that DCE-MRI may be able to depict small tumours in the peripheral gland that lie near the apex of the gland, which is an area that is difficult to biopsy. ITO et al. (2003) compared the visualisation of prostate cancer with DCE-MRI and TRUS with power Doppler using TRUS biopsy as the reference standard. This study showed that the overall sensitivity, specificity and accuracy for cancer visualisation with DCE-MRI (87% 74% and 82% respectively) were better than power Doppler ultrasound (69%, 61% and 68%) but only for peripheral gland tumours. They also noted that reliable detection of central gland tumours (those without a peripheral gland component) was not possible. However, two studies have noted that it is possible to differentiate between tumour and central gland enhancement (TURNBULL et al. 1999; ENGELBRECHT et al. 2003) using complex modelling techniques. Both TURNBULL et al. (1999) and ENGELBRECHT et al. (2003) have described significant differences between carcinoma and BPH in the amplitude of the initial enhancement. In general, cancers have higher amplitude of enhancement when compared to BPH. Additionally, ENGELBRECHT et al. (2003) have recently shown significant differences in the washout patterns between cancers and BPH. What remains unclear is whether this can be done reliably in the clinical setting of a raised PSA level without an abnormality on digital rectal examination and TRUS.

12.4.4.2 Lesion Characterisation

On conventional T_2-weighted MRI, it is not possible to distinguish reliably tumours from other causes of reduced signal in the peripheral gland such as areas of prostatitis or scars (LOVETT et al. 1992). Many studies have shown that there is a high false positive rate or lowered specificity when lesion characterisation is attempted on the basis of hypointensity in the peripheral gland (SCHIEBLER et al. 1989; RIFKIN et al. 1990). NAMIMOTO et al. (1998) noted that it was possible to improve the characterisation of hypointense lesions on T_2-weighted MRI by using subtraction images from DCE-MRI examinations. They showed that both specificity and false positive rates were improved following contrast medium enhancement in 42 patients with hypointense lesions on T_2-weighted MRI with a raised PSA level but without a firm diagnosis of prostate cancer (Table 12.2).

Recently, VAN DORSTEN et al. (2001) compared the ability of DCE-MRI and ^{1}H-MRSI to distinguish acute prostatitis from prostate cancer. They noted that it was not possible to distinguish these entities on the basis of enhancement patterns on DCE-MRI but were able to make the distinction on ^{1}H-MRSI. The distinction of scars from cancers has not been formally evaluated but clinical experience suggests that scars and areas of chronic fibrosis do not enhance in the same manner as cancers in the peripheral gland and seminal vesicles (Fig. 12.8).

Table 12.2. Accuracy of DCE-MRI for the detection of prostate cancer

Author and year	Histological standard	MRI coil type	T_2-weighted MRI				T_1-weighted DCE-MRI			
			Patients (n)	Sensitivity (%)	Specificity (%)	Accuracy (%)	Patients (n)	Sensitivity (%)	Specificity (%)	Accuracy (%)
JAGER et al. (1996)	Prostatectomy	ERC	57	58	81	72	57	74	81	78
NAMIMOTO et al. (1998)	TRUS and needle biopsy	PPA	42	95	57	75	42	86	74	79
TANAKA et al. (1999)	TRUS and needle biopsy	ERC	10	78	95	92	18	100	82	89
OGURA et al. (2001)	Prostatectomy	ERC	–	–	–	–	38	59	88	72
ITO et al. (2003)	TRUS and needle biopsy	PPA	–	–	–	–	31	87	74	82

ERC, endorectal coil; PPA, pelvic phased array; TRUS, transrectal ultrasound.

a

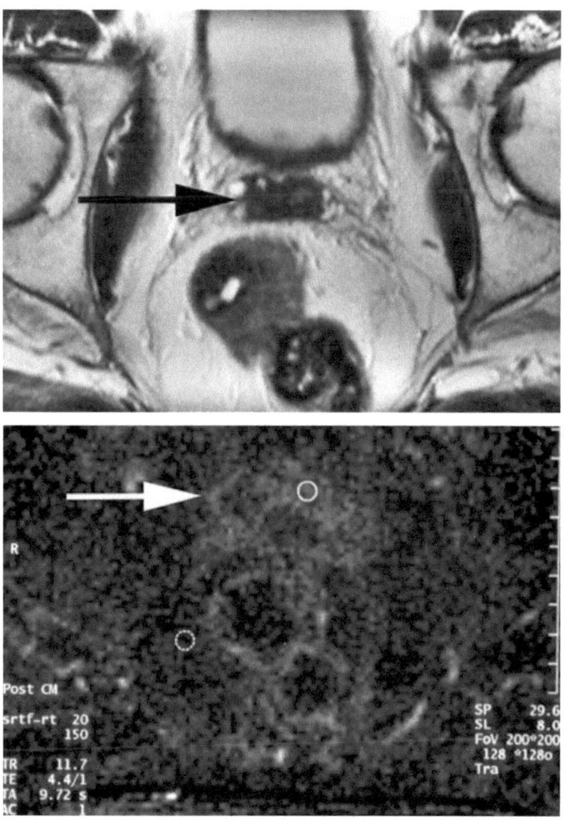

b

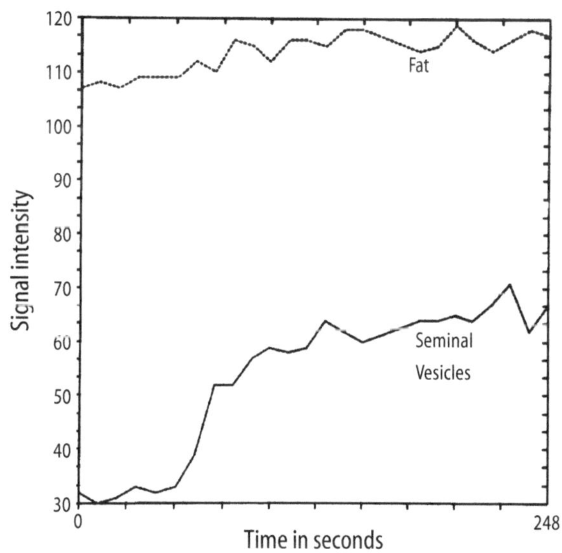

c

Fig. 12.8a–c. Radiation fibrosis of the seminal vesicles. **a** T_2-weighted image obtained 18 months following radiation treatment to the prostate gland. The seminal vesicles are dark (*arrow*) and appear to be infiltrated by tumour. **b** A 90-s subtraction image shows no enhancement in the region of the seminal vesicles (*arrow*). Regions of interest are places in the seminal vesicles and ischiorectal fat. **c** Signal intensity time curve from the regions indicated in the early subtraction image show a benign enhancement pattern

12.4.4.3
Tumour Volume and Staging

Tumour volume is a recognised prognostic indictor in prostate cancer. A disparity between tumour volume on conventional T_2-weighted MRI and pathology is well recognised; tumour volume is often under-estimated (Bezzi et al. 1988; Kahn et al. 1989; McSherry et al. 1991; Quint et al. 1991; Schnall et al. 1991; Sommer et al. 1993; Brawer et al. 1994; Lencioni et al. 1997). This occurs because microscopic tumour in the peripheral zone is not always visible on T_2-weighted images (Carter et al. 1991; Outwater et al. 1992; Lencioni et al. 1997) and poorly differentiated prostate cancer can grow by infiltration, thus causing little architectural distortion or alteration in signal intensity (Schiebler et al. 1989). Limited literature data is available comparing DCE-MRI with T_2-weighted MRI to depict tumour volume (Jager et al. 1996), which suggests that subtraction DCE-MR images may be marginally better than T_2-weighted MRI at depicting the full intraprostatic extent of tumours (Fig. 12.9). There is also evidence that DCE-MRI can improve the accuracy of tumour staging when used in conjunction with T_2-weighted images in patients with equivocal capsular penetration, seminal vesicle invasion and neurovascular bundle involvement (Jager et al. 1997; Ogura et al. 2001). Equivocal capsular penetration and seminal vesicle invasion on T_2-weighted images are well-recognised indications for contrast-enhanced evaluation of prostate cancer (Fig. 12.9).

12.4.4.4
Radiotherapy Planning

The choice of appropriate treatment for patients with prostate cancer remains controversial. The most commonly offered treatments include observation only, radical prostatectomy, radiotherapy, hormone ablation treatment, or a combination. Treatment selection is guided by patient age and general condition, tumour stage and histological grade, serum PSA, and patient and physician preferences. External beam radiotherapy treatment failure is often attributed to the need to limit radiation dose because of the sensitivity of surrounding neighbouring structures including the bladder, bowel and hips. Escalation of dose is one of the major strategies currently being explored to improve local control and overall survival in prostate cancer. Using intensity-modulated radiotherapy (IMRT), a complex 3D dose distribution can be produced to match areas of disease selectively avoiding normal tissue (Nutting et al. 2000; Leibel et al. 2002). IMRT may be used to escalate dose in excess of 80 Gy to the prostate, with a dose constraint on the anterior rectal wall (Zelefsky et al. 2000). To take advantage fully of the opportunity of IMRT, imaging techniques that are able to map functional tumour volume within individual organs are needed (Ling et al. 2000; Rosenman 2001). If it were possible to accurately determine the location of a dominant intraprostatic nodule within the prostate gland, IMRT may allow dose escalation to these nodules with the aim of increasing tumour control with the

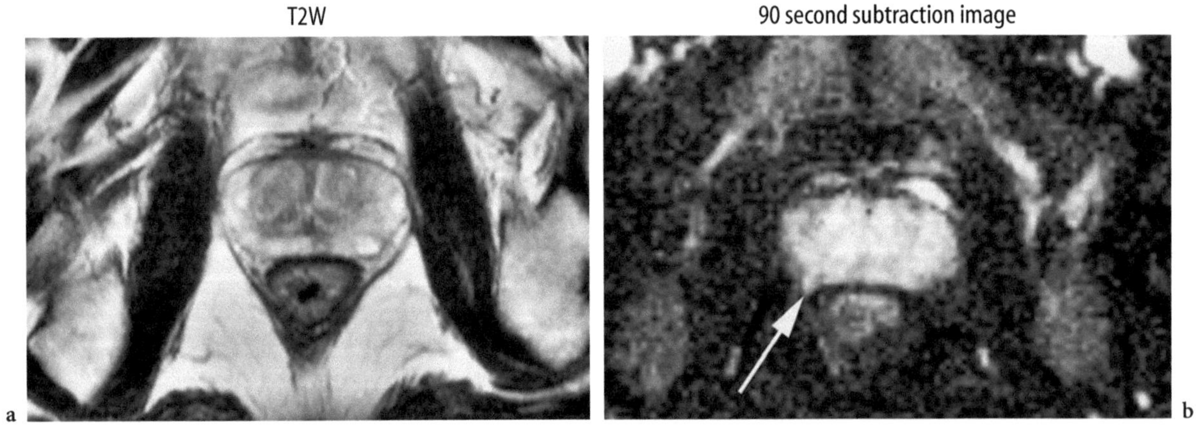

Fig. 12.9a,b. Improved depiction of tumour and capsular penetration on DCE-MRI. (Gleason 3+3, PSA 4.0 ng/ml). TRUS appearances were normal. TRUS-guided biopsy 6 weeks before MRI revealed cancer in the right peripheral gland. **a** T_2-weighted MR image shows a normal appearing peripheral gland with no evidence of a tumour. **b** A 90-s subtraction image demonstrates marked peripheral gland enhancement compatible with tumour infiltration. Extracapsular extension of enhancement is compatible with disease spread beyond the capsule (*arrow*)

benefit of less irradiation to surrounding structures (NUTTING et al. 2002). It has been suggested that DCE-MRI or ^{1}H-MRSI may be able to map functional tumour volume in the prostate gland and thus define the biological tumour volume for irradiation (LING et al. 2000). However, as discussed above, DCE-MRI only adds modestly to tumour localisation (Table 12.2), there is only limited evidence on whether tumour volume definition is improved (JAGER et al. 1996) and as discussed above DCE-MRI has not been shown to be robust in predicting tumour grade. ^{1}H-MRSI appears more promising in this regard, but high resolution ^{1}H-MRSI can only be achieved with the use of an endorectal coil (KURHANEWICZ et al. 1996). Inherent prostate gland distortion associated with endorectal coil usage (HUSBAND et al. 1998) will have to be taken into account if ^{1}H-MRSI is to be used for planning radiotherapy.

12.4.4.5 Monitoring Response to Treatment

Hormone ablation is the preferred treatment choice for patients with advanced disease, but is also used in patients before radiation therapy or prostatectomy. The response of patients to treatment can be assessed by digital rectal examination, by changes in serum PSA levels, TRUS and MRI (PINAULT et al. 1992; SHEARER et al. 1992; CHEN et al. 1996; NAKASHIMA et al. 1997; PADHANI et al. 2001). Clinical evaluations and imaging studies all show significant reductions in both glandular size and tumour volume. Reductions of 10%-52% in prostate glandular volume and 20%-97% in tumour volume have been reported (PINAULT et al. 1992; SHEARER et al. 1992; CHEN et al. 1996; NAKASHIMA et al. 1997; PADHANI et al. 2001). On MRI, the central gland decreased in signal and became more homogenous with treatment and seminal vesicle atrophy has also been noted (SECAF et al. 1991; CHEN et al. 1996; NAKASHIMA et al. 1997; PADHANI et al. 2001). As a result, hormonal ablation also reduced the number of MR detectable tumours (CHEN et al. 1996; NAKASHIMA et al. 1997; PADHANI et al. 2001). This occurs because the peripheral gland showed a decrease in signal intensity thus reducing tumour-peripheral gland contrast. These morphologic appearances are due to distinctive histological changes occurring in patients treated with luteinizing hormone releasing hormone analogues (LH-RHa) (MURPHY et al. 1991; SMITH and MURPHY 1994; CIVANTOS et al. 1996). The histological „LH-RHa effect" is characterised by a reduction in gland size and density, compression of glandular lumina and increased periglandular fibrous tissue.

PADHANI et al. (2001) recently reported that decreases in transfer constant occurred in all prostatic tissues after 3-6 months of hormonal treatment; tumour, median 56%, central gland (40%) and peripheral gland (31%). A typical example is shown in Fig. 12.10. These changes may be explained by the fibrotic changes described histologically (MURPHY et al. 1991; SMITH and MURPHY 1994; CIVANTOS et al. 1996). Additionally, the reduction in transfer constant of prostatic tissues may be related to down-regulation of VEGF production and subsequent apoptosis of immature prostate vessels caused by androgen deprivation (see Sect. 12.2) (BOSTWICK 2000). However, a recent histological study by MATSUSHIMA et al. (1999) appears to contradict this view; their study showed that intratumoral microvessel density (MVD) does not appear to differ in patients treated with neoadjuvant hormonal deprivation compared to untreated patients, but they did show decreased proliferative activity and enhanced apoptosis of prostatic cancer cells.

There is poor documentation on the early vascular effects of radiotherapy as observed by DCE-MRI. Recently, BARKE et al. (2003) have noted that hyperaemia occurs soon after commencing radiotherapy evidenced by an increased permeability surface area product. This confirms the work of HARVEY et al. (2001) who reported on 22 such patients evaluated by functional CT and showed that there was an acute hyperaemic response following radiotherapy to the prostate gland as early as 1-2 weeks following completion of treatment and this remained so after 6–12 weeks. We have recently evaluated 25 patients with DCE-MRI patients 2 years after completion of radiotherapy in whom there is no evidence of tumour recurrence (biochemical or histological). We observed that morphologically the gland has similar appearances to that seen after androgen deprivation, i.e. a small gland with poor zonal differentiation on T_2 weighted images (Fig. 12.11). On DCE-MRI, central gland enhancement was greater than the peripheral gland and kinetic parameters were also statistically higher. A slow rising pattern of enhancement was seen in the majority in the peripheral gland but in only five patients within the central gland. Contrast medium washout was not observed in the peripheral gland and was seen in only one patient within the central gland.

New treatments for prostate cancer and obstructive BPH include pulsed high-energy focused ultrasound, cryosurgical ablation, laser ablation and transurethral thermal ablation using microwaves

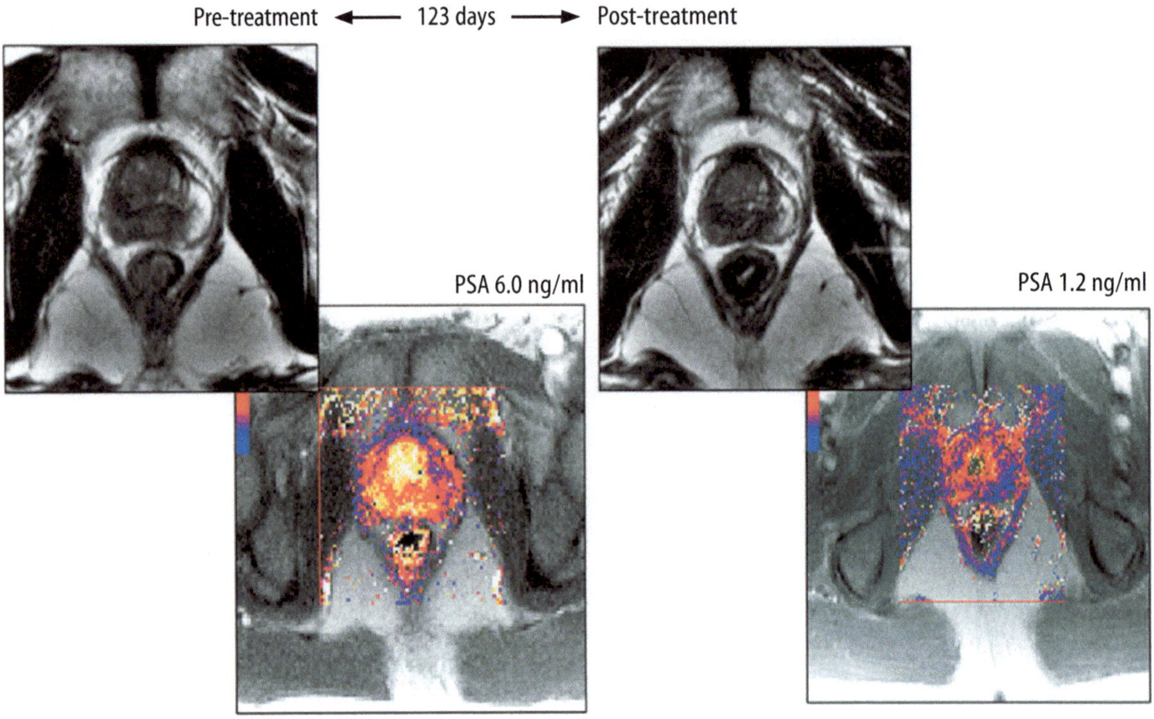

Fig. 12.10. Transfer constant changes after androgen deprivation treatment. A 62-year-old man with prostate cancer (Gleason grade 3+4). Left images: Pre-treatment images (PSA=6 ng/ml), T_2-weighted turbo spin-echo image and transfer constant map (maximum transfer constant=1 min^{-1}). A low signal intensity mass in the right peripheral zone with invasion of the central gland is seen. The peripheral zone in the left side of the gland appears normal. Higher transfer constant levels are seen in the tumour and central gland (0.46 and 1.54 min^{-1}) compared to the peripheral zone (0.17 min^{-1}). Note that some pixels do not display colour because there was a poor fit of the multi-compartment model to the data observed. Right images: Following 123 days of androgen deprivation (PSA 1.2 ng/ml), the glandular volume has reduced by 46%. The tumour and normal peripheral zone are still visible. Transfer constant map shows a decrease in transfer constant both in the tumour and central gland (0.28 and 0.53 min^{-1}). The peripheral zone transfer constant has also reduced to 0.07 min^{-1}. [Images reproduced from Padhani et al. (2001) with kind permission]

(Beerlage et al. 2000). Histopathology of prostatic xenografts has revealed intratumoral haemorrhage, disruption of tumour vasculature, and necrosis in the focus of the ultrasound field (Huber et al. 1999). Histological examination in humans has shown periurethral necrosis following laser treatments for obstructive benign prostatic hyperplasia (Boni et al. 1997). A number of studies have evaluated contrast enhanced MRI in evaluating the effectiveness of such treatments and early results indicate that reductions in enhancement (often called „perfusion defects") closely correlate with the treatment volume (Boni et al. 1997; Huber et al. 1999; Osman et al. 2000).

12.4.4.6 Detecting Relapse

Determining the site of suspected recurrent prostate cancer after radical local treatment can be challenging. A slow rising PSA level may be the only evidence of local treatment failure. Following radical prostatectomy, both TRUS and T_2-weighted MRI are difficult to interpret. The lack of normal landmarks and the presence of scar tissue can lead to diagnostic confusion. Determining the site of recurrence is important because men with isolated local recurrence can benefit from further treatments such as radiation to a prostatectomy resection bed. Endorectal MRI possibly combined with phased array coils has shown some utility (Huch Boni et al. 1995b) and may be useful in those patients with elevated PSA levels and in whom metastatic disease elsewhere has been excluded. Recently, Takeda et al. (2002) have shown that DCE-MRI is able to detect cancer recurrence following a radical prostatectomy even before it can be detected by biopsy. Takeda et al. (2002) studied 16 patients who had a rising PSA level following radical prostatectomy. All patients had an ultrasound-guided biopsy that came back negative and had PSA level below 3.3 ng/ml. In these patients, DCE-MRI indicated local recurrence

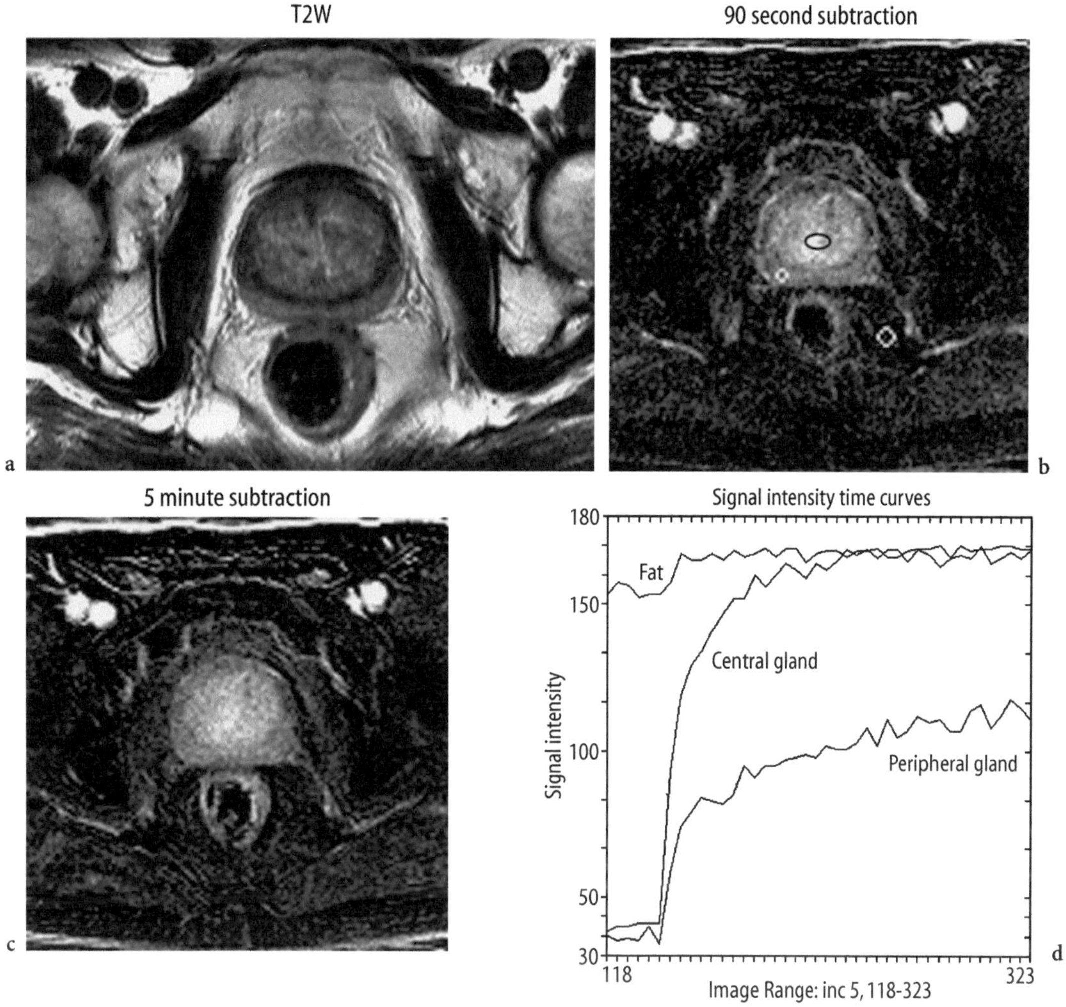

Fig. 12.11a–d. Typical post-radiotherapy prostate gland. A 75-year-old man with prostate cancer (presenting stage T1C, PSA 42 ng/ml, Gleason grade 7). MRI assessment 2 years post-radiotherapy (74 Gy) with PSA 1.1 ng/ml. Current follow-up period 4 years with no biochemical evidence of relapse. **a** T_2-weighted image showing a darkened prostate gland with reduced signal intensity of the peripheral gland. Some zonal differentiation is visible. No tumour relapse is seen. **b** A 90-s subtraction image with regions of interest in the peripheral gland, central gland and ischiorectal fat. **c** A 5-min subtraction image shows low-grade enhancement of the whole prostate. **d** Signal intensity time curve from the regions indicated in the early subtraction image. The peripheral gland enhancement is similar to that observed in the untreated state. The central gland enhancement is brisk but without significant washout

in 13 patients diagnosed on the basis of early, nodular enhancement within the prostatectomy bed. Eight of these 13 patients were treated with local radiation therapy with a fall in PSA level. These results suggest that prostate cancer patients who have a rising PSA level following a radical prostatectomy should undergo an MRI examination first to determine if their cancer has returned. A typical example is shown in Fig. 12.12. What remains to be determined is the sensitivity of DCE-MRI at different levels of serum PSA (particularly when PSA levels are <1 ng/ml or if the PSA doubling time is more than 6 months) and the sensitivity of DCE-MRI in the clinical setting of a rising serum PSA level after radical radiotherapy (Fig. 12.13).

12.4.4.7
Prognostication

Recently, kinetic parameters derived from DCE-MRI have been shown to be able to predict survival of patients with cervix cancers, with tumours of high vascular permeability having an overall poorer prognosis (Hawighorst et al. 1998). Padhani et al.

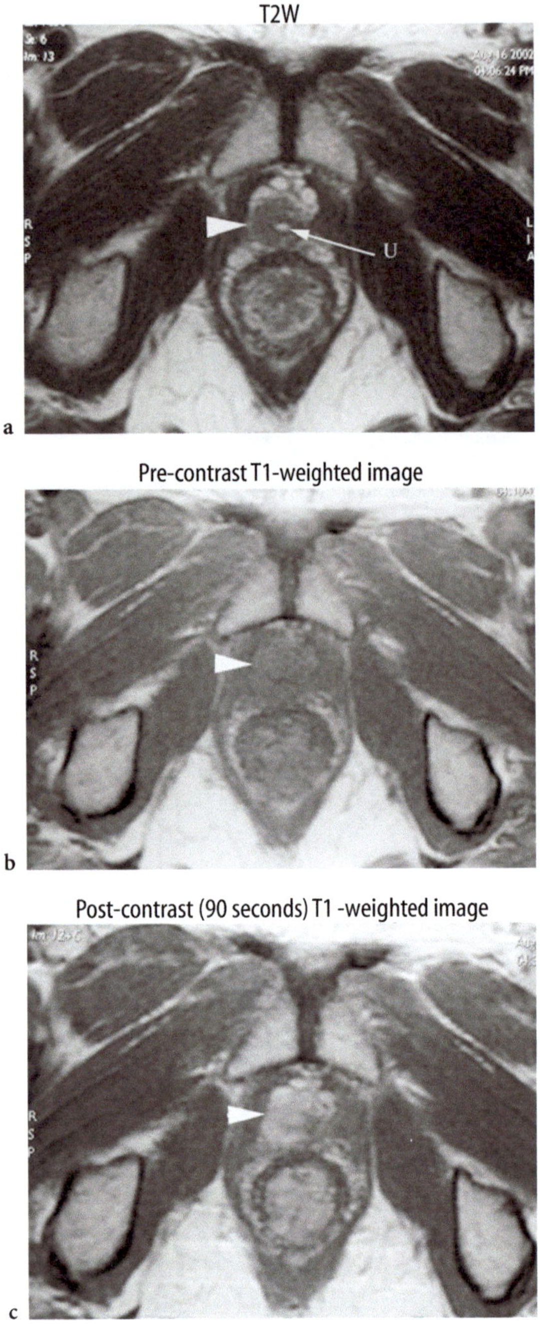

Fig. 12.12a–c. Local tumour recurrence following prostatectomy. A 63-year-old man presenting with raised PSA (9.3 ng/ml) 3 months following radical prostatectomy. Tumour was present at the prostatic apex. Post-operative PSA nadir was 7.3 ng/ml. Bone scan was normal and no enlarged pelvic lymph nodes were seen on MRI. **a,b** T_2- (**a**) and T_1-weighted (**b**) images through the resection bed shows minimal soft tissue (*arrow*) adjacent to the urethra (U) compatible with postoperative scar tissue or residual tumour. **c** T_1-weighted image 90 s following bolus contrast medium administration shows a focal 2.2-cm area of nodular enhancement compatible with local recurrence (*arrow*)

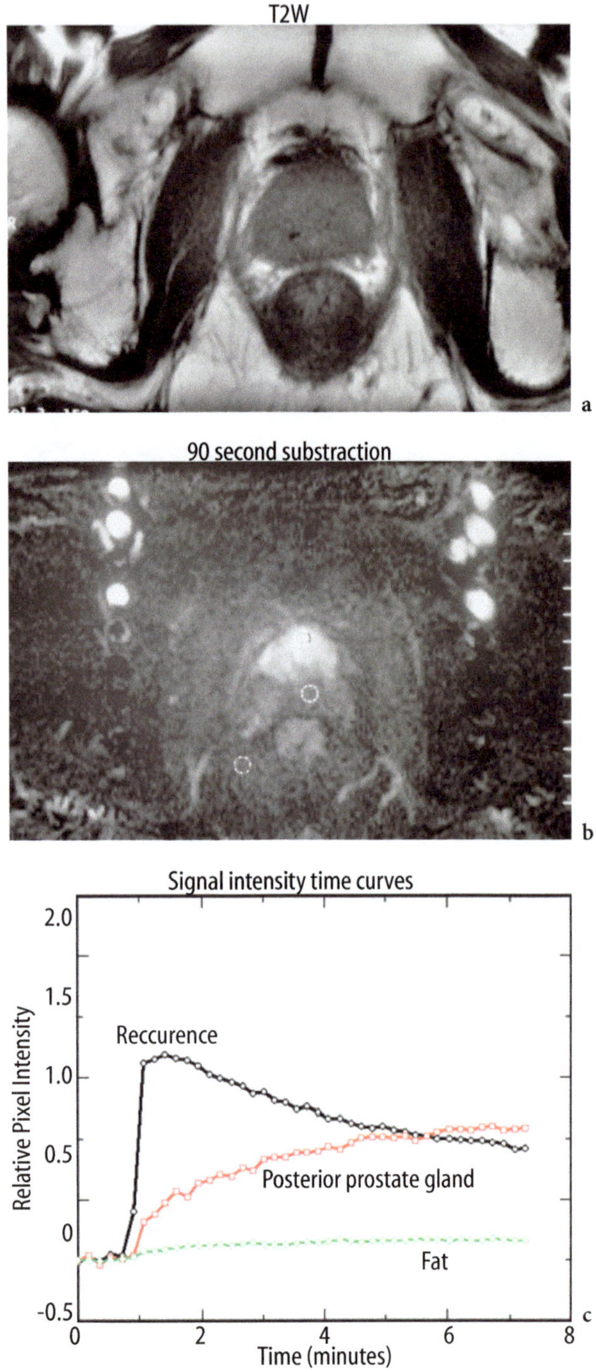

Fig. 12.13a–c. Local tumour recurrence after radical radiotherapy. A 66-year-old man being evaluated for rising PSA (doubling time<6 months) 3.5 years after radical radiotherapy (65 Gy). **a** T_2-weighted image shows a featureless prostate gland with no definite tumour. **b** A 90-s post-contrast subtraction image shows intense enhancement in the anterior gland with high peak and washout on the signal intensity time graph (**c**). This feature is diagnostic of tumour recurrence, which sharply contrasts with the appearances after radiotherapy when there is no evidence of recurrent disease (see Fig. 12.11)

(2002) evaluated whether kinetic parameters derived by DCE-MRI were able to predict disease outcome in patients with prostate cancer after neoadjuvant androgen deprivation and radiation treatment. They noted that no DCE-MRI kinetic parameter was able to predict time to PSA failure in 49 patients at a median follow-up of 3.9 years. They commented that their study lacked adequate power and that a larger patient cohort was needed to explore the role of DCE-MRI in determining treatment outcome in prostate cancer.

12.5 Conclusions

DCE-MRI techniques utilising low molecular weight contrast media have successfully made the transition from methodological development to pre-clinical and clinical validation. DCE-MRI techniques are now rapidly becoming mainstream clinical tools with recognised indications in the imaging of prostate cancer. Current roles include the tumour staging (depiction of capsular penetration and seminal vesicle invasion) and for the detection of suspected tumour recurrence following definitive treatment. Its role in monitoring tumour response to hormonal treatment and radiation remains to be defined. Limitations of the technique include inadequate lesion characterisation particularly differentiating acute prostatitis from cancer in the peripheral gland and distinguishing between benign prostatic hyperplasia and central gland tumours. ^{1}H-MRSI may also have added value in the evaluation of prostate cancer.

Acknowledgements

The support of Cancer Research UK and the Childwick Trust who respectively support the work of the Clinical Magnetic Resonance Research Group at the Royal Marsden Hospital, UK and at the Paul Strickland Scanner Centre, UK is gratefully acknowledged. I would like to personally acknowledge the gracious support and guidance of Professors Janet E. Husband and Martin O. Leach of the Clinical Magnetic Resonance Research Group at the Institute of Cancer Research and Royal Marsden Hospital, UK.

References

Alexander AA (1995) To color Doppler image the prostate or not: that is the question. Radiology 195:11-13

Barbier EL, Lamalle L, Decorps M (2001) Methodology of brain perfusion imaging. J Magn Reson Imaging 13:496-520

Barke A, I'yasov KA, Kiselev VG et al (2003) Dynamic gadolinium-enhanced MR imaging for radiation therapy monitoring in prostate cancer patients. Proceedings of the International Society of Magnetic Resonance in Medicine, 11th scientific meeting, Toronto, p 1463

Beerlage HP, Thuroff S, Madersbacher S et al (2000) Current status of minimally invasive treatment options for localized prostate carcinoma. Eur Urol 37:2-13

Benjamin LE, Golijanin D, Itin A et al (1999) Selective ablation of immature blood vessels in established human tumors follows vascular endothelial growth factor withdrawal. J Clin Invest 103:159-165

Bettencourt MC, Bauer JJ, Sesterhenn IA et al (1998) CD34 immunohistochemical assessment of angiogenesis as a prognostic marker for prostate cancer recurrence after radical prostatectomy. J Urol 160:459-465

Bezzi M, Kressel HY, Allen KS et al (1988) Prostatic carcinoma: staging with MR imaging at 1.5 T. Radiology 169:339-346

Bigler SA, Deering RE, Brawer MK (1993) Comparison of microscopic vascularity in benign and malignant prostate tissue. Hum Pathol 24:220-226

Boni RA, Sulser T, Jochum W et al (1997) Laser ablation-induced changes in the prostate: findings at endorectal MR imaging with histologic correlation. Radiology 202:232-236

Borre M, Offersen BV, Nerstrom B et al (1998) Microvessel density predicts survival in prostate cancer patients subjected to watchful waiting. Br J Cancer 78:940-944

Bostwick DG (2000) Immunohistochemical changes in prostate cancer after androgen deprivation therapy. Mol Urol 4:101-106

Bostwick DG, Iczkowski KA (1998) Microvessel density in prostate cancer: prognostic and therapeutic utility. Semin Urol Oncol 16:118-123

Bostwick DG, Wheeler TM, Blute M et al (1996) Optimized microvessel density analysis improves prediction of cancer stage from prostate needle biopsies. Urology 48:47-57

Bostwick DG, Grignon DJ, Hammond ME et al (2000) Prognostic factors in prostate cancer. College of American Pathologists Consensus Statement 1999. Arch Pathol Lab Med 124:995-1000

Brawer MK, Deering RE, Brown M et al (1994) Predictors of pathologic stage in prostatic carcinoma. The role of neovascularity. Cancer 73:678-687

Bree RL (1997) The role of color Doppler and staging biopsies in prostate cancer detection. Urology 49:31-34

Brix G, Semmler W, Port R et al (1991) Pharmacokinetic parameters in CNS Gd-DTPA enhanced MR imaging. J Comput Assist Tomogr 15:621-628
Brookes JA, Redpath TW, Gilbert FJ et al (1996) Measurement of spin-lattice relaxation times with FLASH for dynamic MRI of the breast. Br J Radiol 69:206-214
Brown G, Macvicar DA, Ayton V et al (1995) The role of intravenous contrast enhancement in magnetic resonance imaging of prostatic carcinoma. Clin Radiol 50:601-606
Buckley DL, Kerslake RW, Blackband SJ et al (1994) Quantitative analysis of multi-slice Gd-DTPA enhanced dynamic MR images using an automated simplex minimization procedure. Magn Reson Med 32:646-651
Carter HB, Brem RF, Tempany CM et al (1991) Nonpalpable prostate cancer: detection with MR imaging. Radiology 178:523-525
Chen M, Hricak H, Kalbhen CL et al (1996) Hormonal ablation of prostatic cancer: effects on prostate morphology, tumor detection, and staging by endorectal coil MR imaging. AJR Am J Roentgenol 166:1157-1163
Cho JY, Kim SH, Lee SE (1998) Diffuse prostatic lesions: role of color Doppler and power Doppler ultrasonography. J Ultrasound Med 17:283-287
Civantos F, Soloway MS, Pinto JE (1996) Histopathological effects of androgen deprivation in prostatic cancer. Semin Urol Oncol 14:22-31
Clements R (2001) Ultrasonography of prostate cancer. Eur Radiol 11:2119-2125
CRC (2001) Cancer research campaign scientific handbook 2000-2002. Cancer Research Campaign, London, UK
Crone C (1963) The permeability of capillaries in various organs as determined by the use of ‚indicator diffusion' method. Acta Physiol Scand 58:292-305
Daldrup HE, Shames DM, Husseini W et al (1998) Quantification of the extraction fraction for gadopentetate across breast cancer capillaries. Magn Reson Med 40:537-543
D'Amico AV, Whittington R, Schnall M et al (1995) The impact of the inclusion of endorectal coil magnetic resonance imaging in a multivariate analysis to predict clinically unsuspected extraprostatic cancer. Cancer 75:2368-2372
Deering RE, Bigler SA, Brown M et al (1995) Microvascularity in benign prostatic hyperplasia. Prostate 26:111-115
Donahue KM, Burstein D, Manning WJ et al (1994) Studies of Gd-DTPA relaxivity and proton exchange rates in tissue. Magn Reson Med 32:66-76
Duque JL, Loughlin KR, Adam RM et al (1999) Plasma levels of vascular endothelial growth factor are increased in patients with metastatic prostate cancer. Urology 54:523-527
Dvorak HF, Brown LF, Detmar M et al (1995) Vascular permeability factor/vascular endothelial growth factor, microvascular hyperpermeability, and angiogenesis. Am J Pathol 146:1029-1039
Eberhard A, Kahlert S, Goede V et al (2000) Heterogeneity of angiogenesis and blood vessel maturation in human tumors: implications for antiangiogenic tumor therapies. Cancer Res 60:1388-1393
Engelbrecht MR, Huisman HJ, Laheij RJ et al (2003) Discrimination of peripheral zone and central gland prostate cancer from normal prostatic tissue using dynamic contrast-enhanced MR imaging. Radiology 229:248–54
EUCAN Cancer Incidence, mortality and prevalence in the European Union (1996 estimates)
Ferrer FA, Miller LJ, Andrawis RI et al (1998) Angiogenesis and prostate cancer: in vivo and in vitro expression of angiogenesis factors by prostate cancer cells. Urology 51:161-167
Flickinger FW, Allison JD, Sherry RM et al (1993) Differentiation of benign from malignant breast masses by time-intensity evaluation of contrast enhanced MRI. Magn Reson Imaging 11:617-620
Frauscher F, Klauser A, Halpern EJ et al (2001) Detection of prostate cancer with a microbubble ultrasound contrast agent. Lancet 357:1849-1850
Frauscher F, Klauser A, Volgger H et al (2002) Comparison of contrast enhanced color Doppler targeted biopsy with conventional systematic biopsy: impact on prostate cancer detection. J Urol 167:1648-1652
Fregene TA, Khanuja PS, Noto AC et al (1993) Tumor-associated angiogenesis in prostate cancer. Anticancer Res 13:2377-2381
Gettman MT, Pacelli A, Slezak J et al (1999) Role of microvessel density in predicting recurrence in pathologic Stage T3 prostatic adenocarcinoma. Urology 54:479-485
Gibbs P, Young BJ, Turnbull LS (2001) Diagnostic utility of perfusion weighted imaging of the prostate. Proceeding of the UK Radiological Congress, London
Greenlee RT, Hill-Harmon MB, Murray T et al (2001) Cancer statistics 2001. CA Cancer J Clin 51:15-36
Gribbestad IS, Nilsen G, Fjosne HE et al (1994) Comparative signal intensity measurements in dynamic gadolinium-enhanced MR mammography. J Magn Reson Imaging 4:477-480
Haggstrom S, Lissbrant IF, Bergh A et al (1999) Testosterone induces vascular endothelial growth factor synthesis in the ventral prostate in castrated rats. J Urol 161:1620-1625
Hall MC, Troncoso P, Pollack A et al (1994) Significance of tumor angiogenesis in clinically localized prostate carcinoma treated with external beam radiotherapy. Urology 44:869-875
Harvey CJ, Blomley MJ, Dawson P et al (2001) Functional CT imaging of the acute hyperemic response to radiation therapy of the prostate gland: early experience. J Comput Assist Tomogr 25:43-49
Hawighorst H, Knapstein PG, Knopp MV et al (1998) Uterine cervical carcinoma: comparison of standard and pharmacokinetic analysis of time-intensity curves for assessment of tumor angiogenesis and patient survival. Cancer Res 58:3598-3602
Hoffmann U, Brix G, Knopp MV et al (1995) Pharmacokinetic mapping of the breast: a new method for dynamic MR mammography. Magn Reson Med 33:506-514
Hricak H, Dooms GC, McNeal JE et al (1987) MR imaging of the prostate gland: normal anatomy. AJR Am J Roentgenol 148:51-58
Huber P, Peschke P, Brix G et al (1999) Synergistic interaction of ultrasonic shock waves and hyperthermia in the Dunning prostate tumor R3327-AT1. Int J Cancer 82:84-91
Huch Boni RA, Boner JA, Lutolf UM et al (1995a) Contrast-enhanced endorectal coil MRI in local staging of prostate carcinoma. J Comput Assist Tomogr 19:232-237
Huch Boni RA, Trinkler F, Pestalozzi D et al (1995b) The value of high resolution MRI for diagnosis of recurrent prostate cancer. Proceedings of the Society of Magnetic Resonance, 3rd scientific meeting, Nice
Huisman HJ, Engelbrecht MR, Barentsz JO (2001) Accu-

rate estimation of pharmacokinetic contrast-enhanced dynamic MRI parameters of the prostate. J Magn Reson Imaging 13:607-614
Husband JE, Padhani AR, MacVicar AD et al (1998) Magnetic resonance imaging of prostate cancer: comparison of image quality using endorectal and pelvic phased array coils. Clin Radiol 53:673-681
Ito H, Kamoi K, Yokoyama K et al (2003) Visualization of prostate cancer using dynamic contrast-enhanced MRI: comparison with transrectal power Doppler ultrasonography. Br J Radiol 76:617–24
Izawa JI, Dinney CP (2001) The role of angiogenesis in prostate and other urologic cancers: a review. CMAJ 164:662-670
Jackson MW, Bentel JM, Tilley WD (1997) Vascular endothelial growth factor (VEGF) expression in prostate cancer and benign prostatic hyperplasia. J Urol 157:2323-2328
Jager GJ, Ruijter ET, van de Kaa CA et al (1996) Local staging of prostate cancer with endorectal MR imaging: correlation with histopathology. AJR Am J Roentgenol 166:845-852
Jager GJ, Ruijter ET, van de Kaa CA et al (1997) Dynamic TurboFLASH subtraction technique for contrast-enhanced MR imaging of the prostate: correlation with histopathologic results. Radiology 203:645-652
Jain RK, Safabakhsh N, Sckell A et al (1998) Endothelial cell death, angiogenesis, and microvascular function after castration in an androgen-dependent tumor: role of vascular endothelial growth factor. Proc Natl Acad Sci USA 95:10820-10825
Joseph IB, Nelson JB, Denmeade SR et al (1997) Androgens regulate vascular endothelial growth factor content in normal and malignant prostatic tissue. Clin Cancer Res 3:2507-2511
Kahn T, Burrig K, Schmitz-Drager B et al (1989) Prostatic carcinoma and benign prostatic hyperplasia: MR imaging with histopathologic correlation. Radiology 173:847-851
Kaiser WA, Zeitler E (1989) MR imaging of the breast: fast imaging sequences with and without Gd-DTPA. Preliminary observations. Radiology 170:681-686
Kurhanewicz J, Vigneron DB, Hricak H et al (1996) Three-dimensional H-1 MR spectroscopic imaging of the in situ human prostate with high (0.24-0.7-cm3) spatial resolution. Radiology 198:795-805
Lee F, Torp-Pedersen S, Littrup PJ et al (1989) Hypoechoic lesions of the prostate: clinical relevance of tumor size, digital rectal examination, and prostate-specific antigen. Radiology 170:29-32
Leibel SA, Fuks Z, Zelefsky MJ et al (2002) Intensity-modulated radiotherapy. Cancer J 8:164-176
Lencioni R, Menchi I, Paolicchi A et al (1997) Prediction of pathological tumor volume in clinically localized prostate cancer: value of endorectal coil magnetic resonance imaging. Magma 5:117-121
Levine AC, Liu XH, Greenberg PD et al (1998) Androgens induce the expression of vascular endothelial growth factor in human fetal prostatic fibroblasts. Endocrinology 139:4672-4678
Liney GP, Turnbull LW, Knowles AJ (1999) In vivo magnetic resonance spectroscopy and dynamic contrast enhanced imaging of the prostate gland. NMR Biomed 12:39-44
Ling CC, Humm J, Larson S et al (2000) Towards multidimensional radiotherapy (MD-CRT): biological imaging and biological conformality. Int J Radiat Oncol Biol Phys 47:551-560
Lovett K, Rifkin MD, McCue PA et al (1992) MR imaging characteristics of noncancerous lesions of the prostate. J Magn Reson Imaging 2:35-39
Matsushima H, Goto T, Hosaka Y et al (1999) Correlation between proliferation, apoptosis, and angiogenesis in prostate carcinoma and their relation to androgen ablation. Cancer 85:1822-1827
McSherry SA, Levy F, Schiebler ML et al (1991) Preoperative prediction of pathological tumor volume and stage in clinically localized prostate cancer: comparison of digital rectal examination, transrectal ultrasonography and magnetic resonance imaging. J Urol 146:85-89
Meyer GE, Yu E, Siegal JA et al (1995) Serum basic fibroblast growth factor in men with and without prostate carcinoma. Cancer 76:2304-2311
Mirowitz SA, Brown JJ, Heiken JP (1993) Evaluation of the prostate and prostatic carcinoma with gadolinium-enhanced endorectal coil MR imaging. Radiology 186:153-157
Moul JW (1999) Angiogenesis, p53, bcl-2 and Ki-67 in the progression of prostate cancer after radical prostatectomy. Eur Urol 35:399-407
Mueller-Lisse UG, Vigneron DB, Hricak H et al (2001) Localized prostate cancer: effect of hormone deprivation therapy measured by using combined three-dimensional 1H MR spectroscopy and MR imaging: clinicopathologic case-controlled study. Radiology 221:380-390
Murphy WM, Soloway MS, Barrows GH (1991) Pathologic changes associated with androgen deprivation therapy for prostate cancer. Cancer 68:821-828
Nakashima J, Imai Y, Tachibana M et al (1997) Effects of endocrine therapy on the primary lesion in patients with prostate carcinoma as evaluated by endorectal magnetic resonance imaging. Cancer 80:237-241
Namimoto T, Morishita S, Saitoh R et al (1998) The value of dynamic MR imaging for hypointensity lesions of the peripheral zone of the prostate. Comput Med Imaging Graph 22:239-245
Newman JS, Bree RL, Rubin JM (1995) Prostate cancer: diagnosis with color Doppler sonography with histologic correlation of each biopsy site. Radiology 195:86-90
Noseworthy MN, Morton G, Wright GA (1999) Comparison of normal and cancerous prostate using dynamic Tl and T2* weighted MRI. Proceedings of the 7th annual meeting of the international society for magnetic resonance in medicine, Philadelphia
Nutting CM, Convery DJ, Cosgrove VP et al (2000) Reduction of small and large bowel irradiation using an optimized intensity-modulated pelvic radiotherapy technique in patients with prostate cancer. Int J Radiat Oncol Biol Phys 48:649-656
Nutting CM, Corbishley CM, Sanchez-Nieto B et al (2002) Potential improvements in the therapeutic ratio of prostate cancer irradiation: dose escalation of pathologically identified tumour nodules using intensity modulated radiotherapy. Br J Radiol 75:151-161
Offersen BV, Borre M and Overgaard J (1998) Immunohistochemical determination of tumor angiogenesis measured by the maximal microvessel density in human prostate cancer. Apmis 106:463-469
Ogura K, Maekawa S, Okubo K et al (2001) Dynamic endorec-

tal magnetic resonance imaging for local staging and detection of neurovascular bundle involvement of prostate cancer: correlation with histopathologic results. Urology 57:721-726

Okihara K, Miki T, Joseph Babaian R (2002) Clinical efficacy of prostate cancer detection using power Doppler imaging in American and Japanese men. J Clin Ultrasound 30:213-221

Osman YM, Larson TR, El-Diasty T et al (2000) Correlation between central zone perfusion defects on gadolinium-enhanced MRI and intraprostatic temperatures during transurethral microwave thermotherapy. J Endourol 14:761-766

Outwater E, Schiebler ML, Tomaszewski JE et al (1992) Mucinous carcinomas involving the prostate: atypical findings at MR imaging. J Magn Reson Imaging 2:597-600

Padhani AR (2002) Dynamic contrast-enhanced MRI in clinical oncology - current status and future directions. J Magn Reson Imaging 16:407-422

Padhani AR, Khoo VS, Suckling J et al (1999) Evaluating the effect of rectal distension and rectal movement on prostate gland position using cine MRI. Int J Radiat Oncol Biol Phys 44:525-533

Padhani AR, Gapinski CJ, Macvicar DA et al (2000) Dynamic contrast enhanced MRI of prostate cancer: correlation with morphology and tumour stage, histological grade and PSA. Clin Radiol 55:99-109

Padhani AR, MacVicar AD, Gapinski CJ et al (2001) Effects of androgen deprivation on prostatic morphology and vascular permeability evaluated with MR imaging. Radiology 218:365-374

Padhani AR, Parker C, Norman A et al (2002) Dynamic contrast-enhanced MRI of prostate cancer for predicting patient outcome. Proceedings of the International Society of Magnetic Resonance in Medicine, 10th Scientific meeting, Honululu

Parker GJ, Tofts PS (1999) Pharmacokinetic analysis of neoplasms using contrast-enhanced dynamic magnetic resonance imaging. Top Magn Reson Imaging 10:130–142

Parker GJ, Suckling J, Tanner SF et al (1997) Probing tumor microvascularity by measurement, analysis and display of contrast agent uptake kinetics. J Magn Reson Imaging 7:564–574

Parker GJ, Suckling J, Tanner SF et al (1998) MRIW: parametric analysis software for contrast-enhanced dynamic MR imaging in cancer. Radiographics 18:497–506

Parker GJ, Baustert I, Tanner SF et al (2000) Improving image quality and T(1) measurements using saturation recovery turboFLASH with an approximate K-space normalisation filter. Magn Reson Imaging 18:157–167

Patel U, Rickards D (1994) The diagnostic value of colour Doppler flow in the peripheral zone of the prostate, with histological correlation. Br J Urol 74:590–595

Phillips ME, Kressel HY, Spritzer CE et al (1987) Normal prostate and adjacent structures: MR imaging at 1.5 T. Radiology 164:381–385

Pinault S, Tetu B, Gagnon J et al (1992) Transrectal ultrasound evaluation of local prostate cancer in patients treated with LHRH agonist and in combination with flutamide. Urology 39:254–261

Quinn SF, Franzini DA, Demlow TA et al (1994) MR imaging of prostate cancer with an endorectal surface coil technique: correlation with whole-mount specimens. Radiology 190:323–327

Quint LE, van Erp JS, Bland PH et al (1991) Carcinoma of the prostate: MR images obtained with body coils do not accurately reflect tumor volume. AJR Am J Roentgenol 156:511–516

Rifkin MD, Zerhouni EA, Gatsonis CA et al (1990) Comparison of magnetic resonance imaging and ultrasonography in staging early prostate cancer. Results of a multi-institutional cooperative trial. N Engl J Med 323:621–626

Rosen BR, Belliveau JW, Buchbinder BR et al (1991) Contrast agents and cerebral hemodynamics. Magn Reson Med 19:285–292

Rosenman J (2001) Incorporating functional imaging information into radiation treatment. Semin Radiat Oncol 11:83–92

Rubin MA, Buyyounouski M, Bagiella E et al (1999) Microvessel density in prostate cancer: lack of correlation with tumor grade, pathologic stage, and clinical outcome. Urology 53:542–547

Sanders H, El-Galley R (1997) Ultrasound findings are not useful for defining stage T1c prostate cancer. World J Urol 15:336–338

Schiebler ML, Tomaszewski JE, Bezzi M et al (1989) Prostatic carcinoma and benign prostatic hyperplasia: correlation of high-resolution MR and histopathologic findings. Radiology 172:131–137

Schnall MD, Imai Y, Tomaszewski J et al (1991) Prostate cancer: local staging with endorectal surface coil MR imaging. Radiology 178:797–802

Secaf E, Nuruddin RN, Hricak H et al (1991) MR imaging of the seminal vesicles. AJR Am J Roentgenol 156:989–994

Shearer RJ, Davies JH, Gelister JS et al (1992) Hormonal cytoreduction and radiotherapy for carcinoma of the prostate. Br J Urol 69:521–524

Silberman MA, Partin AW, Veltri RW et al (1997) Tumor angiogenesis correlates with progression after radical prostatectomy but not with pathologic stage in Gleason sum 5 to 7 adenocarcinoma of the prostate. Cancer 79:772–779

Smith DM, Murphy WM (1994) Histologic changes in prostate carcinomas treated with leuprolide (luteinizing hormone-releasing hormone effect). Distinction from poor tumor differentiation. Cancer 73:1472–1477

Smith DS, Catalona WJ, Herschman JD (1996) Longitudinal screening for prostate cancer with prostate-specific antigen. JAMA 276:1309–1315

Sommer FG, McNeal JE, Carrol CL (1986) MR depiction of zonal anatomy of the prostate at 1.5 T. J Comput Assist Tomogr 10:983–989

Sommer FG, Nghiem HV, Herfkens R et al (1993) Determining the volume of prostatic carcinoma: value of MR imaging with an external-array coil. AJR Am J Roentgenol 161:81–86

Sorensen AG, Tievsky AL, Ostergaard L et al (1997) Contrast agents in functional MR imaging. J Magn Reson Imaging 7:47–55

Strohmeyer D, Rossing C, Bauerfeind A et al (2000) Vascular endothelial growth factor and its correlation with angiogenesis and p53 expression in prostate cancer. Prostate 45:216–224

Swanson MG, Vigneron DB, Tran TK et al (2001) Magnetic resonance imaging and spectroscopic imaging of prostate cancer. Cancer Invest 19:510–523

Takeda M, Akiba H, Yama N et al (2002) Value of multi-sectional fast contrast-enhanced MR imaging in patients with elevated PSA levels after radical prostatectomy. American Roentgen Ray Society, Atlanta. Am J Roentgenol 178(S):97

Tanaka N, Samma S, Joko M et al (1999) Diagnostic usefulness of endorectal magnetic resonance imaging with dynamic contrast-enhancement in patients with localized prostate cancer: mapping studies with biopsy specimens. Int J Urol 6:593–599

Taylor JS, Reddick WE (2000) Evolution from empirical dynamic contrast-enhanced magnetic resonance imaging to pharmacokinetic MRI. Adv Drug Deliv Rev 41:91–110

Thornbury JR, Ornstein DK, Choyke PL et al (2001) Prostate cancer: what is the future role for imaging? AJR Am J Roentgenol 176:17–22

Tofts PS (1997) Modeling tracer kinetics in dynamic Gd-DTPA MR imaging. J Magn Reson Imaging 7:91–101

Tofts PS, Brix G, Buckley DL et al (1999) Estimating kinetic parameters from dynamic contrast-enhanced T(1)-weighted MRI of a diffusable tracer: standardized quantities and symbols. J Magn Reson Imaging 10:223–232

Turnbull LW, Buckley DL, Turnbull LS et al (1999) Differentiation of prostatic carcinoma and benign prostatic hyperplasia: correlation between dynamic Gd-DTPA-enhanced MR imaging and histopathology. J Magn Reson Imaging 9:311–316

Van Dorsten F, Engelbrecht MR, Van de graaf JJ et al (2001) Combined dynamic contrast-enhanced MRI and 1H MR spectroscopy may differentiate prostate carcinoma from chronic prostatitis. Radiology 211(P):585

Vigneron DB, Males RG, Noworolski S et al (1998) 3D MRSI of prostate cancer: correlation with histologic grade. Proceedings of the international society for magnetic resonance in medicine, 6th scientific meeting, Sidney

Walsh K, Sherwood RA, Dew TK et al (1999) Angiogenic peptides in prostatic disease. BJU Int 84:1081–1083

Wefer AE, Hricak H, Vigneron DB et al (2000) Sextant localization of prostate cancer: comparison of sextant biopsy, magnetic resonance imaging and magnetic resonance spectroscopic imaging with step section histology. J Urol 164:400–404

Weidner N, Carroll PR, Flax J et al (1993) Tumor angiogenesis correlates with metastasis in invasive prostate carcinoma. Am J Pathol 143:401–409

Weingartner K, Ben-Sasson SA, Stewart R et al (1998) Endothelial cell proliferation activity in benign prostatic hyperplasia and prostate cancer: an in vitro model for assessment. J Urol 159:465–470

Yu KK, Scheidler J, Hricak H et al (1999) Prostate cancer: prediction of extracapsular extension with endorectal MR imaging and three-dimensional proton MR spectroscopic imaging. Radiology 213:481–488

Zelefsky MJ, Fuks Z, Happersett L et al (2000) Clinical experience with intensity modulated radiation therapy (IMRT) in prostate cancer. Radiother Oncol 55:241–249

Zhong H, De Marzo AM, Laughner E et al (1999) Overexpression of hypoxia-inducible factor 1alpha in common human cancers and their metastases. Cancer Res 59:5830–5835

13 Dynamic Contrast-Enhanced MR Imaging in Musculoskeletal Tumors

June S. Taylor and Wilburn E. Reddick

CONTENTS

Dedication

In memory of Charles B. Pratt, MD (1930–2002), a rare combination of gentleman, scholar, and respected practitioner of oncology. He cared for his patients and his profession, and took the time to teach us why.

J. S. Taylor, PhD
Associate Professor, Department of Radiology, University of Utah, Utah Center for Advanced Imaging Research, 729 Arapeen Drive, Salt Lake City, UT 84108, USA
W. E. Reddick, MD, PhD
Assistant Member, Department of Diagnostic Imaging, St. Jude Children's Research Hospital, 332 North Lauderdale Street, Memphis, TN 38105-2794, USA

13.1 Introduction

13.1.1 Four Oncologic Questions

Radiologists are required to provide information on four clinical questions in imaging primary musculoskeletal neoplasms: diagnosis, staging, response to pre-surgical (or initial) chemotherapy[1], and detection of recurrence. MRI is useful, together with radiography and clinical data, in narrowing the differential diagnosis and establishing the extent of disease. However, it rarely provides a definitive diagnosis for these tumors. MRI is the pre-eminent imaging modality for the remaining three clinical questions (Bloem et al. 1997).

MRI has dramatically improved the pre-operative staging of these tumors. As a result, there has been a major reduction in the incidence of mutilating surgery. Limb-sparing procedures are now commonly used for primary osseous or soft-tissue neoplasms, with resulting improvement in the quality of life for these survivors. Patients with primary osseous and soft-tissue neoplasms have also benefited enormously from developments in initial chemotherapies. Initial chemotherapy is administered prior to tumor resection and has led to improved surgical outcomes and improved survival. For some tumors and/or stages, response to this initial therapy correlates positively with overall survival, in which case measuring this response as precisely and accurately as possible is vital.

MRI is also important in screening for the presence of residual or recurrent tumor. The detection of small recurrent or residual tumor, following initial treatment, poses a challenge. The problem is further complicated by susceptibility artifacts on CT or MR images arising from reconstructive hardware or metallic particles from surgical instruments. In spite

[1] Pre-surgical chemotherapy is sometimes referred to as induction or neoadjuvant therapy.

of this, MRI has proved to have superior sensitivity and specificity to CT in the detection of recurrent musculoskeletal tumor (Berger et al. 2000; Bloem et al. 1997). In patients who have received radiation therapy, radiation-induced soft tissue edema may arise and persist for years. In such cases, dynamic contrast-enhanced MRI (DCE-MRI) may be useful in discriminating between radiation effects, which enhance more slowly, and recurrent tumor, which has a much faster rate of contrast uptake.

13.1.2 MR Imaging of Bone and Soft-Tissue Tumors

A broad consensus on diagnostic assessment of musculoskeletal tumors has emerged over the past decade. Radiographs, clinical data and MRI can narrow the differential diagnosis, and in some cases precisely diagnose the tumor type (Ma 1999; van der Woude et al. 1998a). However, biopsy is still required for accurate diagnosis of musculoskeletal lesions. Protocols for MR imaging and the intricacies of interpretation in diagnosing musculoskeletal lesions have been well described in recent reviews (van der Woude et al. 1998a; Ma 1999; Berger et al. 2000). Table 13.1 shows a basic protocol for imaging musculoskeletal tumors that follows current recommendations (Ma 1999; Berger et al. 2000). Fat suppression is necessary to improve discrimination between fatty marrow components and tumor or red marrow; it improves the sensitivity of detecting edema and fluid collections in the marrow and soft tissues. Some prefer short-tau inversion recover (STIR) images to fat-suppressed T2-weighted spin-echo images, because STIR fat suppression is less susceptible to the effects of local field inhomogeneities; on the other hand, the SNR of STIR images is lower. This choice is also influenced by the field strength and other characteristics of the MR imaging equipment.

Table 13.1. Representative MR imaging procedure for musculoskeletal tumors

Contrast medium	Orientation	Sequence	Contrast weighting
Pre	Axial	Spin echo	T1
Pre	Axial	Fast spin echo	T2 with Fat Sat[a]
Pre	Longitudinal[b]	Spin echo	T1
Pre	Longitudinal	Fast spin echo	T2 with Fat Sat and/or STIR[c]
Post	Axial	Spin echo	T1 with Fat Sat
Post	Longitudinal	Fast spin echo	T1

[a] Fat Sat: fat saturation with chemical-shift-selective presaturation RF pulse; this method cannot be used when prostheses are present.
[b] Longitudinal: coronal or sagittal plane along long axis of bone, including nearest joint in FOV.
[c] STIR: short-tau inversion recovery; also suppresses fat signal.

There has been disagreement among musculoskeletal radiologists on the necessity for contrast administration in MRI examinations of musculoskeletal tumors (Berger et al. 2000; van der Woude and Egmont-Petersen 2001). A case can be made that neither static nor dynamic contrast-enhanced imaging is critical to the diagnosis of these lesions (van der Woude and Egmont-Petersen 2001), because the use of contrast has not been shown to improve the discrimination between benign and malignant tumors, except in cartilaginous tumors. Therefore biopsy is required for any lesions suspicious for malignancy. However, contrast administration should be planned for examinations done for staging, for measuring response to therapy, and for follow-up. DCE-MRI is more discriminating than static contrast MRI for the staging of musculoskeletal lesions, and DCE-MRI may also be necessary to guide the biopsy for optimal tumor sampling. Unless malignancy can be ruled out definitively (as for some bone and soft tissue cysts, lipomas and hemangiomas), it is prudent to plan the use of DCE-MRI in the staging examination of these tumors.

For staging of primary neoplasms of bone and soft tissue, and for surgical planning, including biopsy, MRI is the imaging examination of choice (van der Woude and Egmont-Petersen 2001). It is superior in determining the extent of tumor and the tumor's relation to adjacent muscles and neurovascular structures. It is thus invaluable in planning biopsies and surgical treatment, both of which are critically important in many of these lesions. Staging and treatment of musculoskeletal lesions have progressed to the point that limb-sparing surgery for extremity lesions is frequently undertaken. Biopsies of possibly malignant musculoskeletal tumors are planned with two priorities in mind: first, to obtain the most accurate diagnosis, and second, to avoid compromising the future treatment and prognosis of the patient.

13.1.3 Rationale for DCE-MRI in Advancing Treatment of Musculoskeletal Tumors

DCE-MRI generally refers to a series of T1-weighted images acquired before, during and after contrast injection. It has been successfully performed on imaging systems with fields from 0.5–1.5 T, although higher field strengths clearly are preferable to achieve the best temporal and spatial resolution of tumor. Historically, the data have been collected by 2D spin-echo or gradient-echo imaging of single or multiple slices (VERSTRAETE et al. 1994b; HANNA et al. 1992; REDDICK et al. 1996), and more recently by 3D spoiled gradient echo imaging (KNOPP et al. 2001). 3D imaging has the advantage of more complete coverage of tumor, although temporal sampling times short enough for the question addressed must be maintained. It may also be difficult to include a feeding artery in the 3D slab for simultaneous determination of the dynamic (intensity vs. time) curve in the blood, which improves reproducibility (RIJPKEMA et al. 2001). The temporal sampling required is discussed below for specific applications.

13.1.3.1 Present Applications

Contrast-enhanced MRI, particularly DCE-MRI, is the method of choice for monitoring response to pre-operative chemotherapy in certain malignant musculoskeletal tumors. DCE-MRI is broadly useful for detecting residual or recurrent tumor after surgery (VAN DER WOUDE and EGMONT-PETERSEN 2001; MA et al. 1997; VERSTRAETE and LANG 2000). The experience with DCE-MRI for staging local extent of these tumors and for planning biopsy procedures are discussed below under specific families of musculoskeletal tumors.[2]

The earliest successful application of DCE-MRI to oncologic imaging was in assessing osteosarcoma response to therapy (ERLEMANN et al. 1989; FLETCHER 1991; BONNEROT et al. 1992). Primary bone tumors, in particular, provide a near-ideal system for evidence-based imaging response studies, because at the end of initial chemotherapy the tumor is resected en bloc and the ground truth for tumor response is available from histological analyses of slides matched to the imaging plane(s) (EGMONT-PETERSEN et al. 2000; HANNA et al. 1993). In the decade and a half since the initial results, numerous studies have confirmed the efficacy of DCE-MRI in assessing tumor necrosis in osteosarcoma. There is enormous variability in tumors and therefore it is important to know how best to utilize DCE-MRI for particular clinical questions. Researchers have obtained promising results on the usefulness of DCE-MRI in assessing response to initial chemotherapy for other, rarer musculoskeletal tumors, specifically Ewing's family tumors and soft-tissue sarcomas.

Malignant musculoskeletal tumors resemble normal bone or soft tissues in immunogenic and other properties, but their architecture is different in key ways that can be exploited by DCE-MRI. Their growth and development are chaotropic in varying degrees. This leads to abnormalities in the architecture and permeability of their microvasculature (JAIN 1994, 1996b), and sometimes to remarkable spatial heterogeneity in perfusion of tumor cells. Also, the interstitial space (the extracellular-extravascular space, or EES, compartment) of malignant tumors may be many times larger than normal or even edematous tissues (GULLINO and GRANTHAM 1964), so that uptake into the EES compartment is often a significant part of enhancement in tumors. The distribution of this EES is often inhomogeneous, with some tumor regions having expanded EES and others having a more normal EES.

DCE-MRI can be used to assess the microcirculation and EES of malignant bone sarcoma. It provides a way to discriminate between the uptake of contrast into necrotic or edematous tissue and the leakage of contrast agent from angiogenic microvasculature into the EES of viable tumor. It does this by using either semi-quantitative or quantitative measures of contrast uptake. Separating viable tumor from necrotic tumor and edema is valuable in the staging and biopsy of these lesions. Identifying the fraction of dead tumor just before tumor resection gives a measure of response to initial chemotherapy, and response to therapy has predictive value for overall survival, particularly in patients with local (M0) disease. Moreover, the rate of contrast uptake into various tissue regions appears to serve as a surrogate for the access of drug to these regions, at least in certain tumor types. Basic research in animal tumor models has shown that transient irradiation-induced increases in tumor capillary permeability to cisplatin can be quantified with DCE-MRI (SCHWICKERT et al. 1996). More recent animal studies of solid tumor response to chemotherapy (SU et al. 1999) and to gene therapy (SU et al. 2000) have demonstrated the ability of DCE-MRI to detect vascular changes correlated with response or progression. In clinical observations of numerous series of patients with bone and soft tissue sarcoma, measures of contrast uptake (contrast access) have convincingly demonstrated a relationship with measures and predictions of the tumor's

[2] The staging of musculoskeletal lesions is complex and, with respect to soft-tissue sarcomas, a work in progress. The interested reader is referred to articles by (PEABODY et al. 1998; GEBHARDT 2002).

response to preoperative chemotherapy. The results of these studies have indicated that greater access at the time of presentation, greater decrease in access during therapy, and low access at the completion of preoperative therapy correspond to better response and longer disease-free survival (Verstraete and Lang 2000; Reddick et al. 2001). Consequently, DCE-MRI studies of pre-operative response to therapy should be considered for every patient with a chemotherapy-sensitive musculoskeletal malignancy, for the purposes of adjusting the plan of pre-surgical chemotherapy, revising surgical planning and choosing optimal post-surgical therapy.

13.1.3.2 Future Directions

For chemotherapy to be effective, therapeutic agents must reach target cells (tumor or endothelial) in adequate concentrations and with minimal toxicity to normal tissues (Jain 1996a). Once a blood-borne molecule reaches an exchange capillary, transport across the vessel wall is a function of the surface area of exchange, the transvascular concentration gradient and the interstitial fluid pressure. Tumor interstitial pressure can be much higher than that in normal tissues, to the point of collapsing perfusing microvessels. It was hypothesized that, because of aberrations in the architecture of microvessels and interstitial pressure, some solid tumors may resist drug penetration (Jain 1994). Such *physiological resistance* may play a significant role in treatment failure (Tannock et al. 2002; Jain 2001), and is a different mechanism of resistance to therapy from the better-understood *cellular resistance* phenomenon conferred by the presence of resistance molecules, such as the multiple drug resistance protein Mdr1, in tumor cells. However, it has been difficult to test this hypothesis of physiological resistance in clinical studies, in the absence of non-invasive repeatable measures of drug penetration that sample tumor compartments other than plasma. Now that advances in tumor biology and intelligent drug design are producing many novel therapeutic agents for clinical trials, it is increasingly important to discriminate between *cellular resistance* of tumor cells to drug and *physiological resistance* arising from a failure to expose poorly-perfused tumor regions to sufficient concentrations of drug. This concern is driving the interest of researchers in DCE-MRI as a methodology which can supply new, urgently-needed pharmacokinetic information on the efficiency and heterogeneity of drug access to solid tumors in individual patients.

Optimizing the spatial and temporal sampling rates for DCE-MRI applications is complex, because faster image acquisition rates increase the temporal resolution of the dynamic curve (time-intensity curve), but the signal-to-noise ratio (SNR) and/or spatial resolution of each image is decreased. Likewise, there is the usual trade-off between SNR and better spatial resolution (voxel size) in the DCE images. Current protocols for 2D and 3D DCE-MRI have been published by several European researchers: Shapeero and Vanel (2000); Rijpkema et al. (2001) for 2D and Knopp et al. (2001) for 3D. However, the DCE-MRI protocol should begin prior to contrast injection, as emphasized by Knopp et al. (2001). The sequence used should give tissue signal intensity proportional to contrast concentration in the tissue.

It is important to recognize that reduction of the SNR below a certain threshold will be detrimental to the reliability and sensitivity of the calculations of DCE-MRI measures. Noisy dynamic curves lead to decreases in reliability of both empirical and model estimates, unless methods that are robust to noise are employed. Analyses by region of interest (ROI), which averages together the information from many pixels, can be used to obtain dynamic (time vs. intensity) curves that have higher SNR than curves from individual pixels. This trading of spatial information for SNR is not always possible for all tumors in all applications. A rich area for research is the development of algorithms that can cope with noisy data while maintaining the spatial resolution of DCE images.

13.2 Primary Osseous Sarcomas

13.2.1 Pediatric Osteosarcoma

Osteosarcoma is the most common malignant bone tumor of childhood (60%) (Link and Eilber 1993). Current treatments for these tumors consist of several cycles of initial chemotherapy given before surgical ablation of the osteosarcoma for local control (Fletcher 1997). Preoperative chemotherapy is used to treat systemic disease (micrometastases), provide time for planning surgical therapy (including manufacturing customized prostheses), and allow time for postoperative healing before postoperative chemotherapy is given (Link et al. 1986; Rosen et al. 1979). It is important to note that the period of preoperative chemotherapy can be used to evaluate new chemotherapeutic agents in phase 2 trials.

The use of preoperative chemotherapy has produced a dramatic improvement in prognosis for children with bone sarcoma (LINK et al. 1986). Recent studies show at least 70% of children with osteosarcoma now survive (GRIER 1997). Accurate determination of response is essential in assessing the effectiveness of investigational and standard drug therapies during the pre-surgical treatment window.

There is a paucity of data on primary osteosarcoma in adults. While clinical, histopathologic and prognostic features are well understood for children (under 20) with osteosarcoma, much less is known about these factors for adults. There is a tendency for axial location of primary osteosarcoma in adults (HUVOS 1986; CARSI and ROCK 2002), compared to children. The distribution of histologic variants in adults is similar to those seen in pediatric osteosarcoma (CARSI and ROCK 2002). Negative prognostic factors appear to be metastases prior to or during treatment, large tumor volumes, a pathologic fracture prior to treatment, and inadequate surgical margins. These are not different in kind from negative factors that are known for pediatric osteosarcoma. One possibility is that the poorer outcome in treating primary osteosarcoma in adults is related to the higher fraction of adults presenting with large, advanced stage tumors and the greater fraction of axial tumors, which are much harder to resect cleanly. Reports to date on the response rate for adults to chemotherapy are conflicting (ANTMAN et al. 1998; BACCI et al. 1998; BIELACK et al. 2002). Results of older treatment regimens in adults with primary osteosarcoma have been poor (15%–40% survival rate), and prospective trials of more aggressive multi-agent and multi-modality protocols are needed. European trials have provided evidence of better survival for adults treated with therapy protocols similar to those used in children (BACCI et al. 1998; BIELACK et al. 2002).

13.2.1.1
Rationale for DCE-MRI of Osteosarcoma

Osteosarcoma, like all malignant musculoskeletal tumors, requires biopsy for definitive diagnosis. DCE-MRI is useful for staging and for optimal targeting of tumor biopsy in these large, heterogeneous tumors (GEIRNAERDT et al. 1998; VAN DER WOUDE et al. 1998b).

Evaluating response of bone sarcoma to initial chemotherapy by imaging methods is a challenge. Even osteosarcoma that have become largely necrotic may not shrink significantly, perhaps because of their usually extensive osteoid and bony matrix (WELLINGS et al. 1994). At present, the gold standard for assessing the effects of chemotherapy is histological examination of a representative section of the resected tumor (HUVOS et al. 1977; PICCI et al. 1985; SALZER-KUNTSCHIK et al. 1983; RAYMOND et al. 1987). Response is considered good if there is at least 90% tumor cell necrosis (ROSEN et al. 1982). However, some centrally located tumors may not be resectable and therefore cannot be evaluated by histology.

Tumor necrosis resulting from initial chemotherapy is strongly associated with effective local control. The degree of response is an important prognostic factor that can be used to plan post-surgical treatment and to optimize the timing of surgery (RAYMOND et al. 1987; GLASSER et al. 1992; ROSEN et al. 1979, 1982; PICCI et al. 1994; HUDSON et al. 1990). Response is also an important determinant of the patient's eligibility for limb-salvage procedures without increased risk of local recurrence (PICCI et al. 1994; GHERLINZONI et al. 1992).

This section will focus on two vital issues in the treatment of children with primary osteosarcoma: (1) how to detect response of the primary tumor to initial chemotherapy, and (2) how to identify at initial staging those tumors that are likely to fail pre-surgical therapy. A first approach to answering the response question requires comparing imaging methods with histology. To validate the imaging method for grading response requires the additional step of comparing both histological and imaging methods to the most important outcome measure, disease-free survival, in the same population. Likewise, success in predicting the risk of treatment failure at diagnosis requires, first, comparison of the predictive imaging measure with histopathology and imaging at the end of the initial chemotherapy window, and for full validation, comparison with patient survival.

No satisfactory clinical or static imaging methods exist for determining response of osteosarcoma to cytotoxic drugs before resection (ERLEMANN et al. 1990). Originally, conventional static MR imaging showed promise in response assessment (HOLSCHER et al. 1990, 1992; PAN et al. 1990). However, MR signal intensities of viable and necrotic tumor, edema, and hemorrhage overlap on T2-weighted images (PAN et al. 1990; SANCHEZ et al. 1990). In addition, the use of STIR and contrast-enhanced T1-weighted MR imaging overestimates the extent of viable tumor because of enhancement of nonmalignant reactive tissue (ERLEMANN et al. 1990; ONIKUL et al. 1996; GLAZER et al. 1985). A large study of the effectiveness of CT scanning and conventional MR imaging for evaluating primary tumor response on the basis of several criteria showed that the ability of these imaging modalities to predict histopathology response and outcome was limited (Lawrence et al. 1993).

Other imaging methods that have been used to assess response in bone sarcoma before surgery include conventional radiography, computed tomography, radionuclide scintigraphy, angiography, and ultrasonography, as well as conventional static MR imaging. All proved less than satisfactory in measuring response, as summarized in a recent review (Fletcher 1997). On the other hand, there are now multiple clinical studies from several groups on DCE-MRI for measuring response to initial chemotherapy, and these collectively show that DCE-MRI measures of contrast agent uptake into osteosarcoma provide a reliable, accurate measure of response.

This agreement is satisfying. However, a variety of semi-quantitative and quantitative measures were extracted from the DCE-MR images to arrive at these results, and the multiplicity of measures has created confusion. It is worth comparing the DCE-MRI measures applied to assess response in osteosarcoma studies, as a basis for discussing the pitfalls in generalizing these to other musculoskeletal tumors and other oncologic questions.

13.2.1.2 Acquisition and Analyses of DCE-MRI of Osteosarcoma

Osteosarcoma occur most frequently in the long bones of the extremities and expand in both the intramedullary cavity and the surrounding soft tissue. Therefore, the localization of the tumor is somewhat constrained by the bone. DCE-MRI acquisitions in a single 10-mm thick plane with in-plane resolution of approximately 3 mm^2 can assess response and identify foci of residual tumor with parametric maps, which can be directly compared with histologic maps. Both measuring response and predicting response to therapy prospectively may be done by assessing a single coronal slice, as is the practice in histologic measures, but prediction in particular may prove most accurate when the entire tumor is sampled. For best discrimination, the choice of a temporal sampling rate should adequately sample the fastest accumulation rate. In some of the best-perfused regions of an osteosarcoma, uptake of the contrast agent can equilibrate in as little as 25–30 s; thus, a minimum sampling rate of one image every 13–15 s is required for osteosarcoma. In order to evaluate patient image data using a kinetic model with a plasma term (arterial input function, or AIF) the DCE-MRI would have to sample the dynamic curve at approximately one image every 3 s.

Until recently, the majority of clinical investigations have analyzed DCE-MRI by extracting a variety of empirical parameters as measures. The majority of radiologists must still do so, lacking the expert software to extract the contrast medium concentration–time curve from the DCE-MRI images. The semi-quantitative analyses calculate or estimate parameters that are related to this concentration–time curve: the absolute change in enhancement over some time period, or the relative change (divided by the baseline intensity before contrast injection), or the rate of relative (or absolute) enhancement in the initial period (first 30-60 s) after the contrast injection, or in some few cases the delay in contrast reaching the ROI. The enhancement may or may not be normalized to baseline intensity or compared with other normal tissues in the imaging plane. One of the most popular intensity-based methods takes the ratio of the relative signal intensity enhancement over the time required for the enhancement, to calculate a rate of enhancement (Slope and percentage Slope). This accumulation rate, if it is computed over a short time period (<30 s for osteosarcoma), provides a measure of the rate of contrast agent accumulation in the vasculature and tumor interstitial spaces. Another measure is the maximum enhancement that occurred at any point in the signal intensity curve. The maximum enhancement provides a measure of the total accumulation of contrast agent in the tumor vasculature and interstitial spaces, and this is related to the maximum concentration of contrast agent accumulated in these tumor spaces. Some studies showed it was useful to combine both these measures into a lumped measure, the dynamic magnitude vector (DVM), which is a weighted sum reflecting both the maximum magnitude and the initial rate of enhancement.

More recently, researchers began extracting the contrast medium concentration–time curve from DCE-MR images. This allowed them to apply established compartmental (pharmacokinetic) methods to model k_{ep}, the flux rate constant between EES and plasma, for low-molecular weight (MW) contrast from plasma to tumor EES (related to the slope measures) and the extent to which these agents accumulate in the EES space of the tumor (related to the maximum enhancement measures). Although very complex models can be developed to describe indicator distribution, a simpler two-compartment model may be more appropriate to the limited SNR and spatial resolution of the DCE-MRI images. A conventional two-compartment pharmacokinetic model consisting of plasma volume and EES volume has been used in most studies that have analyzed the distribution of contrast in each image voxel. Tofts et al. (1999) have reported a set of standardized quantities and symbols to be used in estimating kinetic parameters from DCE-MRI (Tofts et al. 1999). The most common parameter reported in the literature is the k_{ep}. As we will show below, the initial rate measures can provide a more or less accurate approximation to the pharmacokinetic parameter $k_{ep.}$

To demonstrate the enhancement-time curves seen, an actual clinical osteosarcoma case (Fig. 13.1) shows these curves for regions seen in a very large tumor of the distal right femur, which was resected and sectioned in the same plane as the MRI studies, for histopathology correlation. According to histopathology, this tumor responded very poorly to presurgical therapy, with less than 10% necrosis at resection. The characteristic shape of the dynamic signal after the contrast arrives at the tumor is shown for four different regions of the tumor (taken from DCE-MR images just prior to surgery). The first region of the tumor (A) is from the central core, which in this tumor is avascular and necrotic. The lack of signal change during DCE-MRI precludes assessment of the pharmacokinetic model variables for this region. However, this characteristic signal pattern identifies the region as necrotic. All of the three remaining regions contained some viable tumor by histology. In the second region (B), an area of soft tissue adjacent to the central necrotic core, the MR signal is low in intensity and slow to enhance. This region corresponds to a semi-necrotic, edematous section of the tumor, which also includes reactive tissues. Pharmacokinetic modeling of the MR signal in this region demonstrates a relatively small transfer rate constant, $k_{ep,}$ which would indicate moderate access to the EES compartment. However, the contrast continues to accumulate throughout the 6 min of the DCE-MRI acquisition, most likely due to an expanded EES compartment. The third tumor region (C) shows more rapid and pronounced enhancement than does region B, as reflected by a larger value of k_{ep}. This well-perfused region of viable tumor with stable microcirculation demonstrates good access and accumulation of contrast agent. Unlike region B, the enhancement (and therefore the contrast concentration) in region C stabilizes within about 2 min of contrast arrival at the tumor and remains fairly constant during the remainder of the acquisition. The fourth region (D) has a very rapid increase in contrast concentration during the 30 s following contrast arrival, and a large degree of enhancement at the end of the measurement. Pharmacokinetic analysis demonstrates a very large k_{ep} for D. The size of this rate constant indicates rapid exchange between the contrast in the vascular and tumor EES compartments, so that the resulting concentration of contrast agent in tumor EES in this region closely follows the plasma concentration. This behavior is characteristic of rapidly proliferating tumor.

Analyzing regions B through D in Fig. 13.1 with each of the techniques listed in Table 13.2 provides

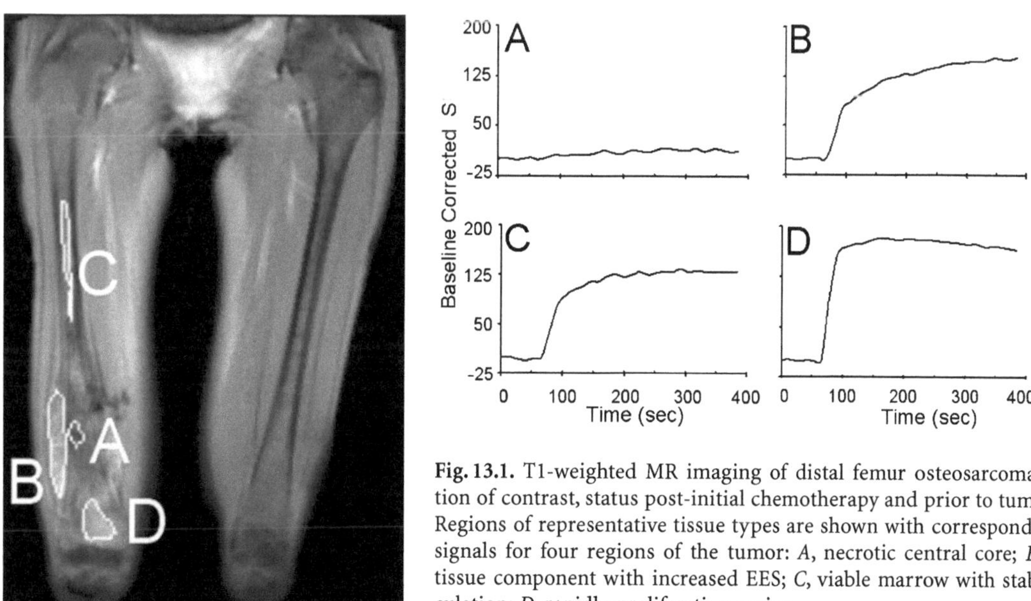

Fig. 13.1. T1-weighted MR imaging of distal femur osteosarcoma after injection of contrast, status post-initial chemotherapy and prior to tumor resection. Regions of representative tissue types are shown with corresponding dynamic signals for four regions of the tumor: *A*, necrotic central core; *B*, viable soft tissue component with increased EES; *C*, viable marrow with stable microcirculation; *D*, rapidly proliferating region

Table 13.2. Results of three different data analysis techniques [% slope, dynamic vector magnitude (DVM), and pharmacokinetic modeling) for the dynamic signals from regions B (soft tissue region – semi-necrotic with expanded EES), C (marrow region – stable microcirculation), and D (rapidly proliferating region – leaky neovasculature) in Fig. 13.1

	Region B Soft Tissue	Region C Marrow	Region D Rapid Proliferation
% Slope (%/min)	32.55	54.92	81.87
DVM (SI/s)	4.24	4.09	7.17
k_{ep} (min^{-1})	1.06	1.55	2.58

a practical comparison of one pharmacokinetic and two empirical measures. Figure B.2 demonstrates the parametric images generated from each of three measures extracted from the same DCE-MRI examination of the osteosarcoma of Figure 13.1. These parametric images demonstrate that each analysis technique provides somewhat different information, depending on the individual measure's sensitivity to characteristics of the dynamic concentration–time curve. Generally, all three measures increase in value from region B to region C to the rapidly proliferating region D. However, the combined measure DVM is slightly higher in B than C primarily due to the continued accumulation of contrast in B. All three measures show that the viable tumor regions have at least moderate access to contrast agent, indicating a good blood supply and effective transfer from plasma to EES for agents of low molecular weight (like GdDTPA)

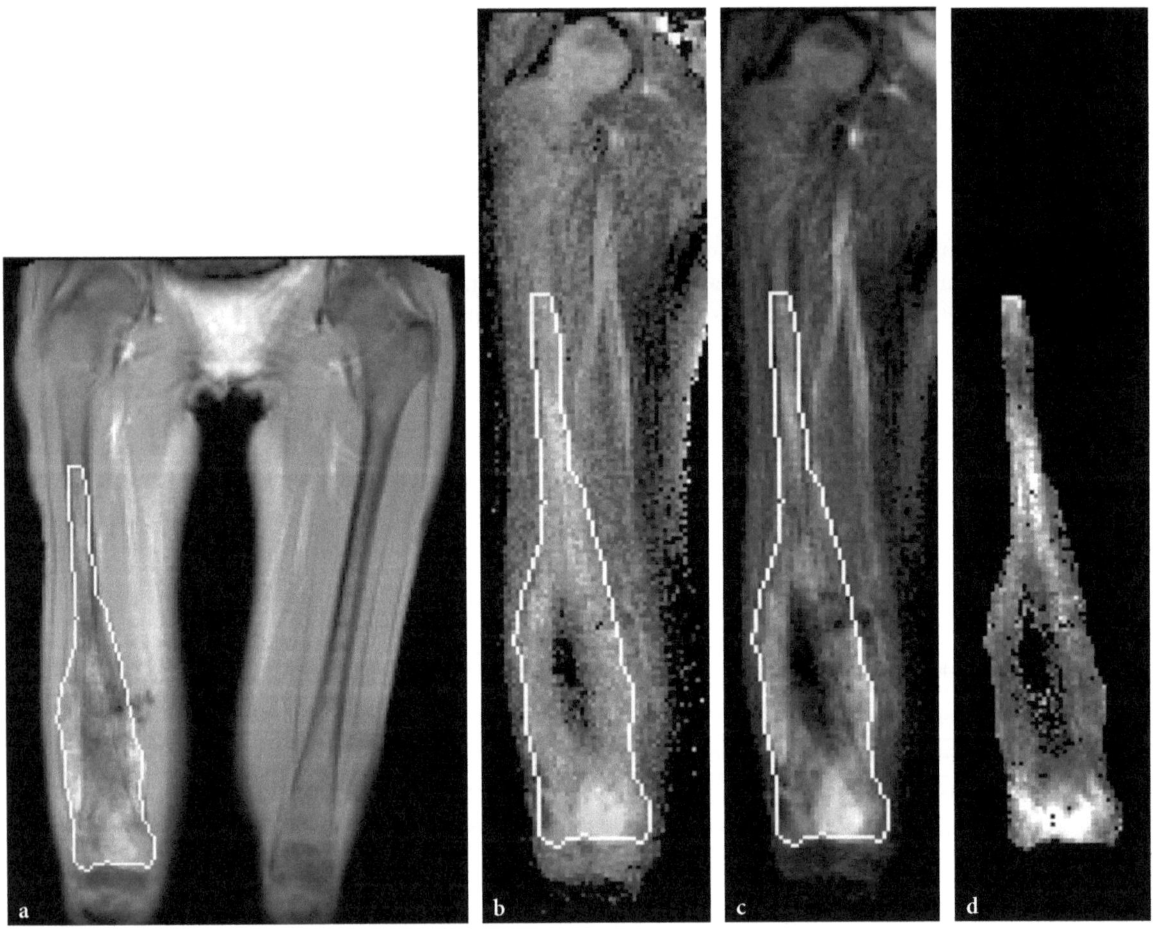

Fig. 13.2a-d. T1-weighted MR imaging of distal femur osteosarcoma after injection of contrast (**a**). Entire tumor is selected as the region of interest. Parametric images calculated from the analysis of the DCE-MRI signals on a pixel-by-pixel basis are shown: **b**, percent slope image; **c**, dynamic vector magnitude (DVM) image; **d**, pharmacokinetic modeling image of k_{ep}

Table 13.3. Predicted DCE-MRI accuracy compared to histological grading of response of bone sarcoma to preoperative chemotherapy

	DCE-MRI measure	Critical value	Accuracy[a]	Sensitivity (non-response)	Specificity (response)
ERLEMANN et al. (1990)	% Slope from linear estimate	60% reduction	86% (18/21)	80% (8/10)	91% (10/11)
BONNEROT et al. (1992)	Factor analysis (FAMIS)	No vascular phase	90% (9/10)	80% (4/5)	100% (5/5)
FLETCHER et al. (1992)	% Slope from linear estimate	44%/min	95% (19/20)	89% (8/9)	100% (11/11)
HANNA et al. (1993)	% Slope from linear estimate	60%/min	89% (8/9)	67% (2/3)	100% (6/6)
REDDICK et al. (1996)	DVM: max slope and amplitude	1.8/min	89% (17/19)	86% (12/14)	100% (5/5)
KAWAI et al. (1997)	% Slope from linear estimate	60%/min	91% (10/11)	86% (6/7)	100% (4/4)
ONGOLO-ZOGO et al. (1999)	% Slope from linear estimate	Increase	92% (11/12)	83% (5/6)	100% (6/6)

[a] Accuracy was calculated using retrospectively selected criteria in all cases, and therefore represents predicted, rather than measured, accuracy.

13.2.1.3 Clinical Application of DCE-MRI to Assess Osteosarcoma Response

Traditionally, studies performed to determine the response of bone sarcoma to preoperative chemotherapy have concentrated primarily on the early contrast uptake phase (initial slopes, $k_{ep)}$, because this can be understood as an assessment of microcirculation. The results of many of these studies have been summarized in Table 13.3 for easy comparison. The clinical application of DCE MRI in bonc sarcoma began with the finding by ERLEMANN et al. (1989) that necrotic areas within tumors enhance less rapidly and less intensely than viable tumor. The DCE-MRI images were analyzed on the basis of the differences in percent slope before and after chemotherapy from a circular ROI encompassing the region exhibiting maximal signal increase on the presentation examination. Within a year, FLETCHER et al. (1992) published an independent verification of these earlier results, using one or more ROIs of various shapes and sizes to allow sampling of as much of the enhancing portion of the tumor as possible. Concurrently, HANNA et al. (1992) reported a refinement of the earlier techniques that increased the spatial resolution by using small contiguous regions, followed later by pixel-by-pixel mapping (HANNA et al. 1993) to detect small foci of residual tumor and to compensate for regional variations in tumor perfusion.

The percent slope analysis approach continues to be used for assessing tumor response. Recently, both KAWAI et al. (1997) and ONGOLO-ZOGO et al. (1999) reported results of DCE-MRI of bone tumors analyzed on the basis of the differences in percent slope before and after chemotherapy, and the results were compared with the histological responses.

Building on evidence that the early initial slope of the DCE-MRI signals provided some surrogate assessment of response in bone sarcoma, Reddick et al. developed a more complete characterization of the DCE-MRI enhancement-time curves (REDDICK et al. 1994) and applied this method to studying response in patients with osteosarcoma (REDDICK et al. 1996). Three useful variables were identified: the initial contrast accumulation rate (ICAR), the delayed contrast accumulation rate (DCAR), and the maximum enhancement (ME) reached in the duration (6.5 min) of the DCE-MRI study. Most of the information appeared to lie in the initial rate and maximum enhancement parameters. This lead to creating a single measure, the dynamic vector magnitude (DVM), which combined the information on how much (ME) contrast entered the region and how rapidly (ICAR) it did so.

An analytic method borrowed from nuclear medicine, factor analysis of medical image sequences (FAMIS), was used by BONNEROT et al. (1992) in a study of ten osteosarcoma patients. FAMIS produces two different factor images from the DCE-MRI images: a "vascular factor" corresponding to early uptake of the contrast (first 1–2 min) which occurs in viable tumor, physes, and arteries; and an "accumulation factor" corresponding to delayed accumulation as the result of inflammation and edema. Note that these factor images correspond well with the two factors (slope and maximum enhancement) used to compute the DVM measure. They also form the basis of a "DCE-MRI light" method suggested by MA (1999), in which images are acquired at contrast injection, and at 30, 60, 90 and 120 s post-injection. In this method, the difference between the images at zero and 30 s post-contrast injection would correspond to the FAMIS "vascular factor" image and the difference between the images at zero and 120 s would correspond to the FAMIS "accumulation factor" image.

A group including Van der Woude, Verstraete and Bloem developed another qualitative method with similarities to the FAMIS approach. They identified three phases of the DCE-MRI intensity-time curves: first pass or wash-in of contrast, equilibration, and wash-out (VAN DER WOUDE et al. 1995, 1998 a,b; VERSTRAETE et al. 1996; VERSTRAETE and LANG 2000). A point-to-point correlation between DCE-MRI and histology revealed that enhancement within 3-6 s after the start of arterial enhancement corresponded to viable tumor and could discriminate between reactive tissue and tumor. Note that this requires at least ultrafast or so-called turbo T1-weighted sequences if multiple slices are to be dynamically imaged (SHAPEERO and VANEL 2000). Visual inspection of such fast "first-pass" images easily discriminated highly perfused viable regions - those enhancing at ≤6 s after contrast arrived in the feeding artery - from edema, normal tissue and necrosis. Response was assessed by the disappearance of these bright regions on the first-pass images at the end of therapy, compared to those from the beginning.

In spite of a notable variety in methods of analyzing clinical DCE-MRI data, studies of substantial series of patients with osteosarcoma for response to initial therapy have consistently reported that regions which enhanced brightly and/or reached near maximum enhancement rapidly (within the first 60-90 s after contrast injection) were correlated with blood vessels and viable tumor regions. For more than a decade, DCE-MRI of osteosarcoma has been performed and validated against histological analyses of en bloc resections following initial chemotherapy. The results of these studies, regardless of the analysis procedure or measure, have demonstrated accuracies of approximately 90% for discrimination of response. The parametric maps produced by pixel-by-pixel analysis methods have enabled the detection of residual disease as small as 3–5 mm^2 (EGMONT-PETERSEN et al. 2000; HANNA et al. 1993). Residual osteosarcoma tends to coalesce in nodules, a pattern which facilitates its detection by DCE-MRI. However, in the rare instances when the osteosarcoma exists as widely scattered individual cells, no MRI method should be expected to provide a satisfactory assessment of the amount of viable tumor.

This process of rigorous validation of DCE-MRI in osteosarcoma has allowed radiologists to distinguish with increased confidence residual viable tissue from surrounding edema, normal tissue and necrosis. However, there are two major pitfalls in interpreting results. With most DCE-MRI analysis methods, fast contrast uptake into newly vascularized tissue, such as immature granulation tissue, may appear as viable tumor. Thus reactive tissue marrow signal changes due to GCSF treatment, for example, can be mistaken for tumor. On the other hand, tumor tissue with low vascularity, such as chondroblastic areas, may appear necrotic.

13.2.1.4
Clinical Application of DCE-MRI to Predict Disease-Free Survival

In most studies, the percentage of tumor necrosis induced by preoperative chemotherapy has been predictive of disease-free survival (PICCI et al. 1994; HUDSON et al. 1990; MEYERS et al. 1992; GLASSER et al. 1992; PROVISOR et al. 1997). There is generally excellent agreement in the results from different researchers when DCE-MRI-based response measures are correlated with histologic measures for bone sarcoma. However, histologic measures of tumor necrosis obviously cannot be used at the time of diagnosis - the optimal point for identifying patients at increased risk of recurrence - and is applicable only to patients with resectable tumors after the completion of preoperative chemotherapy.

A recent retrospective study of DCE-MRI examinations at the time of presentation and at the completion of preoperative chemotherapy was conducted for 31 patients who received protocol-based therapy for non-metastatic osteosarcoma of the extremities (REDDICK et al. 2001). All DCE-MRI examinations were analyzed with a two-compartment pharmacokinetic model, and results were compared with disease-free survival. Disease-free survival was defined as the interval from complete surgical resection to treatment failure or most recent follow-up. The 2-year estimated disease-free survival in this patient cohort was 76.1%±8.0% (SE). At the time of analysis, there had been seven relapses, which occurred 5 months to 4.5 years after surgery. Follow-up intervals ranged from 5 months to 6.0 years with a median of 2.1 years. A total of 17 patients were responders by histologic grading of necrosis, while the remaining 14 patients failed to achieve the required 90% necrosis at resection. For the 31 patients, the median measure of regional contrast access (k_{ep}) was 1.167m in^{-1} at completion of preoperative chemotherapy (mean = 1.25 min^{-1}, standard deviation = 0.40 min^{-1}, range = 0.34 to 2.54 min^{-1}). Histologic assessment of response, based on the degree of necrosis in the resected lesion, was not found to be a prognostic factor of disease-free survival in these patients (p = 0884), using a Cox proportional hazards model. However, the estimated coefficient for k_{ep} at completion of preoperative chemotherapy was significantly correlated to treatment outcome, in a Cox proportional hazards model, with lower regional access (lower k_{ep}) predicting improved outcome. Neither k_{ep} at presentation and nor the change in k_{ep} during therapy were significantly correlated with outcome.

While k_{ep} at completion of preoperative chemotherapy was significantly correlated to treatment outcome, it is preferable to assess patient risk at presentation. Since lower regional access after preoperative therapy was predictive of improved outcome, the relationship between regional access at presentation and change in regional access during therapy was investigated. The linear regression of change in regional access as a function of regional access at presentation had an R-square of 0.85 and is shown in Fig. 13.3

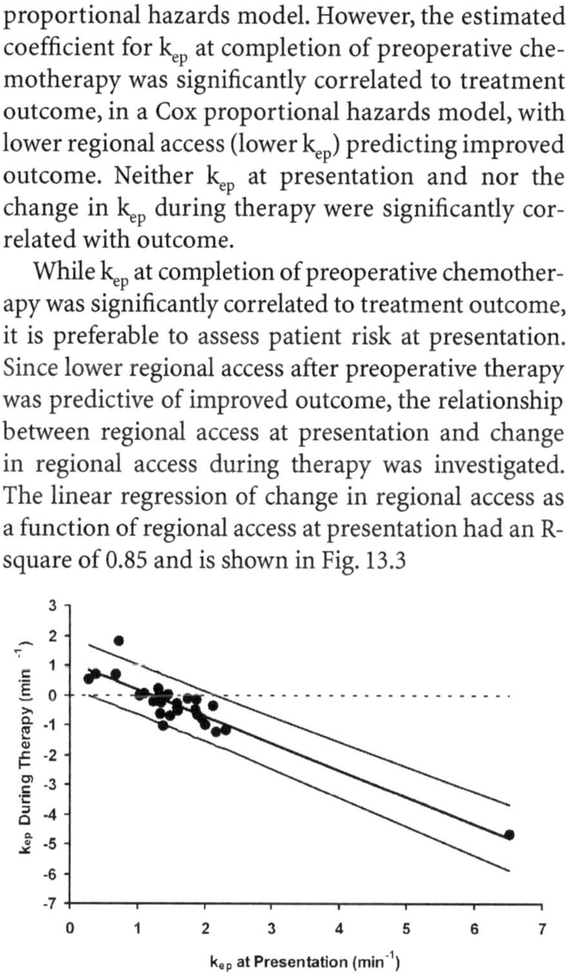

Fig. 13.3. Scatter plot and linear regression analysis demonstrating change in effective regional access (k_{ep}) during preoperative chemotherapy as a significant function of k_{ep} at the time of presentation. Larger effective regional access at the time of presentation corresponds to greater decrease in k_{ep} during therapy

The inverse relationship between regional access (k_{ep}) after initial therapy and improved disease-free survival estimates is consistent with the hypothesis that the transfer rate of low-MW MR contrast agents between the vasculature and the extracellular fluid acts as a surrogate measure of drug delivery. A small value for the k_{ep} after preoperative chemotherapy may indicate that a large proportion of the tumor tissue is necrotic and its perfusing microvasculature is greatly reduced. The regional access measure k_{ep} calculated by pharmacokinetic analysis of the DCE-MRI is a function of both blood flow and vascular permeability of vessels. Tumors with a poorer response to chemotherapy would have larger viable regions, which are highly angiogenic and therefore more permeable to the contrast agent, and such regions would result in a larger value for regional access at the end of therapy. In contrast, larger values for regional access at the time of diagnosis would be expected to correspond with better delivery of the contrast agent, and by inference, improved delivery of low-MW drugs. These larger k_{ep} values were also positively related to response to chemotherapy and to longer disease-free survival estimates, although the latter did not reach significance in this study.

13.2.2 Ewing's Family Tumors (EFTs)

The Ewing's family of tumors includes Ewing's sarcoma and peripheral primitive neuroectodermal tumor (PNET). Classical Ewing's sarcoma and PNET are now identified as the same tumor with variable differentiation, and are treated similarly. Although much more rare than osteosarcoma, Ewing's family tumors are the second most common malignant osseous tumors in children and adolescents. A total of 70% of these tumors occur in patients under the age of 20, with a male preference (Grier 1997). The natural history of Ewing's family tumors appears to be the same in the adult and the pediatric population, although pediatric Ewing's family tumors have been studied much more extensively (Fizazi et al. 1998; Baldini et al. 1999). An interesting feature of these malignancies is that they are extremely rare in patients of African or Asian ancestry, but have similar incidence in other populations (Grier 1997).

The collaborative studies organized by the worldwide pediatric oncology community have greatly improved the outlook for patients with Ewing's family tumors. With modern therapy, more than 60% of patients with local EFT can be cured. However, new therapies for these tumors are still urgently needed, particularly for patients who present with metastases or develop them on therapy. The outcome for the group with disseminated disease is still poor – a recent study finds they have only a 22% relapse-free survival rate at 5 years post diagnosis (Cotterill et al. 2000).

Ewing's family tumors belong to the group of small round cell tumors, which also includes neuroblastoma, rhabdomyosarcoma, and non-Hodgkin's lymphoma. Both Ewing's sarcoma and PNET are characterized by the same histochemical staining profile and demonstrate one of several reciprocal genetic translocations: t(11;22); t(21;22) and t(7;22) (Ambros et al. 1991). Each of these reciprocal translocations fuses part of the EWS gene on chromosome 22 with a transcription factor belonging to the ETS family. This unique translocation is the focus of research to design targeted therapies for these malignancies. A reverse transcriptase-polymerase chain reaction (RT-PCR) assay based on the Ewing's translocation finds tumor cells in patients' bone marrow samples at levels well below those detectable even by microscopic examination. There are ongoing trials to determine the predictive significance of RT-PCR findings at diagnosis (Grier 1997).

Ewing's-PNET can develop in any bone in the body, with flat and long bones almost equally represented. Approximately 50% of EFT occur in the extremities. EFT occur predominantly in the diaphyses, whereas osteosarcoma is more commonly found in the metaphyses. They can occur as soft tissue tumors also, and in this form have no specific distinguishing features on MRI. Negative prognostic factors include metastases at diagnosis, large tumor size/volume, and age over 17 years. The percentage of patients presenting with metastases is usually said to be around 25% (Green 1985), although in two series of adult patients, 29% presented with metastases (Baldini et al. 1999; Fizazi et al. 1998). In a large cooperative study, the incidence of metastases detected at presentation has increased significantly in the post-1991 period, compared to the 1980s, probably due to improved imaging in the latter period (Hense et al. 1999).

Positive prognostic factors include absence of metastases at diagnosis, location of tumor, young age, and good radiological and/or pathological response to pre-surgical chemotherapy (Cotterill et al. 2000; Abudu et al. 1999). Centrally located tumors have worse prognosis. However, the correlation between tumor location and outcome is absent in some studies, and it may not be an independent predictor. The importance of obtaining a good response to initial chemotherapy was underlined by one recent study which tracked actual administered doses of chemotherapeutic agents: those patients receiving higher actual doses were at ten-fold lower risk of metastatic recurrence than those patients who received less dose (Delepine 1997).

Current measures of response are based on radiological or pathological identification of percent necrosis induced by initial chemotherapy in the tumor (Wunder et al. 1998; Aparicio et al. 1998; Abudu et al. 1999; Paulussen et al. 2001; Jenkin et al. 2002). However, for EFT the criterion of response does not override the negative effect of metastases at presentation: children with metastatic Ewing's family tumors have poor chance of survival even when complete response to pre-surgical chemotherapy is observed (Cotterill et al. 2000). Therefore, in studies of response, it is important to analyze the local-disease and metastatic groups separately.

Because disseminated disease is such a key factor in staging these tumors, appropriate imaging protocols for identifying metastases are a critical part of staging for these tumors. EFT tend to metastasize to lung and bone marrow. Imaging protocols for staging therefore include radiographs and CT of the chest, plus bone scintigraphy and bone marrow biopsy, as well as clinical data, radiographs and MRI of the primary tumor (van der Woude et al. 1998b; Ma 1999; Grier 1997).

13.2.2.1 Rationale for DCE-MRI of EFT

MRI is the preferred modality for therapy planning in Ewing's family tumors, for many of the same reasons it is preferred for osteosarcoma therapy planning. Over 50% of EFT occur in the extremities, and limb-salvage surgery is preferred over amputation when possible. MRI shows excellent accuracy in depicting the extension of osseous osteosarcoma and Ewing's tumors to the physis and epiphysis (San Julian et al. 1999).

As with osteosarcoma, the gold standard for response of EFT to pre-operative chemotherapy is histological analysis for the presence of necrosis in resected tumor. The degree of response can be critical in the decision to use limb-sparing surgery. The standard threshold separating good responders from poor responders in EFT is >90% necrosis (≤10% viable tumor) (Rosen et al. 1982; MacVicar et al. 1992). The extra-medullary component of EFT is predominantly cellular and, unlike osteosarcoma, may show a rapid decrease in volume in response to chemotherapy. However, the use of volume change as a criterion for response is fraught with complications even in EFT. An increase in edema or the occurrence of hemorrhage in a responding tumor can be mistaken for non-response. On the other hand, a tumor showing significant shrinkage may still contain active residual tumor cells that make up more than 10% of its volume. Radiography, CT, scintigraphy and static contrast-enhanced MRI have not proved able to determine tumor response accurately (Shapeero and Vanel 2000; Abudu et al. 1999).

13.2.2.2 Clinical Application of DCE-MRI to Assess EFT Response

As with osteosarcoma, the percentage of necrosis induced in Ewing's family tumors by pre-surgical chemotherapy is strongly and positively associated with good patient outcome (both disease-free and overall survival) (Abudu et al. 1999). EFT are rarer than osteosarcoma, and much less attention has been devoted to DCE-MRI studies of these tumors. For over a decade, several European groups have published studies assessing the efficacy of DCE-MRI in measuring response to initial therapy, by comparing DCE-MRI measures of percent necrosis with point-by-point histologic measures (Verstraete et al. 1994a–c; van der Woude et al. 1994, 1995, 1998a,b; Egmont-Petersen et al. 2000; Shapeero and Vanel 2000). However, most reports have grouped the response of osteosarcoma and Ewing's family tumors together, or have covered a mixed group of bone and soft-tissue sarcomas. As has been noted, this can cause confusion because the two tumors differ in their clinical course, radiologic appearance and, not least, in their histologic characteristics (MacVicar et al. 1992). Residual tumor in osteosarcoma tends to coalesce in nodules that are larger than the scattered, microscopic clusters of residual tumor found distributed throughout the soft and bony portions of EFT (MacVicar et al. 1992). This makes it more difficult to determine the percentage of residual tumor in Ewing's sarcoma, compared to osteosarcoma. These researchers have, however, succeeded in showing that DCE-MRI can detect residual tumor volumes of 3–5 mm^2.

As EFT presents a more difficult imaging target for assessing response to initial chemotherapy, it is necessary to validate DCE-MRI for EFT separately from osteosarcoma. This has not yet been done. Only two studies have focused solely on DCE-MRI for measuring response in EFT.

The first is a small study of eight consecutive patients with Ewing's sarcoma, all located in the legs (Egmont-Petersen et al. 2000). No information was given about metastatic disease at diagnosis. All eight underwent DCE-MRI at 0.5 T pre-operatively to assess local response to chemotherapy, and a pathological specimen was available as a gold standard for the imaging results. DCE-MRI studies were performed at 0.5 T with a temporal resolution of 3.3 s per image plane and a spatial resolution of 0.61–3.0 mm^2 in plane by 8-mm section width. To avoid sacrificing spatial resolution in the analysis and to test several pharmacokinetic modeling approaches, the researchers fitted each image pixel to pharmacokinetic models, a technique used by others for osteosarcoma (see above), and produced parametric images for wash-in, wash-out, maximum enhancement, and arrival time of contrast. They carefully matched the histologic macroslice to the parametric images, so that they could make a region mask for viable tumor, which was matched to the parametric images in the same plane, and could calculate statistical measures of correctness for each model in each patient. The two models differed in that one assumed a single global arrival time for the contrast at all regions of the tumor, and the other calculated the arrival time for each pixel as a parameter of the fit.

This study arrived at three salient results. First, comparisons with histology showed that their wash-in parametric image best identified the remnants of viable tumor in these patients. Their wash-in parameter m_1 is proportional to the k_{ep} of the standard terminology (Tofts et al. 1999). The overall sensitivity to tumor ranged from 0.33 to 0.77, with specificities for necrosis of 0.58–0.99. Second, they demonstrated convincingly that the reliability of the wash-in parameter, or k_{ep}, is a strong function of the average size of the residual tumor. Simple pharmacokinetic modeling will yield a k_{ep} that is reliable only when residual "islands" of tumor exceed a certain size – in this study, about 50-250 mm^3. This threshold, however, is not an absolute number but a function of the SNR, the partial-volume effects, and the problems inherent in comparing microscopically thin histologic slices with 4-mm image planes.

Finally, they found that the more complex model (with arrival time as a parameter, not a constant) gave the most accurate fit, but it only worked well when SNR was sufficiently high. They found that the range of variation of the fitted arrival times was ± 10 s within a typical region of interest – a very big range, given that the entire first pass of contrast through the circulation is of the order of 10 s. Moreover, they determined that by making the model more sensitive and therefore more accurate in computing the slope or wash-in parameter of the time–intensity curve, the model also became more sensitive to the noise in the data and therefore less robust. This is an excellent example of the sort of problem that can arise when fitting noisy data. Designing sophisticated algorithms to work around this apparent trade-off between accuracy and robustness is an active area of research in applying pharmacokinetic models to DCE-MR images.

The second, more recent study looked at 21 patients with a very different distribution of EFT: only 43% were in extremities and a large fraction (43%) had metastatic disease at diagnosis (Miller et al. 2001). The single-slice, T1-weighted spoiled gradient echo sequence used had a spatial resolution of 2.4–3.8 mm^2 in plane and 6–10 mm slice thickness. Temporal resolution was 7.6–14 s/image plane. They found that no DCE-MRI parameter showed a significant correlation with disease-free survival (DFS) and progression-free survival (PFS), but that measures of tumor volume from routine MRI did correlate with survival. However, it should be noted first, that this study has an unusually high fraction of patients with metastatic disease, and second, that the 12 patients with local disease were not analyzed separately from the nine with metastases at diagnosis. It is well established that response to standard initial chemotherapy does not correlate with outcome for patients with metastatic disease at or during therapy; these patients do very poorly regardless of response. The fact that nearly half the patients in this study had metastases would be expected to confound the correlation between any response measure and survival. The correlation these researchers found between tumor size at diagnosis and survival measures is probably linked to the presence of metastases. Several studies have reported that the risk of metastases increases with tumor size, and in fact the risk of metastases correlated with tumor size in this study as well.

At week 8 (just prior to surgery), the DCE-MRI parameters ICAR (representing initial contrast uptake rate) and the lumped parameter DVM (which includes both the uptake rate and the maximum enhancement) were beginning to approach significance (p values of 0.05 and 0.06, respectively). It would be of interest to revisit the local-disease population in this study in order to determine the correlation between DCE-MRI and survival. This analysis could not be performed originally because statistical conditions for survival analysis were not met for this group.

13.2.3 Cartilaginous Tumors: Enchondroma vs. Chondrosarcoma

A study of DCE-MRI in 37 patients with cartilaginous tumors (eight enchondroma, 11 osteochondroma and 18 chondrosarcoma) (Geirnaerdt et al. 2000) looked at the utility of DCE-MRI in detecting low-grade chondrosarcoma. Enchondroma and osteochondroma are benign lesions that do not require surgery. In contrast, chondrosarcoma are malignant lesions whose only curative therapy is surgical resection. Although high-grade chondrosarcoma have distinctive malignant radiographic features, differentiation between benign and low-grade malignant cartilaginous tumors is a histological and radiological challenge. The clinical significance of this challenge is increased by the facts that low-grade chondrosarcoma are the more common, and that intralesional procedures increase the risk for local high-grade recurrence and the development of metastases, with a corresponding decrease in overall survival for these patients.

DCE-MRI parameters were correlated with the diagnosis from surgical or biopsy material. Although the data spanned two imaging systems (0.5 T and 1.5 T), with differences in the DCE-MRI timing and coverage, the investigators were able to show that a DCE-MRI pattern of very early exponential enhancement – e.g., tumor enhancement whose rise paralleled that of a nearby artery, but was delayed no more than 10 s after the arterial rise – correlated strongly with the diagnosis of chondrosarcoma. Early enhancement (defined in this study as lesion enhancement starting $\leq$10 s after arterial enhancement) was observed in all chondrosarcoma, none of the enchondroma, and in osteochondroma only in children with unfused growth plates. Using the early enhancement as a sign of malignancy, DCE-MRI identified chondrosarcoma with a sensitivity of 89% and a specificity of 84%. These detection rates are not very satisfactory, and it is not clear that DCE-MRI has any benefits over static contrast-enhanced imaging for distinguishing between enchondroma and low-grade chondrosarcoma.

However, this example illustrates the importance of temporal resolution in certain DCE-MRI applications. In an earlier study, investigators were not able to discriminate between enchondroma and low grade chondrosarcoma, most probably because of insufficient temporal resolution (20 s per image) (Erlemann et al. 1989). However, with a temporal resolution of 3 s per image, Verstraete et al. (1994a) were able to detect differences in the rate (slope) of early enhancement between malignant (two chondrosarcoma) and benign tumors (four osteochondroma, seven enchondroma). The importance of detecting early enhancement at ≤10 s is confirmed in a recent study (Geirnaerdt et al. 2000).

13.3 Primary Soft Tissue Sarcomas

Soft tissue sarcomas are rare malignancies, although the reported incidence is increasing (Shapeero et al. 2002). They are also a very heterogeneous group of cancers, whose pathologic classification has evolved dramatically over the past 30 years. The lineage and genetic features of some malignancies in this group are well understood (e.g., rhabdomyosarcoma), while for others even the cell of origin is a matter of debate. The classification of soft tissue sarcomas thus is something of a work in progress.

Historically, the most common tumor type within this group has been the malignant fibrous histiocytoma (MFH). However, the validity of the MFH classification is being challenged. Two major functions of classification are to associate cancers with specific cell lineages and to identify prognostic groups. An argument has been made that many tumors identified as subtypes of MFH are in fact soft tissue malignancies of distinctly different lines of differentiation, such as leiomyosarcoma or liposarcoma. In a retrospective histologic assessment of MFH subtypes in 100 consecutive patients, 84 could be assigned to specific lines (Fletcher et al. 2001). The most common differentiated sarcomas found were myxofibrosarcoma (29), myogenic sarcoma (30), and myofibroblastic sarcoma (11). For oncologists, the importance of this discussion is the assertion that those tumors now classified as pleomorphic MFH that show myogenic differentiation are significantly more malignant, with shorter time to metastasis and worse outcome, than non-myogenic MFH of comparable stage (Fletcher et al. 2001). This has serious implications for the selection of study populations, and should be kept in mind in the design and interpretation of clinical studies of soft tissue sarcoma.

MRI is the most important imaging modality in the evaluation of soft tissue tumors, again because of its superior soft tissue contrast and the clarity with which it shows the tumor and its relation to tissue anatomic compartments. As is the case for bone sarcomas, biopsy is required for diagnosis of soft tissue malignancy (Crim et al. 1992), although appropriate imaging can narrow the differential.

MRI is critical in staging soft tissue neoplasms (Verstraete and Lang 2000; Berger et al. 2000; Noria et al. 1996; Rao 1993), and should occur before any biopsy or resection is attempted. A case can be made that DCE-MRI is useful in targeting biopsies in some cases, e.g., identification of liposarcomatous degeneration within lipoma (Shapeero et al. 2002), and there is evidence that DCE-MRI can improve accuracy of diagnosis for leiomyosarcoma arising within uterine leiomyoma (see below). Improvements in the accuracy of biopsy planning and resection are needed, because incomplete resection, including contamination from intralesional procedures, is currently the most common cause of amputation for extremity soft tissue tumors (Noria et al. 1996; Kaste et al. 2002; Mankin et al. 1996). The problem of detecting residual tumor after incomplete resection is therefore a major issue with these lesions, and has not received the attention it deserves (Kaste et al. 2002; Noria et al. 1996). However, a number of studies show that static MRI is not sensitive to microscopic disease (Kaste et al. 2002; Mankin et al. 1996), nor is any MRI method, including DCE-MRI, able to discriminate between benign and malignant tissues with sufficient sensitivity (Crim et al. 1992; Verstraete et al. 1996; May et al. 1997). Consequently, decisions regarding resection margins cannot be based on MRI alone.

13.3.1 Clinical Application of DCE-MRI to Distinguish Malignant from Benign Soft Tissue Tumors

There have been several very recent reports of the utility of DCE-MRI in specific diagnostic questions and DCE-MRI-guided biopsy. The first two addressed challenges in distinguishing specific soft tissue malignancies from benign tumors.

The first study looked at the DCE-MRI characteristics of synovial sarcoma (van Rijswijk et al. 2001). Synovial sarcoma classically occurs in soft tissues of extremities, especially near large joints, but can occur anywhere in the body, distant from joint spaces. It typically affects adults in the fourth decade, but half the cases are in children and adolescents. Patients with localized synovial sarcoma have an excellent prognosis, and most are long-term survivors. Synovial sarcomas must be distinguished from hematoma or benign cystic masses, with which they are frequently confused (McCarville et al. 2002). Based on DCE-MRI studies of ten biopsy-proven synovial sarcoma, the most consistent DCE-MRI characteristic was a fast enhancement, occurring in less than 7 s after arterial enhancement (mean time to enhancement ± s.d.=4.4 s ± 2.1 s). Again, fast temporal resolution of 3 s/image is necessary for this application.

The second study addressed the utility of DCE-MRI for distinguishing between benign degenerated leiomyoma (uterine fibroid) and malignant leiomyosarcoma (GOTO et al. 2002). They examined 227 eligible patients, of which ten had uterine leiomyosarcoma (LMS) and 130 had uterine degenerated leiomyoma (DLM). Precontrast MRI and levels of the serum enzyme lactate dehydrogenase (LDH) and its isozymes were measured in all patients. DCE-MRI was obtained in all patients with LMS (ten) and in the 32 patients with DLM in whom elevated LDH levels were observed. All LMS showed enhancement at 60 s after contrast injection; only four of 32 DLM showed enhancement at 60 s. Total LDH and LDH isozyme type 3 were elevated in all ten patients with LMS. The specificity, positive predictive value, negative predictive value, and diagnostic accuracy were 93.8%, 83.3%, 100%, and 95.2% with DCE-MRI alone. Combining DCE-MRI and LDH measures, the corresponding values rose to 100%, 100%, 100%, and 100%. The authors suggest that the combination of information from dynamic MRI and levels of serum LDH (isozymes) appears useful in differentiating between LMS from DLM before treatment.

Finally, investigators at three institutions applied DCE-MRI for biopsy targeting in a total of 40 patients with various soft tissue sarcomas (SHAPEERO et al. 2002). In 25% of these, DCE-MRI defined the foci and extent of the most malignant regions of tumor better than routine static MRI. Results were validated by comparison of MR imaging findings with histologic examination of resected specimens. This study also included findings on the utility of DCE-MRI in assessing response to initial chemotherapy and in follow-up of soft tissue sarcomas, as discussed below.

13.3.2 Potential Clinical Application of DCE-MRI in Soft Tissue Sarcoma

13.3.2.1 Response

Soft tissue sarcoma in general have not had the benefit of the large, multi-center trials that have greatly improved outcome for bone sarcomas in pediatric patients. Although there has been an impression that high-grade soft tissue sarcoma, at least, should receive systemic therapy, data on the efficacy of pre-surgical or post-surgical chemotherapy in these tumors in adults have been lacking. In one small European randomized trial, investigators compared standard local treatment (radical or wide resection plus pre- and post-operative radiation therapy) with local treatment plus initial high-dose chemotherapy. The study population was high-grade soft tissue sarcomas of the extremities and girdles, greater than 5 cm in size, local disease only. Initial chemotherapy resulted in significant increases in disease-free survival at 5 years (GHERLINZONI et al. 1999). Two important new studies have examined the efficacy of initial chemotherapy in soft tissue sarcomas and the role of therapy-induced tumor necrosis as a prognostic factor. Although neither of the two studies had control arms, their results are consistent and provocative.

The first study reviews the 23-year history of evolving neoadjuvant chemotherapy protocols in one major referral center (EILBER et al. 2001). The study population consisted of 309 patients with extremity soft tissue sarcomas of intermediate or high grade, and who had tumors that could be assessed with histopathology. The distribution of sarcomas included 27% liposarcoma, 25% MFH, 16% synovial sarcoma, 7% malignant peripheral nerve sheath tumor, 5% fibrosarcoma, and 4% leiomyosarcoma. They reported that the increase in percentage of patients who achieved a "complete pathological response" (defined as ≥95% necrosis) was correlated with a significant improvement in survival at 5 and 10 years. The authors recommended assessment of pathological necrosis as a valid endpoint for evaluating chemotherapeutic protocols.

The second study reported the results of a 15-year prospective trial of initial intensive chemotherapy in high-risk pelvic and extremity soft tissue sarcomas (HENSHAW et al. 2001). The purpose was to evaluate first, tumor necrosis in response to pre-operative chemotherapy and second, survival rates. The study population was 33 patients with non-metastatic, large, high-grade sarcomas of the pelvis and extremities. The distribution of sarcomas included 52% MFH, 27% liposarcoma, 9% undifferentiated sarcoma, 6% leiomyosarcoma, and 3% each synovial sarcoma and malignant peripheral nerve sheath tumor. Median follow-up from surgery was 5 years.

They reported an overall survival similar to that reported by other studies using post-surgical therapy. While this is a small study with no control arm, it is consistent with the experience in bone sarcoma, where initial chemotherapy has replaced post-surgical therapy, with no deleterious effect on survival. This study produced a disease-free survival of 80% and overall survival of 88% at 5 and 10 years, a very encouraging result. The initial chemotherapy produced a limb-salvage rate of 94%, remarkable in a population who otherwise would have been treated with major amputation and/or high-dose radiation. High rates of necrosis (median tumor necrosis 95%) were observed.

We are aware of only one study on DCE-MRI for measuring response in soft tissue sarcomas. The study was not prospective. The investigators analyzed DCE-MRI studies of 32 patients with newly diagnosed soft tissue sarcomas both before and after initial chemotherapy (within 1–2 weeks of resection). The gold standard for response to chemotherapy was histological determination of the presence of viable tumor on sections in the same plane as the DCE-MRI. The criterion for good response was ≤10% viable tumor cells, adopting the response threshold for bone sarcoma. The DCE-MRI criterion for viable tumor was rapid enhancement, seen 3–10 s after arterial enhancement on subtraction images; this required a temporal resolution of at least 5 s/image, and an in-plane artery to detect the arterial enhancement time curve. The absence of any such foci of enhancement was interpreted as response. DCE-MRI correctly identified all 11 responders from the 21 non-responders. Beyond the fact that one responding tumor was a synovial sarcoma and one non-responder was rhabdomyosarcoma, there are no details on the distribution of soft tissue sarcomas observed, nor on the nature of the initial chemotherapy regimens at the institutions involved. No correlation with survival is presented.

13.3.2.2 Follow-Up

This same study reports on follow-up evaluations in 196 patients for detection of recurrent/residual soft tissue sarcoma (Shapeero et al. 2002). The follow-up protocol was MRI studies every 3 months post-surgery for the first 2 years, at 6-month intervals for the next 5 years, and at yearly intervals thereafter when indicated. Four of the 196 had their initial follow-up MRI only 2 months after resection. All MRI examinations included conventional T1- and T2-weighted imaging, plus static contrast-enhanced imaging following the DCE-MRI study (at 5 min after contrast injection). The sensitivity for tumor detection was 93% (42 of 45 recurrent tumors) and the specificity was 60% (six of ten pseudotumors, or inflammation). All four false-positive DCE-MRI results were due to foci of rapidly enhancing granulation tissue, which matured into slowly enhancing regions within weeks and had completely resolved in 2–6 months. This is a known pitfall, as DCE-MRI detects areas with abundant and very permeable vasculature, and this description fits the angiogenic vascular areas of granulation tissue as well as it fits malignancies. Clinical history and interval comparisons may aid in distinguishing one from the other in follow-up examinations, since granulation tissue will mature but malignant foci will maintain their rapid, early enhancement.

In patients receiving post-operative radiation therapy, the appearance and resolution of radiation effects extends over months and sometimes years, depending on the nature and dose of radiation and the location of the tumor bed (Shapeero et al. 2002). The authors find that these inflammatory changes can often be distinguished from recurrent tumor because the textural features of muscle are preserved on the unenhanced T1-weighted images (Vanel et al. 1994). However, DCE-MRI may be used if necessary to make the call, as radiation-induced changes are late-enhancing relative to recurrent tumor.

13.4 Synopsis

For osteosarcoma and for Ewing's family with local disease, the primary clinical problem is predicting the response of the tumor, so that treatment of individual patients, including limb-sparing surgery, can be planned with confidence (Fletcher 1997). For osteosarcoma, clinical investigations with series of ≤35 patients have convincingly demonstrated the ability of DCE-MRI to measure response to preoperative chemotherapy just prior to tumor removal. These findings should be validated by larger multi-institutional trials. A longer-term goal would be to identify a non-invasive early predictor which could identify good responders not only at the time of surgery, but also earlier in the course of initial chemotherapy. To date, this has not been achieved. More discovery-based research is needed on the temporal evolution of tumor response to therapies, specifically on the evolution of tumor vascular effects.

In EFT, the data are sparser. It is vital to evaluate DCE-MRI just prior to surgery in patients with local EFT disease only, to determine whether it can accurately divide good from poor responders. Evaluating DCE-MRI in these rare tumors will require a common protocol for multi-institutional studies, which is not trivial given the current disparities in imaging platforms and the lack of widely available and validated software for analysis of clinical data.

For soft-tissue sarcoma, one of the most important clinical problems is the same as for bone sarcoma: find an accurate, non-invasive imaging measure of response to systemic therapies. Validating DCE-MRI as a measure of response in soft-tissue sarcomas faces much the same problems as described with EFT. Although soft-tissue sarcomas are much more common than Ewing's family tumors, the plethora of histologic types within this family, combined with ongoing changes in grading and staging these malignancies, produces the same difficulty: a single institution cannot accrue sufficient numbers of well-characterized tumor types in a reasonable period of time. Nevertheless, new systemic therapies may bring much-improved outcomes for these patients. The up-front windows of initial therapy are excellent opportunities to assess the efficacy of agents in the treatment of these tumors. A precise and accurate imaging measure of response would advance our ability to determine treatment efficacy.

DCE-MRI has shown some evidence that it may be helpful in resolving some challenging problems of diagnosis and detection of recurrent local disease on follow-up. More extensive studies are needed to validate these uses. It is encouraging to note that an increasing number of radiologists are beginning to tackle these issues with soft tissue sarcoma.

For both osseous and soft tissue sarcomas, initial chemotherapy protocols can extend for 8–12 weeks prior to surgery, and it would be interesting to determine if interval comparisons between mid-therapy and pre-surgical DCE-MRI may help to resolve some of the questions regarding the discrimination between immature granulation tissue and residual tumor. This could be done efficiently if software were available to automatically register images from sequential examinations in the same patient, so that temporal alterations in regions of interest could be easily and accurately detected.

DCE-MRI has clearly established itself as a valuable imaging method in treating osseous sarcoma, and promises to do the same in the treatment of soft tissue sarcoma. The majority of results to date have come from a small number of dedicated groups who are willing and able to expend considerable resources on coping with the lack of validated software for DCE-MRI image registration and analysis. Wider use of this imaging method in clinical trials and studies will depend on making such software available and refining it to function efficiently in the clinic.

As faster gradients become more widely available, DCE-MR imaging protocols will be able to provide more complete coverage of these tumors without sacrificing temporal or spatial resolution. It will be interesting to see if higher-field magnets achieve superior SNR or spatial resolution in DCE-MR images. Both factors are critical. Reducing partial volume effects is imperative for accuracy of detection in soft-tissue and Ewing's family sarcoma. The SNR constrains the algorithms that can be used successfully for parametric images at the highest spatial and temporal resolution. As noted before, the higher noise levels that can accompany increased spatio-temporal resolution are a major problem for algorithms that fit the dynamic time curves in order to extract either empirical or pharmacokinetic measures. Developing robust algorithms to overcome this is a fertile field for research.

Empirical parameters, such as maximum enhancement, slope and DVM, have several key disadvantages. First, while they are relatively simple to calculate, their magnitudes have no clear meaning in terms of tumor physiology; they cannot be related to, e.g., tumor blood flow, vessel permeability, exchange rates between compartments, or changes in these physiological features. Moreover, they are functions of the field strength, coil sensitivity, coil loading, and imaging sequence. This makes it hard to compare results from studies using different measures, and it makes multi-site studies particularly challenging. One strong point of pharmacokinetic modeling is that measures derived from it are relatively independent of acquisition parameters and can therefore be compared across multiple imaging sites more easily. A second benefit is that, if the pharmacokinetic model is valid, then the DCE-MRI parametric images not only depict viable tumor but also yield insights into regional tumor perfusion and physiology. However, the pharmacokinetic modeling requires some additional imaging, as well as special software and post-processing for analyzing the DCE-MRI. As previously noted, the trade-offs between temporal resolution, spatial resolution and SNR for DCE-MRI are complex. Moreover, DCE-MRI sequences should be optimized to reduce sensitivity to T2* effects. Until pharmacokinetic DCE-MRI protocols are widely available and understood, clinicians will make do with approximate empirical measures, with the understanding that these are not suited for comparison across field strengths and coils of varying sensitivities.

In the absence of widely available, validated software for producing appropriate parametric images from DCE-MRI, the clinicians may follow recent recommendations to include an artery or physis in the tumor imaging plane, use temporal sampling times of 3–5 s/image, and compare intensity–time curves from arteries with those from tumor (Shapeero and Vanel 2000; Shapeero et al. 2002). For imagers not providing intensity–time software, the ROIs in tumor and vessels/physis may be compared on early-uptake vs. late-enhancing images, using a protocol such as that recommended by Ma (1999). Some experienced investigators have successfully interpreted DCE-MRI obtained on different scanners by comparing the rate of change of enhancement to that in a perfusing vessel, thus effectively comparing the dynamic enhancement–time curve in tumor with an approximation to the enhancement–time curve in the blood perfusing the tumor (the so-called arterial input function, or AIF). Although this practice of qualitative comparison has not been prospectively tested, it may be a reasonable procedure. Investigators have recently tested pharmacokinetic modeling of the tumor and the difference made by including one parameter related to the AIF: the local arrival time $t_{0,t}$ of contrast in the tissue (Egmont-Petersen et al. 2000). They found that this improved their initial-rate measures, but only if the SNR of the dynamic curve was moderate to good. It must be noted that this procedure requires careful positioning of the imaging planes as well as fast temporal sampling, at 3–5 s/image, to define the arterial enhancement-time curve. A new study from the Nijmwegen group elegantly demonstrates that it is possible to automate selection of appropriate pixels for defining the AIF, and that incorporating this AIF information into the pharmacokinetic model significantly improves the reproducibility of the k_{ep} measure. They performed duplicate examinations of a series of 11 patients with solid tumors (six head/neck tumors, three prostate carcinoma, two brain tumors). When the AIF for the individual patient examination was used, average k_{ep} values for tumors in replicate examinations were not significantly different in ten of 11 patients; k_{ep} was not replicable for only one of 11 patients. In contrast, the more usual procedure of using a global value for the AIF (see, for example, Knopp et al. 2001) resulted in statistically significant differences between k_{ep} values from duplicate DCE-MRI examinations for six of 11 patients.

Important research questions still exist about the utility of DCE-MRI in musculoskeletal sarcoma and especially regarding optimal imaging protocols and analysis. It remains to be established in which tumors and for which clinical questions particular analysis assumptions are justified. Likewise, there is not yet consensus on details of the pharmacokinetic modeling algorithms: how to define the values of important input variables to be used, or how to overcome the difficulties inherent in fitting noisy data in modeling intensity-time curves from DCE-MR image data, to name just two.

There is agreement, however, on the fact that until validated, user-friendly and imager-independent software for analysis and presentation of DCE-MRI results is widely available, the potential of DCE-MRI will not be fully developed and realized.

It has been hypothesized that DCE-MRI-derived variables are good predictors of response in bone sarcoma because they measure the regional microcirculatory access of intravenous contrast agent across individual tumors. To the extent this is true, these DCE-MRI variables may be an appropriate surrogate for access of the chemotherapeutic agent to the tumor. Molecular factors such as size, charge and binding groups modulate the access of molecules to the tumor EES, so the DCE-MRI of the future will not necessarily be limited to the low-MW contrast media now in clinical use. There is considerable debate about the optimal molecular size for the contrast media to be used. It is likely that contrast media with a range of molecular weights, some with binding moieties for particular receptors, will be designed to address a variety of clinical questions.

This is a time of great productivity in drug development. Entirely new kinds of agents are being designed and readied for clinical testing. One class of new agent – anti-angiogenic therapeutics – has numerous new candidates for cancer therapy already in Phase I clinical trials. Anti-angiogenic agents generally target tumor vasculature and require very different trial models from traditional cytotoxic cancer agents. The paradigms are still evolving for the way that anti-angiogenic agents act on tumor vasculature, alone or in concert with traditional therapies (including cytotoxic drugs and irradiation) (Jain 2001). Jain (2001) has suggested that anti-angiogenic therapy might increase the efficiency of the abnormal tumor vasculature, increasing the delivery of cytotoxic drugs or oxygen. Or, conversely, anti-angiogenic therapy might make endothelial cells more sensitive to cytotoxic therapies.

A noninvasive measure of drug access could facilitate the identification of physiological factors underlying the resistance of some tumors to chemotherapy and could also provide a window into the state of the vascular in an individual tumor. This would be a significant advance toward the goal of customized drug delivery to ensure adequate exposure to chemotherapy in the EES of all tumors and a valuable addition to tumor imaging in studies involving anti-angiogenic agents. The hope of refining DCE-MRI measures into tools which provide not only clinical utility but insight into alterations in tumor microcirculation drives the current interest in applying pharmacokinetic models to interpret DCE-MR images.

Despite the technical questions to be resolved, there is good reason to hope that these novel dynamic imaging models can provide important new measures that reflect the range of biological variation within and between musculoskeletal sarcoma and move us toward the ultimate goal of tailoring cancer therapies to the risk profiles of individual patients.

References

Abudu A, Davies AM, Pynsent PB, Mangham DC, Tillman RM, Carter SR, Grimer RJ (1999) Tumour volume as a predictor of necrosis after chemotherapy in Ewing's sarcoma. J Bone Joint Surg [Br] 81:317–322

Ambros IM, Ambros PF, Strehl S, Kovar H, Gadner H, Salzer-Kuntschik M (1991) MIC2 is a specific marker for Ewing's sarcoma and peripheral primitive neuroectodermal tumors. Cancer 67:1886-1893

Antman K, Crowley J, Balcerzak SP, Kempf RA, Weiss RB; Clamon GH, Baker LH (1998) A Southwest Oncology Group and Cancer and Leukemia Group B phase II study of doxorubicin, dacarbazine, ifosfamide, and mesna in adults with advanced osteosarcoma, Ewing's sarcoma, and rhabdomyosarcoma. Cancer 82:1288–1295

Aparicio J, Munarriz B, Pastor M, Vera FJ, Castel V, Aparisi F, Montalar J, Badal MD, Gomez-Codina J, Herrnaz C (1998) Long-term followup and prognostic factors in Ewing's sarcoma. A multivariate analysis of 116 patients from a single institution. Oncology 55:20–26

Bacci G, Ferrari S, Donati D, Longhi A, Bertoni F, DiFiore M, Comandone A, Cesari M, Campanacci M (1998) Neoadjuvant chemotherapy for osteosarcoma of the extremity in patients in the fourth and fifth decade of life. Oncol Rep 5:1259–1263

Baldini EH, Demetri GD, Fletcher CD, Foran J, Marcus KC, Singer S (1999) Adults with Ewing's sarcoma/primitive neuroectodermal tumor: adverse effect of older age and primary extraosseous disease on outcome. Ann Surg 230:79–86

Berger FH, Verstraete KL, Gooding CA, Lang P (2000) MR imaging of musculoskeletal neoplasm. Magn Reson Imaging Clin North Am 8:929–951

Bielack SS, Kempf-Bielack B, Delling G, Exner GU, Flege S, Helmke K, Kotz R, Salzer-Kuntschik M, Werner M, Winkelmann W, Zoubek A, Jurgens H, Winkler K (2002) Prognostic factors in high-grade osteosarcoma of the extremities or trunk: an analysis of 1,702 patients treated on neoadjuvant cooperative osteosarcoma study group protocols. J Clin Oncol 20:776–790

Bloem JL, van der Woude H-J, Geirnaerdt MJA, Hogendoorn PCW, Taminiau AHM, Hermans Jo (1997) Does magnetic resonance imaging make a difference for patients with musculoskeletal sarcoma? Br J Radiol 70:327–337

Bonnerot V, Charpentier A, Frouin F, Kalifa C, Vanel D, Di Paola R (1992) Factor analysis of dynamic magnetic resonance imaging in predicting the response of osteosarcoma to chemotherapy. Invest Radiol 27:847–855

Carsi B, Rock MG (2002) Primary osteosarcoma in adults older than 40 years. Clin Orthop 397:53–61

Cotterill SJ, Ahrens S, Paulussen M, Jurgens HF, Voute PA, Gadner H, Craft AW (2000) Prognostic factors in Ewing's tumor of bone: analysis of 975 patients from the European Intergroup Cooperative Ewing's Sarcoma Study Group. J Clin Oncol 18:3108–3114

Crim JR, Seeger LL, Yao L, Chandnani V, Eckardt JJ (1992) Diagnosis of soft-tissue masses with MR imaging: can benign masses be differentiated from malignant ones? Radiology 185:581–586

Delepine N, Delepine G, Cornille H, Voisin MC, Brun B, Desbois JC (1997) Prognostic factors in patients with localized Ewing's sarcoma: the effect on survival of actual received drug dose intensity and of histologic response to induction therapy. J Chemother 9:352–363

Egmont-Petersen M, Hogendoorn PC, van der Geest RJ, Vrooman HA, van der WH, Janssen JP, Bloem JL, Reiber JH (2000) Detection of areas with viable remnant tumor in postchemotherapy patients with Ewing's sarcoma by dynamic contrast-enhanced MRI using pharmacokinetic modeling. Magn Reson Imaging 18:525–535

Eilber FC, Rosen G, Eckardt J, Forscher C, Nelson SD, Selch M, Dorey F, Eilber FR (2001) Treatment-induced pathologic necrosis: a predictor of local recurrence and survival in patients receiving neoadjuvant therapy for high-grade extremity soft tissue sarcomas. J Clin Oncol 19:3203-3209

Erlemann R, Reiser MF, Peters PE, Vasallo P, Nommensen B, Kusnierz-Glaz CR, Ritter J, Roessner A (1989) Musculoskeletal neoplasms: static and dynamic Gd-DTPA-enhanced MR imaging. Radiology 171:767–773

Erlemann R, Sciuk J, Bosse A, Ritter J, Kusnierz-Glaz CR, Peters PE, Wuisman P (1990) Response of osteosarcoma and Ewing sarcoma to preoperative chemotherapy: assessment with dynamic and static MR imaging and skeletal scintigraphy. Radiology 175:791–796

Fizazi K, Dohollou N, Blay JY, Guerin S, Le Cesne A, Andre F, Pouillart P, Tursz T, Nguyen BB (1998) Ewing's family of tumors in adults: multivariate analysis of survival and long-term results of multimodality therapy in 182 patients. J Clin Oncol 16:3736–3743

Fletcher BD (1991) Response of osteosarcoma and Ewing sarcoma to chemotherapy: imaging evaluation. AJR 157:825–833

Fletcher BD (1997) Imaging pediatric bone sarcomas: diagnosis and treatment-related issues. Radiol Clin North Am 35:1477–1494

Fletcher CD, Gustafson P, Rydholm A, Willen H, Akerman M (2001) Clinicopathologic re-evaluation of 100 malignant fibrous histiocytomas: prognostic relevance of subclassification. J Clin Oncol 19:3045–3050

Fletcher BD, Hanna SL, Fairclough DL, Gronemeyer SA (1992) Pediatric musculoskeletal tumors: use of dynamic contrast-enhanced MR imaging to monitor response to chemotherapy. Radiology 184:243–248

Gebhardt MC (2002) What's new in musculoskeletal oncology. J Bone Joint Surg [Am] 84-A:694–701

Geirnaerdt MJA, Bloem JL, van der Woude H-J, Taminiau AHM, Nooy MA, Hogendoorn PCW (1998) Chondroblastic osteosarcoma: characterisation by gadolinium-enhanced MR imaging correlated with histopathology. Skeletal Radiol 27:145–153

Geirnaerdt MJ, Hogendoorn PC, Bloem JL, Taminiau AH, van der Woude HJ (2000) Cartilaginous tumors: fast contrast-enhanced MR imaging. Radiology 214:539–546

Gherlinzoni F, Picci P, Bacci G, Campanacci D (1992) Limb sparing versus amputation in osteosarcoma: correlation between local control, surgical margins, and tumor necrosis – Instituto Rizzoli experience. Ann Oncol 3:23–27

Gherlinzoni F, Bacci G, de Paoli A et al (1999) Adjuvant chemotherapy for high grade soft tissue sarcomas of the extremities in adult patients: results of the Italian randomized trial. Acta Orthop Scand 289:46–47

Glasser DB, Lane JM, Huvos AG, Marcove RC, Rosen G (1992) Survival, prognosis, and therapeutic response in osteogenic sarcoma. Cancer 69:698–708

Glazer HS, Lee JKT, Levitt RG, Heiken JP, Ling D, Totty WG, Balfe DM, Emani B, Wasserman TH, Murphy WA (1985) Radiation fibrosis: differentiation from recurrent tumor by MR imaging. Radiology 156:721–726

Goto A, Takeuchi S, Sugimura K, Maruo T (2002) Usefulness of Gd-DTPA contrast-enhanced dynamic MRI and serum determination of LDH and its isozymes in the differential diagnosis of leiomyosarcoma from degenerated leiomyoma of the uterus. Int J Gynecol Cancer 12:354–361

Green DM (1985) Diagnosis and management of malignant solid tumors in infants and children. Martinus Nijhoff Publishing, Boston

Grier HE (1997) The Ewing family of tumors. Ewing's sarcoma and primitive neuroectodermal tumors. Pediatr Clin North Am 44:991–1004

Gullino PM, Grantham FH (1964) The vascular space of growing tumors. Cancer Res 24:1727–1732

Hanna SL, Parham DM, Fairclough DL, Meyer WH, Le AH, Fletcher BD (1992) Assessment of osteosarcoma response to preoperative chemotherapy using dynamic FLASH gadolinium-DTPA-enhanced magnetic resonance mapping. Invest Radiol 27:367–373

Hanna SL, Reddick WE, Parham DM, Gronemeyer SA, Taylor JS, Fletcher BD (1993) Automated pixel-by-pixel mapping of dynamic contrast-enhanced MR images for evaluation of osteosarcoma response to chemotherapy: preliminary results. J Magn Reson Imaging 3:849–853

Hense HW, Ahrens S, Paulussen M, Lehnert M, Jurgens H (1999) Factors associated with tumor volume and primary metastases in Ewing tumors: results from the (EI)CESS studies. Ann Oncol 10:1073–1077

Henshaw RM, Priebat DA, Perry DJ, Shmookler BM, Malawer MM (2001) Survival after induction chemotherapy and surgical resection for high-grade soft tissue sarcoma. Is radiation necessary? Ann Surg Oncol 8:484–495

Holscher HC, Bloem JL, Nooy MA, Taminiau AHM, Eulderink F, Hermans J (1990) The value of MR imaging in monitoring the effect of chemotherapy on bone sarcomas. AJR 154:763–769

Holscher HC, Bloem JL, Vanel D, Hermans Jo, Nooy MA, Taminiau AHM, Henry-Amar M (1992) Osteosarcoma: chemotherapy-induced changes at MR imaging. Radiology 182:839–844

Hudson M, Jaffe M, Jaffe N, Ayala AG, Raymond AK, Carrasco CH, Wallace S, Murray JA, Robertson R (1990) Pediatric osteosarcoma: therapeutic strategies, results, and prognostic factors derived from a 10-year experience. J Clin Oncol 8:1988–1997

Huvos AG (1986) Osteogenic sarcoma of bones and soft tissues in older persons. A clinicopathologic analysis of 117 patients older than 60 years. Cancer 57:1442–1449

Huvos AG, Rosen G, Marcove RC (1977) Primary osteogenic sarcoma: pathologic aspects in 20 patients after treatment with chemotherapy, en bloc resection, and prosthetic bone replacement. Arch Pathol Lab Med 101:14–18

Jain RK (1994) Barriers to drug delivery in solid tumors. Sci Am 271:58–65

Jain RK (1996a) 1995 Whitaker lecture: delivery of molecules, particles, and cells to solid tumors. Ann Biomed Eng 24:457–473

Jain RK (1996b) Delivery of molecular medicine to solid tumors. Science 271:1079–1080

Jain RK (2001) Normalizing tumor vasculature with anti-angiogenic therapy: a new paradigm for combination therapy. Nature Med 7:987–989

JenkinRD, Al Fawaz I, Shabanah M, Allam A, Ayas M, Khafaga Y, Memon M, Rifai S, Schultz H, Younge D (2002) Localised Ewing sarcoma/PNET of bone – prognostic factors and international data comparison. Med Pediatr Oncol 39:586–593

Kaste SC, Hill A, Conley L, Shidler TJ, Rao BN, Neel MM (2002) Magnetic resonance imaging after incomplete resection of soft tissue sarcoma. Clin Orthop 397:204–211

Kawai A, Sugihara S, Kunisada T, Uchida Y, Inoue H (1997) Imaging assessment of the response of bone tumors to preoperative chemotherapy. Clin Orthop Relat Res 337:216–225

Knopp MV, Giesel FL, Marcos H, von Tengg-Kobligk H, Choyke P (2001) Dynamic contrast-enhanced magnetic resonance imaging in oncology. Top Magn Reson Imaging 12:301–308

Lawrence JA, Babyn PS, Chan HSL, Thorner PS, Pron GE, Krajbich IJ (1993) Extremity osteosarcoma in childhood: prognostic value of radiologic imaging. Radiology 189:43–47

Link MP, Eilber F (1993) Osteosarcoma. In: Pizzo PA, Poplack DG (eds) Principles and practice of pediatric oncology, 2nd edn. Lippincott, Philadelphia

Link MP, Goorin AM, Miser AW, Green AA, Pratt CB, Belasco JB, Pritchard J, Malpas JS, Baker AR, Kirkpatrick JA, Ayala AG, Shuster JJ, Abelson HT, Simone JV, Vietti TJ (1986) The effect of adjuvant chemotherapy on relapse-free survival in patients with osteosarcoma of the extremity. N Engl J Med 314:1600–1606

Ma LD (1999) Magnetic resonance imaging of musculoskeletal tumors: skeletal and soft tissue masses. Curr Probl Diagn Radiol 28:29–62

Ma LD, Frassica FJ, McCarthy EF, Bluemke DA, Zerhouni EA (1997) Benign and malignant musculoskeletal masses: MR imaging differentiation with rim-to-center differential enhancement ratios. Radiology 202:739–744

MacVicar AD, Olliff JF, Pringle J, Pinkerton CR, Husband JE (1992) Ewing sarcoma: MR imaging of chemotherapy-induced changes with histologic correlation. Radiology 184:859–864

Mankin HJ, Mankin CJ, Simon MA (1996) The hazards of the biopsy, revisited. Members of the Musculoskeletal Tumor Society. J Bone Joint Surg [Am] 78:656–663

May DA, Good RB, Smith DK, Parsons TW (1997) MR imaging of musculoskeletal tumors and tumor mimickers with intravenous gadolinium: experience with 242 patients. Skeletal Radiol 26:2–15

McCarville MB, Spunt SL, Skapek SX, Pappo AS (2002) Synovial sarcoma in pediatric patients. AJR Am J Roentgenol 179:797–801

Meyers PA, Heller G, Healy J, Huvos AG, Lane JM, Marcove RC, Applewhite A, Vlamis V, Rosen G (1992) Chemotherapy for non-metastatic osteogenic sarcoma: the Memorial Sloan-Kettering experience. J Clin Oncol 10:5–15

Miller SL, Hoffer FA, Reddick WE, Wu S, Glass JO, Gronemeyer SA Haliloglu M, Nikoranov AY, Xiong X, Pappo AS (2001) Tumor volume or dynamic contrast-enhanced MRI for prediction of clinical outcome of Ewing sarcoma family of tumors. Pediatr Radiol 31:518–523

Noria S, Davis A, Kandel R, Levesque J, O'Sullivan B, Wunder J, Bell R (1996) Residual disease following unplanned excision of soft-tissue sarcoma of an extremity. J Bone Joint Surg [Am] 78:650–655

Ongolo-Zogo P, Thiesse P, Sau J, Desuzinges C, Blay JY, Bonmartin A, Bochu M, Philip T (1999) Assessment of osteosarcoma response to neoadjuvant chemotherapy: comparative usefulness of dynamic gadolinium-enhanced spin-echo magnetic resonance imaging and technetium-99 m skeletal angioscintigraphy. Eur Radiol 9:907–914

Onikul E, Fletcher BD, Parham DM, Chen G (1996) Accuracy of MR imaging for estimating intraosseous extent of osteosarcoma. AJR 167:1211–1215

Pan G, Raymond AK, Carrasco CH, Wallace S, Kim EE, Shirkhoda A, Jaffe N, Murray JA, Benjamin RS (1990) Osteosarcoma: MR imaging after preoperative chemotherapy. Radiology 174:517–526

Paulussen M, Ahrens S, Dunst J, Winkelmann W, Exner GU, Kotz R, Amann G, Dockhorn-Dworniczak B, Harms D, Muller-Weihrich S, Welte K, Kornhuber B,Janka-Schaub G, Gobel U, Treuner J, Voute PA, Zoubek A, Gadner H, Jurgens H (2001) Localized Ewing tumor of bone: final results of the cooperative Ewing's Sarcoma Study CESS 86. J Clin Oncol 19:1818–1829

Peabody TD, Gibbs CP Jr, Simon MA (1998) Evaluation and staging of musculoskeletal neoplasms. J Bone Joint Surg Am 80:1204–1218

Picci P, Bacci G, Campanacci D, Gasparini M, Pilotti S, Cerasoli S, Bertoni F, Guerra A, Capanna R, Albisinni U (1985) Histologic evaluation of necrosis in osteosarcoma induced by chemotherapy: regional mapping of viable and nonviable tumor. Cancer 56:1515–1521

Picci P, Sangiorgi L, Rougraff BT, Neff JR, Casadei R, Campanacci M (1994) Relationship of chemotherapy-induced necrosis and surgical margins to local recurrence in osteosarcoma. J Clin Oncol 12:2699–2705

Provisor AJ, Ettinger LJ, Nachman JB, Krailo MD, Makley JT, Yunis EJ, Huvos AG, Betcher DL, Baum ES, Kisker CT, Miser JS (1997) Treatment of nonmetastatic osteosarcoma of the extremity with preoperative and postoperative chemotherapy: a report from the Children's Cancer Group. J Clin Oncol 15:76–84

Rao BN (1993) Nonrhabdomyosarcoma in children: prognostic factors influencing survival. Semin Surg Oncol 9:524–531

Raymond AK, Chawla SP, Carrasco CH et al (1987) Osteosarcoma chemotherapy effect: a prognostic factor. Semin Diagn Pathol 4:212–236

Reddick WE, Langston JW, Meyer WH, Gronemeyer SA, Steen RG, Chen G, Taylor JS (1994) Discrete signal processing of dynamic contrast-enhanced MR imaging: statistical validation and preliminary clinical application. J Magn Reson Imaging 4:397–404

Reddick WE, Bhargava R, Taylor JS, Meyer WH, Fletcher BD (1996) Dynamic contrast-enhanced MR imaging evaluation of osteosarcoma response to neoadjuvant chemotherapy. J Magn Reson Imaging 5:689–694

Reddick WE, Wang S-C, Xiong X, Glass JO, Wu S, Kaste SC, Pratt CB, Meyer WH, Fletcher BD (2001) Dynamic magnetic resonance imaging of regional contrast access as an additional prognostic factor in pediatric osteosarcoma. Cancer 91:2230–2237

Rijpkema M, Kaanders JHAM, Joosten FBM, van der Kogel AJ, Heerschap A (2001) Method for quantitative mapping of dynamic MRI contrast agent uptake in human tumors. J Magn Reson Imaging 14:457–463

Rosen G, Marcove RC, Caparros B, Nirenberg A, Kasloff C, Huvos AG (1979) Primary osteogenic sarcoma: the rationale for preoperative chemotherapy and delayed surgery. Cancer 43:2163–2177

Rosen G, Caparros B, Huvos AG et al (1982) Preoperative chemotherapy for osteogenic sarcoma: selection of postoperative adjuvant chemotherapy based on the response of the primary tumor to preoperative chemotherapy. Cancer 49:1221–1230

Salzer-Kuntschik M, Delling G, Beron G, Sigmund R (1983) Morphological grades of regression in osteosarcoma after polychemotherapy – study COSS 80. J Cancer Res Clin Oncol 106:21–24

Sanchez RB, Quinn SF, Walling A, Estrada J, Greenberg H (1990) Musculoskeletal neoplasms after intraarterial chemotherapy: correlation of MR images with pathologic specimens. Radiology 174:237–240

San Julian M, Aquerreta JD, Benito A, Canadell J (1999) Indications for epiphyseal preservation in metaphyseal malignant bone tumors of children: relationship between image methods and histological findings. J Pediatr Orthop 19:543–548

Schwickert HC, Stiskal M, Roberts TPL, van Dijke CF, Mann JS, Muhler A, Shames DM, Demsar F, Disston A, Brasch RC (1996) Contrast-enhanced MR imaging assessment of tumor capillary permeability: effect of irradiation on delivery of chemotherapy. Radiology 198:893–898

Shapeero LG, Vanel D (2000) Imaging evaluation of the response of high-grade osteosarcoma and Ewing sarcoma to chemotherapy with emphasis on dynamic contrast-enhanced magnetic resonance imaging. Semin Musculoskelet Radiol 4:137–146

Shapeero LG, Vanel D, Verstraete KL, Bloem JL (2002) Fast magnetic resonance imaging with contrast for soft tissue sarcoma viability. Clin Orthop 212–227

Su M-Y, Wang Z, Nalcioglu O (1999) Investigation of longitudinal vascular changes in control and chemotherapy-treated tumors to serve as therapeutic efficacy predictors. J Magn Reson Imaging [Suppl] 9:128–137

Su M-Y, Taylor JA, Villarreal LP, Nalcioglu O (2000) Prediction of gene therapy-induced tumor size changes by the vascularity changes measured using dynamic contrast-enhanced MRI. Magn Reson Imaging 18:311–317

Tannock IF, Lee CM, Tunggal JK, Cowan DSM, Egorin MJ (2002) Limited penetration of anticancer drugs through tumor tissue: a potential cause of resistance of solid tumors to chemotherapy. Clin Cancer Res 8:878–884

Tofts PS, Brix G, Buckley DL, Evelhoch JL, Henderson E, Knopp MV, Larsson HBW, Lee T-Y, Mayr NA, Parker GJM, Port R, Taylor JS, Weisskoff RM (1999) Estimating kinetic parameters from dynamic contrast-enhanced T1-weighted MRI of a diffusible tracer: standardized quantities and symbols. J Magn Reson Imaging 10:223–232

Van der Woude HJ, Bloem JL, Taminiau AHM, Nooy MA, Hogendoorn PCW (1994) Classification of histopathologic changes following chemotherapy in Ewing's sarcoma of bone. Skeletal Radiol 23:501–507

Van der Woude HJ, Bloem JL, Verstraete KL, Taminiau AHM, Nooy MA, Hogendoorn PCW (1995) Osteosarcoma and Ewing's sarcoma after neoadjuvant chemotherapy: value of dynamic MR imaging in detecting viable tumor before surgery. AJR 165:593–598

Van der Woude HJ, Bloem JL, Hogendoorn PCW (1998a) Preoperative evaluation and monitoring chemotherapy in patients with high-grade osteogenic and Ewing's sarcoma: review of current imaging modalities. Skeletal Radiol 27:57–71

Van der Woude HJ, Bloem JL, Pope TL Jr (1998b) Magnetic resonance imaging of the musculoskeletal system, part 9. Primary tumors. Clin Orthop 347:272–286

Van der Woude HJ, Egmont-Petersen M (2001) Contrast-enhanced magnetic resonance imaging of bone marrow. Semin Musculoskelet Radiol 5:21–33

Van Rijswijk CS, Hogendoorn PC, Taminiau AH, Bloem JL (2001) Synovial sarcoma: dynamic contrast-enhanced MR imaging features. Skeletal Radiol 30:25–30

Vanel D, Shapeero LG, De Baere T, Gilles R, Tardivon A, Genin J, Guinebretiere JM (1994) MR imaging in the follow-up of malignant and aggressive soft-tissue tumors: results of 511 examinations. Radiology 190:263–268

Verstraete KL, Lang P (2000) Bone and soft tissue tumors: the role of contrast agents for MR imaging. Eur Radiol 34:229–246

Verstraete KL, de Deene Y, Roels H, Dierick A, Uyttendaele D, Kunnen M (1994a) Benign and malignant musculoskeletal lesions: dynamic contrast-enhanced MR imaging – parametric "first-pass" images depict tissue vascularization and perfusion. Radiology 192:835–843

Verstraete KL, Dierick A, de Deene Y, Uyttendaele D, Vandamme F, Roels H, Kunnen M (1994b) First-pass images of musculoskeletal lesions: a new and useful diagnostic application of dynamic contrast-enhanced MRI. Magn Reson Imaging 12:687–702

Verstraete KL, van der Woude H-J, Hogendoorn PCW, de Deene Y, Kunnen M, Bloem JL (1996) Dynamic contrast-enhanced MR imaging of musculoskeletal tumors: basic principles and clinical applications. J Magn Reson Imaging 6:311–321

Verstraete KL, Vanzieleghem B, de Deene Y, Palmans H, de Greef D, Kristoffersen DT, Uyttendaele D, Roels H, Hamers J, Kunnen M (1994c) Static, dynamic and first-pass MR imaging of musculoskeletal lesions using gadodiamide injection. Acta Radiol 35:1–10

Wellings RM, Davies AM, Pynsent PB, Carter SR, Grimer RJ (1994) The value of computed tomographic measurements in osteosarcoma as a predictor of response to adjuvant chemotherapy. Clin Radiol 49:19–23

Wunder JS, Paulian G, Huvos AG, Heller G, Meyers PA, Healey JH (1998) The histologic response to chemotherapy as a predictor of the oncological outcome of operative treatment of Ewing sarcoma. J Bone Joint Surg [Am] 80:1020–1033

14 Dynamic Contrast-Enhanced MRI in the Liver

Alan Jackson and David A. Nicholson

CONTENTS

14.1 Introduction

As little as 10 years ago routine magnetic resonance imaging (MRI) of the liver for the investigation of clinical disease was considered to be of limited utility despite the massive growth of the technique in the investigation of diseases of the musculoskeletal and central nervous system. Since that time improvements in technology, particularly fast acquisition techniques, have led to the routine use of MRI for abdominal and in particular for liver imaging (Rummeny and Marchal 1997; Van Beers et al. 1997; Bartolozzi et al. 1999). As with any organ system successful application of MRI techniques is dependent on optimising the contrast between normal and pathological tissues. Conventional T1 and T2 weighted sequences can now be acquired rapidly, often in a single breath hold using fast spin echo (FSE), half-Fourier single shot fast spin echo (HASTE) sequences and fast and ultrafast spoiled T1 weighted gradient echo sequences (Catasca and Mirowitz 1994; Siewert et al. 1994; Reinig 1995; Spritzer et al. 1996). The ability to routinely suppress signal from fat on T1 weighted images using frequency selective fat suppression pre-pulses has also improved the ability of MRI to demonstrate the liver and associated pathology. The value of MRI in the liver is significantly enhanced by the wide range of contrast agents available to increase the distinction between normal and pathological tissues (Morana et al. 2002). A wider range of contrast agents is available for use in the hepatobiliary system than for any other application and the range of contrast agents constantly expands with both positive and negative enhancement agents available. These contrast media can be classified into three main categories depending on their bio distribution mechanism. Conventional intravenous contrast agents accumulate in the vascular space and extracellular extra vascular space (EES) by a passive distribution mechanism, the second class of agents are actively taken up by hepatocytes and are excreted to variable extents in the biliary tract, the third group are actively accumulated by the reticuloendothelial system (RES). A number of agents are available within each of these three classes and are increasingly used routinely in the clinical setting.

Although dynamic imaging techniques are routinely used in only a small number of clinical applications, some form of dynamic contrast enhanced examination is a normal part of many diagnostic imaging investigations of the liver (Bartolozzi et al. 1999; Morana et al. 2002). The most common approach is the use of multi-phasic imaging (see below) during contrast enhancement which is commonly performed on computed tomography (CT) but which is increasingly used

A. Jackson, MBChB (Hons), PhD, FRCP, FRCR
Professor, Imaging Science and Biomedical Engineering, The Medical School, University of Manchester, Stopford Building, Oxford Road, Manchester, M13 9PT, UK
D. A. Nicholson, Bmed Sci, FRCR
Consultant in Gastrointestinal Radiology, Hope Hospital, Stott Lane, Salford, M6 8HD, UK

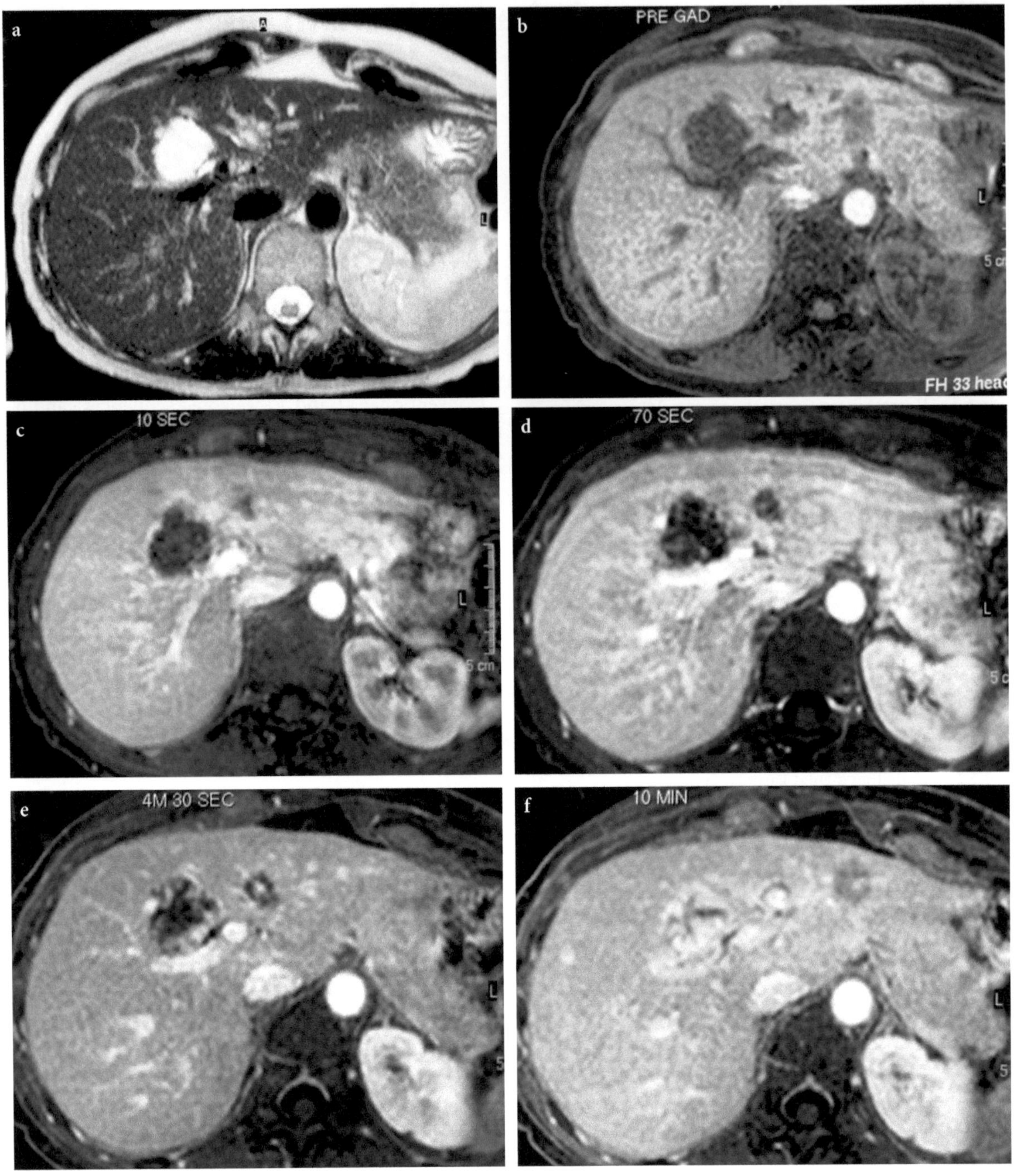

Fig. 14.1a–f. Patient with solitary liver lesion for characterisation. T2 TSE sequence pre gadolinium (**a**) shows a Segment 5 high signal lesion measuring 5×3 cm, containing a few septations. Wave T1 SENSE pre-gadolinium (**b**) shows the lesion to be low signal. Dynamic post intravenous gadolinium scans (**c–f**; 10 s, 135 s, 4.5 min, 10 min) show peripheral globular enhancement of this lesion with subsequent 'filling in' typical of a haemangioma

in routine clinical MRI (Yu et al. 2000; Semelka et al. 2001; Kanematsu et al. 2002; Hussain et al. 2003; Noguchi et al. 2003) (see Fig. 14.1). The common use of dynamic enhancement studies reflects the unique vascular anatomy of the liver, which receives both arterial blood from the systemic circulation via the hepatic arteries and a larger venous blood flow from the bowel and spleen via the hepatic portal system. The importance of this dual blood supply in the investigation of hepatic diseases cannot be overstated. Some lesions, particularly those that arise from normal liver tissue, will also receive dual blood supply. Others will be isolated from the portal venous or arterial supplies and will exhibit vascular enhancement patterns that reflect their vascular anatomy. This gives rise to differential temporal enhancement patterns in a wide range of hepatic diseases that can be usefully used for diagnosis and classification. More importantly, it allows the identification of lesions within the hepatic parenchyma that may not be evident on routine post contrast scans.

In this chapter we will initially describe the available contrast media for use in hepatic investigations, describe the technique of multiphasic imaging and review the common diseases of the liver and their enhancement characteristics. The remainder of the chapter will describe the application of pharmacokinetic models for the analysis of contrast distribution mechanisms in the liver and hepatic disease.

14.2 Reticuloendothelial System (RES) Contrast Agents

Kupffer cells in the normal liver will take up particulate matter from the circulation and this property has been utilised to produce a group of RES-specific contrast agents. These consist of suspensions of superparamagnetic iron oxide particles (SPIO) or ultra-small superparamagnetic iron oxide particles (USPIO) (Fretz et al. 1989; Weissleder et al. 1989; Weissleder et al. 1990). These agents exhibit strong T1 relaxation properties, and due to susceptibility differences to their surroundings also produce a strongly varying local magnetic field, which enhances T2 relaxation. Pathological tissues which do not contain reticuloendothelial cells will maintain their normal signal intensity. Currently available agents include Endorem and Feridex [Ferrumoxides. (USAN). SPIO. Ami-25, dextran-coated)], Resovist (Ferucarbotran, carboxy-dextran coated iron oxide nanoparticles), Sinerem (USPIO) and Combidex (ferumoxtran-10, AMI-227) (for manufacturers' details see Table 14.1). There is considerable evidence that these contrast agents increased tumour to liver contrast and improve the detection of focal liver lesions on T2 weighted images (Stark et al. 1988; Fretz et al. 1990). They also offer specific diagnostic support for the characterisation of some types of focal liver lesion such as focal nodular hyperplasia (Figs. 14.2, 14.3).

Table 14.1. Contrast media available for use in hepatic investigations – manufacturers' details

Product	Manufacturer	Location
Endorem	Guerbet S.A.	France
Feridex	Advanced Magnetics Ltd	MA, USA
Resovist	Schering AG	Germany
Sinerem	Guerbet S.A.	France
Combidex	Advanced Magnetics Ltd	MA, USA
Tesla Scan	Amersham Health	UK
Multihance	Bracco Diagnostics	Milan, Italy
Eovist	Schering AG	Germany

14.3 Hepatobiliary Contrast Agents

Hepatobiliary contrast agents utilise hepatic excretion mechanisms to produce hepatocyte uptake and biliary excretion of paramagnetic ions by binding them to appropriate ligands. Two agents, Mangafodipir trisodium (Mn-DPDP, Teslascan) and Gadobenate dimeglumine (Gd-BOPTA, Multihance) are commercially available and a third, gadoxetic acid (Gd-EOB-DTPA, Eovist) is in phase III clinical trials (see Tables 14.1 and 14.2). Mangafodipir is the first manganese complex that has been used as contrast agent in clinical trials. The Mn2+ ion is a powerful T1 relaxation agent and the molecule demonstrates both biliary and renal excretion causing positive enhancement of normal liver tissue (Rofsky and Earls 1996; Morana et al. 2002). Its chemical similarity to vitamin B6 is cited as the reason for the hepatocyte uptake, but it is likely that some of the liver accumulation of paramagnetic manganese is due to the metabolism of the parent compound with release of free Mn^{2+} ions, which are known to accumulate in hepatocytes. The maximum tissue enhancement is seen at approximately 20 min and lasts for up to 4 h. Normal liver parenchyma shows significant enhancement and tumours of non-hepatocytic origin show little or no enhancement resulting in improved demonstration of lesions (King et al. 2002). Biliary tree enhancement allows contrast enhanced MR cholangiography (Carlos et al. 2002) (Fig. 14.4) after infusion of Mangafodipir (Murakami et al. 1996; Padovani et al. 1996; Schima et al. 1997; Wang et al. 1997). However uptake of Mangafodipir has also been observed in hepatic metastasis from non-functioning endocrine tumours of the pancreas. The agent also provides no information for differentiation between benign and malignant neoplasms (Coffin et al. 1999).

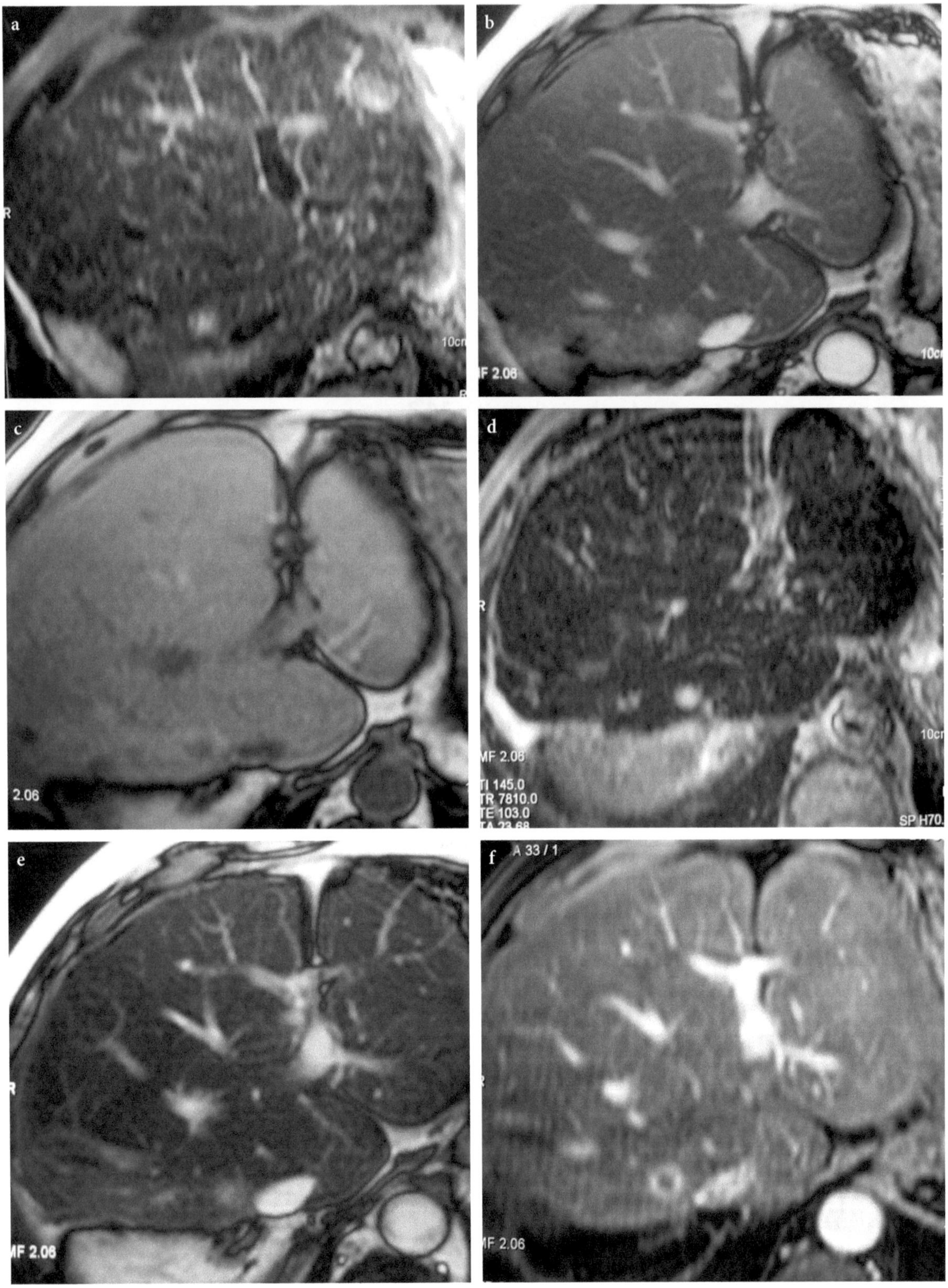

Fig. 14.2a-f. Previous partial right hepatectomy resection for colorectal metastasis. Follow-up scan shows a 4- to 5-cm posterior elliptical collection at site of previous surgery consistent with biloma. Lying anterior to the biloma is a solitary 1-cm metastasis just lateral to IVC within the residual segment 8 of the liver, close to resection margin. The metastasis is very difficult to identify on the T2 True FISP (**b**) and T1 FLASH 2D (**c**). An area of slight increased signal is seen on the T2 gradient echo (TRIM) (**a**) sequence. Following infusion of a SPIO (Endorem) the metastasis is seen with much greater confidence on T2 weighted sequences (**a, b**). With dynamic imaging following an intravenous agent (Gadolinium) the metastasis shows the typical peripheral ring enhancement, further characterising the lesion as malignant (**c**)

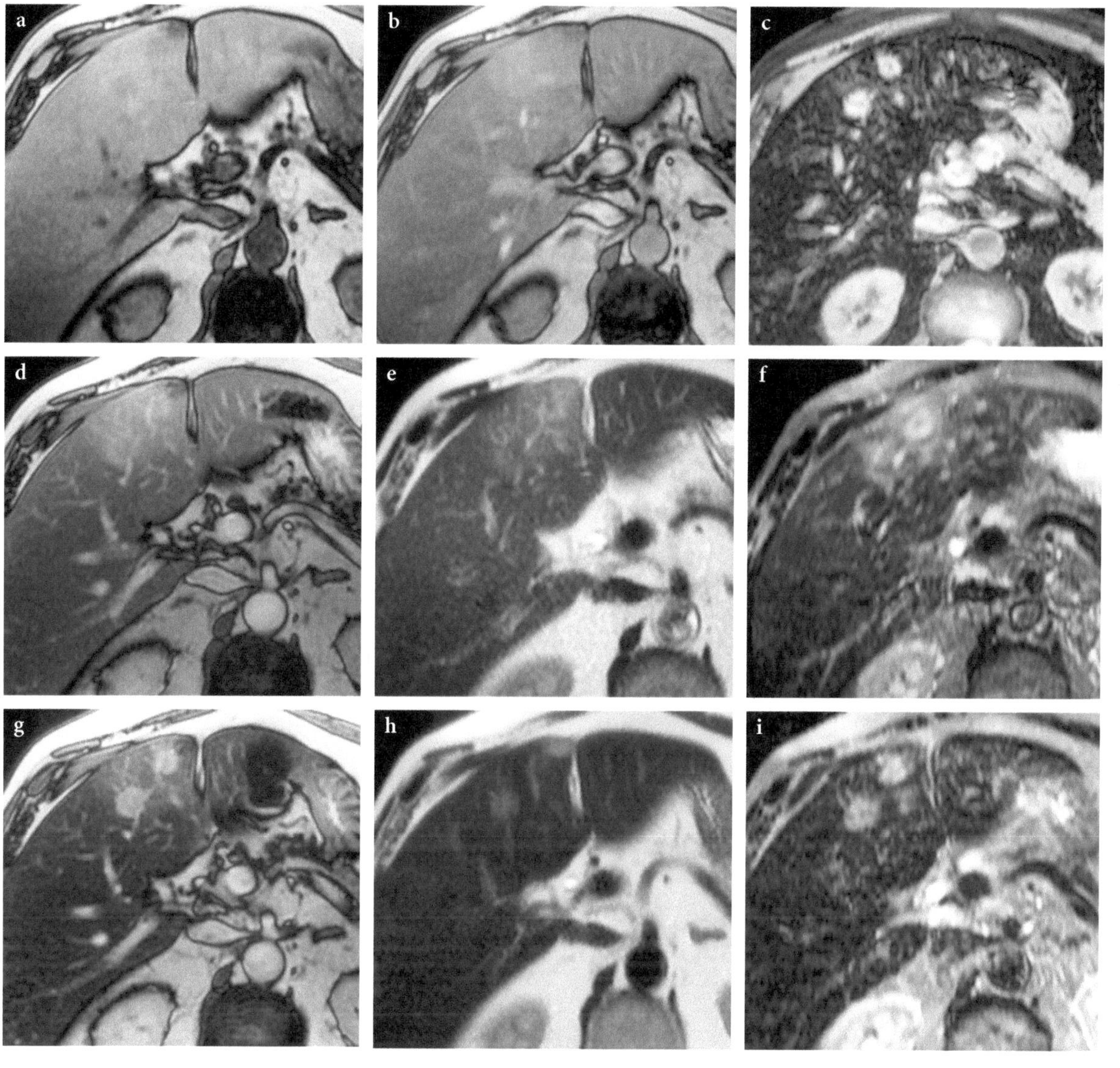

Fig. 14.3a–f. Patient with hepatic metastases from colonic cancer; for assessment of surgical resection. Metastases are not evident on T1 weighted sequences (**a**). T2 weighted sequences (**d**, true FISP; **e**, HASTE; **f**, gradient echo) show variable visibility of metastases best seen with a gradient echo sequence. Following SPIO (Endorem) the reduced signal from normal liver makes the lesions more conspicuous on the T2 (**g**), true FISP (**h**) and T2 gradient echo(**i**) sequences. Additionally a further metastases is appreciated in the periphery of the right hemi-liver significantly altering clinical management. Best seen on post Endorem T2 true FISP. T1 weighted images post IV gadolinium (**b**) do not demonstrate the lesions but T2* weighted flash following intravenous gadolinium and SPIO (**c**) shows lesions with high conspicuity

Gadolinium dimeglumine (Gd-BOPTA) was the fifth Gadolinium chelate to become available on the market. Although it has similar properties in many ways to other intravenous small molecular weight agents such as gadopentetate dimeglumine it is metabolised differently since a fraction of the administered dose is taken up specifically by functioning hepatocytes (Spinazzi et al. 1999; Morana et al. 2002). Several workers have suggested that this agent combines the advantages of conventional gadolinium enhanced MR imaging with the properties of liver specific contrast agents for lesion detection since the addition of a late delayed phase into the imaging protocol will provide images with enhancement of normal liver parenchyma (Morana et al. 2002). Gd-BOPTA has indeed been shown to improve the detection of liver lesions including metastatic disease (Manfredi et al. 1999; Petersein et al. 2000; Pirovano et al. 2000; Pena et al. 2001; Kirchin and Spinazzi 2002). However,

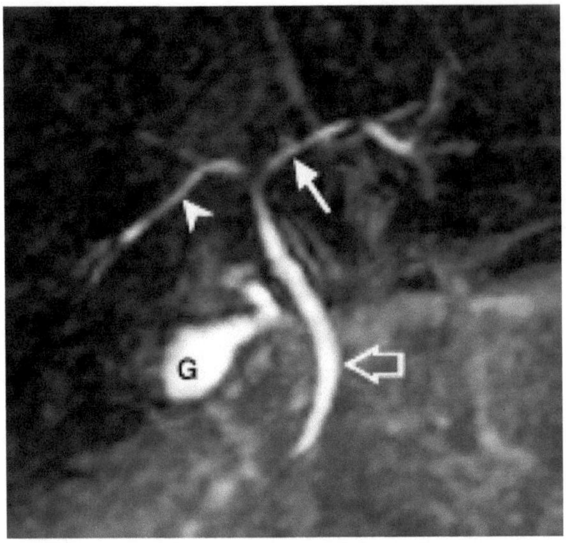

Fig. 14.4. MR contrast enhanced cholangiogram 20 min after administration of Gd-EOB-DTPA [from Carlos et al (2002)]. *G*, gallbladder; *open arrow*, common bile duct; *arrow head*, right hepatic duct; *solid arrow*, left hepatic duct. [Reproduced from Carlos et al (2002)]

Table 14.2. Contrast media for liver imaging. [Adapted from Medcyclopaedia, Amersham Health]

Name of compound	Central moiety	Relaxivity	Distribution	Indication	Development stage	Trademark
Mangafodipir trisodium MN-DPDP, Manganese dipyroxyl diphosphate	Mn^{2+}	r1=2.3, r2=4.0, B0=1.0 T	Hepatobiliary, pancreatic, adrenal	Liver lesions	Approved	Teslascan
Gadobenate dimeglumine, Gd-BOPTA	Gd^{3+}	r1=4.6, r2=6.2, B0=1.0 T	Intravascular, extracellular, hepatobiliary	Neuro/whole body, liver lesions	Approved	Multihance
Gadoxetic acid, Gd-EOB-DTPA	Gd^{3+}	short T1-relaxation time	Hepatobiliary	Liver lesions	Phase III	Eovist
Fe-HBED	Fe^{2+}		Hepatobiliary	Liver lesions		
Fe-EHPD	Fe^{2+}		Hepatobiliary	Liver lesions		
Liposomes, paramagnetic	Gd^{3+}		RES-directed	Liver lesions		
Ferrum oxid. (USAN), SPIO, Ami-25, dextran-coated	Fe^{2+}/Fe^{3+}	r1=40.0, r2=160, B0=0.47T, Xm=0.4	RES-directed	Liver lesions and control	On sale	Endorem, Feridex
Ferrum oxid. (USAN), SPIO, Ami-25, dextran-coated	Fe^{2+}/Fe^{3+}	r1=40.0, r2=160, B0=0.47T, Xm=0.4	RES-directed	Liver lesions and control	On sale	Endorem, Feridex
Ferrixan, Carboxy-dextran coated iron oxide nanoparticles, SHU 555A	Fe^{2+}	r1=25.4, r2=151	RES-directed	Liver lesions	Phase III	Resovist
USPIO, AMI-227	Fe^{3+}/Fe^{2+}	r1=25, r2=160, B0=0.47T, Xm=0.34, r1=23.3, r2=48.9, B0=0.47T	Vascular, lymph v. hepatocyte (AG-USPIO)	MR-angiography vascular staging of RES-directed liver diseases	Phase II/III, lymph nodes	Sinerem, Combidex
Magnetic starch microspheres	Mr^{2+}/Mr^{3+}	r1=27.6, r2=183.7, B0=1.0T	RES-directed	Liver lesions, spleen	Preclinical	
PION, polycrystalline iron oxide nanoparticles (larger particles = DDM 128, PION-ASF)	Fe^{2+}/Fe^{3+}	T2*enhanced, r2/r1=4.4, r2/r1=7	RES-directed lymph v. hepatocyte	Liver lesions, MR lymphography	Preclinical	

the proportional signal change resulting from hepatic uptake of Gd-BOPTA is far less than that seen with superparamagnetic contrast agents and direct comparison studies have recommended the use of ferrumoxides in preference to Gd-BOPTA (DEL FRATE et al. 2002).

Another gadolinium chelate-based compound currently in phase three clinical trials is gadoxetic acid (Gd-EOB-DTPA, Eovist). This compound exploits the carrier molecule used by hepatocytes for the uptake of bilirubin (MORANA et al. 2002). Although gadoxetic acid is excreted almost exclusively by the biliary route via hepatocyte uptake it differs from other hepatophilic molecules by being intensely hydrophilic and having low levels of protein binding (SCHMITZ, WAGNER, SCHUH-MANN-GIAMPIERI et al. 1997; SCHMITZ, WAGNER, SCHUHMANN-GIAMPIERI et al. 1997; SCHUHMANN-GIAMPIERI, MAHLER, ROLL et al. 1997). These give it a favourable tolerance for intravenous injection similar to that seen with other, renally excreted, compounds. Approximately two thirds of the agent is eliminated via the hepatic biliary route and the resulting reduction in T1 relaxation times allows its use for contrast enhanced cholangiography (KAPOOR et al. 2002; MORANA et al. 2002; VITELLAS et al. 2002; VITELLAS et al. 2002). The maximum increase in liver parenchyma signal is seen approximately 20 min after injection and enhancement lasts approximately 2 h. During the arterial and portal phases the dynamic enhancement patterns seen are similar to those with conventional gadolinium chelates so that the agent combines the benefit of both groups of compounds (VOGL et al. 1996).

14.4 Extracellular Fluid Space Agents

The most commonly used contrast agents for hepatic MRI are small molecular weight gadolinium chelates, which act non-specifically to enhance the vascular and extracellular extravascular fluid spaces. These intravenous contrast agents are in routine use in human MRI examinations and act through the paramagnetic property of gadolinium to reduce T1 reactivity and to a far lesser extent T2 reactivity, thus increasing tissue signal intensity on T1 weighted images and causing smaller signal reductions on T2 weighted images. As we have seen in previous chapters one of the potential benefits of extracellular fluid space agents is the ability to quantify the pharmacokinetic distribution of the agent and to develop surrogate markers of physiological tissue parameters such as blood vessel density, blood flow and endothelial permeability surface area product (see Chaps. 5 and 6). This approach can also be taken in liver imaging but suffers from a number of specific complications particularly related to the presence of a dual vascular supply and the severity of respiratory motion which will be discussed in detail below. For these reasons many dynamic studies in the liver have been performed using multiple sequential breath hold acquisitions. This approach, commonly called multiphasic imaging, produces a series of distinct multi-slice data acquisitions at separate time points during the circulation of the contrast media. Although it is less elegant in conception than continuous dynamic acquisition techniques it has enormous clinical value enabling improved identification and classification of liver disease on the basis of variations in vascular enhancement patterns.

14.5 Multiphasic Imaging Techniques

Multiphasic imaging to document the distribution of a bolus injection of contrast agent is probably the most important component of CT and MRI examinations of the abdomen, particularly for the identification and characterisation of liver lesions. Multiphasic imaging is commonly performed on CT because of the lower cost and its equal or superior ability to detect extra hepatic disease. Some workers recommend routine use of MRI instead of or, more commonly, in addition to dynamic helical CT techniques for routine investigation of some diseases, particularly where hepatocellular carcinoma is suspected (SEMELKA et al. 1994; SEMELKA et al. 2001; NOGUCHI et al. 2003). However, despite the dominance of CT in clinical imaging applications MRI is increasingly widely used and is the investigation of choice where contraindications to CT contrast media exist, where the liver shows diffuse fatty infiltration or where lesion identification or diagnosis is equivocal on preliminary CT.

Multiphasic MRI scanning protocols most commonly employ a multi-slice or volume spoiled gradient echo technique which allows high spatial resolution imaging of the entire region of interest during a single breath hold (BARTOLOZZI et al. 1999;

Lavelle et al. 2001; Hussain et al. 2003). The use of volume acquisitions, although not widespread, provides higher spatial resolution and improved anatomical detail, particularly if fat suppression can also be achieved. These high spatial resolution arterial and venous phase images also allow reconstruction of MR angiograms and venograms, which are valuable in surgical planning applications. Images are typically obtained during four physiological phases: (1) pre-contrast; (2) arterial, prior to enhancement of the portal venous system; (3) portal, when contrast accumulates within the sinusoids; and (4) delayed, which demonstrates extracellular collection of the contrast media (Figs. 14.5, 14.6). As stated above, studies using Gd-BOPTA can take advantage of its hepatic excretion mechanism by the addition of a further delayed scan to demonstrate hepatocytic enhancement (Manfredi et al. 1999; Schima et al. 1999; Pena et al. 2001; Morana et al. 2002). Although the timing of the portal and delayed phases can be approximated the timing of the arterial phase is more critical since the duration of the bolus of contrast during its first passage through the vasculature is relatively short (Van Beers et al. 1999). Many workers use a test injection of gadolinium chelate to estimate the timing parameters for initial arterial phase scans (Lavelle et al. 2001) and automated bolus detection techniques have also be applied (Hussain et al. 2003). The necessity for this degree of accuracy remains contentious however and some workers have found little difference between the use of a fixed delay and a tailored delay in the timing of initial scans (Materne et al. 2000). However, other workers studying hypervascular tumours have stressed the need for careful timing of the arterial phase (Van Beers et al. 1999) and some have found value in the use of a double arterial phase dynamic acquisition in these circumstances (Yoshioka et al. 2002). High resolution arterial phase acquisitions also allow identification of arterial abnormalities (Fig. 14.7)

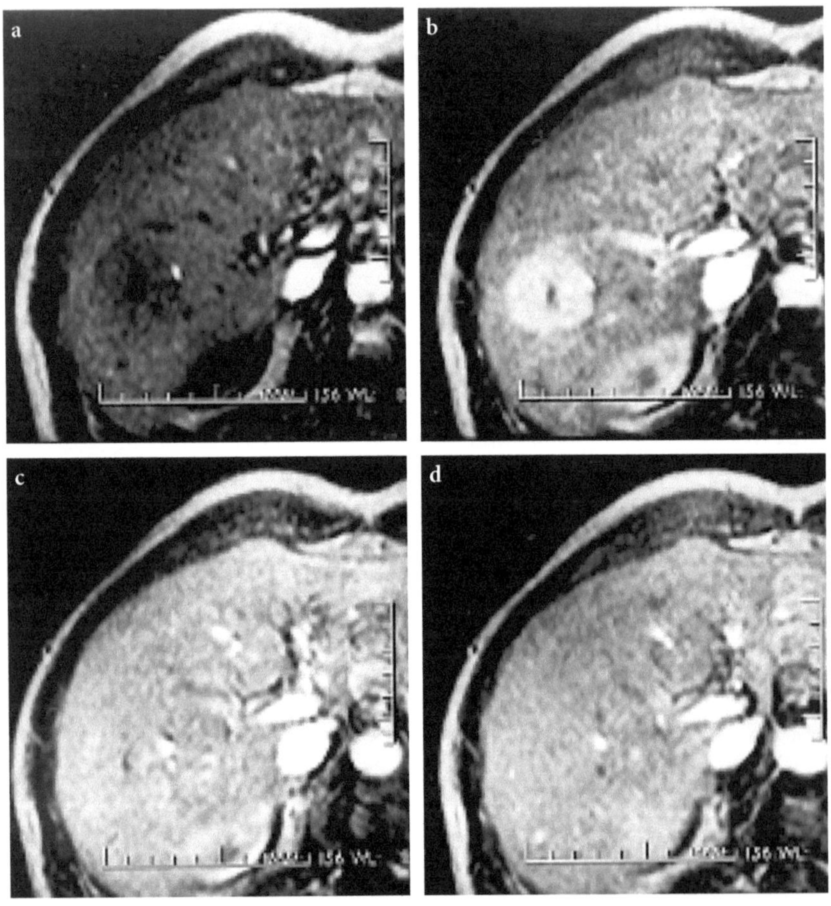

Fig. 14.5a–d. Focal nodular hyperplasia. Dynamic gadolinium-enhanced MR study with a spoiled T1-weighted GRE technique (**a**, precontrast; **b**, arterial phase; **c**, portal venous phase; **d**, delayed phase) shows clear-cut enhancement of the lesion in the arterial phase, with sparing of the central scar. In the portal venous phase and the delayed phase, the nodule is isointense to liver. In particular, no delayed enhancement of the central scar is seen. [Reproduced from Bartolozzi et al. (1999)]

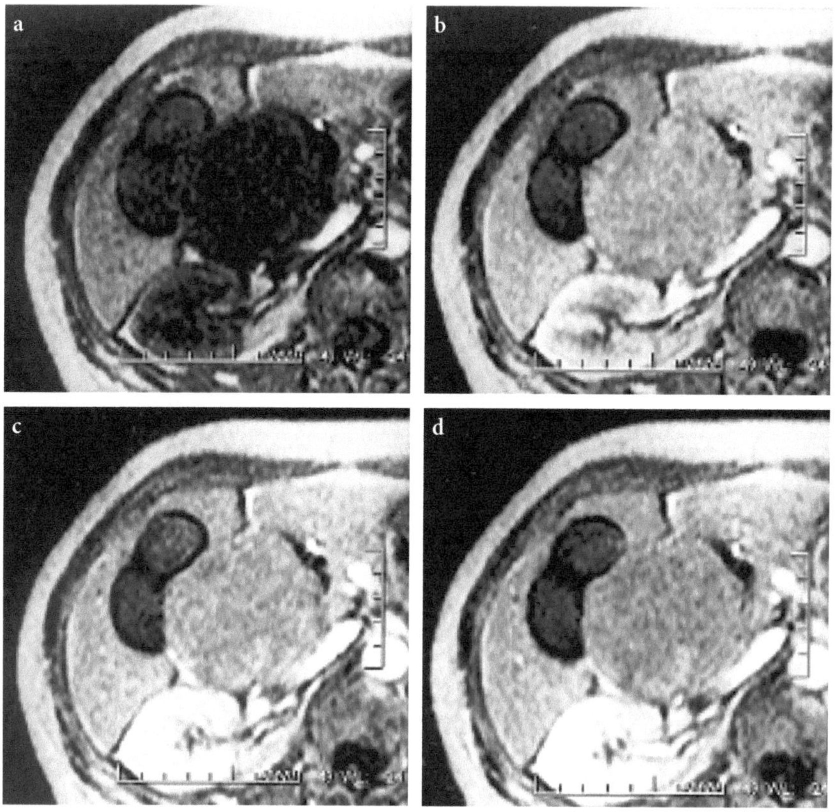

Fig. 14.6a–d. Metastasis from carcinoid tumour. Dynamic gadolinium-enhanced study with a spoiled T1-weighted GRE technique (**a**, precontrast; **b**, arterial phase; **c**, portal venous phase; **d**, delayed phase) shows substantial lesion enhancement in the arterial phase. In the portal venous phase and the delayed phase, the mass is nearly isointense to liver. **d** Contrast-enhanced T1-weighted SE image obtained 5 min after injection shows the lesion to be mostly isointense to liver, with small irregular areas of persistent enhancement. [Reproduced from BARTOLOZZI et al (1999)]

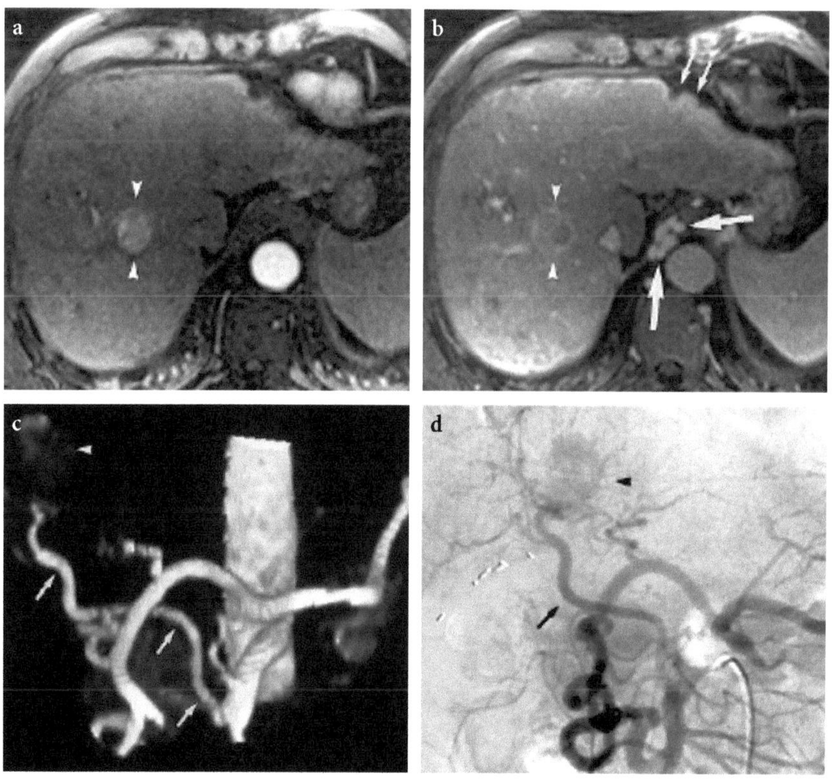

Fig. 14.7a–d. Replaced right hepatic artery arising from the superior mesenteric artery in a 58-year-old man with cirrhosis. **a,b** Transverse 3D contrast-enhanced source MR images (4.2/1.8, 12° flip angle) acquired during the (**a**) arterial and (**b**) portal venous phases of enhancement show a 1.5-cm enhancing mass (*arrowheads*) in the right hepatic lobe. The liver has a nodular contour [*small arrows* in (**b**)], and there are oesophageal varices [*large arrows* in (**b**)]. **c** Coronal volume rendered MR angiogram constructed from the transverse 3D source images, as shown in (**a**), shows the replaced right hepatic artery (*arrows*) feeding the mass (*arrowhead*). **d** DSA image after selective superior mesenteric artery injection confirms a replaced right hepatic artery (*arrow*) feeding the faintly opacified tumour (*arrowhead*). [Reproduced from LAVELLE et al (2001)]

14.6 Liver Diseases

A wide range of the hepatic abnormalities require imaging investigation (Bartolozzi et al. 1999; Morana et al. 2002). This section will concentrate on those disorders where multiphasic imaging techniques are of value in diagnosis and management.

14.6.1 Benign Liver Tumours

14.6.1.1 Haemangiomas

Haemangiomas are the most common solid hepatic lesion and are found in 4%–20% of autopsies (Hanafusa et al. 1995; Hanafusa et al. 1997). They may be divided into capillary and cavernous subtypes (Whitney et al. 1993; Morana et al. 2002). Capillary haemangiomas tend to be smaller and represent 40% of haemangiomas below 1 cm in diameter. Cavernous haemangiomas may present in patients at any age but are most commonly seen in premenopausal women. Lesions can range in size from a few millimetres to over 20 cm and those over 10 cm in diameter are designated giant haemangiomas. The lesions are multiple in 10%–20% of cases. The presence of the blood within the haemangioma vascular spaces gives a characteristic high signal on T2 weighted images which is seen in few other hepatic lesions although it can occur in some hypervascular metastasis such as islet cell tumour, phaeochromocytoma, carcinoid, renal cell carcinoma and sarcoma. A central hypointense area, often referred to as a scar, which represents central fibrous tissue may also typically be seen in larger cavernous haemangiomas. Multiphasic contrast enhanced imaging demonstrates a typical pattern in capillary haemangioma with signal intensities that are isointense to the arterial system in all phases (Jeong et al. 1999)(see Fig. 14.1). In cavernous haemangiomas multiphasic MRI may reveal similar behaviour or peripheral nodular enhancement which may or may not progress to involve central regions of the tumour may be seen (Whitney et al. 1993; Bartolozzi et al. 1999).

14.6.1.2 Focal Nodular Hyperplasia

Focal nodular hyperplasia (FNH) is the second most common benign lesion of the liver and is present in 3%–5% of the population (Nguyen et al. 1999). It is most commonly seen in women of childbearing age. The lesions represent normal liver tissue but show abnormal or absent portal vein supply and dysplastic biliary drainage. A central area of scarring is seen in some cases (approximately 30%) and is the focus of the dysplastic biliary system. The pathogenesis of the lesion is unclear although it has been suggested that it is developmental and associated with pre-existing arterial vascular anomalies. Focal nodular hyperplasia typically appears almost isointense to normal liver parenchyma on all non-enhanced sequences. The central area of scarring is hypointense on T1weighted images but may be hyperintense on T2 weighted images due to development of new vessels. Multiphasic imaging classically shows early arterial phase enhancement with a rapid washout phase reflecting the hypervascular nature of these regenerative nodules (see Fig. 14.5). The central scar may demonstrate prominent enhancement during the portal and delayed phases at a time when the peripheral nodules are isointense to liver, this is more commonly seen in large lesions (Choi and Freeny 1998; Bartolozzi et al. 1999; Carlson et al. 2000; Morana et al. 2002). Where diagnostic doubt exists the use of RES-specific contrast media will produce a decrease in signal intensity on T2 weighted images due to the presence of active reticuloendothelial cells within the peripheral portion of the lesion.

14.6.1.3 Hepatic Adenoma

Hepatic adenomas are benign neoplasms usually found in women with a history of oral contraceptive usage (Grazioli et al. 2001). They are also associated with type one glycogen storage disease and to a lesser extent with chronic hepatic iron deposition. The lesions consists of cells larger than normal hepatocytes which contain large amounts of glycogen and lipid. They receive systemic arterial supply but no portal venous supply and have no bile ducts. Kupffer cells are commonly found in the lesions but usually have little or no functional ability. Lesions may be large and multiple and can cause significant clinical problems due to internal haemorrhage and rupture. The MRI appearances are extremely variable and areas of increased signal on T1 weighted images may be seen due to fat and haemorrhage while the areas of low intensity can be seen due to necrosis calcification and previous haemorrhage. Most are predominantly hypointense on T2 weighted images. Multiphasic contrast enhanced MRI demonstrates early arterial phase enhancement which rapidly becomes iso- or hypointense in the portal phase (Bartolozzi et al. 1999; Ichikawa et al. 2000; Morana et al. 2002).

14.6.2 Malignant Liver Tumours

14.6.2.1 Hepatocellular Carcinoma (HCC)

Hepatocellular carcinoma is the commonest primary hepatic malignancy (80%–90%) representing over 5% of all cancers . There are 500,000 to one million new cases each year worldwide (Bruix et al. 2001). It is commoner in developing countries but its incidence is rising in the West. It is commonly associated with liver cirrhosis particularly secondary to high alcohol consumption. HCC develops from dysplastic nodules and there are three steps in development; regenerative nodule; dysplastic nodule (low grade; high grade with focus of HCC) and small HCC (Wanless 1996). The lesions have been classified as expanding, spreading or multifocal and can be classified from grade I to IV based on histological criteria (Edmondson and Steiner 1954). Normal contrast enhanced MRI shows non-specific variable signal intensity changes although most (80%) are hyperintense on T2 weighted images. Signal intensity on T1 weighted images correlates with histological grade with high signal intensity seen more commonly in well-differentiated disease. A low intensity rim may be seen on T1 weighted images due to the presence of a well-defined capsule surrounding the tumour (Grazioli et al. 1999). Where this is present it is associated with improved prognosis and allows differentiation from metastatic deposits and cholangiocarcinoma. The vascular supply of the tumour is typically from the systemic arterial supply and HCC typically shows a peak of contrast enhancement during the arterial phase of multiphasic contrast enhanced studies. Multiphasic imaging is of utmost importance for the detection of small HCCs since they are often isointense on other pulse sequences and on portal and late phase images. A relationship between tumour grade and peak contrast enhancement has been demonstrated with a lower peak enhancement relating to smaller and better differentiated lesions (Yamashita et al. 1994). Poorly differentiated lesions may not show this early peak enhancement and can be difficult to distinguish from other tumour types.

HCC typically spreads by vascular invasion along the hepatic portal vein and dynamic contrast enhanced MR studies allow identification of such embolic growth due to the early arterial phase of enhancement of the intravenous tumour mass.

Hepatocellular carcinomas contain Kupffer cells and functioning hepatocytes so that liver specific contrast agents will produce enhancement (Tanaka et al. 1996). All HCCs demonstrate enhancement with Mn-DPDP and enhancement is greater in patients with well-differentiated tumours (Rofsky and Earls 1996). Hepatocytic uptake has also been demonstrated with the Gd-BOPTA and has again been shown to be greater in well-differentiated lesions (Manfredi et al. 1999). Signal characteristics of HCCs following administration of RES-specific contrast agents vary depending on the number of Kupffer cells within the lesion (Lim et al. 2001), although most tumours do not show a significant decrease in signal intensity. The signal intensity of normal liver does of course decrease improving the contrast noise ratio and improving lesion identification. There have been many studies addressing the sensitivity and specificity of different imaging approaches for the identification of HCC. Contrast enhanced multiphasic imaging certainly appears to be of significant value in the detection of small tumours whilst the use of ferrumoxides alone, or in combination with conventional multiphasic imaging with a gadolinium chelate provide significant improvements in diagnostic accuracy for larger tumours (Asahina et al. 2003).

14.6.2.2 Cholangiocellular Carcinoma

Cholangiocarcinoma is the second most common primary liver tumour after HCC representing 10% of all primary hepatic malignancies. MRI demonstrates peripheral high signal on T2 weighted images with a central hypointense area due to scar tissue. Multiphasic contrast enhanced studies showed progressive moderate enhancement in the peripheral part of the tumour. Although the morphological pattern of enhancement is similar to that seen in haemangiomas the rate of enhancement is far slower.

14.6.2.3 Metastatic Disease

Metastatic deposits are by far the most common malignant lesions seen in the liver and metastatic spread to the liver is one of the most important disease patterns in clinical oncology. As many as 50% of patients dying from malignancy will have metastatic disease in the liver at autopsy and the presence and extent of liver metastasis are a major prognostic feature in a broad range of tumour types. The role of radiological investigations is to identify the presence of metastatic lesions, to distinguish them from less sinister benign abnormalities that may also be present and to identify their location to determine whether or not the disease is treatable by resection or partial hepatectomy.

MRI is extremely efficient at differentiating between metastasis and benign lesions such as haemangiomas or cysts (Mitchell, Saini, Weinreb et al. 1994). Most liver metastatic deposits will appear hypointense on T1 weighted images and hyperintense on T2. Although the signal intensity on T2 weighted images is usually less than typically seen in haemangiomas and benign cystic lesions there may be confusion particularly where metastatic deposits have areas of central necrosis or cyst formation (Bartolozzi et al. 1999). Some endocrine tumour deposits may also demonstrate extremely high signal on T2 weighted images (these include carcinoid, islet cell tumour, renal carcinoma, thyroid carcinoma and phaeochromocytoma) (Morana et al. 2002). Metastatic disease from malignant melanoma will show different signal characteristics due to the presence of paramagnetic melanin within the tumour cells. This produces deposits which may be hyperintense on T1 weighted images and hypointense on T2.

Metastatic deposits may demonstrate heterogeneous signal intensity due to central areas of liquefactive necrosis or haemorrhage. These areas will show high signal on T2 weighted images often associated with a decrease in signal intensity on T1. This appearance, described as a target lesion, is highly indicative of malignancy and therefore of metastatic disease.

Multiphasic contrast enhanced imaging provides valuable additional information concerning the characterisation and extent of metastatic disease (see Fig. 14.6). Most metastatic lesions are hypovascular and show little or no enhancement during the arterial phase. Enhancement occurs predominantly in the peripheral component of the tumour with no enhancement being seen in central necrotic areas, even on late phase images. These typical enhancement patterns are not however seen in all metastatic liver deposits and a significant number will be hypervascular (these include carcinoid, islet cell tumour, renal carcinoma, thyroid carcinoma, phaeochromocytoma, melanoma and the breast carcinoma) with rapid enhancement during the arterial phase often associated with apparent hypointensity on later phases due to rapid washout of the contrast media from the hypervascular, high flow peripheral component of the tumour.

Metastatic lesions typically have a very variable vascular supply and are capable of parasitising arterial input from both the systemic arterial and hepatic portal systems. The vascular supply may therefore represent any admixture of these two sources, which can give rise to difficulties in the interpretation of enhancement patterns from multiphasic studies. Despite this there is considerable evidence that multiphasic contrast enhanced imaging does enhance the detection of focal liver lesions when compared with unenhanced studies. However, hepatic biliary and RES-targeted agents provide a more reliable method for enhancing the contrast between pathological areas and normal liver and significantly improve the detection of liver metastasis and the accuracy of surgical planning.

14.6.3 Monitoring of Tumour Therapy

Resection or ablation of liver tumours, particularly metastatic lesions and hepatocellular carcinoma have become increasingly common. There is good evidence that resection of solitary or lobar metastases produces significant improvement in survival in a number of tumour types particularly in patients with colorectal malignancy. Partial hepatectomy and surgical resection offer the most aggressive option but increasingly these lesions are being treated by image guided approaches using chemical or thermal ablation or by intra-arterial injection of embolic or chemotherapeutic agents. Contrast enhanced multiphasic MRI is a valuable tool for the evaluation of tumour response and can demonstrate residual viable neoplastic tissue. This is of particular value in hypervascular lesions were peripheral areas of rapidly enhancing tissue seen in the arterial phase will reliably identify residual tumour. However, many metastatic lesions will not be demonstrated in this way since they are by nature hypovascular and identification of recurrent tumour in these cases can be extremely difficult since the MR characteristics of residual or recurrent tumour are similar to those seen in normal liver tissue during the recovery phase from thermal or chemotherapeutic treatments.

14.6.4 Diffuse Liver Diseases

Diffuse liver diseases such as fatty infiltration, iron deposition and cirrhosis produce changes in morphology and signal characteristics which can be characterised by MRI. Although these do not represent malignant disease processes cirrhosis in particular is of importance since it is a common association with hepatocellular carcinoma. Cirrhosis does not significantly alter the T1 or T2 relaxation times of liver although it will produce the classic morphological changes of caudate and left hepatic lobe enlargement and distortion and compression of intrinsic hepatic vessels. Extrahepatic findings such as ascites, splenomegaly and enlargement of the hepatic portal vein and its tributaries may indicate portal hypertension and collateral varices are easily detected.

14.7 Dynamic Contrast Enhanced Imaging

The use of true dynamic contrast enhanced imaging techniques in the liver has been remarkably limited with only a small handful of papers attempting to apply these methods in hepatic disease. In part this reflects the excellent diagnostic information that can be gained by simple multiphasic enhancement studies. However, there can be no doubt that standard analysis techniques for dynamic contrast enhanced studies such as have been presented in Chaps. 5 and 6 would have potential major advantages in a number of applications. The ability to identify blood volume, blood flow and contrast transfer coefficients could provide valuable clinical information, avoiding misinterpretation of enhancement patterns which might result from the temporal under sampling of a multiphasic study. More importantly the production of reproducible objective quantitative markers would be of potential value in diagnosis, treatment monitoring and therapeutic trials.

Unfortunately, there are a number of significant technical problems to be addressed when applying dynamic contrast enhanced techniques in the liver. Firstly, it must be appreciated that the liver lies directly below the diaphragm and is therefore extremely subject to respiratory physiological motion. Pharmacokinetic modelling depends upon inherent assumptions that signal changes observed in the time course data represent the same volume of tissue throughout the study. This means that movement effects within the time course data are highly undesirable. Secondly, standard pharmacokinetic models applied to the majority of human tissues will not suffice in the liver which has a dual vascular supply from the systemic arterial circulation and from the hepatic portal venous system. In order to correctly calculate the contrast transfer coefficient (K^{trans}) and EES distribution volume (v_e) it is essential that the proportional contribution of these vascular inputs to each voxel of tissue should be known a priori.

14.7.1 Respiratory Motion

A number of potential strategies are available to compensate for the errors introduced by respiratory motion in dynamic contrast enhanced studies of the liver. One potential approach is to minimise the movement by encouraging shallow breathing at rest and further reduce the impact by the acquisition of relatively thick slices, which will be affected to a lesser extent by small respiratory excursions. In practice this is clearly unsatisfactory and although it may be used to generate data from large regions of interest it is entirely unsuitable for the production of parametric images. Respiratory gating can be routinely applied to most dynamic MRI sequences. Unfortunately, the use of respiratory gating significantly affects the temporal resolution that can be obtained from dynamic studies. Indeed, the respiratory rate is so low in comparison to the temporal distribution dynamics of intravenous contrast agents that respiratory gating does not offer a feasible solution. The use of data acquisition restricted to a single breath hold has been described. This approach removes the effect of respiratory artefact completely but has a number of major disadvantages. Firstly, many patients will be too unwell to maintain a breath hold for a prolonged period of time. Secondly, and most importantly, breath hold acquisitions can only acquire information concerning the distribution of contrast entering the liver via the hepatic artery, which occurs within the first 10–15 s of contrast passage. The portal vein inflow is delayed since contrast must first pass through the spleen or bowel and typically does not occur for at least 10–15 s after the onset of the arterial input function. To measure both vascular input functions would therefore require a breath hold of at least 40 s which is not feasible in the majority of clinical cases. Breath hold techniques can therefore only address pharmacokinetic parameters related to the distribution of hepatic arterial contrast media prior to the arrival of contrast in the hepatic portal vein circulation. Although this can provide useful information it effectively excludes any estimation of pharmacokinetic parameters relating to hepatic portal venous inflow. Respiratory motion can be minimised by the use of navigator echo techniques although this has not currently been described in relationship to dynamic contrast enhanced imaging within the liver. Navigator echo techniques will not suffer from the major restrictions on temporal resolution associated with respiratory gating but do have other technical limitations which may prove problematic. A further potential solution is the use of rapid volume acquisitions designed to minimise blurring due to motion occurring during the acquisition itself combined with post-processing coregistration techniques which can be applied to correct respiratory motion within the dynamic dataset. This approach places considerable demands on the acquisition sequence and on the hardware. In addition post-processing correction of motion artefact is likely to lead to the loss of tissue at the periphery of the imaging volume requiring significant spatial oversampling that will further restrict sequence design. However, despite these problems post-acquisition coregistration offers a potentially acceptable solution with the additional advantages that it could, potentially, deal with distortions of liver tissue in a more principled way than the other available techniques by the use of non-affine transformations. In practice only a small subset of these approaches have so far been explored in the literature.

14.7.2 Modelling a Dual Vascular Input

The presence of two vascular input functions for the majority of voxels in the image generates significant problems that limit the application of standard pharmacokinetic models (Fig. 14.8). One approach is to

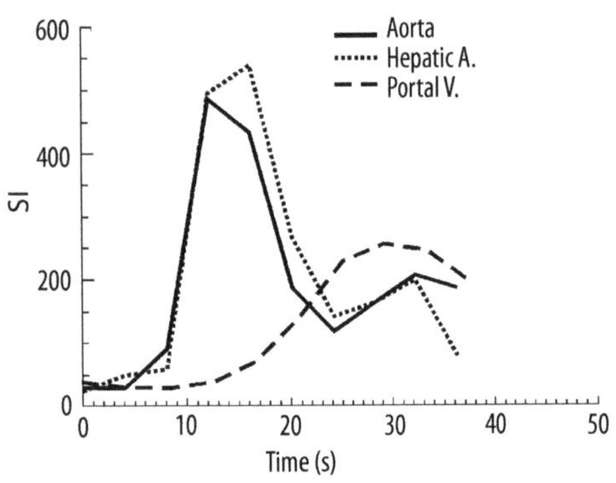

Fig. 14.8. Signal intensity changes over time following an intravenous injection of contrast media. The three *curves* represent signal changes in the aorta, hepatic artery and portal vein. Note the delay and smoothing of the curve within the portal vein compared to the arterial curves from aorta and hepatic artery

deal only with data collected during the early phase of the arterial circulation prior to the onset of significant portal vein contrast delivery. This approach can be combined with breath hold acquisitions to provide reproducible objective parameters for both contrast transfer coefficient and blood volume. However, the loss of information concerning the portal venous supply is a significant problem since abnormalities of portal perfusion are a common feature of many diseases. One approach to this is to modify the existing pharmacokinetic models to deal explicitly with the existence of a dual vascular input function. It is important to be able to distinguish the proportional vascular supply to each voxel from each of the two potential sources. However, in practice this is less difficult than it might at first appear since there is a very significant delay between the arrival of contrast in the arterial system and the later arrival of contrast in the hepatic portal vein. This delay, which is in the region of 10–20 s, allows relatively accurate identification of contrast enhancement contributions arising from the two separate vascular components.

14.8 Quantification of Liver Perfusion

The main application of dynamic contrast enhanced imaging in the liver has been for the study of liver perfusion in normal and pathological liver tissue. Much of this work has been done by the Brussels group who have a particular interest in liver perfusion changes occurring in cirrhosis. It has been shown previously that changes in hepatic vascular resistance occur as a result of chronic liver diseases and particularly in the cirrhotic liver. This increase in vascular resistance decreases the portal fraction of liver perfusion and changes in hepatic perfusion contribute directly to the deterioration of hepatic function by decreasing blood-hepatocyte exchange. The decrease in portal perfusion is partially compensated by an increase of arterial inflow and morphological changes occur with capillarisation of the sinusoids, deposition of collagen in the extra vascular spaces and the formation of basal lamina. Reliable non-invasive methods for the measurement of hepatic perfusion may therefore have significant clinical value in diffuse liver disease.

In 1999 Scharf et al (Scharf et al. 1999) described a method for measurement of liver blood flow based on dynamic contrast enhanced T1 weighted imaging in pigs. The technique used a very simple single compartment single input model and assumed a simple exponential decay for plasma contrast concentration. Although recognising the presence of a complex vascular supply to liver tissue the authors modelled this as a single vascular input function with a variable delay. Despite the simplicity of the modelling approach estimates of liver perfusion acquired before and after partial portal vein occlusion showed excellent correlation (r=0.89, p<0.001) with independent thermal dilution technique measurements.

In 2000 MATERNE et al. (2000b) described a method for non-invasive quantification of liver perfusion using single slice dynamic contrast enhanced CT scans. The method they described has been used in subsequent CT studies and has also been modified and validated for use with MRI (MATERNE et al. 2000a,2000b; MATERNE et al. 2002a,2002b). The original CT technique used 10-mm thick axial images at the level of the hepatic portal vein acquiring a single slice every 3 s. A dynamic contrast bolus was administered at a rate of 7 ml/s with a total injection time of 7 s and this was followed by a saline flush of 30 ml given at the same rate. Images were analysed using a dual input one compartment model with first order rate constants to reflect the fact that the liver receives its blood supply from both the systemic and portal circulations and therefore has two inflow rate constants. The mathematical equation for the compartmental model is:

$$dC_L(t)/dt = k_{1a}C_a(t) + k_{1p}C_p(t) - k_2C_{L(t)} \qquad (1)$$

where C_a(t), Cp(t) and CL(t) represent the concentration in the aorta, portal vein and liver compartments, respectively, k_{1a} represents the aortic inflow rate constant, k_{1p} the portal venous inflow rate constant and k_2 the outflow rate constant.

Solving for C_L(t) and adding two delay parameters, $ô_a$ and $ô_p$ which represent the transit time from the aorta and portal vein to the liver region of interest then:

$$C_L(t) = \int_0^t [k_{1a}C_a(t'-\tau_a) + k_{1p}C_p(t'-\tau_p)]e^{-k_2(t-t')}dt' \qquad (2)$$

Since contrast media do not enter red blood cells the time series values were divided by one minus the large vessel haematocrit. The delays $ô_a$ and $ô_p$ were defined to be equal and were fixed at the delay between the first nonzero value of the arterial inflow curve and the first nonzero value of the liver parenchyma curve. Equation 2 was fitted to derive the three parameters k_{1a}, k_{1p} and k_2.

From these data portal perfusion can be expressed as the fraction of total liver perfusion:

$$\text{Portal perfusion fraction} = 100.k_{1p}/(k_{1a}+k_{1p})$$

Apparent liver perfusion is defined as k_{1a}+k_{1p}. The distribution volume of the contrast agent is calculated as:

$$\text{Distribution volume} = (k_{1a}+k_{1p})/k_2$$

And the mean transit time (MTT) is calculated as:

$$\text{MTT} = 1/k$$

Initial validation studies of this technique in rabbits showed excellent correlation between measured values and total liver perfusion (r=0.92), arterial perfusion (r=0.81) and portal perfusion (r=0.85) when compared to microsphere measurements. Application of this technique in ten patients with liver disease demonstrated flow values in keeping with those expected from more invasive measurement techniques (MATERNE et al. 2000b). This group went on to use the technique in patients with chronic liver disease and were able to demonstrate significant reductions in liver perfusion and increases in arterial fraction and mean transit time in patients with cirrhosis compared to normal volunteers and patients with diffuse non-cirrhotic liver disease (VAN BEERS et al. 2001) . The same group has also validated a modified version of the same technique for use with T1 weighted MRI (MATERNE et al. 2002b). This study, performed in rabbits, used the same pharmacokinetic model combined with a technique for the calibration of signal intensity changes observed in response to contrast bolus passage. Comparison with microsphere measurements again showed good correlation (Fig. 14.9). A later study comparing measurements in patients with cirrhosis and diffuse liver disease compared this technique with Doppler ultrasound measurements of portal velocity, portal flow, congestion index, right hepatic artery resistance index and modified hepatic index (ANNET et al. 2003). Interestingly all MR derived flow parameters (except distribution volume were significantly different between patients with and without cirrhosis and there was a significant correlation between all flow parameters and measured portal pressure (p<0.02) (Fig. 14.10). Apparent arterial (p=0.024) and portal (p<0.001) perfusion, portal fraction (p<0.001), and mean transit time (p=0.004) were significantly correlated with the Child-Pugh index (PUGH et al. 1973). In

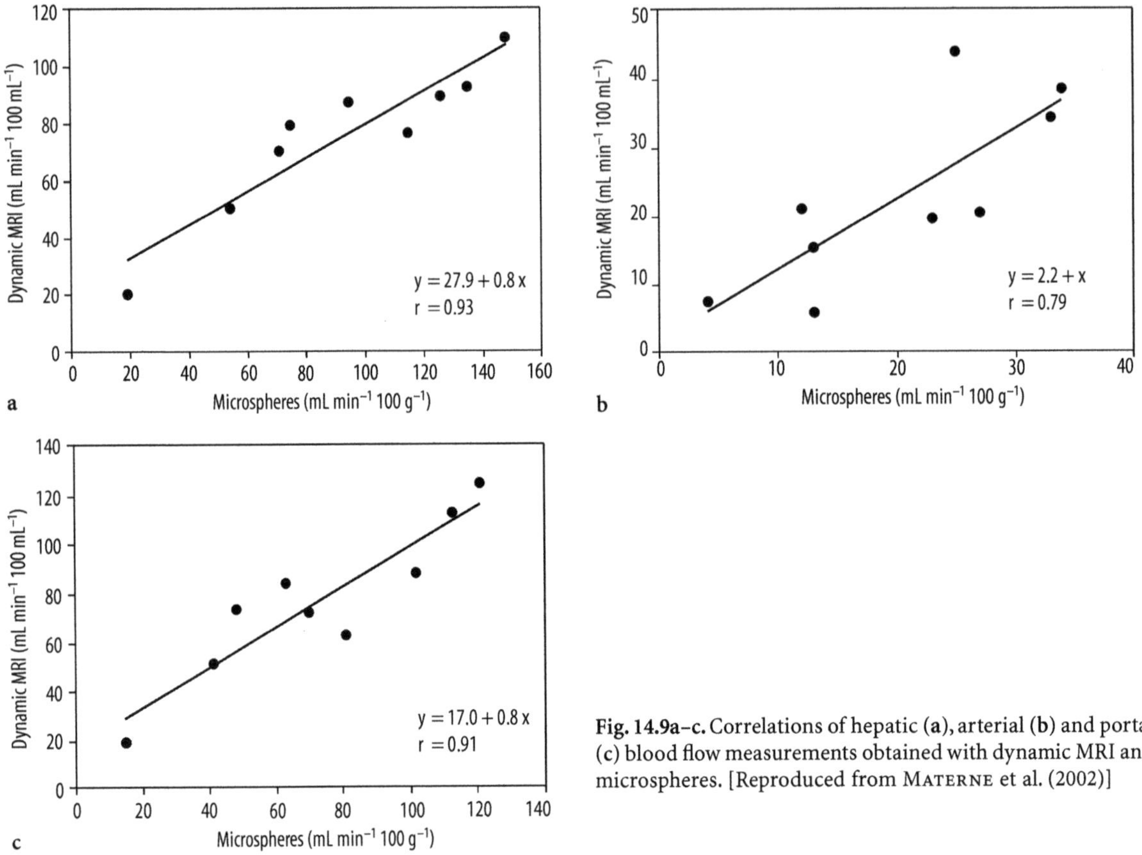

Fig. 14.9a–c. Correlations of hepatic (**a**), arterial (**b**) and portal (**c**) blood flow measurements obtained with dynamic MRI and microspheres. [Reproduced from MATERNE et al. (2002)]

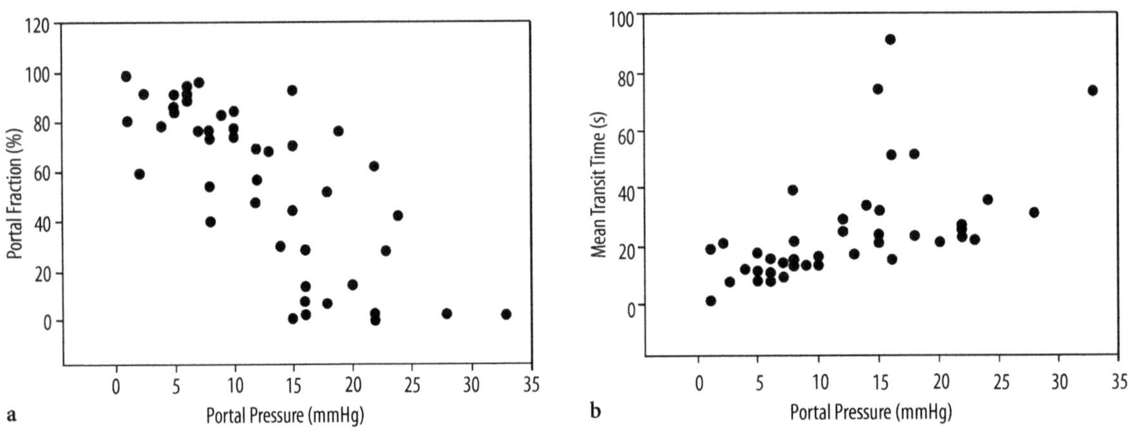

Fig. 14.10a,b. Scatterplots depict substantial correlations (**a**) between portal fraction of liver perfusion and portal pressure (r 0.769, p 0.001) and (**b**) between mean transit time and portal pressure (r 0.721, p 0.001). Portal fraction of liver perfusion and mean transit time are calculated with the dual-input one-compartmental model from the MR imaging data. [Reproduced from ANNET et al (2003)]

comparison flow parameters measured with Doppler ultrasound showed little or no correlation with portal pressure or Child-Pugh class. This elegant group of studies shows that dynamic contrast enhanced techniques combined with simple pharmacokinetic analyses can give useful clinical data in diffuse liver disease and that the problems of respiratory motion can be minimised by the use of simple acquisition strategies. Nonetheless, it is clear that this approach could benefit from further methodological improvements. The use of a single thick slice combined with spontaneous breathing is likely to be problematic if the techniques are to be taken into clinical use and effectively precludes the production of parametric images. It is likely that a significant amount of partial volume averaging occurred within the regions of interest used for the studies and this is supported by the requirement to extract in different temporal components of the portal venous input function from different regions of interest within the dataset (VAN BEERS et al. 2001). In addition 10-mm thick slices are clearly suboptimal for many applications and whole liver coverage is desirable. Nonetheless this important group of papers highlight the potential clinical value of pharmacokinetic analysis of dynamic contrast enhanced MRI studies in liver disease.

Very few groups have attempted to apply classic pharmacokinetic analyses to malignant diseases within the liver. HANNEKE et al. (2003) have used a basic contrast distribution model to estimate the rate constant k_{ep} between EES and blood plasma in patients with colorectal liver metastasis. This study compared the reproducibility of these measurements when values were calculated using a vascular input function from the aorta versus a vascular input function from the spleen. Reproducibility was higher using the modified splenic vascular input function, although it was clear that arterial input functions obtained in this study were of poor quality due to inflow artefacts. No attempt was made to address the question of how the measurements were affected by the choice between systemic and portal venous arterial input functions. MORGAN et al. (2003) have recently published a clinical trial of the vascular endothelial growth factor receptor tyrosine kinase inhibitor PTK787/ZK in patients with metastatic colorectal cancer. These workers used 10-mm thick slices with rapid acquisition times to avoid image blurring. Images were acquired every 3 s for 5 min in a coronal oblique plane to include both the aorta and the tumour. Images were acquired during spontaneous respiration. In addition to the dynamic imaging a proton density sequence was acquired to enable estimation of baseline T1 values and to allow calculation of contrast agent concentration changes against time. Contrast related changes in signal intensity were expressed as the bidirectional transfer constant Ki (ml/100 g/min). Because of considerable variability in baseline values of Ki, relative values in comparison to baseline were used. The vascular input function data was derived from the aorta on the assumption that metastatic deposits will have a predominantly arterial supply. In this study rapid reduction in enhancement was seen within 26–33 h after the first administration of drug. The study also showed evidence of dose dependent changes in Ki with larger changes seen as the maximum achievable concentration of drug was approached. Patients who responded to the drug showed proportionately greater reduction in enhancement compared to patients who progressed on treatment. This study shows that relatively simple scanning techniques combined with basic pharmacokinetic analyses can reliably demonstrate biological effects in a clinical situation.

Our group took an alternative approach to pharmacokinetic analysis of contrast distribution in hepatic tumours (JACKSON et al. 2002; JAYSON et al. 2002). In previous publications we have described a new technique for simultaneous calculation of blood volume and transfer coefficient images from large 3D data sets in patients with brain tumours (LI et al. 2000, 2003). The novel feature of the model is that it uses only data collected during the first passage of the bolus of contrast media through the target tissue so that data acquisition is extremely fast compared with conventional methods. Since data acquisition can be performed in a single breath hold we applied the method to a group of 14 patients with hepatic mass lesions including cavernous haemangioma (n=2) colorectal adenocarcinoma (n=8), ovarian serious carcinoma (n=2) and hepatocellular carcinoma (n=2) (JACKSON et al. 2002). The pharmacokinetic model has been described in detail elsewhere (see Chap. 6). Briefly it is based on the assumption that back-flow of contrast from the extravascular extracellular space can be ignored during the first passage of the contrast bolus through the vasculature. This allows decomposition of the contrast concentration time course data to produce two separate data sets reflecting the changes occurring within the intravascular and extravascular extracellular spaces. The data can be used to generate pixel by pixel parametric maps of regional blood volume (rBV) and transfer coefficient, which is designated k_{fp} to identify as the result of a first pass analysis. This study gave rise to high quality parametric maps of k_{fp} and rBV (Figs. 14.11, 14.12) and demonstrated clustering of values for different disease groups (Fig. 14.13). More importantly, the method produced highly reproducible estimates of both k_{fp} and rBV with variance ratios of 0.134 and 0.11, respectively, in a group of five patients who underwent repeat scanning (Fig. 14.14). A major problem with this approach is that it cannot deal explicitly with the portal vascular supply since the delay in contrast arrival in the portal blood flow compared to arterial blood exceeds the time available within a breath hold acquisition. Early enhancement of the portal vein can however be reliably seen and voxels within the liver can be identified based on the time at which they show enhancement. This allows exclusion of pixels showing predominantly portal enhancement from the analy-

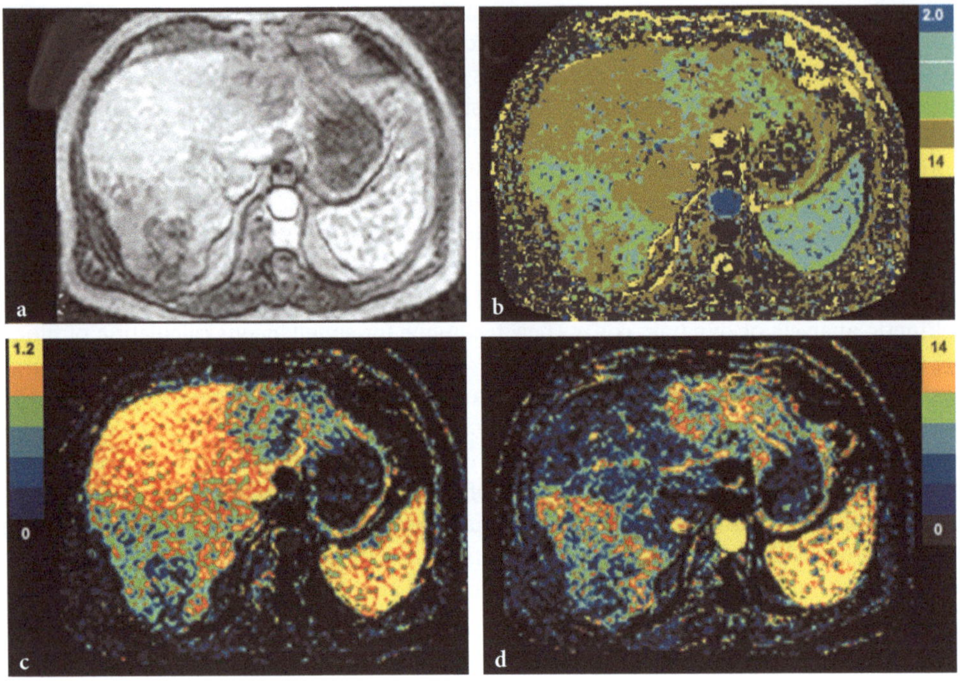

Fig. 14.11a–d. Transverse contrast enhanced image from dynamic series (**a**) and maps of T0 (**b**), kfp (**c**) and rBV (**d**) in a patient with metastatic colonic carcinoma (patient 4). Metastatic deposits are seen in the right and left lobes. The T0 map shows early contrast arrival compared to normal liver in both metastases. Maps of kfp (**c**) and rBV (**d**) show a peripheral rim of high kfp and rBV in both metastases with low values in the tumour centre. This tumour rim shows kfp values that appear lower than those of normal liver parenchyma and rBV values that appear lower

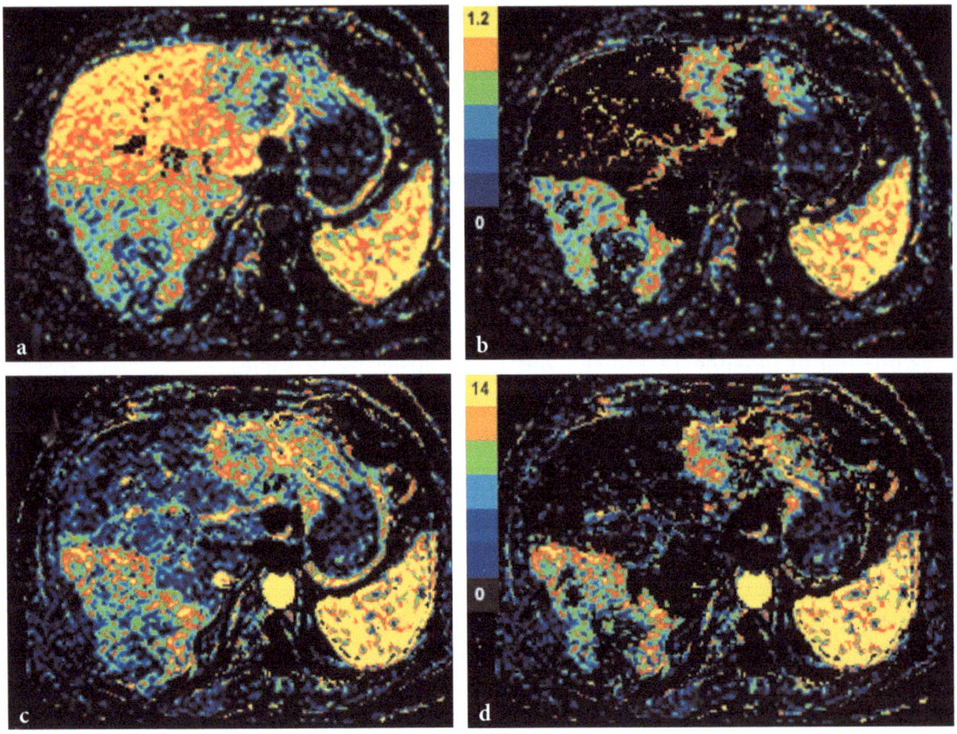

Fig. 14.12a–d. Transverse sections of kfp (**a,b**) and rBV (**c,d**) in the same patient as in Fig. 14.11. The images on the *right* are generated from the original images on the left by exclusion of pixels with T0 values in keeping with a portal venous blood supply (T0>10 s). This removes areas of erroneously elevated kfp and reduced rBV

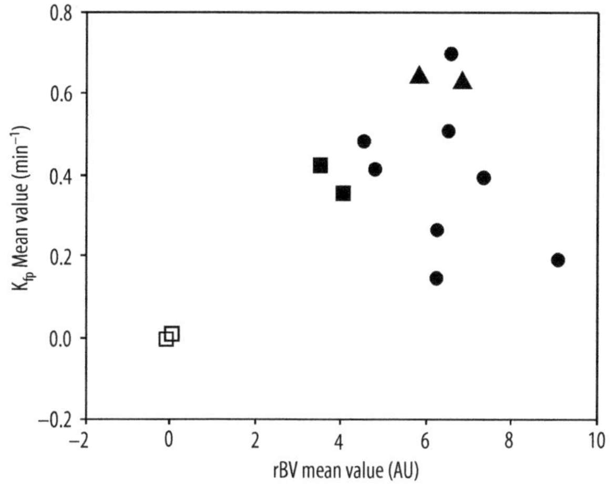

Fig. 14.13. Median values of kfp and rBV from all patients. Patient numbers correspond to Table 14.2. Diagnoses are cavernous haemangioma (*open circles*), metastatic colonic carcinoma (*open squares*), metastatic serous carcinoma of the ovary (*circles*), HCC in cirrhotic liver (*open triangle*), HCC in normal liver (*triangle*). *AU*, arbitrary units

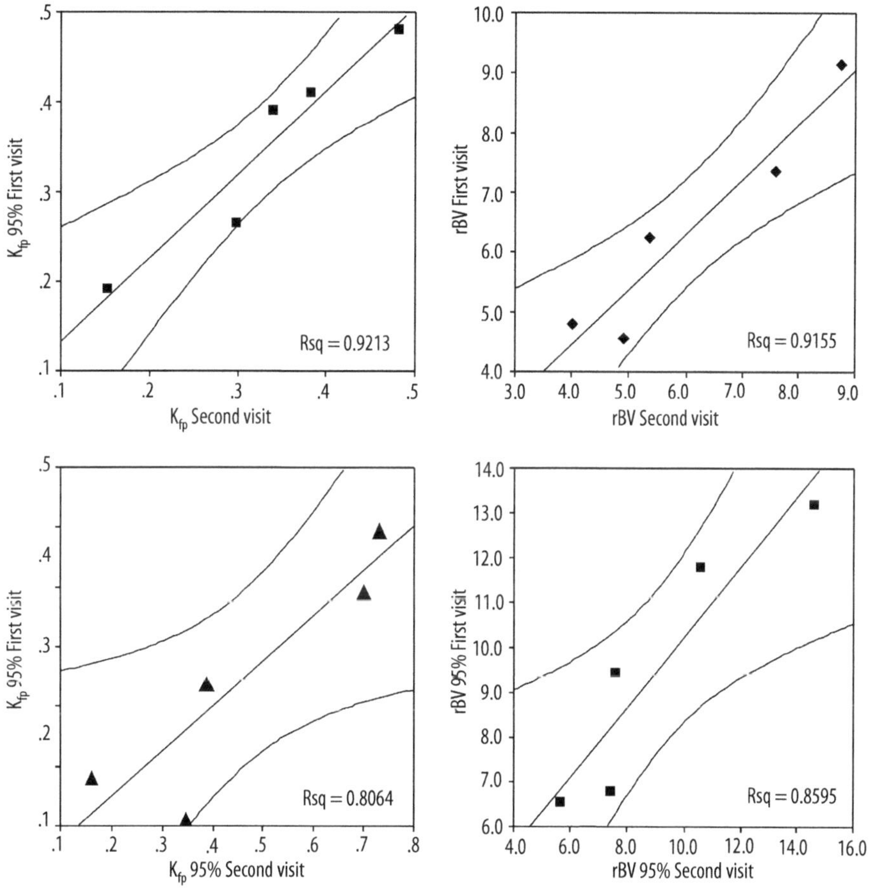

Fig. 14.14. Plots showing reproducibility of kfp (*top*) and rBV (*bottom*) values represent median values (*left*) and median of the upper 5th centile of measurements (*right*)

sis. Within the liver tumours in this study enhancing voxels showed a significant arterial supply in all cases but it must be appreciated that the transfer coefficient values produced in this study reflect only the transfer coefficient between systemic capillaries and tumour interstitial space. We have also used this technique in a study of an anti-VEGF antibody in patients with abdominal malignancy. Tumours showed significant decreases in k_{fp} within 48 h of drug administration and a clear dose response was seen with smaller decreases, which were less well sustained, in patients on the lowest dose regime. Since the technique uses multi-sliced acquisitions to cover a large area of interest it was also possible to spatially co-register these parametric maps with both baseline CT and positron emission tomography images of labelled drug.

14.9 Conclusions

The liver represents a unique biological system for the use of dynamic contrast enhanced MRI. The importance of variations in blood flow distribution in diffuse liver disease and the potential diagnostic value of dynamic contrast enhanced imaging in focal diseases are clear. The routine uptake of multiphasic imaging techniques confirms the clinical value of dynamic acquisition approaches in the management and monitoring of liver disease and the paucity of published data on formal dynamic acquisition techniques and pharmacokinetic analyses is surprising. It may be that clinicians do not perceive the need for more complex analysis techniques when multiphasic imaging approaches more than adequately address their diagnostic requirements. Nonetheless it is clear that objective quantitative surrogate markers of microvascular structure in liver disease have potential value in clinical management, therapeutic monitoring and clinical trials. The major hurdles in the development of routine dynamic contrast enhanced methodologies are the development of methods to deal effectively with respiratory motion and the development of formal pharmacokinetic models which can deal with the presence of a mixed dual arterial input function in a principled manner. Some of these restrictions will be addressed with the development of faster imaging techniques such as parallel imaging which will allow rapid acquisition of large volumes of data. Increased sophistication in image coregistration methodologies may address the problems of respiratory motion and the development of modified pharmacokinetic models is feasible assuming that adequate quality data can be acquired. It seems reasonable to predict that pharmacokinetic analysis of dynamic contrast enhanced images of the liver will become far more common and, if acquisition and analysis problems are overcome, may replace routine multiphasic imaging approaches.

References

Annet L, Materne R, Danse E, Jamart J, Horsmans Y, Van Beers BE (2003) Hepatic flow parameters measured with MR imaging and Doppler US: correlations with degree of cirrhosis and portal hypertension. Radiology 229:409–414

Asahina Y, Izumi N, Uchihara M, Noguchi O, Ueda K, Inoue K, Nishimura Y, Tsuchiya K, Hamano K, Itakura J, Himeno Y, Koike M, Miyake S (2003) Assessment of Kupffer cells by ferumoxides-enhanced MR imaging is beneficial for diagnosis of hepatocellular carcinoma: comparison of pathological diagnosis and perfusion patterns assessed by CT hepatic arteriography and CT arterioportography. Hepatol Res 27:196–204

Bartolozzi C, Lencioni R, Donati F, Cioni D (1999) Abdominal MR: liver and pancreas. Eur Radiol 9:1496–1512

Bruix J, Sherman M, Llovet JM, Beaugrand M, Lencioni R, Burroughs AK, Christensen E, Pagliaro L, Colombo M, Rodes J (2001) Clinical management of hepatocellular carcinoma. Conclusions of the Barcelona-2000 EASL conference. European Association for the Study of the Liver. J Hepatol 35(3):421–430

Carlos RC, Branam JD, Dong Q, Hussain HK, Francis IR (2002) Biliary imaging with Gd-EOB-DTPA: is a 20-minute delay sufficient. Acad Radiol 9:1322–1325

Carlson SK, Johnson CD, Bender CE, Welch TJ (2000) CT of focal nodular hyperplasia of the liver. AJR Am J Roentgenol 174(3):705–712

Catasca JV, Mirowitz SA (1994) T2-weighted MR imaging of the abdomen: fast spin-echo vs conventional spin-echo sequences. AJR Am J Roentgenol 162(1):61–67

Choi CS, Freeny PC (1998) Triphasic helical CT of hepatic focal nodular hyperplasia: incidence of atypical findings. AJR Am J Roentgenol 170:391–395

Coffin CM, Diche T, Mahfouz A, Alexandre M, Caseiro-Alves F, Rahmouni A, Vasile N, Mathieu D (1999) Benign and malignant hepatocellular tumors: evaluation of tumoral enhancement after mangafodipir trisodium injection on MR imaging. Eur Radiol 9:444–449

del Frate C, Bazzocchi M, Mortele KJ, Zuiani C, Londero V, Como G, Zanardi R, Ros PR (2002) Detection of liver metastases: comparison of gadobenate dimeglumine-enhanced and ferumoxides-enhanced MR imaging examinations. Radiology 225:766–772

Edmondson HA, Steiner PE (1954) Primary carcinoma of the liver: a study of 100 cases among 48,900 necropsies. Cancer 7(3):462–503

Fretz CJ, Elizondo G, Weissleder R, Hahn PF, Stark DD, Ferrucci JT (1989) Superparamagnetic iron oxide-enhanced MR imaging: pulse sequence optimization for detection of liver cancer. Radiology 172:393–397

Fretz CJ, Stark DD, Metz CE, Elizondo G, Weissleder R, Shen JH, Wittenberg J, Simeone J, Ferrucci JT (1990) Detection of hepatic metastases: comparison of contrast-enhanced CT, unenhanced MR imaging, and iron oxide-enhanced MR imaging. AJR Am J Roentgenol 155:763–770

Grazioli L, Olivetti L, Fugazzola C, Benetti A, Stanga C, Dettori E, Gallo C, Matricardi L, Giacobbe A, Chiesa A (1999) The pseudocapsule in hepatocellular carcinoma: correlation between dynamic MR imaging and pathology. Eur Radiol 9:62–67

Grazioli L, Federle MP, Brancatelli G, Ichikawa T, Olivetti L, Blachar A (2001) Hepatic adenomas: imaging and pathologic findings. Radiographics 21:877-92; discussion 892-4

Hanafusa K, Ohashi I, Himeno Y, Suzuki S, Shibuya H (1995) Hepatic hemangioma: findings with two-phase CT. Radiology 196:465–469

Hanafusa K, Ohashi I, Gomi N, Himeno Y, Wakita T, Shibuya H (1997) Differential diagnosis of early homogeneously enhancing hepatocellular carcinoma and hemangioma by two-phase CT. J Comput Assist Tomogr 21:361–368

Hussain HK, Londy FJ, Francis IR, Nghiem HV, Weadock WJ,

Gebremariam A, Chenevert TL (2003) Hepatic arterial phase MR imaging with automated bolus-detection three-dimensional fast gradient-recalled-echo sequence: comparison with test-bolus method. Radiology 226:558–566

Ichikawa T, Federle MP, Grazioli L, Nalesnik M (2000) Hepatocellular adenoma: multiphasic CT and histopathologic findings in 25 patients. Radiology 214:861–868

Jackson A, Haroon H, Zhu XP, Li KL, Thacker NA, Jayson G (2002) Breath-hold perfusion and permeability mapping of hepatic malignancies using magnetic resonance imaging and a first-pass leakage profile model. NMR Biomed 15(2):164–173

Jayson GC, Zweit J, Jackson A, Mulatero C, Julyan P, Ranson M, Broughton L, Wagstaff J, Hakannson L, Groenewegen G, Bailey J, Smith N, Hastings D, Lawrance J, Haroon H, Ward T, McGown AT, Tang M, Levitt D, Marreaud S, Lehmann FF, Herold M, Zwierzina H (2002) Molecular imaging and biological evaluation of HuMV833 anti-VEGF antibody: implications for trial design of antiangiogenic antibodies. J Natl Cancer Inst 94:1484–1493

Jeong MG, Yu JS, Kim KW, Jo BJ, Kim JK (1999) Early homogeneously enhancing hemangioma versus hepatocellular carcinoma: differentiation using quantitative analysis of multiphasic dynamic magnetic resonance imaging. Yonsei Med J 40:248–255

Kanematsu M, Semelka RC, Matsuo M, Kondo H, Enya M, Goshima S, Moriyama N, Hoshi H (2002) Gadolinium-enhanced MR imaging of the liver: optimizing imaging delay for hepatic arterial and portal venous phases--a prospective randomized study in patients with chronic liver damage. Radiology 225:407–415

Kapoor V, Peterson MS, Baron RL, Patel S, Eghtesad B, Fung JJ (2002) Intrahepatic biliary anatomy of living adult liver donors: correlation of mangafodipir trisodium-enhanced MR cholangiography and intraoperative cholangiography. AJR Am J Roentgenol 179:1281–1286

King LJ, Burkill GJ, Scurr ED, Vlavianos P, Murray-Lyons I, Healy JC (2002) MnDPDP enhanced magnetic resonance imaging of focal liver lesions. Clin Radiol 57:1047–1057

Kirchin MA, Spinazzi A (2002) Low-dose gadobenate dimeglumine-enhanced MRI in the detection and characterization of focal liver lesions. Acad Radiol 9 Suppl 1:S121–126

Lavelle MT, Lee VS, Rofsky NM, Krinsky GA, Weinreb JC (2001) Dynamic contrast-enhanced three-dimensional MR imaging of liver parenchyma: source images and angiographic reconstructions to define hepatic arterial anatomy. Radiology 218(2):389–394

Li KL, Zhu XP, Waterton J, Jackson A (2000) Improved 3D quantitative mapping of blood volume and endothelial permeability in brain tumors. J Magn Reson Imaging 12:347–357

Li KL, Zhu XP, Checkley DR, Tessier JJ, Hillier VF, Waterton JC, Jackson A (2003) Simultaneous mapping of blood volume and endothelial permeability surface area product in gliomas using iterative analysis of first-pass dynamic contrast enhanced MRI data. Br J Radiol 76:39–50

Lim JH, Choi D, Cho SK, Kim SH, Lee WJ, Lim HK, Park CK, Paik SW, Kim YI (2001) Conspicuity of hepatocellular nodular lesions in cirrhotic livers at ferumoxides-enhanced MR imaging: importance of Kupffer cell number. Radiology 220:669–676

Manfredi R, Maresca G, Baron RL, Cotroneo AR, De Gaetano AM, De Franco A, Pirovano G, Spinazzi A, Marano P (1999) Delayed MR imaging of hepatocellular carcinoma enhanced by gadobenate dimeglumine (Gd-BOPTA). J Magn Reson Imaging 9:704–710

Materne R, Horsmans Y, Jamart J, Smith AM, Gigot JF, Van Beers BE (2000) Gadolinium-enhanced arterial-phase MR imaging of hypervascular liver tumors: comparison between tailored and fixed scanning delays in the same patients. J Magn Reson Imaging 11:244–249

Materne R, Van Beers BE, Smith AM, Leconte I, Jamart J, Dehoux JP, Keyeux A, Horsmans Y (2000) Non-invasive quantification of liver perfusion with dynamic computed tomography and a dual-input one-compartmental model. Clin Sci (Lond) 99:517–525

Materne R, Annet L, Dechambre S, Sempoux C, Smith AM, Corot C, Horsmans Y, Van Beers BE (2002) Dynamic computed tomography with low- and high-molecular-mass contrast agents to assess microvascular permeability modifications in a model of liver fibrosis. Clin Sci (Lond) 103:213–216

Materne R, Smith AM, Peeters F, Dehoux JP, Keyeux A, Horsmans Y, Van Beers BE (2002) Assessment of hepatic perfusion parameters with dynamic MRI. Magn Reson Med 47:135–142

Mitchell DG, Saini S, Weinreb J, De Lange EE, Runge VM, Kuhlman JE, Parisky Y, Johnson CD, Brown JJ, Schnall M et al (1994) Hepatic metastases and cavernous hemangiomas: distinction with standard- and triple-dose gadoteridol-enhanced MR imaging. Radiology 193:49–57

Morana G, Grazioli L, Schneider G, Testoni M, Menni K, Chiesa A, Procacci C (2002) Hypervascular hepatic lesions: dynamic and late enhancement pattern with Gd-BOPTA. Acad Radiol 9 Suppl 2:S476–479

Morana G, Grazioli L, Testoni M, Caccia P, Procacci C (2002) Contrast agents for hepatic magnetic resonance imaging. Top Magn Reson Imaging 13:117–150

Morgan B, Thomas AL, Drevs J, Hennig J, Buchert M, Jivan A, Horsfield MA, Mross K, Ball HA, Lee L, Mietlowski W, Fuxuis S, Unger C, O'Byrne K, Henry A, Cherryman GR, Laurent D, Dugan M, Marmé D, Steward WP (2003) Dynamic contrast-enhanced magnetic resonance imaging as a biomarker for the pharmacological response of PTK787/ZK 222584, an inhibitor of the vascular endothelial growth factor receptor tyrosine kinases, in patients with advanced colorectal cancer and liver metastases: results from two phase I studies. J Clin Oncol 21:3955–3964

Murakami T, Baron RL, Peterson MS, Oliver JH 3rd, Davis PL, Confer SR, Federle MP (1996) Hepatocellular carcinoma: MR imaging with mangafodipir trisodium (Mn-DPDP). Radiology 200(1):69–77

Nguyen BN, Fléjou JF, Terris B, Belghiti J, Degott C (1999) Focal nodular hyperplasia of the liver: a comprehensive pathologic study of 305 lesions and recognition of new histologic forms. Am J Surg Pathol 23:1441–1454

Noguchi Y, Murakami T, Kim T, Hori M, Osuga K, Kawata S, Kumano S, Okada A, Sugiura T, Nakamura H (2003) Detection of hepatocellular carcinoma: comparison of dynamic MR imaging with dynamic double arterial phase helical CT. AJR Am J Roentgenol 180:455–460

Padovani B, Lecesne R, Raffaelli C, Chevallier P, Drouillard J, Bruneton JN, Lambrechts M, Gordon P (1996) Tolerability and utility of mangafodipir trisodium injection (MnDPDP) at the dose of 5 mumol/kg body weight in detecting focal liver tumors: results of a phase III trial using an infusion technique. Eur J Radiol 23:205–211

Pena CS, Saini S, Baron RL, Hamm BA, Morana G, Caudana R, Giovagnoni A, Villa A, Carriero A, Mathieu D, Bourne MW, Kirchin MA, Pirovano G, Spinazzi A (2001) Detection of malignant primary hepatic neoplasms with gadobenate dimeglumine (Gd-BOPTA) enhanced T1-weighted hepatocyte phase MR imaging: results of off-site blinded review in a phase-II multicenter trial. Korean J Radiol 2(4):210–215

Petersein J, Spinazzi A, Giovagnoni A, Soyer P, Terrier F, Lencioni R, Bartolozzi C, Grazioli L, Chiesa A, Manfredi R, Marano P, Van Persijn Van Meerten EL, Bloem JL, Petre C, Marchal G, Greco A, McNamara MT, Heuck A, Reiser M, Laniado M, Claussen C, Daldrup HE, Rummeny E, Kirchin MA, Pirovano G, Hamm B (2000) Focal liver lesions: evaluation of the efficacy of gadobenate dimeglumine in MR imaging--a multicenter phase III clinical study. Radiology 215:727–736

Pirovano G, Vanzulli A, Marti-Bonmati L, Grazioli L, Manfredi R, Greco A, Holzknecht N, Daldrup-Link HE, Rummeny E, Hamm B, Arneson V, Imperatori L, Kirchin MA, Spinazzi A (2000) Evaluation of the accuracy of gadobenate dimeglumine-enhanced MR imaging in the detection and characterization of focal liver lesions. AJR Am J Roentgenol 175:1111–1120

Pugh RN, Murray-Lyon IM, Dawson JL, Pietroni MC, Williams R (1973) Transection of the oesophagus for bleeding oesophageal varices. Br J Surg 60:646–649

Reinig JW (1995) Breath-hold fast spin-echo MR imaging of the liver: a technique for high-quality T2-weighted images. Radiology 194(2):303–304

Rofsky NM, Earls JP (1996) Mangafodipir trisodium injection (Mn-DPDP). A contrast agent for abdominal MR imaging. Magn Reson Imaging Clin N Am 4(1):73–85

Rummeny EJ, Marchal G (1997) Liver imaging. Clinical applications and future perspectives. Acta Radiol 38(4 Pt 2):626–630

Scharf J, Zapletal C, Hess T, Hoffmann U, Mehrabi A, Mihm D, Hoffmann V, Brix G, Kraus T, Richter GM, Klar E (1999) Assessment of hepatic perfusion in pigs by pharmacokinetic analysis of dynamic MR images. J Magn Reson Imaging 9:568–572

Schima W, Petersein J, Hahn PF, Harisinghani M, Halpern E, Saini S (1997) Contrast-enhanced MR imaging of the liver: comparison between Gd-BOPTA and Mangafodipir. J Magn Reson Imaging 7:130–135

Schima W, Saini S, Petersein J, Weissleder R, Harisinghani M, Mayo-Smith W, Hahn PF (1999) MR imaging of the liver with Gd-BOPTA: quantitative analysis of T1-weighted images at two different doses. J Magn Reson Imaging 10:80–83

Schmitz SA, Wagner S, Schuhmann-Giampieri G, Krause W, Bollow M, Wolf KJ (1997) Gd-EOB-DTPA and Yb-EOB-DTPA: two prototypic contrast media for CT detection of liver lesions in dogs. Radiology 205:361–366

Schmitz SA, Wagner S, Schuhmann-Giampieri G, Krause W, Wolf KJ (1997) A prototype liver-specific contrast medium for CT: preclinical evaluation of gadoxetic acid disodium, or Gd-EOB-DTPA. Radiology 202:407–412

Schuhmann-Giampieri G, Mahler M, Roll G, Maibauer R, Schmitz S (1997) Pharmacokinetics of the liver-specific contrast agent Gd-EOB-DTPA in relation to contrast-enhanced liver imaging in humans. J Clin Pharmacol 37(7):587–596

Semelka RC, Shoenut JP, Ascher SM, Kroeker MA, Greenberg HM, Yaffe CS, Micflikier AB (1994) Solitary hepatic metastasis: comparison of dynamic contrast-enhanced CT and MR imaging with fat-suppressed T2-weighted, breath-hold T1-weighted FLASH, and dynamic gadolinium-enhanced FLASH sequences. J Magn Reson Imaging 4(3):319–323

Semelka RC, Martin DR, Balci C, Lance T (2001) Focal liver lesions: comparison of dual-phase CT and multisequence multiplanar MR imaging including dynamic gadolinium enhancement. J Magn Reson Imaging 13:397–401

Siewert B, Müller MF, Foley M, Wielopolski PA, Finn JP (1994) Fast MR imaging of the liver: quantitative comparison of techniques. Radiology 193:37–42

Spinazzi A, Lorusso V, Pirovano G, Kirchin M (1999) Safety, tolerance, biodistribution, and MR imaging enhancement of the liver with gadobenate dimeglumine: results of clinical pharmacologic and pilot imaging studies in nonpatient and patient volunteers. Acad Radiol 6:282–291

Spritzer CE, Keogan MT, DeLong DM, Dahlke J, MacFall JR (1996) Optimizing fast spin echo acquisitions for hepatic imaging in normal subjects. J Magn Reson Imaging 6:128–135

Stark DD, Weissleder R, Elizondo G, Hahn PF, Saini S, Todd LE, Wittenberg J, Ferrucci JT (1988) Superparamagnetic iron oxide: clinical application as a contrast agent for MR imaging of the liver. Radiology 168:297–301

Tanaka M, Nakashima O, Wada Y, Kage M, Kojiro M (1996) Pathomorphological study of Kupffer cells in hepatocellular carcinoma and hyperplastic nodular lesions in the liver. Hepatology 24:807–812

Van Beers BE, Gallez B, Pringot J (1997) Contrast-enhanced MR imaging of the liver. Radiology 203(2):297–306

Van Beers BE, Materne R, Lacrosse M, Jamart J, Smith AM, Horsmans Y, Gigot JF, Gilon R, Pringot J (1999) MR imaging of hypervascular liver tumors: timing optimization during the arterial phase. J Magn Reson Imaging 9:562–567

Van Beers BE, Leconte I, Materne R, Smith AM, Jamart J, Horsmans Y (2001) Hepatic perfusion parameters in chronic liver disease: dynamic CT measurements correlated with disease severity. AJR Am J Roentgenol 176:667–673

van Laarhoven HW, Rijpkema M, Punt CJ, Ruers TJ, Hendriks JC, Barentsz JO, Heerschap A (2003) Method for quantitation of dynamic MRI contrast agent uptake in colorectal liver metastases. J Magn Reson Imaging 18:315–320

Vitellas KM, El-Dieb A, Vaswani KK, Bennett WF, Fromkes J, Ellison C, Bova JG (2002) Using contrast-enhanced MR cholangiography with IV mangafodipir trisodium (Teslascan) to evaluate bile duct leaks after cholecystectomy: a prospective study of 11 patients. AJR Am J Roentgenol 179:409–416

Vitellas KM, Enns RA, Keogan MT, Freed KS, Spritzer CE, Baillie J, Nelson RC (2002) Comparison of MR cholangiopancreatographic techniques with contrast-enhanced cholangiography in the evaluation of sclerosing cholangitis. AJR Am J Roentgenol 178:327–334

Vogl TJ, Kümmel S, Hammerstingl R, Schellenbeck M, Schumacher G, Balzer T, Schwarz W, Müller PK, Bechstein WO, Mack MG, Söllner O, Felix R (1996) Liver tumors: comparison of MR imaging with Gd-EOB-DTPA and Gd-DTPA. Radiology 200:59–67

Wang C, Ahlström H, Ekholm S, Fagertun H, Hellström M, Hemmingsson A, Holtås S, Isberg B, Jonnson E, Lönnemark-Magnusson M, McGill S, Wallengren NO,

Westman L (1997) Diagnostic efficacy of MnDPDP in MR imaging of the liver. A phase III multicentre study. Acta Radiol 38:643–649

Wanless IR (1996) Nodular regenerative hyperplasia, dysplasia, and hepatocellular carcinoma. Am J Gastroenterol 91:836–837

Weissleder R, Stark DD, Engelstad BL, Bacon BR, Compton CC, White DL, Jacobs P, Lewis J (1989) Superparamagnetic iron oxide: pharmacokinetics and toxicity. AJR Am J Roentgenol 152:167–173

Weissleder R, Elizondo G, Wittenberg J, Rabito CA, Bengele HH, Josephson L (1990) Ultrasmall superparamagnetic iron oxide: characterization of a new class of contrast agents for MR imaging. Radiology 175:489–493

Whitney WS, Herfkens RJ, Jeffrey RB, McDonnell CH, Li KC, Van Dalsem WJ, Low RN, Francis IR, Dabatin JF, Glazer GM (1993) Dynamic breath-hold multiplanar spoiled gradient-recalled MR imaging with gadolinium enhancement for differentiating hepatic hemangiomas from malignancies at 1. 5 T. Radiology 189:863–870

Yamashita Y, Fan ZM, Yamamoto H, Matsukawa T, Yoshimatsu S, Miyazaki T, Sumi M, Harada M, Takahashi M (1994) Spin-echo and dynamic gadolinium-enhanced FLASH MR imaging of hepatocellular carcinoma: correlation with histopathologic findings. J Magn Reson Imaging 4:83–90

Yoshioka H, Takahashi N, Yamaguchi M, Lou D, Saida Y, Itai Y (2002) Double arterial phase dynamic MRI with sensitivity encoding (SENSE) for hypervascular hepatocellular carcinomas. J Magn Reson Imaging 16:259–266

Yu JS, Kim KW, Jo BJ, Jeong MG, Kim JK, Hahm JK, Lee JT, Yoo HS (2000) Test-bolus injection for optimization of arterial phase imaging during contrast-enhanced hepatic MR imaging. Yonsei Med J 41:459–467

Applications

15 Use of Dynamic Contrast-Enhanced MRI in Multi-Centre Trials with Particular Reference to Breast Cancer Screening in Women at Genetic Risk

Martin O. Leach

CONTENTS

15.1 Introduction

This chapter considers issues concerned with developing multi-centre trials using dynamic contrast-enhanced MRI studies. As techniques have been considered in other chapters, emphasis is placed on issues that relate to trials, and particularly their implementation across centres. Both diagnostic and therapeutic trials are considered, although as yet most experience arises from diagnostic trials. Trials that have been reported are considered, and the UK study of magnetic resonance as a method of screening women at genetic risk of breast cancer (MARIBS) using dynamic contrast-enhanced MRI is taken as an example. Issues of organisation, instrumentation, quality assurance and analysis are considered.

M. O. Leach, PhD, FInstP
Joint Director, Cancer Research UK, Clinical Magnetic Resonance Research Group, Institute of Cancer Research and Royal Marsden Hospital, Downs Road, Sutton, Surrey, SM2 5PT, UK

15.2 Multi-Centre Trials

15.2.1 New Diagnostic Techniques

Development and evaluation of new techniques often occurs initially at single centres. Where new approaches are developed at a university or hospital, the centre evaluating the technique is often the same centre that developed the approach. This has the benefit of maximising the expertise in the technique, and is often an essential part of the interactive process of developing and optimising a new clinical technique. Those involved are likely to be advocates of the approach, and the utility established in such a single-centre evaluation may not be representative of the effectiveness of an approach across a range of centres. Manufacturers may also initially pilot a new approach at a single centre, in this case because of the strong continuing interaction required to optimise development. Such a strong interaction allows resources to be focussed, and may lead to scientific publications, assisting the manufacturer's role in alerting the community to new methods and equipment. Often this preliminary stage is then followed by a stage of more widespread evaluation, defining the role of the technique at a number of centres representing the range of clinical applications and potential purchasers, in some cases leading to further modifications in the technique.

An important issue affecting many preliminary studies is that the clinical conditions examined may not be representative of the final target group. One example would be testing diagnostic methods that might be used for screening for breast cancer on symptomatic patients with more advanced disease than would be typical of a screening population. While this is a reasonable approach in defining utility for more advanced disease, and in developing a technique, it is important to ensure that data obtained for one purpose are not inappropriately utilised to infer utility for more demanding applications.

15.2.2 Multi-Centre Evaluation

While these initial single-centre studies may define the potential utility of a technique, increasing emphasis is being placed on defining the impact of new technologies and approaches on healthcare outcome in the target group. This type of study often requires multiple centres to provide the numbers required for statistical power, and also ensures a representative evaluation of the technique, with a range of expertise more typical of clinical practice. Identification of impact on outcome (and definition of any associated morbidity) is central to evaluating the clinical impact, which may not be directly determined from the immediate diagnostic value.

Although there are many examples of diagnostic evaluation studies, for example comparison of different diagnostic modalities, being performed within a single centre, an objective assessment often has to be built on evaluation of several such studies. These studies may still be influenced by advocacy of a particular technique, and are unlikely to provide as robust an evaluation as a formal multi-centre evaluation.

Screening studies are a particular example of a question that usually requires data from multi-centre studies to define utility. Not only is symptomatic disease often not representative of screen detected disease, in many applications the prevalence of the disease in the screened population is low, requiring large studies to establish a significant result. The studies set up to evaluate the efficacy of breast cancer screening using X-ray mammography provide an example of the complexity of such studies, and identify some of the issues involved in multi-centre trials (Moss and Chamberlain 1996; National Institutes of Health Consensus Statement 1997). While this form of screening clearly identifies women at an earlier stage than would otherwise be the case, the impact on health outcome remains a matter for debate, in part due to the morbidity arising from the radiotherapy treatments used during the period covered by the studies.

15.2.3 Therapeutic Trials

Early stage clinical trials of new therapies have traditionally been carried out at single centres (Phase I/II). Recently Phase I trials have begun to include hypothesis testing elements, rather than concentrating on toxicity and establishing maximum tolerated doses. With these changes there is interest at pharmaceutical companies in performing such studies at two centres, to aid recruitment and provide increased experience. With novel approaches in cancer therapeutics, where treatment may increasingly be tailored to the individual's genome, there may be an increased need to broaden the base, and therefore catchment, of such trials. This trend may develop with other diseases where there is considerable individual variation. Phase III trials are multicentre, requiring coordination and agreed standards across centres. To date, in cancer, these trials have generally used solid tumour volume response as the radiological endpoint, graded using WHO or RECIST criteria (Miller et al. 1981; Therasse et al. 2000). In these studies there has been little cross-site quality assurance or diagnostic protocol standardisation. New therapeutic agents may lead to tumour stasis, but may have other effects on tumour metabolism or function that can be detected by MRI. One example is the effects of anti-angiogenic or anti-vascular treatments, where MR dynamic contrast agent measurements have shown particular promise in demonstrating drug action. Multi-centre studies using these approaches will require much greater standardisation and quality assurance than has hitherto been necessary. Applications of MR to assessing therapeutic response in breast cancer have recently been reviewed (Leach 2002) as has the use of MRI to evaluate angiogenic changes (Leach 2001). The potential for using MRI to assess response in clinical trials has also been evaluated (Harms 2001; Julian 2001).

15.3 Dynamic Contrast Agent Studies

15.3.1 T1-Weighted Methods

Contrast agents used with MRI have predominantly been based on gadolinium chelates, providing a positive contrast on T1-weighted images. Their initial application was to demonstrate areas of blood–brain barrier breakdown, as a method of identifying and classifying CNS lesions such as those from multiple sclerosis, or from cancer. More recently in cancer their use has extended to the evaluation of other solid tumours, aiding discrimination of active disease from fibrosis, necrosis and normal tissues. They are used in other applications to identify perfusion defects, and to increase the sensitivity of MR angiography. In tumour studies, in addition to morphological assessment of the enhanced region, there has been interest in evaluating the dynamics of contrast uptake and wash out, which can be related to physiological parameters by the use of appropriate physiological models.

Initially observations were related to the shape of the uptake and washout curve obtained from a region of interest (Kaiser and Zeitler 1989), from a time series of T1-weighted images, using this as an additional radiological descriptor (Heywang-Köbrunner 1990; Kuhl et al. 1999). This has been shown to be of particular value in breast cancer diagnosis and assessment. Both descriptive and calculated parameters have been developed to characterise these curves (Tofts et al. 1999). More recently there has been interest in deriving physiological parameters, firstly by fitting the curve from a region of interest to an appropriate model (Tofts and Kermode 1991; Hittmair et al. 1994; Tofts et al. 1995; 1999; Kuhl et al. 1999), and then by performing pixel-wise fitting of a time series of images to a model. This leads to the calculation of maps of the parameters generated by the fitting process (Knopp et al. 1994; Parker et al. 1997, 1998; Hayes et al. 2002). Applications of these maps include identification of areas of abnormality, characterising heterogeneity in tumours, assessing response to treatment. Parameters include K^{trans}, which reflects perfusion and vascular permeability, and extracellular, extravascular volume (v_e) (Tofts et al. 1999). In some cases similar techniques may provide information on vascular volume (Li et al. 2000; Zhu et al. 2000). Parameter maps give rise to the question of how best to analyse such information, and how to relate several different parameters, that may be generated in the same study.

While descriptive parameters from T1-weighted images have been shown to be helpful, they are not readily transportable, and are affected by a range of factors including specific sequence parameters, instrumental parameters, inherent tissue T1 relaxation times, built in image processing. These can vary markedly between MR systems, and some may vary with hardware and software revision, or routine maintenance. Thus there are significant problems to be addressed in generalising such techniques across several centres for multi-centre trials.

Analysis methods based on model fitting require the concentration of contrast in the tissue to be calculated. This involves certain assumptions, such as the relaxivity of the contrast agent in plasma, and is calculated either based on an assumption that T1 relaxation change, and hence contrast agent concentration, is proportional to the change in signal intensity; or more accurately is based on methods that directly measure T1 relaxation time. The former approach is liable to bias between tissues having different intrinsic T1 relaxation times, as well as from non-linearities between T1 relaxation change and signal intensity. Application of such techniques to multi-centre trials also requires considerable attention to transferability and quality assurance, but this is aided by the considerable analysis and evaluation required to implement such techniques.

15.3.2 T2*-Weighted Methods

Assessment of the first-pass bolus of contrast agent, resulting in transient susceptibility changes close to capillaries, as the bolus passes, and resultant loss of signal in areas of perfusion on T2*-weighted images, provides further information on local blood volume and perfusion (Ostergaard et al. 1996). This has been shown to be of value in differentiating benign from malignant breast lesions (Kuhl et al. 1997; Kvistad et al. 1999). This technique has been used to evaluate regional brain perfusion and blood volume, and more recently has been applied to the study of extra-cranial tumours. The technique demands high temporal resolution, ideally 1–2 s per time point, so it has usually not been associated with methods evaluating K^{trans}. However, recently several approaches combining both T1-weighted and T2*-weighted imaging to obtain a wider range of parameters that characterise tumours have been reported (Baustert et al. 1998; Barbier et al. 1999; Vonken et al. 2003). These techniques are also being employed to assess response to treatment. As yet there are no reports of the techniques being applied in multi-centre trials, but the general principles are similar to those for T1-weighted studies.

15.4 Current Multi-Centre Studies Using Dynamic Contrast-Enhanced MRI

15.4.1 Breast Cancer

Although a number of multi-centre studies of breast cancer diagnosis or screening are in progress using dynamic contrast agent MRI, few have published details of the protocol and methodology to be employed. Heywang-Köbrunner et al. (2001) have reported a trial conducted at 11 centres using Siemens 1.0-T or 1.5-T scanners to improve standardisation and optimise interpretation guidelines for dynamic contrast-enhanced MRI. This study employed an 87-s 3D fast low-angle shot (FLASH) sequence repeated once before and five times after a standardised bolus of 0.2 mmol Gd-DTPA/kg. Imaging findings were correlated retrospectively with histopathology in 512 histologically correlated lesions. By setting specificity thresholds of 30%, 50% and 64%–71%, sensitivities of respectively 98%, 97% or 96% at 1.0 T and 96%, 93% and 86% at 1.5 T were reported. The best results were obtained by combining up to five wash in or wash out descriptors.

The UK study of contrast-enhanced magnetic resonance imaging as a method of screening women at genetic risk of breast cancer (MARIBS) has published its rationale (Brown et al. 2000b), study protocol (Leach 1997; Brown et al. 2000c) and radiological measurement and assessment protocol (Brown et al. 2000a). Much of this has also been included in a report of the International Working Group on Breast MRI (1999) which includes details of other studies in progress at the time of the report. The MARIBS protocol is also summarised in a review of MR in breast screening (Leach and Kessar 2002) and in a recent update reporting progress to date and comparing reported detection rates in similar studies (Leach and MARIBS Advisory Group 2002).

Kuhl and colleagues (2000; Kuhl 2003) have reported initial results from a trial of MRI screening in women diagnosed as or suspected of carrying a breast cancer susceptibility gene. This is a single-centre study, and together with other similar single-centre studies is reviewed in Leach and MARIBS Advisory Group (2002).

A further study is applying and evaluating a method of breast cancer diagnosis based on the use of three time points (the 3TP method) (Furman-Haran et al. 1998; Weinstein et al. 1999). Recently a multi-centre study of breast cancer screening has commenced at nine centres in Italy (Podo et al. 2002), with 102 participants recruited so far. The study includes participants at 1 in 2 risk of being mutation carriers, from age 25 (women) and 50 (men) with no upper age limit, and includes individuals with a previous history of breast cancer. Imaging is based on T1-weighted 3D spoilt gradient echo images acquired coronally or axially with a matrix of 128×256 coronally. MRI is compared with X-ray mammography and ultrasound. Out of 119 screening measurements, eight cancers have been detected, with five being invasive ductal or lobular, and three being ductal or lobular cancer in situ. Five of these occurred in patients with a previous history of breast cancer. Of the eight cancers detected, only one was seen on X-ray mammography and ultrasound.

Multi-centre studies investigating dynamic contrast-enhanced MRI in breast cancer may be divided into those considering morphological features alone, those considering dynamic contrast alone and those considering both morphology and contrast enhancement. This classification aids consideration of aspects important for multi-centre trials.

The use of the morphological features of tumours, observed on contrast-enhanced MRI images at specific times following injection, to determine a diagnosis was introduced by Heywang and colleagues (1986). Similar approaches, in some cases utilising fat suppression techniques, have been used in a number of studies (Harms et al. 1993; Allgayer et al. 1993; Fischer et al. 1993; Greenstein Orel et al. 1995; Tesoro-Tess et al. 1995; Orel 2000) showing high sensitivity (88%–100%), but often lower specificity (37%–89%). While early studies used 2D imaging techniques, more recent work has used 3D imaging sequences, in several cases accompanied by interleaved alternate breast imaging (requiring switching of coil elements to maximise sensitivity), which allows smaller fields of view, optimising acquisition time and spatial resolution (Greenman et al. 1998). 3D techniques have intrinsically longer acquisition times than 2D approaches, but allow all of one (or two) breasts to be assessed, of particular importance in diagnostic assessments and in screening.

The measurement of the shape of the contrast curve obtained from an ROI was introduced by KAISER and ZEITLER (1989), and has been widely used (HEYWANG-KÖBRUNNER 1990; KUHL et al. 1999). It provides strong independent diagnostic power. Using simple descriptors of the shape of the wash-out curve. KUHL and colleagues (1999) in Bonn have reported a sensitivity of 91% and specificity of 83% for cancer detection. Time resolution varies, from about 10 s or less for single-slice approaches to 90 s for 3D volume measurements (OREL and SCHNALL 1999; BROWN et al. 2000b). Initial studies have used curves derived from regions of interest (ROI) for analysis, allowing a number of empirical descriptors of the contrast curve to be defined (KUHL et al. 1999; BROWN et al. 2000b). More sophisticated approaches use model fitting, in some cases accompanied by quantitative imaging approaches, either on an ROI basis, or calculated pixel by pixel.

Many investigators combine the information from morphology and contrast kinetics. GREENSTEIN OREL and colleagues (1994) in Philadelphia reported on this approach in 1994, showing that addition of morphology to kinetic data improved discrimination of benign and malignant disease. Morphology is particularly helpful in discriminating fibroadenoma, some of which demonstrate tumour like contrast kinetics. The MARIBS study (BROWN et al. 2000b) includes a primary 3D screening assessment, with 90 s time resolution, allowing both dynamic and morphological assessment. Equivocal cases are recalled for a further high-time-resolution 2D imaging study to provide higher-time-resolution dynamic data to aid specificity.

Morphological parameters are recorded based on a predetermined set of descriptors, as is the spatial pattern of contrast uptake (BROWN et al. 2000b). The shape of the contrast curve is similarly described, and several qualitative parameters are calculated to describe contrast wash-in. All of these factors are assigned scores which are summed to give a numerical estimate of likely malignancy. A diagnostic decision is based on the radiologist's experience rather than the score, which is currently being evaluated in a symptomatic cohort. All MR results are double read blind, as are the comparison X-ray mammograms. This standardisation of reporting is an important aspect of the standardisation required for a multi-centre trial. Recently the International Working Group on Breast MRI has used a similar but more detailed categorisation of morphological and dynamic features to develop a lexicon of MR descriptors for diagnostic reporting (INTERNATIONAL WORKING GROUP ON BREAST MRI 1999; IKEDA et al. 2001), which will be helpful in future studies.

15.4.2 Multi-Centre Trials in Other Conditions

BARKHOF et al. (1997) have considered the requirements for multi-centre trials in multiple sclerosis, identifying the need to establish observer variability over multiple centres, as well as improve quantification methods and compare the different techniques in a multi-centre longitudinal fashion in order to include variation caused by both scanner and segmentation techniques, in addition to biological activity. BARKHOF et al. (1993) report a database developed for recording serial brain MRI results suitable for multiple sclerosis multi-centre trials. NYLAND et al. (1996) report on a randomised, double-blind, placebo controlled multi-centre study at eight centres in Norway to evaluate the efficacy and safety of 4.5 and 9.0 MIU recombinant human interferon alfa-2a (Roferon-A) given thrice weekly in patients with relapsing-remittent multiple sclerosis. The primary objective is to determine new disease activity analysed by monthly MRI with gadodiamide (Gd-DTPA-BMA, Omniscan).

PARODI et al. (2002) have investigated the intra- and inter-observer agreement variability of a locally developed Growing Region Segmentation Software (GRES), comparing them with those obtained using manual contouring (MC) in MS lesions seen on proton-density-weighted images (PDWI) and on Gd-DTPA-BMA enhanced T1-weighted images. The authors report that the intra- and inter-observer agreements were significantly greater for GRES compared with MC ($p<0.0001$ and $p=0.0023$, respectively) for PDWI, while no difference was found between GRES an MC for Gd-T1WI. The intra-observer variability for GRES was significantly lower on both PDWI ($p=0.0001$) and Gd-T1WI ($p=0.0067$), whereas for MC the same result was found only for PDWI ($p=0.0147$). These data indicated that this implementation of GRES reduces both the intra- and the inter-observer variability in assessing the area of MS lesions on PDWI and might prove useful in multi-centre studies.

A number of multi-centre studies have reported on the utility and acceptability of contrast agents (for example ASLANIAN et al. 1995, 1996; WANG et al. 1997; SAINI et al. 2000). Although these studies require a degree of standardisation, they do not employ dynamic contrast analysis and are aimed at demonstrating efficacy of the contrast agent rather than addressing a diagnostic or therapeutic question.

Several multi-centre studies have examined the utility of ^{1}H magnetic resonance spectroscopy for the diagnosis and evaluation of brain tumours (Sijens et al. 1995; Negendank and Sauter 1996; Negendank et al. 1996), and to examine the neurological complications of AIDS (Paley et al. 1996). This required standardisation of measurement parameters, selection of placement of region of interest, and analysis. A multi-centre study of ^{31}P magnetic resonance spectroscopy is currently in progress (Arias-Mendoza et al. 2000).

A statistical evaluation to determine the sample size is an essential first step, taking account of the population, likely potential accrual rate, maximum possible measurement or treatment capacity taking account of return visits, estimating drop-out and acceptability of the study design to prospective patients. Poor entry into trials is a major problem, and this may be acerbated by high drop-out in studies that are measurement intensive. Realistic estimates of accrual, and of instrument access, can be difficult to obtain, and in practice are often affected by local or national policy changes during the course of a study. However, good estimates, adequate funding and clear local agreement, are particularly important to multi-centre trials.

15.5 Standardisation and Issues to Be Resolved in a Multi-Centre Trial Design

Taking the MARIBS design as an example, given that more details are available and published than for most other multi-centre trials using dynamic contrast agent evaluation, in this section the major issues to be addressed in trial design are described, together with consideration of their relative importance with respect to single-centre trials. The following section will then consider approaches to tackling each issue.

15.5.1 Defining the Scientific Question and Design

The question posed dictates the trial design, the test required, and the power required of the study. A decision as to whether a study is single- or multi-centre has major implications for design and funding, and is likely to be dictated by prevalence of the condition and likely recruitment at individual centres, the context of the question (for example evidence that a new technique has diagnostic and clinical potential might be addressed at a single centre, establishing that the technique is robust and can be used routinely in a general hospital setting, as a change in practice would require a multi-centre trial), issues of regulation and pharmaceutical licensing, and the level of confidence required by a pharmaceutical manufacturer before committing significant funding to further development or Phase III trials. One or more control arms may be required and comparison of diagnostic tests may be required.

15.5.2 Ethical Approval

Approval by the appropriate ethical committees is a prerequisite for any research study. This is an area where multi-centre trials are considerably more complex than single-centre studies. In the UK, until recently, full consideration and approval of a study was required by each local ethical committee involved. For the MARIBS study, this involved many committees, with one individual measurement centre potentially having to submit applications to many local ethics committees if recruiting from a number of hospitals. Recently this system has been streamlined, with the establishment of regional multi-centre research ethics committees (MRECs). If one such committee approves a study, the same protocol has still to be submitted to individual local research ethics committees (LRECs), but they are guided by the MREC decision.

15.5.3 Determining an Imaging Protocol

The imaging protocol must address the scientific question. For dynamic contrast studies the measurement endpoints required will determine the protocol. Issues to be considered include:

- Is morphology required, what image weightings are required for any non-contrast aspects of the examination (e.g. T2-weighted images), what spatial resolution is desirable (including slice thickness), what FOV is required?
- For the dynamic contrast component, what spatial resolution (including slice thickness) is required, what FOV, is a 3D examination (or complete organ

coverage) required for each time point, what temporal resolution is required?
- For a dynamic study, what type of information is required? This will define the type of sequences to be used.
- The simplest form of study will obtain information before and at a time point after contrast, providing little functional information other than the uptake of contrast. Dynamic studies with time resolutions of the order of 90 s provide information on the change in signal intensity over a number of time points, providing some of the dynamic information characterising washout shape referred to above, and allow a number of qualitative parameters to be defined. These measurements can be made quantitative by incorporating sequences that allow contrast agent concentration to be calculated, allowing these parameters to be put on a quantitative basis, and providing absolute contrast concentration.
- Higher time resolution studies, including sequences designed for quantitative studies, allow the image data to be fitted to pharmacokinetic models of contrast uptake, allowing parameters to be obtained that describe aspects of the tissue physiology, delivery of the agent or descriptors of the contrast kinetics. Time resolutions of the order of 10 s have been used for T1-weighted studies (shorter time resolution has been used in some studies), or approaching 1–2 s for T2*-weighted studies. Many of these studies have used 2D imaging, although as instrumentation improves there is interest in performing 3D measurements.

The final imaging protocol is likely to involve compromises, both in the number and range of measurements, the resolution and volume coverage, and the temporal resolution. In a multi-centre trial, the capabilities and the practicality of implementation at different sites must also be considered.

15.5.4 Equipment Issues

Given an ideal imaging protocol from a scientific point of view, the next issue to be considered is the practicality of implementing it on the equipment available for the study. This should be easiest for a single-centre study, where the investigators are very familiar with the equipment and its capabilities. However, a number of issues still arise, which are a subset of those faced by a multi-centre trial.

15.5.4.1 Issues for Single-Centre Trials

Choice of field strength, imaging coils and patient set up – these must be appropriate to the trial.

Does the manufacturer provide the sequences required for the trial? This may be a particular problem when quantitative measurements are required, or when faster than usual measurements are needed, or when the protocol calls for an unusual combination of information, for example interleaved T1 and T2*-weighted information (D'Arcy et al. 2002). If a non-standard sequence is required the investigators may need to prepare it themselves, with all the required testing and validation. They will need to persuade the manufacturer to implement it, or they will need to transfer it from another academic site (which again may require manufacturer's agreement). To develop and install a sequence (other than for minor modifications) it is likely that the user will require access to the pulse sequence development language and facilities, have staff with the required know-how, and have a degree of support from the manufacturer, usually with a research agreement. Persuading a manufacturer to tailor and provide a non-standard sequence has recently become very difficult, due to the requirements for good manufacturing practice, and satisfying medical equipment regulatory bodies such as the Medical Devices Directorate in the UK and the Federal Food and Drugs Administration in the USA, and legislation such as the EU Medical Devices Directive. Manufacturers are unwilling to commit themselves to doing this, and further are requiring complex legal indemnities to be agreed before transferring such non product sequences to clinical sites. While clinical research centres may have the expertise and resources to deal with these issues, they present more of a problem (and a drain on staff time) at non-expert centres. This is also an impediment in multi-centre trials, as an academic site producing a new sequence may meet similar risks in transferring sequences, and may also need a contractual framework clearly identifying intended use, limits on liability etc. These barriers to medical research are now significant, and require revision of international medical equipment approval mechanisms to reduce adverse impact on medical research.

The user needs to ascertain whether the equipment will allow the measurements required. One example is a common practice in performing quantitative measurement sequences, where the signal acquisition parameters for a number of sequences are fixed, so that the numbers obtained with one sequence can be used as a reference for subsequent sequences (e.g. proton density sequences used as a reference for T1-weighted sequences, to allow rapid measurement of T1 relaxation times). Some manufacturers do not support this facility, and automatically reoptimise some or all of transmit amplifier/attenuator settings, receiver amplifier/attenuator settings, receive ADC set up and image scaling factors and filter factors each time a sequence is loaded.

Once the sequences have been defined, if any non-standard processing (other than that provided by the manufacturer) is required, programs may need to be developed and run off-line. Again the user is likely to need access to the image file structure (usually requiring manufacturer's agreement), the image file store on the imaging device, and if the program needs to be run on a separate workstation, the means to export the data in a way that can be read. All of these steps can pose problems, where provision was not made at specification and purchase of the imaging equipment.

Equipment performance may need to be monitored via a quality assurance programme to ensure that equipment performance variation does not introduce unacceptable variance in the measurements, and that the location and conduct of measurements are themselves not the cause of variance.

Users should review routine maintenance and any upgrades of hardware or software critically. It is not unusual for such activities to vary the status of the equipment in a way that adversely affects a clinical study, and upgrades can remove or change sequences in a way that is not advised or expected.

If execution of a study requires any special equipment or software modifications, programs or datasets, the user should ensure that a mechanism for reinstating them on top of the manufacturer's rebuild is available, in the event of, for example, a disc crash or operating system corruption.

15.5.4.2 Issues for Multi-Centre Trials

Multi-centre trials involve all of the above issues, but in addition the issues posed by arrangements and level of expertise at the different centres, the possibility of different equipment and software (model, revision level, manufacturer) have a major impact on the design of, and requirements to support, trials.

In all cases there will be a need to ensure that staff are trained in the protocol and in the analysis of the data, including use of any specialised software. Equipment will need to be assessed to ensure comparable performance at the different centres, and over time. It is advisable to have a central quality assurance resource, that will monitor equipment performance, and diagnostic performance, during the study. This will help identify and resolve potential problems, thereby improving the quality of the study.

15.5.4.2.1 Studies with Equipment from One Manufacturer

A number of multi-centre diagnostic trials have been designed using equipment from one manufacturer. Examples include the evaluation of dynamic contrast breast MRI (Heywang-Köbrunner et al. 2001), and a study examining ^{1}H MRS in the brain (Negendank and Sauter 1996). Usually single-manufacturer studies reduce the problems attached to sequence selection and provision, and to data sharing and transfer for analysis. Significant issues may remain if different models or releases of equipment and software are involved. Support by the manufacturer for the protocol and study can considerably reduce the burden on the study co-ordinating centre. However the investigators should remember that scientific responsibility resides with them and that despite their best intentions, manufacturers can make mistakes. The investigators need to confirm that sequences and analysis programs do perform as intended.

15.5.4.2.2 Studies with Equipment from Several Manufacturers

Fewer studies have used equipment from multiple manufacturers, including detailed analysis and quantitative approaches. Two examples are the MARIBS study (Brown et al. 2000b,c) and the multi-centre study of ^{31}P MR spectroscopy in cancer (Arias-Mendoza et al. 2000) which has involved advanced decoupled spectroscopy including extending the instrumentation routinely available. An international workshop reported on requirements for standardisation of measurements using magnetic resonance spectroscopy (Leach et al. 1994). When equipment from different manufacturers is used, it is important to ensure that the planned protocol on each instrument is as close as possible to the imaging protocol for the study. This can require considerable understanding of the peculiarities of each instrument, and it is advisable for an expert in each type of hardware to be available to the study. Often descriptors and adjustable parameters vary between machines, and there may not be a one to one relationship. The closest approximation must be identified, taking into account the implications of changes made to sequences. There may be differences in the way sequential repetitions of sequences can be run, and the results stored (of importance in dynamic contrast measurements) and there may be other issues affecting relative normalisation of sequence set-up (as discussed above).

Analysis and data storage may vary between manufacturers, including the ability to store regions of interest, contrast uptake curves, and the capability to regenerate them if required. It may be necessary to transfer data between sites, or to common independent processing software at the user site, or to a co-ordinating centre. Access to transfer routes, and information on the data structure, can be an issue. A central coordinating site is unlikely to have close working arrangements with all manufacturers, requiring some issues to be solved by a lead site for a given manufacturer.

The image information from different manufacturers may (and does) vary. Issues include different image scale factors, leading to different apparent enhancements between manufacturers, which can give manufacturer-dependent ranges for empirical pharmacokinetic parameters; different image processing and filters, which may or may not be accessible to the user; different number ranges and ADC set-up. Analysis and evaluation protocols need to take account of these issues.

Suitable quality assurance protocols and calibrations will be required to address these issues.

15.5.5 Data Analysis

In addition to ensuring that data are obtained in a consistent way and identifying and addressing differences between equipment, it is necessary to determine how the data are to be evaluated.

Morphological information may be assessed by normal radiological review. However, a consistent and robust reporting is required, and it is likely that this may need to be recorded in an evaluable form. It is therefore desirable to establish terminology, and the importance attached to given characteristics, at the outset. Often, if the technique is new, experience will be limited, and independent double reading, together with some independent quality assurance process, will be advantageous.

If dynamic data are to be obtained from a region of interest, criteria for selecting a region, and parameters to be assessed or measured need to be established.

If pixel-wise calculation of empirical or quantitative parameters is to be performed, this needs to be done in a consistent and robust way, with identification of any assumptions or approximations. The methods of analysing these parameter maps have to be defined and applied consistently, with appropriate quality assurance.

Provision for retaining and backing up the data, together with ensuring confidentiality and security need to be established.

15.5.6 Publication Policy

It is advisable for multi-centre studies to have a publication policy to define authorship issues at the outset.

15.6 Addressing Issues in Multi-Centre Trials Using Dynamic Contrast Agents, with Reference to the MARIBS Study

Based on the issues identified above, the approach taken in the MARIBS study is described, as an example.

15.6.1 Scientific Question and Study Design

The study was designed to address the question of whether dynamic contrast-enhanced MRI was superior to X-ray mammography in detecting and diagnosing breast cancer in women at high genetic risk. The target group was women below the age of 50, where X-ray mammography has limitations. Based on the estimated sensitivity of MRI (based on symptomatic studies) and X-ray mammography in this age group, it was originally estimated that some 1500 women at 50% risk of carrying BRCA1, BRCA2 or TP53 gene mutations needed to be accrued. This meant adopting a comparative rather than randomised trial design, as this type of design required the smallest numbers, due to the relatively small number of known mutation carriers. However, this also meant that mortality could not be used as an endpoint. The statistical basis for the trial design has been reported (Brown et al. 2000c). This accrual required a multi-centre design to accrue sufficient women, and to provide sufficient imaging capacity. It also was necessary to use a range of MR imaging equipment and manufacturers, to take account of instruments available at the different recruiting centres. Some 22 genetics and MRI centres are contributing to the study. In the event, due to limitations in recruitment at genetics centres, and availability of imaging resources, the overall accrual has been reduced to 950 women and a total of 3300 scans, which should detect a difference between X-ray mammography and MRI at the 1% significance level with 70% power, and at the 5% significance level with 90% power (Leach and MARIBS Advisory Group 2002).

15.6.2 Ethical Approval

Based on the protocol, approval was sought originally at the Royal Marsden Hospital Research Ethics Committee, and subsequently at all referring and imaging centres. Due to changes in requirements, and later recruitment of some centres, multi-centre ethics approval was also obtained from the North Thames Multi-Centre Research Ethics Committee (MREC).

15.6.3 The Imaging Protocol

The objective of the study was to investigate dynamic contrast-enhanced MRI in comparison with the standard technique of X-ray mammography. As a screening investigation it was necessary to evaluate both breasts, maximising the sensitivity for detection of small lesions, whilst providing adequate resolution to define them. This implied using dedicated breast coils and a field strength of 1.0 T or 1.5 T. In order to maximise sensitivity, a double dose (0.2 mmol/kg of Gd-DTPA) of contrast was used, delivered by bolus injection (about 10 s). In addition to maximising sensitivity, it was important to optimise specificity, to minimise unnecessary follow-up or biopsy. This suggested including both morphological and dynamic evaluation, maximising spatial resolution to improve structural definition, and minimising time resolution to improve characterisation of the contrast dynamics. The protocol therefore includes high-resolution 3D scans prior to and after the dynamic contrast sequence (0.89*0.66 mm resolution) and a lower resolution dynamic 3D sequence before and after contrast (1.33*1.33 mm resolution), both with 2.5-mm slice thickness. The lower resolution 3D sequence provides dynamic enhancement information with a time resolution of 90 s, in line with much published information on using dynamic contrast uptake curves as a discriminant in breast cancer diagnosis. Images are taken in the coronal plane to minimise the sequence duration for a given field of view by allowing an asymmetrical field of view. The dynamic T1-weighted sequence is preceded by a proton density sequence with identical timing but a 6° rather than 35° flip angle. This allows T1 relaxation times for tissues before and during contrast enhancement to be calculated. Figure 15.1 shows the full set of image data in an example of a screen detected cancer. Figure 15.2 shows a graph of the dynamic uptake curve obtained in regions of interest in fat, parenchymal tissue and tumour. This protocol could be applied with little modification to a wide range of 1.0 T and 1.5 T instruments, although high-specification instruments could have employed better time resolution or obtained higher resolution.

While the above (Visit A) protocol provided the primary screening measurement, it was recognised that the 90-s time resolution might limit the specificity of dynamic contrast measurements. Where findings were equivocal, a second visit 2 weeks later (Visit B) would be performed. This was designed to provide higher temporal resolution in equivocal lesions, providing 10-s time resolution for 2D slices through lesions of interest. The protocol contained the same pre- and post-contrast high-resolution 3D images, but now uses a 2D sequence with up to five slices, preceded again with a proton density sequence for the same slices, to follow the dynamic contrast uptake. Again the protocol could be implemented on a wide range of scanners. The full protocol for both visits has been published (Brown et al. 2000a).

15.6.4 Quality Assurance

To ensure that the sequence operated accurately, and provided the correct contrast, a quality assurance protocol was devised. This incorporates a routine QA measurement to be performed on a phantom provided to the centre, that is tailored to fit within the specific breast coil at that centre. This contains a material of known T1 relaxation time and allows T1 relaxation time, signal to noise, and coil homogeneity to be measured (Hayes et al. 1998, 1999).

An additional more detailed test assessment has been designed to be conducted by a study physicist

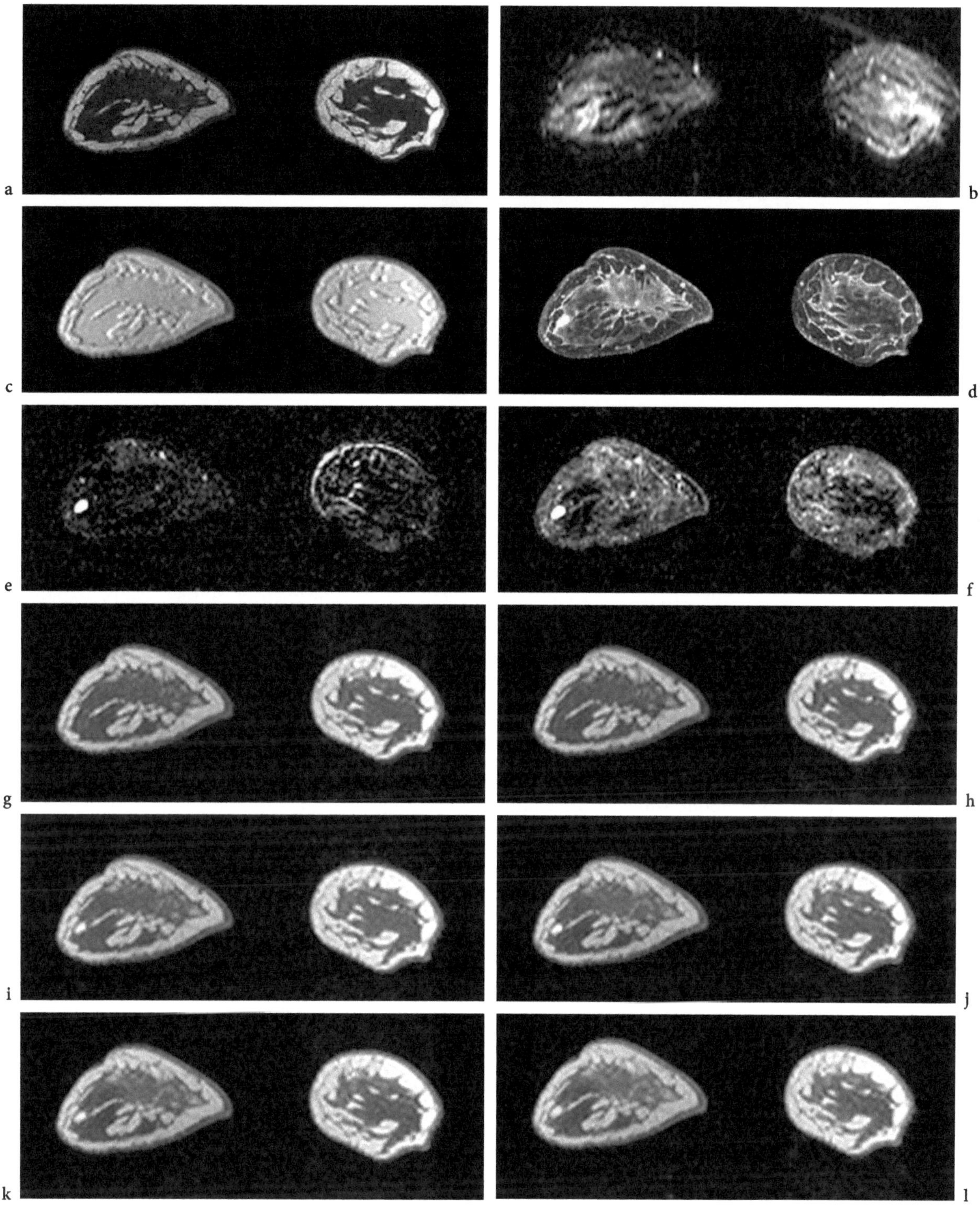

Fig. 15.1a-l. Images from the "Visit-A" screening protocol from a 35-year-old woman with an MR screening detected lesion considered suspicious on the Visit-A scans. A subsequent "Visit-B" scan confirmed a suspicious time intensity curve and cytology following a fine needle aspirate confirmed carcinoma. Images show the initial high-resolution T1-weighted scan (**a**, pre-contrast T1-weighted high resolution); a T2-weighted image (**b**, T2-weighted); proton density-weighted image (**c**, proton density-weighted); six of the seven dynamic contrast images at 90-s intervals (starting at –90, 0, 90, 180, 270, 450 s, with contrast commencing at 0 seconds) (**g–l**, 1st–5th dynamic, 7th dynamic), the post-contrast high-resolution fat suppressed image (**d**, T1-weighted fat-suppressed post-contrast), early subtraction image (180–0 s) (**e**, early subtraction), late subtraction image [450 s (–90 s)] (**f**, late subtraction)

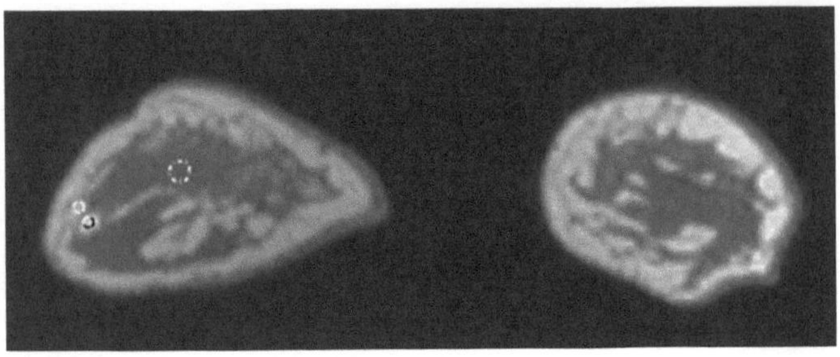

a

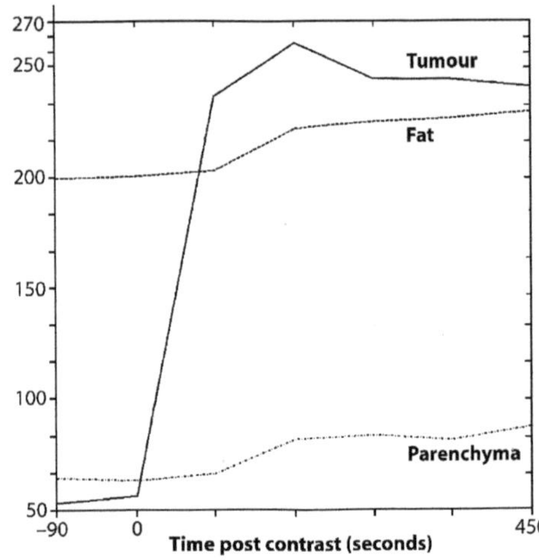

b

Fig. 15.2ab. **a** Selected regions of interest shown on the 3rd dynamic post-contrast image selected in tumour, parenchyma and fat in the patient shown in Fig. 15.1. **b** Signal intensity versus time curves for the three selected regions

after installation of the study sequences, twice yearly thereafter, or after equipment upgrades or modifications. This includes T1 measurements, checking the contrast response of the sequences and providing T1 calibration curves to allow the quantitative measurements to be corrected for the effects of slice profile. In addition, slice profile is measured, together with spatial resolution.

These quality assurance tests aim to strike a balance between ensuring each MR system is performing the protocol properly, by periodic detailed assessments; providing a routine check on performance to alert local staff and the coordinating centre to any problems; and avoiding undue time on the equipment. Some centres, particularly those with no active developmental research, have difficulty scheduling these QA sessions, a problem that could be reduced with more explicit funding of machine time.

15.6.5 Implementation of Sequences

The sequences used in the study were implemented on the Siemens Vision 1.5-T MR system at the Royal Marsden Hospital, Sutton, by modifying standard sequences, and tested extensively. The sequences were modified to run on 1.0-T Siemens systems, with the help of the manufacturer who installed them at non-research sites. Standard sequences were modified for GE sites, again with some support from the manufacturer. Standard sequences were available for Philips systems, although additional steps were needed to provide calibration between proton density and T1-weighted images, this being applied at the processing stage. Again the manufacturer provided support for this. The sequence was also implemented on a Marconi scanner, with support from the company. One of the aims of this study was to use sequences that were transferable and close to standard sequences.

For some multi-centre studies using dynamic contrast agents, it may be desirable to use more advanced approaches, that may not be closely based on a standard sequence. In such cases it will be necessary either to base the studies at capable research sites with the capability and explicit funding to implement the approach, or to ensure that the study uses similar systems such that pulse sequences and processing are directly compatible. Alternatively, a major process of securing appropriate manufacturer's support (to an agreed timetable) will be required, which may require influence greater than that wielded by an individual academic research centre. Currently the US NCI is considering reaching such agreements with manufacturers, and funding the necessary costs, to attain objectives that are not otherwise practicable.

15.6.6 Equipment Issues

Many issues relating to different manufacturers' equipment have been discussed above and addressed in sequence design and in pulse programming. A number of equipment issues can have an important bearing on the quality of measurements. In the MARIBS study, the major determinants in selecting equipment were that there should be a dedicated breast coil, that the field strength should be 1.0 T or 1.5 T, that the system should have shielded gradients.

The latter requirement was an important determinant of image quality and speed. For many dynamic contrast studies, uniformity of the transmit coil will be important, as this can affect the measured signal intensity, and adversely affect quantitative measurements. Assessing this over the area to be imaged is an important part of quality assurance and validation.

Generation of slice-selective pulses often varies between manufacturers, and may also show relative changes with slice thickness. It is advisable to assess this, and ensure that evaluation takes account of it. A relatively poor slice profile will reduce the contrast of T1-weighted images, compared with systems having a better slice profile, thereby reducing contrast sensitivity. Gradient amplitude, slew rate, gradient screening, eddy current corrections and imaging coil design (which can provide a source of unshielded eddy currents) can affect both the speed of equipment and the quality of images, and should be assessed by standard quality assurance tests. High transmitter power can allow shorter RF pulses, which may benefit imaging speed.

The receive chain, including the way the ADC is normalised, the properties of in-line filters, and the use of image processing, can vary widely between manufacturers, and should again be controlled for in multi-centre studies.

Data formats and the media available for recording data are another major source of incompatibility between systems. Although DICOM in principle provides a format that should translate, in practice this may only be available on certain output routes, which may not be those most convenient for a multi-centre trial. It is likely that a capability to read the internal file formats of different systems will be required, which may require agreements with manufacturers. This is a specific problem it was necessary for us to overcome in the MARIBS study, and we are grateful to the manufacturers for their support in achieving this. Processing software from manufacturers also varies widely, and if this is to be used, care needs to be taken to ensure that it works as described and that the users understand the description. Some standardisation may be required to allow for different gain factors, or image value offsets between manufacturers. It is also important to check for dynamic studies that the software is correctly identifying the timing of sequences. While manufacturers are beginning to introduce analysis packages for functional MR information, these are likely to depend on particular acquisition strategies, and may not translate well to a trans-manufacturer trial.

15.6.7 Data Analysis

Many of the issues to be considered have been dealt with above. Double reading, blinded to the first reader, and ideally at another centre, has been included in the MARIBS study. This provides a variety of checks on the process and is recommended. Investigation of problem cases and a random sample of cases is also undertaken by the MARIBS study radiologist, and is recommended. In the MARIBS study, a symptomatic cohort has also been evaluated, providing additional experience to radiologists, and providing for a separate evaluation of the radiological process. A quality assurance evaluation is in process, circulating complete data sets to the participating radiologists. Separate test cases have been prepared for radiologists familiar with different machines. A quantitative analysis of study data, allowing comparison with the empirical analysis conducted at each centre, is planned, and will provide an interesting comparison, as well as, in principle, more standardised data. This approach is likely to be necessary in studies monitoring new anti-angiogenic or anti-vascular agents.

15.7 Conclusions

Dynamic contrast-enhanced MRI methods provide a powerful tool for detection, diagnosis and evaluation of several diseases. They have a growing role in the assessment of therapies, particularly new treatments directed at vascular processes. While much has been achieved in developing and demonstrating techniques, with the continuing advances in instrumentation, there is clearly room for considerable further development. For these applications to have widespread use there is a need for further academic development, and the support from manufacturers, regulators, pharmaceutical companies and research funding organisations to support implementation of techniques across different platforms.

Acknowledgements
The contribution of the MARIBS Study Advisory Group, staff and collaborating centres, to the development and prosecution of the MARIBS study is gratefully acknowledged. The study is supported by UK Medical Research Council and the National Health Service Research and Development Board.

References

Allgayer B, Lukas P, Loos W et al (1993) The MRT of the breast with 2D spin-echo and gradient-echo sequences in diagnostically problematic cases. Rofo Fortschr Geb Rontgenstr Neuen Bildgeb Verfahr 158:423–427

Arias-Mendoza F, Brown TR, Schwarz A et al (2000) Preliminary results of a multi-institutional trial to demonstrate clinical predictive value of in vivo localized 31P MR spectroscopy data in human non-Hodgkin's lynphoma. ISMRM, Berkeley, 1:98. Proceedings of the 8th International Society of Magnetic Resonance in Medicine

Aslanian V, Lemaignen H, Bunouf P et al (1995) Clinical evaluation of the tolerability of gadodiamide, a new nonionic contrast agent in MRI of the central nervous system. J Radiol 76:431–434

Aslanian V, Lemaignen H, Bunouf P et al (1996) Evaluation of the clinical safety of gadodiamide injection, a new nonionic MRI contrast medium for the central nervous system: a European perspective 11. Neuroradiology 38:537–541

Barbier EL, den Boer JA, Peters AR et al (1999) A model of the dual effect of gadopentate dimeglumine on dynamic brain MR images. J Magn Reson Imaging 10:242–253

Barkhof F, Filippi M, Miller DH et al (1997) Strategies for optimizing MRI techniques aimed at monitoring disease activity in multiple sclerosis treatment trials 7. J Neurol 244:76–84

Barkhof F, Thompson AJ, Kappos L et al (1993) Database for serial magnetic resonance imaging in multiple sclerosis 13. Neuroradiology 35:362–366

Baustert IC, Padhani A, Revell P et al (1998) From qualitative to quantitative measurements of permeability, leakage space and relative blood volume. International Society of Magnetic Resonance in Medicine, Berkeley, Ca 1655, Proceedings of the ISMRM 6th annual meeting, Sydney

Brown J, Buckley D, Coulthard A et al (2000a) Magnetic resonance imaging screening in women at genetic risk of breast cancer: imaging and analysis protocol for the UK multicentre study. Magn Reson Imaging 18:765–776

Brown J, Coulthard A, Dixon AK et al (2000b) Rationale for a national multi-centre study of magnetic resonance imaging screening in women at genetic risk of breast cancer. Breast 9:72–77

Brown J, Coulthard A, Dixon AK et al (2000c) Protocol for a national multi-centre study of magnetic resonance imaging screening in women at genetic risk of breast cancer. Breast 9:78–82

D'Arcy JA, Collins DJ, Rowland IJ et al (2002) Applications of sliding window reconstruction with cartesian sampling for dynamic contrast enhanced MRI. NMR Biomed 15:174–183

Fischer U, von Heyden D, Vosshenrich R et al (1993) Signal characteristics of malignant and benign lesions in dynamic 2D-MRT of the breast. Rofo Fortschr Geb Rontgenstr Neuen Bildgeb Verfahr 158:287–292

Furman-Haran E, Grobgeld D, Margalit R et al (1998) Response of MCF7 human breast cancer to tamoxifen: evaluation by the three-time-point, contrast-enhanced magnetic resonance imaging method. Clin Cancer Res 4:2299–2304

Greenman RL, Lenkinski RE, Schnall MD (1998) Bilateral imaging using separate interleaved 3D volumes and dynamically switched multiple receive coil arrays. Magn Reson Med 39:108–115

Greenstein Orel S, Schnall MD, Livolsi VA et al (1994) Suspicious breast lesions: MR imaging with radiologic-pathalogic correlation. Radiology 190:485–493

Greenstein Orel S, Schnall MD, Powell CM et al (1995) Staging of suspected breast cancer: effect of MR imaging and MR-guided biopsy. Radiology 196:115–122

Harms SE (2001) Integration of breast MRI in clinical trials. J Magn Reson Imaging 13:830–836

Harms SE, Flamig DP, Hesley KL et al (1993) MR imaging of the breast with rotating delivery of excitation off resonance: clinical experience with pathologic correlation. Radiology 187:493–501

Hayes C, Utting J, Leach MO et al (1998) A breast phantom for quality assurance in a multi-centre screening trial. International Society of Magnetic Resonance in Medicine, Berkeley, CA, 2:929. Proceedings of the 6th scientific meeting of the ISMRM

Hayes C, Liney G, Leach MO (1999) Quality assurance in the UK multi-centre study of MRI screening for breast cancer. Proceedings of the 7th scientific meeting ISMRM, Philadelphia, vol 2, p 1072. International Society of Magnetic Resonance in Medicine, Berkeley, CA, 2:1072. Proceedings of the 7th scientific meeting ISMRM, Philadelphia

Hayes C, Padhani AR, Leach MO (2002) Assessing changes in tumour vascular function using dynamic contrast-enhanced magnetic resonance imaging. NMR Biomed 15:154–163

Heywang-Kobrunner SH (1990) Contrast enhanced MRI of the breast. Karger, Munich

Heywang-Kobrunner SH, Bick U, Bradley WG Jr et al (2001) International investigation of breast MRI: results of a multicentre study (11 sites) concerning diagnostic parameters for contrast-enhanced MRI based on 519 histopathologically correlated lesions. Eur Radiol 11:531–546

Heywang SH, Hahn D, Schmidt H et al (1986) MR imaging of the breast using gadolinium-DTPA. J Comput Assist Tomogr 10:199–204

Hittmair K, GomiscekG, Langenberger K, Recht M, Imhof H, Kramer J et al (1994) Method for the quantitative assessment of contrast agent uptake in dynamic contrast-enhanced MRI. Magn Reson Med 31:567–571

Ikeda DM, Hylton NM, Kinkel K et al (2001) Development, standardization, and testing of a lexicon for reporting contrast-enhanced breast magnetic resonance imaging studies. J Magn Reson Imaging 13:889–895

International Working Group on Breast MRI (1999) Technical report of the International Working Group on Breast MRI. J Magn Reson Imaging 10:980–981

Julian TB (2001) MRI: a role in clinical trials. J Magn Reson Imaging 13:837–841

Kaiser WA, Zeitler E (1989) MR imaging of the breast: fast imaging sequences with and without Gd-DTPA. Preliminary observations. Radiology 170:681–686

Knopp MV, Brix G, Junkermann HJ et al (1994) MR mammography with pharmacokinetic mapping for monitoring of breast cancer treatment during neoadjuvant therapy. Magn Reson Imaging Clin North Am 2:633–658

Kuhl CK (2003) High-risk screening: multi-modality surveillance of women at high risk of breast cancer (proven or suspected carriers of a breast cancer susceptibility gene). J Exp Clin Cancer Res 21:103–106

Kuhl CK, Bieling H, Gieseke J et al (1997) Breast neoplasms:

T2* susceptibility-contrast, first-pass perfusion MR imaging. Radiology 202:87–95
Kuhl CK, Mielcareck P, Klaschik S et al (1999) Dynamic breast MR imaging: are signal intensity time course data useful for differential diagnosis of enhancing lesions? Radiology 211:101–110
Kuhl CK, Schmutzler RK, Leutner CC et al (2000) Breast MR imaging screening in 192 women proved or suspected to be carriers of a breast cancer susceptibility gene: preliminary results. Radiology 215:267–279
Kvistad KA, Lundgren S, Fjosne H et al (1999) Differentiating benign and malignant breast lesions with T2*-weighted first pass perfusion imaging. Acta Radiol 40:45–51
Leach MO and MARIBS Advisory Group Study Advisory Committe (2002) The UK national study study of magnetic resonance imaging as a method of screening for breast cancer (MARIBS). J Exp Clin Cancer Res 21:107–114
Leach MO (1997) Protocol 97PRT/4: National study of magnetic resonance imaging to screen women at genetic risk of breast cancer. Lancet, http://www.thelancet.com/info/info.isa?n1=authorinfo&n2=Protocol+review&uid=1187
Leach MO (2001) Application of magnetic resonance imaging to angiogenesis in breast cancer. Breast Cancer Res 3:22–27
Leach MO (2002) Assessing response to treatment in breast cancer using magnetic resonance. J Exp Clin Cancer Res 21:111
Leach MO, Kessar P (2002) Breast MRI and screening. In: Warren R, Coulthard A (eds) Breast MRI in practice. Dunitz, London, pp 227–236
Leach MO, Arnold D, Brown TR et al (1994) International workshop on standardization in clinical magnetic resonance spectroscopy measurements: proceedings and recommendations. Acad Radiol 1:171–186
Li KL, Zhu XP, Waterton J et al (2000) Improved 3D quantitative mapping of blood volume and endothelial permeability in brain tumours. J Magn Reson Imaging 12:347–357
Miller AB, Hoogstraten B, Staquet M (1981) Reporting results of cancer treatment. Cancer 47:207–214
Moss S, Chamberlain J (1996) Screening for cancer of the breast. In: Chamberlain J, Moss S (eds) Evaluation of cancer screening. Springer, Berlin Heidelberg New York, pp 33–53
National Institutes of Health Consensus Statement (1997) Breast cancer screening for women aged 40–49, 21–23 Jan 1997. National Cancer Institute, Bethesda MD
Negendank W, Sauter R (1996) Intratumoral lipids in 1H MRS in vivo in brain tumors: experience of the Siemens cooperative clinical trial. Anticancer Res 16:1533–1538
Negendank WG, Sauter R, Brown TR et al (1996) Proton magnetic resonance spectroscopy in patients with glial tumors: a multicenter study. J Neurosurg 84:449–458
Nyland H, Myhr KM, Lillas F et al (1996) Treatment of relapsing-remittent multiple sclerosis with recombinant human interferon-alfa-2a: design of a randomised, placebo-controlled, double blind trial in Norway. Mult Scler 1:372–375
Orel SG (2000) MR imaging of the breast. Radiol Clin North Am 38:899–913
Orel SG, Schnall MD (1999) High risk screening working group report. J Magn Reson Imaging 10:995–1005
Ostergaard L, Sorenson AG, Kwong KK et al (1996) High resolution measurement of cerebral blood flow using extravascular tracer bolus passages, part II. Experimental comparison and preliminary results. Magn Reson Med 36:726–736
Paley M, Cozzone PJ, Alonso J et al (1996) A multicenter proton magnetic resonance spectroscopy study of neurological complications of AIDS. AIDS Res Hum Retroviruses 12:213–222
Parker GJ, Suckling J, Tanner SF et al (1997) Probing tumor microvascularity by measurement, analysis and display of contrast agent uptake kinetics. J Magn Reson Imaging 7:564–574
Parker GJ, Suckling J, Tanner SF et al (1998) MRIW: parametric analysis software for contrast-enhanced dynamic MR imaging in cancer. Radiographics 18:497–506
Parodi RC, Sardanelli F, Renzetti P et al (2002) Growing Region Segmentation Software (GRES) for quantitative magnetic resonance imaging of multiple sclerosis: intra- and inter-observer agreement variability: a comparison with manual contouring method. Eur Radiol 12:866–871
Podo F, Sardanelli F, Canese R et al (2002) The Italian multicentre project on evaluation of MRI and other imaging modalities in early detection of breast cancer in subjects at high genetic risk. J Exp Clin Cancer Res 21:115–124
Saini S, Sharma R, Baron RL et al (2000) Multicentre dose-ranging study on the efficacy of USPIO ferumoxtran-10 for liver MR imaging. Clin Radiol 55:690–695
Sijens PE, Knopp MV, Brunetti A et al (1995) 1H MR spectroscopy in patients with metastatic brain tumors: a multicenter study. Magn Reson Med 33:818–826
Tesoro-Tess JD, Amoruso A et al (1995) Microcalcifications in clinically normal breast – the value of high field, surface coil, Gd-DTPA-enhanced MRI. Eur Radiol 5:417–422
Therasse P, Arbuck SG, Eisenhauer EA et al (2000) New guidelines to evaluate the response to treatment in solid tumors. J Natl Cancer Inst 92:205–216
Tofts P, Kermode AG (1991) Measurement of the blood-brain barrier permeability and leakage space using dynamic MR imaging. I. Fundamental Concepts. Magn Reson Med 17:357–367
Tofts PS, Berkowitz B, Schnall MD (1995) Quantitative analysis of dynamic Gd-DTPA enhancement in breast tumors using a permeability model. Magn Reson Med 33:564–568
Tofts PS, Brix G, Buckley DL et al (1999) Estimating kinetic parameters from dynamic contrast-enhanced T(1)-weighted MRI of a diffusable tracer: standardized quantities and symbols. J Magn Reson Imaging 10:223–232
Vonken EPA, van Osch MJP, Bakkar CJG et al (2003) Simultaneous quantitative cerebral perfusion and Gd-DTPA extravasation measurement with dual-echo dynamic susceptibility contrast MRI. Magn Reson Med 43:820–827
Wang C, Ahlstrom H, Ekholm S et al (1997) Diagnostic efficacy of MnDPDP in MR imaging of the liver. A phase III multicentre study 6. Acta Radiol 38:643–649
Weinstein D, Strano S, Cohen P et al (1999) Breast fibroadenoma: mapping of pathophysiologic features with three-time-point, contrast-enhanced MR imaging-pilot study. Radiology 210:233–240
Zhu XP, Li KL, Kamaly-Asl ID et al (2000) Quantification of endothelial permeability, leakage space and blood volume in brain tumours using combined T1 and T2* contrast-enhanced dynamic MR. J Magn Reson Imaging 11:575–585

16 Applications of Dynamic Contrast-Enhanced MRI in Oncology Drug Development

Gordon C. Jayson and John C. Waterton

CONTENTS

16.1 Introduction

Approximately one third of the American and European population will develop cancer at some time in their lives. The incidence of cancer increases with age, and among middle-aged people cancer is the single greatest cause of mortality. Currently the major treatment options are surgery, radiotherapy, cytotoxic chemotherapy and hormonal modulation. Despite the tremendous improvements in cancer treatment over the past few decades, survival rates for many cancers are still poor and cytotoxic chemotherapy is usually accompanied by significant toxicity. There remains a huge unmet medical need for better cancer therapies.

Entirely novel anti-cancer drugs are under development. With the explosion in our understanding of the molecular biology of cancer, hundreds of potential molecular targets have been identified for anti-cancer drugs beyond the traditional antiproliferative agents. Potential drug targets have been identified in receptors, enzymes and associated biochemical pathways, in areas such as angiogenesis and tumour perfusion, the cell cycle, apoptosis, invasion and growth factors. Targets identified in angiogenesis and the tumour vasculature (Cristofanilli et al. 2002) include the signalling pathways responsible for the growth of new blood vessels, together with factors required for the survival and structural integrity of immature endothelium. Vascular targets are particularly attractive, since effects on a small number of endothelial cells may affect the nutrition of a large number of tumour cells and, because the cancer cell itself is not targeted, the problem of resistance may be reduced. Medicinal chemists and molecular biologists have been highly successful at devising candidate drugs with good activity in preclinical testing against targets in the tumour vasculature. Dozens of such molecules are now in clinical trial, and hundreds more are in pre-clinical evaluation. These candidate drugs include orally bioavailable, rationally designed, small organic molecules, as well as biological agents that would typically require parenteral administration, for example neutralising antibodies.

The clinical development of these newer types of agent poses new challenges. Unlike antiproliferative drugs, they will not inevitably show acute dose-limiting toxicity. In addition the anti-angiogenic compounds are likely to be cytostatic rather than cytotoxic, and thus tumour stabilisation rather than response is a more probable outcome, at least with monotherapy. Considerable ingenuity may be required to devise a development programme that effectively identifies the best candidate drug molecule

G. C. Jayson, PhD, FRCP
Cancer Research UK, Department of Medical Oncology, Christie Hospital NHS Trust, Wilmslow Road, Withington, Manchester, M20 6DB, UK
J. C. Waterton, PhD, FRSC
Enabling Science & Technology, AstraZeneca, Alderley Park, Macclesfield, Cheshire, SK10 4TG, UK

at the optimum dose for full Phase III clinical testing, probably in a variety of solid tumours, at different tumour stages, both in monotherapy and combination therapy. In this chapter we will illustrate the incorporation of DCE-MRI into drug development with examples of anti-angiogenic and anti-vascular agents. For instance a large number of drugs have been targeted against the angiogenic cytokine vascular endothelial growth factor (VEGF) and its signalling pathways. This cytokine is a principal mediator of perfusion and vascular permeability. The ability of DCE-MRI to detect changes in these parameters (Jayson et al. 2002) provides powerful biomarkers for rapid evaluation of the acute pharmacology of these newer agents in clinical trials.

16.2 What Does the Drug Developer Need to Know?

16.2.1 Phase III

The ultimate aim of the drug developer is to gain approval for the molecule from the regulatory agencies in Europe (EMEA), Japan (MHLW), the United States (FDA) and other territories, so that the drug can be made available for oncologists to prescribe for their patients. A key component of the regulatory submission is the findings of prospective randomised phase III clinical trials where the new treatment is compared in a randomised controlled trial against the best available current therapy, involving typically many hundreds or thousands of patients, in tens or hundreds of centres spread across several different countries. The most powerful evidence that can be provided to regulators is statistically significant evidence of clinical benefit, for example improvements in overall survival, progression free survival, or quality of life. Surrogate endpoints can also provide supporting evidence provided they have been validated, i.e. have been shown to be predictive of the clinical outcome of the disease (how a patient feels, functions or survives). While tumour dimensions (Therasse et al. 2000) measured by imaging may be considered a surrogate, no acute biomarker or pharmacodynamic measure from imaging has yet been accepted as a valid surrogate endpoint in oncology. In the future DCE-MRI biomarkers may provide validated surrogate endpoints for use in phase III, allowing shorter or smaller trials than those that use clinical endpoints such as survival alone, thus permitting regulatory authorities to approve drugs more quickly. However the validation of surrogate endpoints is an immense task (Lesko et al. 2001). The development and validation of DCE-MRI endpoints in this phase III setting has barely begun and is not the focus of this chapter. Given the choice of targets and candidate drugs, one of the greatest challenges for the drug developer is deciding which molecule to take into phase III, at which dose and schedule, in which tumours and perhaps in which combinations. The cost of a failed phase III programme is immense. Thousands of patients could be exposed to ineffective therapy, while the hundreds of millions of dollars wasted could, instead, have been spent on bringing another, more effective, treatment to patients. A failed phase III programme cannot be repeated.

16.2.2 Phase I/II

Before a molecule enters phase III cancer trials a considerable body of data must be acquired in smaller-scale phase I/II clinical trials, together with toxicology and efficacy studies in animals, in vitro pharmacology and other studies. At some point in the development of a new drug the compound will be administered to patients for the first time. These phase I clinical trials have several aims, which include identification of the toxicities associated with the agent and the maximum tolerated dose. The aim is to choose doses that can be taken forwards, firstly into phase II clinical trials, each of which tests a defined dose and schedule in a specific clinical situation and ultimately into phase III.

The conventional design of phase I clinical trials is to recruit cohorts of patients who receive the new drug over a few weeks. If that cohort tolerates the drug without developing significant toxicity then a new cohort is treated at a higher dose. Again, if there is no significant toxicity then a further dose is opened and so on until a dose limiting toxicity is identified. But what happens if there is no acute dose limiting toxicity, as is often the case, for example, with some humanised monoclonal antibodies? In that situation, the maximum tolerated dose and therefore also the doses and schedules for further clinical evaluation cannot be identified. A further issue is that when these apparently non-toxic drugs are combined with conventional agents marked toxicities have been recorded in certain situations (Kuenen et al. 2002).

One approach to these problems has been to incorporate biomarkers that rapidly demonstrate a relevant pharmacological action of the drug, into early clinical trials. These are sometimes described as pharmacodynamic endpoints, although strictly that term should be reserved for pharmacological responses coupled to pharmacokinetics of drug that is being administered. For instance agents that inhibit particular DNA repair enzymes can be assessed for their ability to inhibit the biological activity of the target enzyme in tumour biopsy specimens taken from patients on treatment. Alternatively various imaging biomarkers have been used to characterise the vasculature before and after administration of anti-angiogenic (Jayson et al. 2002; Eder et al. 2002; Medved et al. 2002; Thomas et al. 2001) and anti-vascular (Dowlati et al. 2002; Evelhoch et al. 2002; Galbraith et al. 2002) agents. These imaging methodologies have the aims of confirming that the putative drug has appropriate biological activity, identifying the lowest doses that have the required biological effect and more recently, have started to give mechanistic insights into why certain drugs are only partially effective. This is potentially very important as development of an ineffective drug can be halted if the desired effects on the biomarker are not seen in early clinical trials. Ineffective compounds (and worthless drug targets) can be abandoned early after few patient exposures, while for effective compounds a biologically effective dose can be established early for further testing. The principal value of a biomarker (from, for example, DCE-MRI) to the drug developer is thus to help select the right compound, at the best dose in phase I/II trials, and which to abandon.

DCE-MRI, in which imaging is performed at the same time as contrast agent administration, has been widely used in studies to understand the biology of the tumour vasculature. While some consensus recommendations have appeared (Evelhoch et al. 2000; Leach et al. 2003; Tofts et al. 1999) there remain significant issues in translating these approaches into reliable tools which can support robust decision-making in early drug development. The issues will be reviewed in this chapter. The main topics include the design of the DCE-MRI acquisition and analysis protocol, the validity and reproducibility of the measurement and the complexity introduced by tumour heterogeneity.

16.3 Considerations in Study Design – Physics, Image Informatics and Validation

16.3.1 Which DCE-MRI Parameter to Measure

The use of DCE-MRI merely implies that images will be acquired at defined intervals during the uptake and elimination of a contrast agent. For an effective compound given at the top of the dose-response curve, the radiologist's qualitative assessment may reveal an effect of the drug, but this non-quantitative approach lacks statistical power especially in early trials where effects on the tumour, and on the images, at lower doses may be small. For practical drug development some quantitative image analysis is essential. The aim is to provide measurements (or maps) of parameters which reflect tumour perfusion and which can be analysed to give a quantitative assessment of the pharmacology. Several analyses, with different levels of analytical complexity, are available to characterise DCE-MRI data. Analysis may simply provide a summary of the MR signal intensity changes or it may attempt to derive underlying physiologic parameters such as blood flow by fitting a pharmacokinetic model of contrast agent uptake and washout to the data. Note however, that although terms such as "vascular permeability" and "blood flow" are commonly encountered in the DCE-MRI literature, an accurate and absolute determination of these parameters (in $mm^{-2}.sec^{-1}$ or $ml.g^{-1}.min^{-1}$) in tumours, from DCE-MRI data, is quite difficult.

Several semi-quantitative non-pharmacokinetic analyses based on signal intensity changes are available, for example maximum enhancement, gradient or alternatively the time for signal to reach 90% of its maximum value. A limitation of such approaches is that they may be difficult to compare between centres and between studies because signal intensity changes depend on scanner and pulse sequence in a complex way. In principle, a more robust approach would be to convert signal intensity changes (via effects on T_1 and with simplifying assumptions) to tissue contrast agent concentration. A useful non-pharmacokinetic measurement of such data, which still captures information from the entire uptake-washout curve, is the initial area under the contrast agent time-concentration curve (IAUC). Alternatively, pharmacokinetic modelling can be used to estimate the transfer constant (K^{trans}) for contrast agent movement out of the vasculature and into the tissue extravascular extracellular space. Both IAUC and K^{trans} can be used to derive a "hypervascularised tumour volume" by counting the voxels with values above a pre-determined threshold, e.g. the mean value in a control tissue such as skeletal muscle. These two parameters have the merit that they are measured in absolute units (mM.s and s^{-1} respectively) and so, in principle, should provide data that are comparable between pulse sequences and scanners. Additionally, they have been widely used in animals and humans, and there is now considerable experience on a number of centres on reproducibility and, in animals, there is some evidence of a dose–response relationship (Checkley et al. 2003; Robinson et al. 2003). More complex pharmacokinetic models attempt to separate the contribution of intrinsic physiological parameters such as flow, endothelial permeability and vascular volume from the DCE-MRI signal. Such models make greater demands of the image data than K^{trans} or IAUC analyses either because they require rapid imaging in order to obtain flow measurements from initial uptake and are difficult to perform in three dimensions, require very high signal-to-noise ratio so are difficult to analyse voxelwise, or require the use of investigational high-molecular-weight contrast media (Pradel et al. 2003). In the future complex pharmacokinetic analyses may provide additional information on anti-cancer drug efficacy in clinical trials but currently they are difficult to implement and have not yet been demonstrated to provide additional value in clinical drug development.

A clinical trial should have a clear primary endpoint established prospectively, with known (and good) statistical power and with potential confounds understood and controlled. In particular, it is essential to know that the DCE-MRI technique is robust and reproducible in the proposed setting. It is also important to estimate the expected size of the effect at the top and near the bottom of the dose–response curve, and for novel drugs this is best measured through animal experiments. It is particularly important to the drug developer that the expected direction of change is certain, otherwise the study has no statistical power. For example, v_e, the extravascular extracellular volume fraction, might either increase or decrease with effective antivascular or antiangiogenic cancer therapies, depending for example on changes in tumour oedema. It is therefore of limited value as a primary imaging endpoint, although it may provide interesting mechanistic data.

16.3.2 Contrast Media

Contrast-enhanced MR protocols measure the change in MR signal in the tumour following an intravenous bolus of a contrast medium such as gadoterate (Dotarem), gadodiamide (Omniscan), gadopentetate (Magnevist) or gadoteridol (ProHance). These are all hydrophilic gadolinium chelates with molecular weight around 500 Da. They remain extracellular but rapidly cross permeable endothelia and diffuse through the tumour interstitium. The pharmacokinetics of these agents can be estimated from the model of Tofts et al. (1991), a two-compartment open model with mean distribution and elimination half-lives of the order of 10 and 100 min, respectively, the latter representing renal clearance. These different contrast agents differ in their charge and protein binding and so in principle might show differences in their equilibration between blood plasma and the tumour interstitium. This possibility does not appear to have been exploited in the study of tumour biology or response to therapy; however, it would be prudent in a clinical trial of a new therapeutic agent to ensure that each patient receives the same contrast agent.

Higher molecular weight MR contrast agents such as P792 (Vistarem) (Pradel et al. 2003; Ruehm et al. 2002) offer an interesting alternative as their tissue penetration is modulated by larger gaps in the endothelium than low molecular weight contrast media, and their renal clearance is slower. Potentially evaluation of vascular permeability with high molecular weight compounds might provide a more useful dynamic range of uptake parameters than has been seen with the lower molecular weight compounds. Protocols combining two contrast media potentially permit the separate contributions of permeability, vascular volume and flow to be estimated. However, such agents are themselves currently still in clinical development, and from the point of view of the oncology drug developer, their use remains largely a future prospect, because of ethical and regulatory difficulties in performing trials involving both an investigational diagnostic and an investigational therapeutic drug.

16.3.3 Design of the DCE-MRI Acquisition Protocol

For most studies. it is desirable to have the possibility of implementing the study protocol in a multicentre trial, possibly with different MRI instrumentation at each site. Repeat scanning must be possible and acceptable to patients, usually with sufficient time for contrast medium washout. Techniques must be feasible in the relevant anatomic locations, including maybe locations affected by physiologic motion (e.g. bowel, lung), organs with special patterns of blood supply [e.g. liver (Jackson et al. 2002), lung], and locations suffering magnetic field heterogeneity. Ideally, techniques should be feasible both in newly presenting disease and also after surgical or radio- or chemotherapeutic intervention.

A choice has to be made between slower 3D and faster 2D protocols. Conventional 3D DCE-MRI protocols normally use T_1-weighted MRI, have relatively poor time resolution (typically ≥5 s) and are therefore inadequate to measure contrast medium pharmacokinetics during first pass. "Fast" protocols may use T_2*-weighted or T_1-weighted MRI and have sufficient time resolution (typically <5 s) to measure contrast medium pharmacokinetics during first pass, allowing blood flow and vascular volume to be estimated as well as IAUC or K^{trans}. Another attraction of these fast acquisition sequences is that they allow the investigator to evaluate drugs in parts of the body usually affected by motion artefact such as the thorax; however, limitations in the gradient performance of many older scanners in existence would normally limit them to 2D acquisition.

Typical trials of antivascular and anti-angiogenic agents require comparison of DCE-MRI before and after therapy over periods of hours to weeks (Jayson et al. 2002). In order to obtain valid estimates of tumour response, it is important that the same region of tumour is imaged on each patient visit. For this reason 3D protocols, which provide data over the whole tumour, are often preferred to single slice protocols. Although some MRI instruments now allow fairly "fast" protocols in 3D, some cancer hospitals lack machines with such capability, so that the drug developer must choose between a more complete assessment of perfusion from "fast" 2D protocols, or alternatively, greater confidence from conventional 3D protocols in which follow-up scans are truly comparable, albeit with some vulnerability to motion artefact. Furthermore, if imaging is to be performed over a few weeks then it is appropriate to study tumours that are not growing too rapidly as comparison of data from different time points becomes very difficult.

Since contrast medium uptake into the tumour is driven by the plasma concentration, it is desirable that this is standardised, for example by use of a power injector. The analysis should also control for variations in contrast medium pharmacokinetics (e.g. renal clearance) between patients or in the same patient before and after treatment. Thus it is critical to know the concentration of contrast medium that enters a tumour if the pharmacokinetic analysis is to be accurate and comparable between patients. This input function may be measured from the images. For "fast" protocols a true arterial input function is required, but for "slow" protocols the venous concentration from, for example the sagittal sinus may provide an acceptable approximation. Alternatively a "normal" tissue such as skeletal muscle, spleen or the choroid plexus may be employed as a control.

Assuming that contrast agent concentration is to be measured, a valid measurement of T_1 is essential. Variable flip angle spoiled gradient echo approaches are fast and provide good coverage in 3D but are very sensitive to RF inhomogeneity and should ideally be validated in every coil and every relevant spatial location in the scanners used in the study. Saturation-recovery and inversion-recovery sequences are more robust but very time-consuming in 3D.

In summary a number of methodologies have been developed. The more complex protocols, requiring more advanced hardware capabilities and more complex analysis algorithms, may provide additional insights into biology but may be less easy to compare between institutions. Thus for phase I trials, where one or two institutions participate, complex analytical techniques can be used. For multicentre trials less demanding technologies should be used but it is essential that adequate standardisation and quality control are performed between centres, perhaps by a specialist imaging contract research organisation, with analysis in a central reading centre, if comparable data are to be obtained.

16.3.4 Design of the DCE-MRI Analysis Protocol

For a DCE-MRI image processing and analysis (informatics) protocol, the steps include the definition of the region of interest (ROI) by a process of image segmentation; definition of the arterial or vascular input function, calculation of pre-contrast T_1, and calculation of K^{trans} or IAUC. If the analysis is performed voxelwise, the individual data points may be noisy, leading to fitting errors in compartmental models; however, if analysis is confined to the total ROI then it is not possible to account for heterogeneity. There are several sources of variation in DCE-MRI of which two major sources of bias and subjectivity must be considered. Firstly, if the segmentation is performed on post-contrast images, the ROI may vary with drug efficacy, which is clearly an undesirable situation. It may therefore be appropriate to define ROIs on co-registered independent contrast images, such as T_2 weighted images, in conjunction with pre-and post-contrast T_1 weighted images. The ROIs should be drawn by a radiologist familiar with oncological MRI who should ideally be blinded to visit and dose. This blinding may be difficult to arrange in the context of a phase I trial where it is desirable to analyse each patient's data in real time: one solution is to perform the measurement unblinded as the trial progresses and then a second blinded read by a different radiologist at the end of the study. If there are multiple lesions, for example multiple hepatic metastases, then either all lesions must be measured or alternatively, index lesion(s) must be identified, with a procedure in place to ensure that the same lesion(s) are measured in follow-up scans.

A second potential source of bias is determination of arrival time of contrast agent in the arteries supplying the tumour: again this should be performed objectively by software or by a blinded radiologist.

16.3.5 Spatial Heterogeneity Within the Tumour

Tumour heterogeneity poses considerable difficulty. Tumours with large necrotic cores will show very different average values in K^{trans} in comparison with small well-vascularised tumours, even though values of K^{trans} in the angiogenic rim may be identical. On the other hand if the core is excluded from the analysis there is a risk of biasing assessments of drug efficacy. Analysis of a single region of interest covering the entire tumour will obscure differences between rim and core. On the other hand voxelwise analyses may be time consuming, and, if a pharmacokinetic analysis is employed, noisy data may not fit to a pharmacokinetic curve from the compartmental model, particularly in the presence of motion artefact. Vascular targeting agents in particular have dramatically different effects in different regions of tumour (Fig. 16.1), producing deep reductions in gadolinium uptake in the core while leaving the rim hardly affected. The tools for the statistical analysis of the histograms of parameter values from such maps require further development.

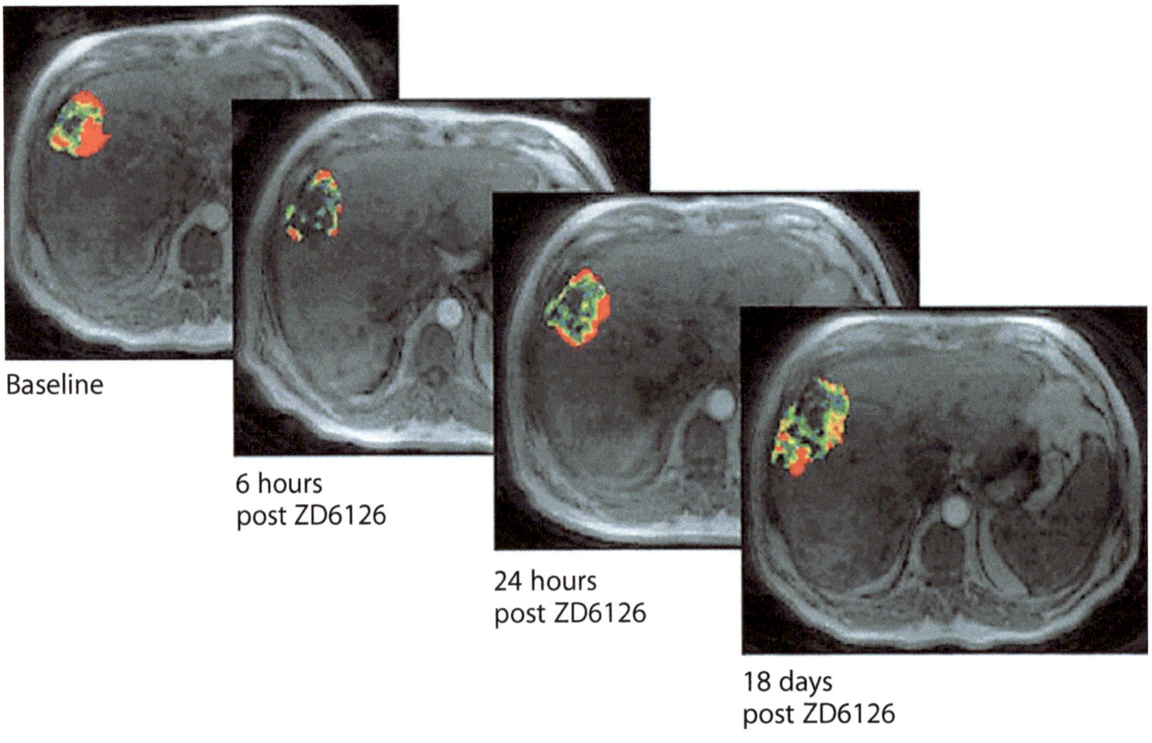

Fig. 16.1. Map of measurement of IAUC measurement in a liver metastasis from colon cancer Magnetic resonance images were used to determine tumour IAUC before (*left*) and at 6 h, 24 h, and 18 days after (*right*) the patient was treated with the vascular targeting agent ZD6126. One representative slice from the 3D sets is shown. The areas of *red* and *blue* represent high and low IAUC in the tumour, respectively (Evelhoch et al. 2002)

Overall the heterogeneity of tumour biology within and between patients has largely not been taken into account when incorporating imaging strategies into early phase clinical trials. This contrasts with the standardisation of normal organ function that are entry criteria for most phase I trials, perhaps explaining why we are able to see dose–response effects with regard to toxicity but sometimes not biological or pharmacodynamic phenomena.

16.3.6 The Validity of the Measurement

It is critical to assess the validity of DCEMRI endpoints as biomarkers or surrogates for the desired parameter. Validation has several aspects which should be distinguished.

The strongest validation is *validation against outcome* (sometimes referred to as predictive validity, an aspect of criterion validity). This is achieved if we can show, for example that tumours with high K^{trans} are always associated with a worse prognosis, and that interventions which reduce K^{trans} in particular patients also improve prognosis in a dose-dependent way in the same patients. If an endpoint has shown predictive validity in a number of large studies, it may be considered by regulatory authorities as a validated surrogate endpoint. A weaker, but still important, validation is *validation against histopathology* (sometimes referred to as content validity). This is achieved if we can show, for example, that tumours with high K^{trans} tend to have a high microvessel density (MVD). Histopathological validation is important if an endpoint is to be employed as a biomarker in Phase I/II. A number of studies have attempted to make these comparisons, and the data are summarised in Table 16.1. Although some studies have not confirmed that there is a statistical association between particular MRI parameters and the vascular

Table 16.1. Studies assessing validity of DCE-MRI with respect to vascular agents

		Tumour	MRI parameter	p Value
MVD	Buadu et al. 1996	Breast	Gradient	0.001
	Stomper et al. 1997	Breast	Time interval	0.02
	Hawighorst et al. 1997a,b, 1998	Cervix	Amplitude	<0.001
			K^{trans}	<0.05
	Buckley et al. 1997	Breast	Time Interval	0.002
	Hulka et al. 1997	Breast	K^{trans}	<0.01
	Tynninen et al. 1999	Brain	Time interval	0.01
	Ikeda et al. 1999	Breast	K^{trans}	0.01
	Matsubayashi et al. 2000	Breast	Time interval	<0.001
VEGF	Hawighorst 1997, 1998a,b	Cervix	Amplitude	NS
			K^{trans}	NS
	Knopp et al. 1999	Breast	K^{trans}	0.05
	Matsubayashi et al. 2000	Breast	Time interval	0.008
pO_2	Cooper et al. 2000	Cervix	Amplitude	0.001
			Gradient	0.071
			Amplitude/gradient	NS
Metastasis	Nagashima et al. 2002	Breast	Time interval	<0.05
Response	Baba et al. 1997	Head and neck	Amplitude	
	Barentsz et al. 1998	Bladder	Amplitude	<0.05
	Hoskin et al. 1999	Head and neck	Amplitude	0.004
			Time interval	NS
	Devries et al. 2001	Rectal	Perfusion	<0.001
Survival	Hawighorst et al. 1999	Cervix	K^{trans}/gradient	<0.05

Studies are listed according to the clinical parameter that was compared with the MRI study. **Content validity:** *MVD*, microvascular density; *VEGF*, immunohistochemical assessment of amount of VEGF present; *pO_2*, partial pressure of oxygen in tissue. **Predictive validity:** *metastasis*, risk of developing metastasis; *response*, likelihood of responding to anti-cancer treatment; *survival*, duration of survival from diagnosis. **MRI parameters:** *gradient*, maximum gradient of uptake of gadolinium; *time interval*, interval from injection to a defined percentage of maximum gadolinium uptake; *K^{trans}*, transfer constant; *amplitude*, maximum amount of enhancement by contrast agent; *perfusion*, semi-quantitative assessment of blood flow.

or tumour measurement there are sufficient studies reported to support the provisional incorporation of MRI into the early clinical evaluation of anti-angiogenic compounds. Nevertheless validation studies are to some extent confounded by the difficulty incurred by comparing histological evaluation, measured at the micrometre level, with DCE-MRI measurements where resolution is measured in millimetres.

A third aspect of validation is the computer systems validation. ICH GCP (good clinical practice) demands that the computer system and software used for the analysis are validated. In this context "validation" refers to the documentation, audit trails and change control procedures used to provide assurance that the analysis is performed correctly on every scan without bugs or errors. Computer systems validation does not of course of itself indicate that the endpoint is appropriate to use.

16.3.7 The Reproducibility and Statistical Power of the Measurement

If there is reliance upon DCE-MRI as a biomarker then it is critical to establish how reproducible the measurement is. For example, if an experimental agent causes a 20% decline in a vascular parameter measured by MRI but the day-to-day variation in that parameter is 25% then there is unlikely to be sufficient statistical power to determine whether the drug is active. There are many sources of variation in the DCE-MRI experiment due to biological variation, random error, and systematic error. Figure 16.2 shows a hierarchy of sources of variation, from the basic design of the study and tumour type, through to the image informatics. Even this is an oversimplification as many of the terms interact. Figure 16.2 represents an enormous notional nested analysis of variance which, although never performed in practice, does provide a framework for thinking about reproducibility and variation. In practice, in any particular study, the components of variation which are of most interest are of course the effect of drug, together with the same-tumour same-analysis different-scan reproducibility (marked with an asterisk in Fig. 16.2). If possible, reproducibility studies for each patient should be built into any study that relies upon DCE-MRI as a biomarker, for example by performing two baseline studies before administering the drug (Leach et al. 2003). It is however important to consider all the factors summarised in Fig. 16.2, both in order to minimise variability and avoid confounds, and also when comparing values of reproducibility quoted from different studies and centres. The appendix shows a statistical treatment of reproducibility whereby we can calculate statistical power and thus whether a new therapeutic agent has an effect in a single individual or a group beyond the observed day-to-day variation in the technique in the absence of intervention.

Reproducibility studies have been performed for a number of tumours. Data show that these studies can most easily be performed in newly presenting supratentorial glioma, in which the tumour is contained in the rigid skull, while studies in the upper abdomen and thorax have been compromised, to some extent, by respiratory motion artefact. For instance we have compared three different analytical techniques in patients with glioma (Fig. 16.3). These data show that the coefficients of variation for the gradient, time to 90% saturation and K^{trans} parameters were 17.9%, 7.1% and 7.7%. Using a two-way mixed effect model, the estimated percentage change needed in these parameters to be 95% confident that a therapeutic effect was in excess of that produced by day-to-day variation, one would need to record 20.2%, 5.2% and 6.2% changes in the parameters, respectively (Jackson et al. 2003). As expected the reproducibility of these studies in the abdomen is lower than that in the brain. When liver metastases were studied over an 8-h period using a first pass method to measure K^{trans}, the coefficient of variation was 11% and the percentage change required to prove drug activity in this setting was 15% (Jackson et al. 2002). Currently techniques are under development to measure vascular parameters through DCE-MRI in the thorax, the major hurdle being the movement associated with breathing and the heart beat.

16.4 Clinical Considerations in Study Design

As one of the principal uses of DCE-MRI has been to investigate changes in the vasculature it is appropriate to consider the results of imaging studies in the context of an anti-angiogenic agent.

Angiogenesis is the process of new blood vessel formation that is disordered in many pathophysiological conditions including cancer. In experimental models inhibition of angiogenesis has been associated with tumour growth restraint and even shrinkage while clinical studies of anti-angiogenic agents

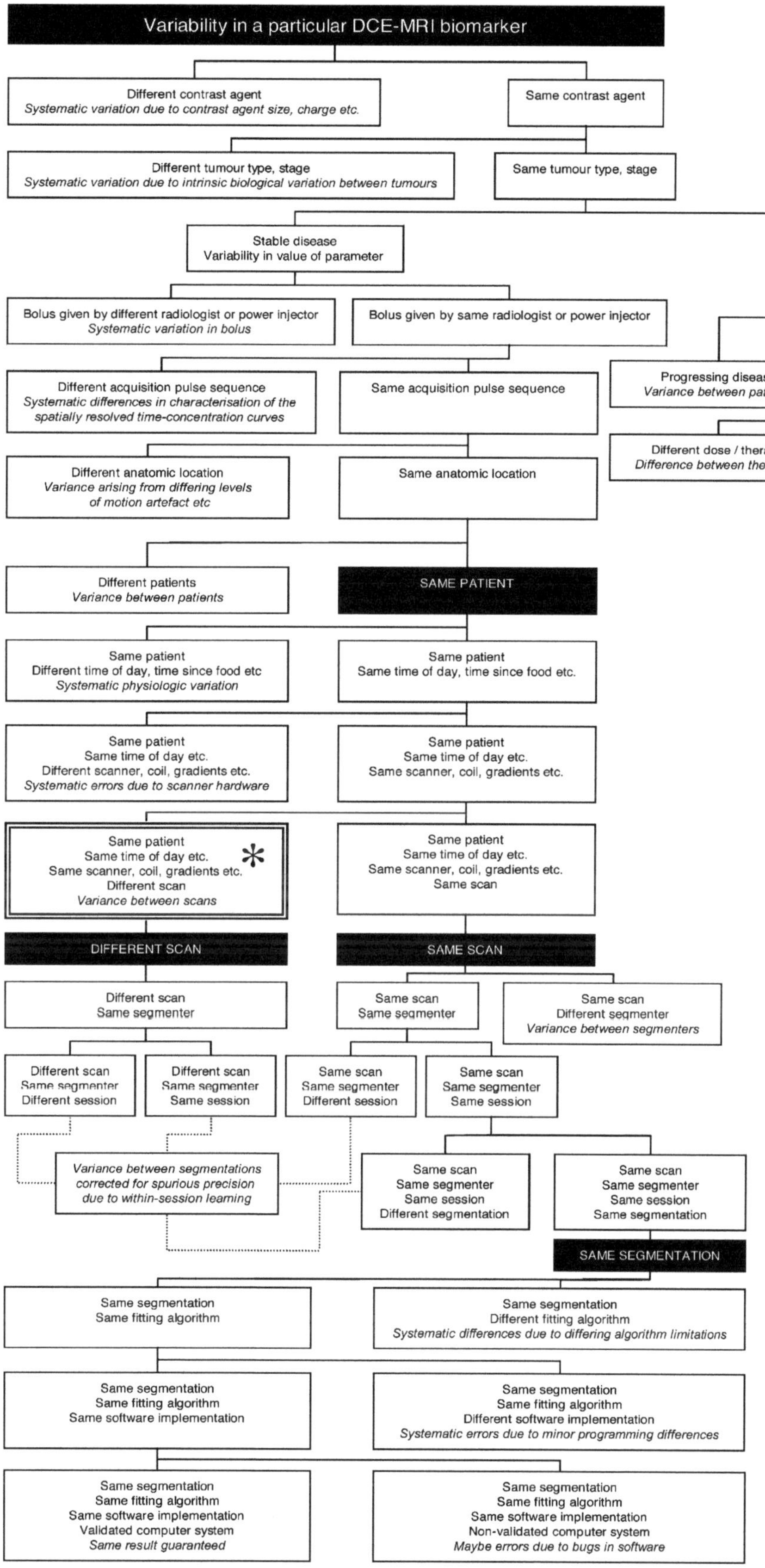

Fig. 16.2. A hierarchical approach to sources of variation in DCE-MRI in cancer. The most commonly considered component of reproducibility is marked with an *asterisk*

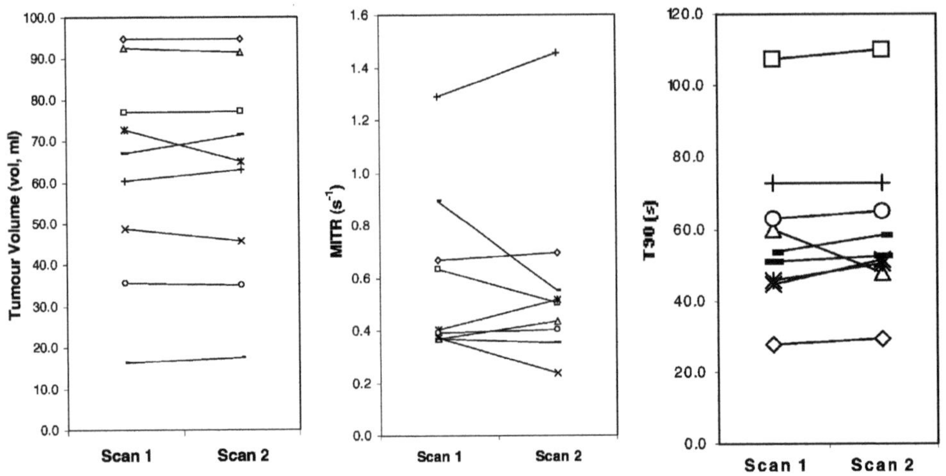

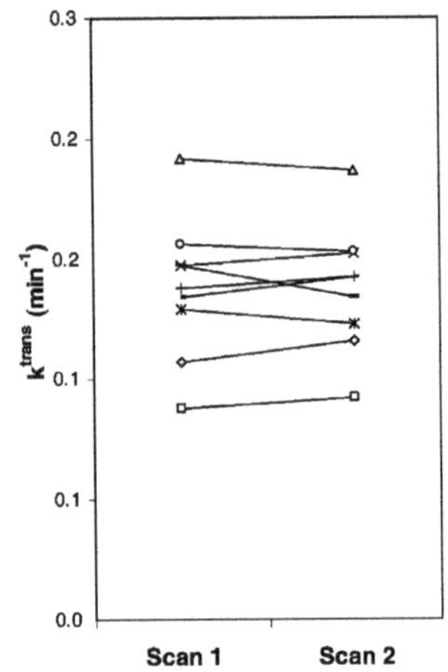

Fig. 16.3. Reproducibility of different DCE-MRI vascular parameters. Changes in measured parameters: **a** Tumour volume (*vol*). **b** MITR, the maximum gradient of uptake of contrast agent. **c** T90, the time to 90% of maximum contrast agent-induced signal change. **d** K^{trans} over the 2-day study period. Symbols represent individual patients. [Reproduced with permission from Jackson et al. (2003)]

have shown some signs of anti-tumour activity. VEGF is one of the most potent angiogenic cytokines and therefore has been selected as the target for a number of anti-angiogenic drugs (Hasan and Jayson 2001). Strategies for inhibiting this cytokine have involved antibodies against either the cytokine or signalling receptor or small molecule inhibitors of the tyrosine kinase domain of the receptor. K^{trans} (which depends on the permeability-endothelial surface area product) has been extensively investigated in the context of VEGF inhibitors, since VEGF controls vascular permeability and therefore measurements of K^{trans} should reflect the biological activity of VEGF. Potentially therefore, we can use K^{trans} (and other compartmental modelling parameters) as a biomarker in the evaluation of this class of compound.

One of us has completed a phase I evaluation of a humanised monoclonal antibody in which we monitored the distribution of the antibody by labelling the drug with ^{124}I and positron emission tomography. These data were then related to the DCE-MRI derived measurements of K^{trans}. The data showed that there was heterogeneity at multiple levels in the tumours including baseline K^{trans} measurements, intra- and inter-tumoral reductions in K^{trans} after treatment, drug distribution and drug clearance, even when tumour deposits in the same patient were compared (Jayson et al. 2002). An important implication of these data was that plasma pharmacokinetics were not representative of tumour pharmacokinetics (although they did compare favourably with normal organ clearance) and that imaging, to some extent, can overcome this problem by looking at the tumour directly.

These problems have manifested themselves in other clinical trials where imaging has been used to study vascular parameters (Galbraith et al. 2001, 2002; Herbst et al. 2002). At present the best interpretation of the data is that they show a threshold effect rather than a dose response effect.

In other words trialists are reporting that as higher doses are given one eventually encounters a dose at which the majority of patients manifest the desired change in the DCE-MRI endpoint. Below that dose the majority of MRI studies do not show the anticipated change. In the anti-VEGF antibody trial the lowest dose level, 0.3 mg/kg, was significantly inferior, in terms of its effect on K^{trans}, than the three higher doses (1, 3 and 10 mg/kg). However, there was no dose response relationship in the three higher dose levels. In the evaluation of combretastatin A4 phosphate, a vascular targeting agent, the initial report of the data suggested a threshold effect at a minimum of 52 mg/m^2 (Galbraith et al. 2001). Thus, to date, we have developed methodologies that identify the minimum effective dose rather than the optimum biologically effective dose. Clearly if we are to use DCE-MRI to guide us in the development of anti-angiogenic and anti-vascular agents we need to modify clinical trial design so that the degree of heterogeneity is addressed. This might then allow us to identify dose response curves and thus the optimum biologically active dose for further study.

16.5 Clinical Trial Design

Two possible approaches would address the impact of tumour heterogeneity on tumour imaging studies in phase I clinical trials. In addition to the classical cohort based phase I study we could recruit a further cohort of patients who are treated using an intra-patient dose escalation regimen. In this case we could focus on a particular tumour mass in a patient who is then treated with increasing doses of a drug. Imaging studies would be performed at each dose level and the response compared. This approach would control for the biological and functional differences discussed above, although it would be sensitive to a drug that significantly altered tumour behaviour. Evidence in favour of such an approach could be taken from our most recent investigation of the anti-VEGF antibody (Jayson et al. 2002), in which a more profound reduction in K^{trans} (Fig. 16.4) was recorded when the antibody concentration was higher rather than when the concentration was lower, suggesting that within each patient there was a concentration–response effect. Whether this would translate into a dose–response

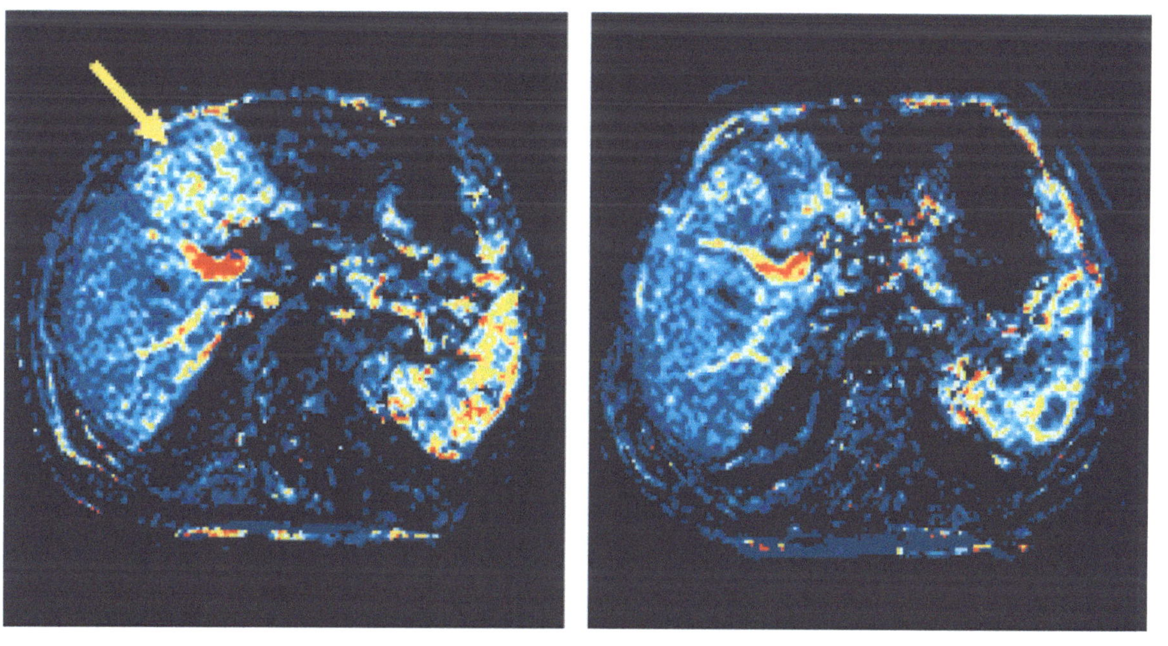

Fig. 16.4. Colour enhanced DCE-MRI measurement of K^{trans} in a liver metastasis from colon cancer. Magnetic resonance images were used to determine tumour K^{trans} before (*left*) and after (*right*) patients were treated with the humanised anti-vascular endothelial cell growth factor (VEGF) antibody HuMV833. Representative images are shown: *Left*, the maps showing the K^{trans} of a metastasis (*yellow arrow*) in the left lobe of the liver in a patient before receiving HuMV833 (1 mg/kg). *Right*, the maps showing the K^{trans} of the same metastasis from the same patients 48 h after receiving HuMV833. The areas of *green* and *blue* represent high and low K^{trans}, respectively. *Red* and *yellow* pixels represent artifactually high measurements in the hepatic vein. [Reproduced with permission from Jayson et al. (2002)]

effect using the intra-patient dose escalation strategy is unknown.

A second approach to dealing with tumour heterogeneity in imaging studies is the inclusion of larger more precisely defined entry criteria in the hope that this would equalise the biology of the tumours included at baseline. Evidence in favour of this was presented in a recent study (Morgan et al. 2003) in which 26 of 56 patients were selected from two phase I studies of a VEGF receptor tyrosine kinase inhibitor. These patients had colorectal cancer metastasis in the liver and, by measuring K^{trans} in these patients, the authors reported a dose–response effect. Whether the selection criteria for this retrospective selection were appropriate will be confirmed in prospective trials ongoing at the moment. However, to date, this is the only trial where a dose–response relationship has been observed.

16.6 DCE-MRI in Comparison with Other Imaging Modalities

From the drug development perspective, DCE-MRI is just one of many ways to provide a biomarker; a signal that provides evidence of an acute, maybe pharmacodynamic, response to the investigational agent. Biomarkers might be sought in principle in biofluids (e.g. blood, urine), in accessible non-tumour tissue (e.g. skin), from immunohistochemistry of tumour biopsies or from imaging. Imaging has the advantage that pharmacological responses in the tumour itself can be evaluated and perhaps compared with responses in non-target tissue. Unlike biopsy, imaging measurements can be repeated at the same site, although in the case of DCE-MRI normally it would be necessary to wait at least a few hours between examinations for the previous contrast medium dose to be eliminated. Biomarkers have been sought from four major imaging modalities radioisotope imaging (PET, SPECT, scintigraphy), contrast-enhanced ultrasound, contrast-enhanced CT, and magnetic resonance. MR approaches include spectroscopy (MRS) "conventional" (non-contrast) MRI, arterial spin tagging MRI, as well as contrast-enhanced MRI.

In addition to existing technologies new PET probes are under development and again co-registration methodologies will allow us to relate DCE-MRI parameters to other tumour measurements, principally through PET. These include measurements of tumour metabolism (^{18}FDG) (Herbst et al. 2002), tumour DNA synthesis (^{124}IUDR) (Blasberg et al. 2000) or tumour apoptosis (annexin V) (Hofstra et al. 2000). The next critical step is to work out how to relate, or co-register, the images so that for instance one could determine the extent to which drug distribution (positron emitting isotope labelled drug) (Jayson et al. 2002) influences changes in vascular parameter (DCE-MRI) and vice versa. Certainly the software now exists to co-register these studies and data will emerge shortly.

Although ultrasound methods have been developed for the assessment of vascularity these studies have been hampered by the operator dependency of the study and the difficulty in obtaining images when flow rates are very low, which may obscure determination of drug effects.

The incorporation of CT scanners into PET scanners has led to an increased use of dynamic CT in the evaluation of anti-angiogenic and anti-vascular compounds. Early data suggest that the measurements are reproducible and clearly this technology is widely available. A further advantage is that the relationship between signal and contrast agent concentration is linear, unlike MRI. The problem is that early clinical studies of anti-angiogenic compounds require multiple imaging assessments of the tumour. As contrast enhanced dynamic CT necessarily utilises ionising radiation there will be a limit on the number of studies that can be performed in any one patient.

Conventional (non-contrast) MRI can provide measurements of parameters such as mobile proton density (PD), the relaxation times T_1, T_2 or T_2* and the apparent self-diffusion coefficient of tissue water (ADC). These parameters have shown interesting responses to intervention in animal and human tumour studies. However, interpretation is difficult as they depend in a complex way on tumour oedema, necrosis, interstitial space and deoxyhaemoglobin, and for this reason they are probably insufficiently evaluated to use as a primary endpoint in a clinical trial. However, parameters such as T_2* and ADC may be useful as secondary endpoints and it may be feasible to measure one or more of these parameters during the precontrast part of a DCE-MRI protocol.

An alternative MRI technique, arterial spin labelling, measures blood flow into the tumour. Arterial spin labelling is an attractive approach since it does not require the administration of any contrast medium and can be repeated rapidly, since no contrast medium needs to be eliminated. It relies on the inversion of magnetisation in the arteries supplying the tumour. In prospective trials, where patients may have tumour in many different anatomic locations it might be difficult to establish a protocol to provide adequate arterial spin tagging measurements of tumour blood blow in all subjects and perhaps for that reason it has been little used to date.

16.7 Summary

DCE-MRI has been used in the early clinical trial evaluation of a number of anti-angiogenic and anti-vascular compounds. Data have shown that the techniques used are reproducible and have some validity. Clinical trial results have shown that this methodology can be used to identify the minimum effective dose, but to date there is only limited information to show that this methodology can identify the optimum biologically active dose. A possible explanation for this is that tumour heterogeneity obscures any dose–response relationship and we have suggested two clinical trial design strategies that could be used to circumvent these problems. In the absence of any changes in DCE-MRI one would be reluctant to develop an anti-vascular or anti-angiogenic drug further. Yet the ultimate arbiter for drug development and registration is survival advantage in phase III trials. Thus at present DCE-MRI provides useful information in early clinical trial drug development. Whether this biomarker information can be incorporated into late drug development strategies remains to be proven.

References

Baba Y, Furusawa M, Murakami R, Yokoyama T, Sakamoto Y, Nishimura R, Yamashita Y, Takahashi M, Ishikawa T (1997) Role of dynamic MRI in the evaluation of head and neck cancers treated with radiation therapy. Int J Radiat Oncol Biol Phys 37:783–787

Barentsz JO, Berger-Hartog O, Witjes JA, Hulsbergen-van der Kaa C, Oosterhof GO, VanderLaak JA, Kondacki H, Ruijs SH (1998) Evaluation of chemotherapy in advanced urinary bladder cancer with fast dynamic contrast-enhanced MR imaging. Radiology 207:791–797

Blasberg RG, Roelcke U, Weinreich R, Beattie B, von Ammon K, Yonekawa Y, Landolt H, Guenther I, Crompton NE, Vontobel P, Missimer J, Maguire RP, Koziorowski J, Knust EJ, Finn RD, Leenders KL (2000) Imaging brain tumor proliferative activity with [124I]iododeoxyuridine. Cancer Res 60:624–635

Buadu LD, Murakami J, Murayama S, Hashiguchi N, Sakai S, Masuda K, Toyoshima S, Kuroki S, Ohno S (1996) Breast lesions: correlation of contrast medium enhancement patterns on MR images with histopathologic findings and tumor angiogenesis. Radiology 200:639–649

Buckley DL, Drew PJ, Mussurakis S, Monson JR, Horsman A (1997) Microvessel density of invasive breast cancer assessed by dynamic Gd-DTPA enhanced MRI. J Magn Reson Imaging 7:461–464

Checkley D, Tessier J, Kendrew J et al (2003) Dynamic contrast-enhanced MRI as a marker of biological effect: a study of a human prostate tumor xenograft treated with the VEGF signaling inhibitor ZD6474. Br J Cancer 89:1889–1895.

Cooper RA, Carrington BM, Loncaster JA et al. (2000) Tumour oxygenation levels correlate with dynamic contrast-enhanced magnetic resonance imaging parameters in carcinoma of the cervix. Radiother Oncol 57:53–59

Cristofanilli M, Charnsangavej C, Hortobagyi GN (2002) Angiogenesis modulation in cancer research: novel clinical approaches. Nat Rev Drug Discov 1:415–426

Devries AF, Griebel J, Kremser C, Judmaier W, Gneiting T, Kreczy A, Ofner D, Pfeiffer KP, Brix G, Lukas P (2001) Tumor microcirculation evaluated by dynamic magnetic resonance imaging predicts therapy outcome for primary rectal carcinoma. Cancer Res 61:2513–2516

Dowlati A, Robertson K, Cooney M, Petros WP, Stratford M, Jesberger J, Rafie N, Overmoyer B, Makkar V, Stambler B, Taylor A, Waas J, Lewin JS, McCrae KR, Remick SC (2002) A phase I pharmacokinetic and translational study of the novel vascular targeting agent combretastatin a-4 phosphate on a single-dose intravenous schedule in patients with advanced cancer. Cancer Res 62:3408–3416

Eder JP, Supko JG, Clark JW, Puchalski TA, Garcia-Carbonero R, Ryan DP, Shulman LN, Proper J, Kirvan M, Rattner B, Connors S, Keogan MT, Janicek MJ, Fogler WE, Schnipper L, Kinchla N, Sidor C, Phillips E, Folkman J, Kufe DW (2002) Phase I clinical trial of recombinant human endostatin administered as a short intravenous infusion repeated daily. J Clin Oncol 20:3772–3784

Evelhoch J, Brown T, Chenevert T et al. (2000) Consensus recommendation for acquisition of dynamic contrasted-enhanced MRI data in oncology. Proc Int Soc Magn Reson Med Eighth Scientific Meeting and Exhibition, p 1439

Evelhoch J, Lorusso P, Delproposto Z et al. (2002) Dynamic contrast-enhanced MRI evaluation of the effects of ZD6126 on tumor vasculature in a phase I clinical trial. Proc Int Soc Magn Reson Med Tenth Scientific Meeting and Exhibition, p 2095

Federal Register (1997) ICH harmonised tripartite guideline for good clinical practice, vol. 62, pp 25691-25709

Galbraith SM, Lodge MA, Taylor NJ et al. (2001) Combretastatin A4 phosphate (CA4P) reduces tumor blood flow in animals and man, demonstrated by MRI. Proc Amer Soc Clin Oncol 20:278 (abstract)

Galbraith SM, Rustin GJ, Lodge MA, Taylor NJ, Stirling JJ, Jameson M, Thompson P, Hough D, Gumbrell L, Padhani AR (2002) Effects of 5,6-dimethylxanthenone-4-acetic acid on human tumor microcirculation assessed by dynamic contrast-enhanced magnetic resonance imaging. J Clin Oncol 20:3826–3840

Hasan J, Jayson GC (2001) VEGF antagonists. Expert Opin Biol Ther 1:703–718

Hawighorst H, Knapstein PG, Weikel W, Knopp MV, Zuna I, Knof A, Brix G, Schaeffer U, Wilkens C, Schoenberg SO, Essig M, Vaupel P, van Kaick G (1997) Angiogenesis of uterine cervical carcinoma: characterization by pharmacokinetic magnetic resonance parameters and histological microvessel density with correlation to lymphatic involvement. Cancer Res 57:4777–4786

Hawighorst H, Knapstein PG, Knopp MV, Weikel W, Brix G, Zuna I, Schönberg SO, Essig M, Vaupel P, van Kaick G (1998) Uterine cervical carcinoma: comparison of standard and pharmacokinetic analysis of time-intensity curves for assessment of tumor angiogenesis and patient survival. Cancer Res 58:3598–3602

Hawighorst H, Weikel W, Knapstein PG, Knopp MV, Zuna I, Schönberg SO, Vaupel P, van Kaick G (1998) Angiogenic activity of cervical carcinoma: assessment by functional magnetic resonance imaging-based parameters and a histomorphological approach in correlation with disease outcome. Clin Cancer Res 4:2305–2312

Hawighorst H, Knapstein PG, Knopp MV, Vaupel P, van Kaick G (1999) Cervical carcinoma: standard and pharmacokinetic analysis of time-intensity curves for assessment of tumor angiogenesis and patient survival. MAGMA 8:55–62

Herbst RS, Hess KR, Tran HT, Tseng JE, Mullani NA, Charnsangavej C, Madden T, Davis DW, McConkey DJ, O'Reilly MS, Ellis LM, Pluda J, Hong WK, Abbruzzese JL (2002) Phase I study of recombinant human endostatin in patients with advanced solid tumors. J Clin Oncol 20:3792–3803

Herbst RS, Mullani NA, Davis DW, Hess KR, McConkey DJ, Charnsangavej C, O'Reilly MS, Kim HW, Baker C, Roach J, Ellis LM, Rashid A, Pluda J, Bucana C, Madden TL, Tran HT, Abbruzzese JL (2002) Development of biologic markers of response and assessment of antiangiogenic activity in a clinical trial of human recombinant endostatin. J Clin Oncol 20:3804–3814

Hofstra L, Liem IH, Dumont EA, Boersma HH, van Heerde WL, Doevendans PA, De Muinck E, Wellens HJ, Kemerink GJ, Reutelingsperger CP, Heidendal GA (2000) Visualisation of cell death in vivo in patients with acute myocardial infarction. Lancet 356:209–212

Hoskin, P.J., Saunders, M.I., Goodchild, K. et al (1999) Dynamic contrast enhanced magnetic resonance scanning as a predictor of response to accelerated radiotherapy for advanced head and neck cancer. Br J Radiol 72:1093–1098

Hulka CA, Edmister WB, Smith BL, Tan L, Sgroi DC, Campbell T, Kopans DB, Weisskoff RM (1997) Dynamic echo-planar imaging of the breast: experience in diagnosing breast carcinoma and correlation with tumor angiogenesis. Radiology 205:837–842

Ikeda O, Yamashita Y, Morishita S, Kido T, Kitajima M, Okamura K, Fukuda S, Takahashi M (1999) Characterization of breast masses by dynamic enhanced MR imaging. A logistic regression analysis. Acta Radiol 40:585–592

Jackson A, Haroon H, Zhu XP, Li KL, Thacker NA, Jayson G (2002) Breath-hold perfusion and permeability mapping of hepatic malignancies using magnetic resonance imaging and a first-pass leakage profile model. NMR Biomed 15:164–173

Jackson A, Jayson GC, Li KL, Zhu XP, Checkley DR, Tessier JJ, Waterton JC (2003) Reproducibility of quantitative dynamic contrast-enhanced MRI in newly presenting glioma. Br J Radiol 76:153–162

Jayson GC, Zweit J, Jackson A et al. (2002) Molecular imaging and biological evaluation of HuMV833 anti-VEGF antibody: implications for trial design of antiangiogenic antibodies. J Natl Cancer Inst 94:1484–1493

Knopp MV, Weiss E, Sinn HP et al. (1999) Pathophysiologic basis of contrast enhancement in breast tumors. J Magn Reson Imaging 10:260–266

Kuenen BC, Rosen L, Smit EF et al. (2002) Dose-finding and pharmacokinetic study of cisplatin, gemcitabine, and SU5416 in patients with solid tumors. J Clin Oncol 20:1657–1667

Leach MO, Brindle KM, Evelhoch JL et al. (2003) Assessment of anti-angiogenic and anti-vascular therapeutics using Magnetic Resonance Imaging: recommendations for appropriate methodology for clinical trials. Proc Int Soc Magn Reson Med Eleventh Scientific Meeting and Exhibition, p 1268

Lesko LJ, Atkinson AJ Jr (2001) Use of biomarkers and surrogate endpoints in drug development and regulatory decision making: criteria, validation, strategies. Annu Rev Pharmacol Toxicol 41:347–366

Matsubayashi R, Matsuo Y, Edakuni G, Satoh T, Tokunaga O, Kudo S (2000) Breast masses with peripheral rim enhancement on dynamic contrast-enhanced MR images: correlation of MR findings with histologic features and expression of growth factors. Radiology 217:841–848

Medved M, Maceneany P, Karxzmar H et al. (2002) Contrast agent dynamics as an early marker of antiangiogenic agent action. Proc Int Soc Magn Reson Med Tenth Scientific Meeting, p 2099

Morgan B, Thomas AL, Drevs J et al. (2003) Dynamic contrast-enhanced magnetic resonance imaging as a biomarker for the pharmacological response of PTK787/ZK 222584, an inhibitor of the vascular endothelial growth factor receptor tyrosine kinases, in patients with advanced colorectal cancer and liver metastases: results from two phase I studies. J Clin Oncol 21:3955–3964

Nagashima T, Suzuki M, Yagata H et al. (2002) Dynamic-enhanced MRI predicts metastatic potential of invasive ductal breast cancer. Breast Cancer 9:226–230

Pradel C, Siauve N, Bruneteau G et al. (2003) Reduced capillary perfusion and permeability in human tumour xenografts treated with the VEGF signalling inhibitor ZD4190: an in vivo assessment using dynamic MR imaging and macromolecular contrast media. Magn Reson Imaging 21:845–851

Robinson SP, Mcintyre SJO, Checkley D et al. (2003) Tumour dose response to the anti-vascular agent ZD6126 assessed by magnetic resonance imaging. Br J Cancer 88:1592-1597

Ruehm SG, Christina H, Violas X Corot C, Debatin JF (2002) MR angiography with a new rapid-clearance blood pool agent: Initial experience in rabbits. Magn Reson Med 48:844–851

Stomper PC, Winston JS, Herman S, Klippenstein DL, Arredondo MA, Blumenson LE (1997) Angiogenesis and dynamic MR imaging gadolinium enhancement of malignant and benign breast lesions. Breast Cancer Res Treat 45:39–46

Therasse P, Arbuck SG, Eisenhauer EA et al (2000) New guidelines to evaluate the response to treatment in solid tumors. European Organization for Research and Treatment of Cancer, National Cancer Institute of the United States, National Cancer Institute of Canada. J Natl Cancer Inst 92:205–216

Thomas A, Morgan B, Drevs J et al (2001) Pharmacodynamic results using dynamic contrast enhanced magnetic resonance imaging, of 2 phase 1 studies of the VEGF inhibitor PTK787/ZK 222584 in patients with liver metastases from colorectal cancer. Proc Amer Soc Clin Oncol 20:abstr 279

Tofts PS, Kermode AG (1991) Measurement of the blood-brain barrier permeability and leakage space using dynamic MR imaging. 1. Fundamental concepts. Magn Reson Med 17:357–367

Tofts PS, Brix G, Buckley DL et al. (1999) Estimating kinetic parameters from dynamic contrast-enhanced T(1)-weighted MRI of a diffusable tracer: standardized quantities and symbols. J Magn Reson Imaging 10:223–232

Tynninen O, Aronen HJ, Ruhala M et al. (1999) MRI enhancement and microvascular density in gliomas. Correlation with tumor cell proliferation. Invest Radiol 34:427–434

Glossary of Terms

Biomarker
A characteristic (e.g. from blood sample, biopsy, or imaging) that is objectively measured and evaluated as an indicator of normal biological processes, pathogenic processes or pharmacologic responses to a therapeutic intervention and expected to predict clinical benefit or harm based on epidemiologic, therapeutic, pathophysiologic or other evidence. An efficacy biomarker might measure changes occurring at the molecular target for the drug, in an associated signalling pathway, or consequent changes at the cellular level (e.g. proliferation, apoptosis), or in local physiologic response (e.g. perfusion, permeability), before a clinical response becomes apparent. K^{trans} may provide a biomarker for anti-vascular drugs.

Central Reading Centre
A specialist laboratory, possibly a CRO, responsible for collecting images from all the sites participating in the trial and processing them in a uniform manner, e.g. defining regions of interest (voxels identified as tumour, or non-tumour control, by a process of image segmentation), determining arrival time of contrast bolus in the image, and deriving parameters such as K^{trans}.

Clinical Endpoint
A measure of how the patient feels, functions, benefits or survives.

CRO
A Contract Research Organisation. Specialist imaging CROs often play a key role in supporting the drug developer in the management and quality control of multi-centre trials using imaging, and may provide a central reading centre.

Candidate Drug
A molecule which in pre-clinical testing shows a combination of pharmacological, toxicological and pharmacokinetic properties suitable for entry into Drug Development.

Drug Development
Starting from a candidate drug, the clinical studies (Phase I/II/III) and other activities (toxicology, formulation, manufacturing etc.) necessary for regulatory approval.

EMEA
The regulatory authorities in Europe (the European Medicines Evaluation Agency), Japan (the Ministry of Health, Labour and Welfare) and the United States (the Food and Drug Administration) responsible for the approval of new drugs.

Endpoint
A status of the patient that constitutes the 'endpoint' of a patient's participation in a clinical study and that is used as the final outcome. Examples of clinical endpoints are death or tumour progression. Endpoint is sometimes used in a more general sense (as in "pharmacodynamic endpoint", "surrogate endpoint") indicating a variable to be used for evaluating the trial objectives, whether it constitutes a true 'Endpoint' or not.

GCP
Good Clinical Practice (GCP) is an international ethical and scientific quality standard for designing, conducting, recording and reporting trials that involve the participation of human subjects. ICH has issued guidelines on GCP.

ICH
The International Conference on Harmonisation of Technical Requirements for Registration of Pharmaceuticals for Human Use [ICH (Federal Register 1997)] is a project that brings together the regulatory authorities of Europe, Japan and the United States and experts from the pharmaceutical industry in the three regions to discuss scientific and technical aspects of product registration.

Image Informatics
The algorithms and technology used to transform raw image data into medical information. In the case of DCE-MRI, it includes data formatting, storage, and transmission; image registration, segmentation and visualisation, modelling of contrast agent pharmacokinetics, and significance testing.

Pharmacodynamic Response
An acute pharmacological response (e.g. blood pressure) whose magnitude follows directly the pharmacokinetics of the drug eliciting the response. Sometimes used more loosely to describe any acute response to drug treatment. Plasma VEGF concentration may have a pharmacodynamic effect on vascular permeability.

Pharmacokinetics
The time-course of absorption, distribution, metabolism and elimination of a drug molecule in each tissue and organ in vivo.

Phase I Trial
A clinical trial to study the safety and tolerability and/or pharmacokinetic and/or pharmacodynamic properties of a drug.

Phase II Trial
A clinical trial in patients performed to assess safety, tolerability and efficacy, and/or:

- to demonstrate a pharmacological effect
- to demonstrate a dose–response relationship
- to identify suitable doses for the Phase III studies.

Phase III Trial
A clinical trial in patients to confirm safety and efficacy for regulatory approval.

Statistical Power
The statistical power in a clinical trial is the probability of declaring a compound effective in a significance test using the given trial design, if the compound is, in fact, effective. The principal considerations are the magnitude of the effect, the reproducibility of the measurement and the number of patients.

Surrogate Endpoint
A surrogate is a measurement that reflects disease activity, and is a substitute for the disease process, and is expected to predict clinical benefit. Change in tumour size is often used as a surrogate endpoint. A surrogate endpoint may be considered validated when there is compelling evidence, ideally from large clinical studies with different kinds of intervention, that the endpoint predicts clinical outcome and that a change in the endpoint either during the natural history of the disease or as a result of intervention predicts a changed outcome. It may be recognised by the Regulatory Authorities as a specific expression of the disease in patients.

Validated Computer System
A measurement performed according to a procedure that ensures accuracy and which provides a documented audit trail so that analysed data can be verified against raw data. In particular it is a requirement of ICH GCP that "When using electronic trial data handling and/or remote electronic trial data systems, the [drug developer] should...ensure and document that the electronic data processing system(s) conforms to the sponsor's established requirements for completeness, accuracy, reliability, and consistent intended performance (i.e. validation)".

Appendix

Calculation of Reproducibility

Errors in DCE-MRI may arise from many different experimental or physiological sources (Fig. 16.2). Such errors can be assessed either through a single experiment with nested analysis of variance in which all possible sources of error are systematically varied, or through a series of studies in which each possible source of error is varied in turn. The latter approach is more practical and more commonly adopted. For each source of error the standard deviation (SD), or the coefficient of variation (CoV), can be assessed

$$CoV = \frac{\sigma}{\mu} \times 100\%$$

where σ and μ are the standard deviation and mean, respectively. Typically CoV or SD are assessed from a series of repeated measures in different patients or images, so that for example, for same-observer-different-scan reproducibility, a study is performed in a number of patients, in which a small number of measurements (often as few as two) is obtained from each patient, i, on different occasions. For each patient unbiased (although imprecise) SD, $\sigma\sigma_i$, can be calculated (provided that Bessel's correction is used). The overall test-retest CoV for a group of N patients is then

$$\sqrt{\sum_i \left(\frac{\sigma_i}{\mu_i}\right)^2 / N}$$

Use of the CoV is most appropriate when $\sigma\sigma_i$ is proportional to $\mu\mu_i$. If $\sigma\sigma_i$ is independent of $\mu\mu_i$, it is appropriate to compare $\sigma\sigma_i$ values directly. Point estimates of CoV are imprecise unless very large numbers of patients are sampled: in this case it is also helpful to calculate the 95% confidence upper bound:

$$\text{CoV upper bound} = \sqrt{\frac{(N-1)\cdot(\text{CoV})^2}{\chi^2_{0.975,N-1}}}$$

Calculation of Statistical Power

If a drug is studied in a group of N patients, we need to calculate the power of the study to detect a given effect in the DCE-MRI parameter. The statistical power can be calculated using statistical programs such as SAS and is given by

$$1-\beta = 1 - PROBT\left\{TINV(1-\alpha,\ 2N-2),\ (2N-2),\ \frac{\delta}{\sigma\sqrt{2/N}}\right\}$$

where 1-β is the power, or probability of declaring a drug active, with $P < \alpha$ in a one-sided Student's t-test, if the true effect is to change the DCE-MRI measurement by δ; PROBT (*x*, *df*, *nc*) is the probability that an observation from Student's *t*-distribution with degrees of freedom, *df*, and non-centrality parameter *nc* is <*x*; and TINV (*p*, *df*) is the *p*th quantile from a Student's *t*-distribution with degrees of freedom *df*.

Significance of Change in an Individual Patient

In a study *n* people undergo two measurements that are close together in time giving values y_{1i} and . If we define $di = y_2i - y_1i$ as the difference between the measurements and then define the mean difference as

$$\bar{d} = \frac{1}{n}\sum_{i=1}^{n} d_i,$$

and the standard deviation of the differences as

$$S_d = \sqrt{\frac{1}{n-1}\sum_{i=1}^{n}(d_i - \bar{d})^2},$$

then a 95% [100(1-α%)] prediction interval (*PI*) for a future *d* is:

$$PI_{95} = \bar{d} \pm t_{0.025,n-1} S_d$$

giving the upper and lower confidence limits on the range of differences expected between measurements (with, for example, no treatment effects) (d_L, d_U).

In general a 100(1-α%) prediction interval is defined as

$$PI_{100(1-\alpha\%)} = \bar{d} \pm t_{\alpha/2,n-1} S_d,$$

where *t* can be obtained from standard statistical tables, e.g. $t_{0.025,5} = 2.57$.

In the actual study, if observed *d*s, calculated as $y'_2 - y'_1$ (i.e. the difference between pre- and post-treatment), are outside the interval (d_L, d_U), we deem them unlikely in the context of an 'inactive intervention' and thus attribute this to the intervention.

Subject Index

T

List of Contributors

Mei-Lin W. Ah-See, MD
Paul Strickland Scanner Centre
Mount Vernon Hospital
Rickmansworth Road, Northwood
Middlesex HA6 2RN
UK

Robert C. Brasch, MD
Professor of Radiology and Pediatrics
Center for Pharmaceutical and Molecular Imaging
Department of Radiology
University of California San Francisco
513 Parnassus Avenue
San Francisco, CA 94143-0628
USA

David L. Buckley, PhD
Lecturer in Imaging Science
Imaging Science and Biomedical Engineering
University of Manchester
Stopford Building
Oxford Road
Manchester M13 9PT
UK

Fernando Calamante MD, PhD
Radiology and Physics Unit
Institute of Child Health
University College London
30 Guilford Street
London WC1N 1EH
UK

Jeffrey L. Evelhoch, PhD
Director, Structural Imaging
World Wide Clinical Technology
Pfizer Global Research & Development
2800 Plymouth Road
Mail Stop: 50-M129
Ann Arbor, MI 48105
USA

Laure S. Fournier, MD
Center for Pharmaceutical and Molecular Imaging
Department of Radiology
University of California San Francisco
513 Parnassus Avenue
San Francisco, CA 94143-0628
USA

Ingrid S. Gribbestad, MD
SINTEF Unimed
MR Center
7465 Trondheim
Norway
and
Cancer Clinic
St. Olavs University Hospital
7006 Trondheim
Norway

Kjell I. Gjesdal, MD
Ullevaal University Hospital
0407 Oslo
Norway

Mari H.B. Hjelstuen, MD
Central Hospital of Stavanger
4068 Stavanger
Norway

Alan Jackson, MBChB (Hons), PhD, FRCP, FRCR
Professor, Imaging Science
and Biomedical Engineering
The Medical School
University of Manchester
Stopford Building
Oxford Road
Manchester M13 9PT
UK

Gordon C. Jayson, FRCP, PhD
Senior Lecturer in Medical Oncology
Cancer Research UK
Department of Medical Oncology
Christie Hospital NHS Trust
Wilmslow Road, Withington
Manchester M20 6 DB
UK

Michael W. Knopp, MD, PhD
Division of Imaging Research
Department of Radiology
Ohio State University
657 Means Hall
1654 Upham Dr.
Columbus, OH 43210-1250
USA

Martin O. Leach, PhD, FInstP
Joint Director, Cancer Research UK
Clinical Magnetic Resonance Research Group
Institute of Cancer Research and Royal Marsden Hospital
Downs Road
Sutton, Surrey SM2 5PT
UK

Kah Loh Li, PhD
Department of Radiology
University of California, San Francisco
4150 Clement Street
San Francisco, CA 94121
USA

Steinar Lundgren, MD
Cancer Clinic
St. Olavs University Hospital
7006 Trondheim
Norway

Nina A. Mayr, MD
Professor, Director Radiation Oncology Center
Department of Radiology
Oklahoma University Health Sciences Center
University Hospital
1200 N. Everett Drive
Rm. BNP 603
Oklahoma City, OK 73190
USA

D. Scott McMeekin, MD
Department of Obstetrics and Gynecology
Oklahoma University Health Sciences Center
Oklahoma City, OK 73190
USA

Joseph F. Montebello, MD
Radiation Oncology Center
Oklahoma University Health Sciences Center
Oklahoma City, OK 73190
USA

Chrit T. W. Moonen, PhD
Imagerie Moléculaire et Fonctionnelle:
de la Physiologie à la Thérapie
ERT CNRS/ Universite Victor Segalen Bordeaux 2
146 rue Leo Saignat, Case 117
33076 Bordeaux Cedex
France

David A. Nicholson, Bmed Sci, FRCR
Consultant in Gastrointestinal Radiology
Hope Hospital
Stott Lane
Salford, M6 8HD
UK

Gunnar Nilsen, MD
Molde Hospital
6400 Molde
Norway

Michael D. Noseworthy, PhD
Departments of Medical Imaging and Medical Biophysics
University of Toronto
Hospital for Sick Children Research Scientist
The Hospital for Sick Children
555 University Ave.
Toronto, Ontario M56 1X8
Canada

Anwar Padhani, MRCP, FRCR
Consultant Radiologist and Lead in MRI
Paul Strickland Scanner Centre
Mount Vernon Hospital
Rickmansworth Road, Northwood
Middlesex HA6 2RN
UK

Geoffrey J. M. Parker, PhD
Research Fellow in Imaging Science
Imaging Science and Biomedical Engineering
University of Manchester
Stopford Building
Oxford Road
Manchester M13 9PT
UK

Michael Pedersen, PhD
Imagerie Moléculaire et Fonctionnelle:
de la Physiologie à la Thérapie
ERT CNRS/ Universite Victor Segalen Bordeaux 2
146 rue Leo Saignat, Case 117
33076 Bordeaux Cedex
France
and
MR Research Center, Institute of Experimental Clinical Research
Aarhus University Hospital, Skejby
8200 Aarhus N
Denmark

Wilburn E. Reddick, MD, PhD
Assistant Member, Department of Diagnostic Imaging
St. Jude Children's Research Hospital
332 North Lauderdale Street
Memphis, TN 38105-2794
USA

Timothy P. L. Roberts, PhD
Professor, Department of Medical Imaging
University of Toronto
Canada Research Chair in Imaging Research
Fitzgerald Building
150 College St., Rm 88
Toronto, Ontario M5S 3E2
Canada

June S. Taylor, PhD
Associate Professor
Department of Radiology
University of Utah
Utah Center for Advanced Imaging Research
729 Arapeen Drive
Salt Lake City, UT 84108
USA

Peter van Gelderen, PhD
Laboratory for Molecular and Functional Imaging
NINDS, NIH
Bethesda, 20892 MD
USA

John C. Waterton, PhD, FRSC
Enabling Science & Technology
AstraZeneca
Alderley Park
Macclesfield, Cheshire, SK10 4TG
UK

Dee H. Wu, PhD
Department of Radiology
Oklahoma University Health Sciences Center
Oklahoma City, OK 73190
USA

William T.C. Yuh, MD, MSEE
Radiation Oncology Center
Oklahoma University Health Sciences Center
Oklahoma City, OK 73190
USA

Xiao Ping Zhu, MD, PhD
MR Unit and Department of Radiology, VA Medical Center
University of California, San Francisco
4150 Clement Street
San Francisco, CA 94121
USA

MEDICAL RADIOLOGY Diagnostic Imaging and Radiation Oncology

Titles in the series already published

Medical Radiology Diagnostic Imaging and Radiation Oncology

Titles in the series already published

Magnetic Resonance Angiography
2nd Revised Edition
Edited by I. P. Arlart,
G. M. Bongratz, and G. Marchal

Pediatric Chest Imaging
Edited by Javier Lucaya
and Janet L. Strife

Applications of Sonography in Head and Neck Pathology
Edited by J. N. Bruneton
in collaboration with C. Raffaelli
and O. Dassonville

Imaging of the Larynx
Edited by R. Hermans

3D Image Processing
Techniques and
Clinical Applications
Edited by D. Caramella
and C. Bartolozzi

Imaging of Orbital and Visual Pathway Pathology
Edited by W. S. Müller-Forell

Pediatric ENT Radiology
Edited by S. J. King
and A. E. Boothroyd

Radiological Imaging of the Small Intestine
Edited by N. C. Gourtsoyiannis

Imaging of the Knee
Techniques and Applications
Edited by A. M. Davies
and V. N. Cassar-Pullicino

Perinatal Imaging
From Ultrasound
to MR Imaging
Edited by Fred E. Avni

Radiological Imaging of the Neonatal Chest
Edited by V. Donoghue

Diagnostic and Interventional Radiology in Liver Transplantation
Edited by E. Bücheler, V. Nicolas,
C. E. Broelsch, X. Rogiers,
and G. Krupski

Radiology of Osteoporosis
Edited by S. Grampp

Imaging Pelvic Floor Disorders
Edited by C. I. Bartram
and J. O. L. DeLancey
Associate Editors: S. Halligan,
F. M. Kelvin, and J. Stoker

Imaging of the Pancreas
Cystic and Rare Tumors
Edited by C. Procacci
and A. J. Megibow

High Resolution Sonography of the Peripheral Nervous System
Edited by S. Peer and G. Bodner

Imaging of the Foot and Ankle
Techniques and Applications
Edited by A. M. Davies,
R. W. Whitehouse,
and J. P. R. Jenkins

Radiology Imaging of the Ureter
Edited by F. Joffre, Ph. Otal,
and M. Soulie

Imaging of the Shoulder
Techniques and Applications
Edited by A. M. Davies and J. Hodler

Radiology of the Petrous Bone
Edited by M. Lemmerling
and S. S. Kollias

Interventional Radiology in Cancer
Edited by A. Adam, R. F. Dondelinger,
and P. R. Mueller

Duplex and Color Doppler Imaging of the Venous System
Edited by G. H. Mostbeck

Multidetector-Row CT of the Thorax
Edited by U. J. Schoepf

Functional Imaging of the Chest
Edited by H.-U. Kauczor

Radiology of the Pharynx and the Esophagus
Edited by O. Ekberg

Radiological Imaging in Hematological Malignancies
Edited by A. Guermazi

Imaging and Intervention in Abdominal Trauma
Edited by R. F. Dondelinger

Multislice CT
2nd Revised Edition
Edited by M. F. Reiser, M. Takahashi,
M. Modic, and C. R. Becker

Intracranial Vascular Malformations and Aneurysms
From Diagnostic Work-Up
to Endovascular Therapy
Edited by M. Forsting

Radiology and Imaing of the Colon
Edited by A. H. Chapman

Coronary Radiology
Edited by M. Oudkerk

Dynamic Contrast-Enhanced Magnetic Resonance Imaging in Oncology
Edited by A. Jackson, D. L. Buckley,
and G. J. M. Parker

Imaging in Treatment Planning for Sinonasal Diseases
Edited by R. Maroldi and P. Nicolai

Clinical Cardiac MRI
With Interactive CD-ROM
Edited by J. Bogaert,
S. Dymarkowski, and A. M. Taylor

Focal Liver Lesions
Detection, Characterization,
Ablation
Edited by R. Lencioni, D. Cioni,
and C. Bartolozzi

Multidetector-Row CT Angiography
Edited by C. Catalano
and R. Passariello

Springer

RADIATION ONCOLOGY

Lung Cancer
Edited by C.W. Scarantino

Innovations in Radiation Oncology
Edited by H. R. Withers and L. J. Peters

Radiation Therapy of Head and Neck Cancer
Edited by G. E. Laramore

Gastrointestinal Cancer – Radiation Therapy
Edited by R.R. Dobelbower, Jr.

Radiation Exposure and Occupational Risks
Edited by E. Scherer, C. Streffer, and K.-R. Trott

Radiation Therapy of Benign Diseases
A Clinical Guide
S. E. Order and S. S. Donaldson

Interventional Radiation Therapy Techniques – Brachytherapy
Edited by R. Sauer

Radiopathology of Organs and Tissues
Edited by E. Scherer, C. Streffer, and K.-R. Trott

Concomitant Continuous Infusion Chemotherapy and Radiation
Edited by M. Rotman and C. J. Rosenthal

Intraoperative Radiotherapy – Clinical Experiences and Results
Edited by F. A. Calvo, M. Santos, and L.W. Brady

Radiotherapy of Intraocular and Orbital Tumors
Edited by W. E. Alberti and R. H. Sagerman

Interstitial and Intracavitary Thermoradiotherapy
Edited by M. H. Seegenschmiedt and R. Sauer

Non-Disseminated Breast Cancer
Controversial Issues in Management
Edited by G. H. Fletcher and S.H. Levitt

Current Topics in Clinical Radiobiology of Tumors
Edited by H.-P. Beck-Bornholdt

Practical Approaches to Cancer Invasion and Metastases
A Compendium of Radiation Oncologists' Responses to 40 Histories
Edited by A. R. Kagan with the Assistance of R. J. Steckel

Radiation Therapy in Pediatric Oncology
Edited by J. R. Cassady

Radiation Therapy Physics
Edited by A. R. Smith

Late Sequelae in Oncology
Edited by J. Dunst and R. Sauer

Mediastinal Tumors. Update 1995
Edited by D. E. Wood and C. R. Thomas, Jr.

Thermoradiotherapy and Thermochemotherapy
Volume 1:
Biology, Physiology, and Physics
Volume 2:
Clinical Applications
Edited by M.H. Seegenschmiedt, P. Fessenden, and C.C. Vernon

Carcinoma of the Prostate
Innovations in Management
Edited by Z. Petrovich, L. Baert, and L.W. Brady

Radiation Oncology of Gynecological Cancers
Edited by H.W. Vahrson

Carcinoma of the Bladder
Innovations in Management
Edited by Z. Petrovich, L. Baert, and L.W. Brady

Blood Perfusion and Microenvironment of Human Tumors
Implications for Clinical Radiooncology
Edited by M. Molls and P. Vaupel

Radiation Therapy of Benign Diseases
A Clinical Guide
2nd Revised Edition
S. E. Order and S. S. Donaldson

Carcinoma of the Kidney and Testis, and Rare Urologic Malignancies
Innovations in Management
Edited by Z. Petrovich, L. Baert, and L.W. Brady

Progress and Perspectives in the Treatment of Lung Cancer
Edited by P. Van Houtte, J. Klastersky, and P. Rocmans

Combined Modality Therapy of Central Nervous System Tumors
Edited by Z. Petrovich, L. W. Brady, M. L. Apuzzo, and M. Bamberg

Age-Related Macular Degeneration
Current Treatment Concepts
Edited by W. A. Alberti, G. Richard, and R. H. Sagerman

Radiotherapy of Intraocular and Orbital Tumors
2nd Revised Edition
Edited by R. H. Sagerman, and W. E. Alberti

Modification of Radiation Response
Cytokines, Growth Factors, and Other Biolgical Targets
Edited by C. Nieder, L. Milas, and K. K. Ang

Radiation Oncology for Cure and Palliation
R. G. Parker, N. A. Janjan, and M. T. Selch

Clinical Target Volumes in Conformal and Intensity Modulated Radiation Therapy
A Clinical Guide to Cancer Treatment
Edited by V. Grégoire, P. Scalliet, and K. K. Ang

Advances in Radiation Oncology in Lung Cancer
Edited by Branislav Jeremić

MIX
Papier aus verantwortungsvollen Quellen
Paper from responsible sources
FSC® C105338

If you have any concerns about our products,
you can contact us on
ProductSafety@springernature.com

In case Publisher is established outside the EU,
the EU authorized representative is:
Springer Nature Customer Service Center GmbH
Europaplatz 3, 69115 Heidelberg, Germany

Printed by Libri Plureos GmbH
in Hamburg, Germany